MAMMALIAN AMINO ACID TRANSPORT

Mechanisms and Control

MAMMALIAN AMINO ACID TRANSPORT
Mechanisms and Control

Edited by

Michael S. Kilberg
University of Florida
Gainesville, Florida

and

Dieter Häussinger
Medizinische Universitätsklinik
Freiburg im Breisgau, Germany

Springer Science+Business Media, LLC

Library of Congress Cataloging in Publication Data

Mammalian amino acid transport: mechanisms and control / edited by Michael S. Kilberg and Dieter Häussinger.
p. cm.
Includes bibliographical references and index.

1. Amino acids—Physiological transport—Congresses. I. Kilberg, Michael S. II. Häussinger, D. (Dieter), date.
QP561.M35 1993 92-37711
599'.0875—dc20 CIP

Cover art courtesy of Cell Press, Cambridge, Massachusetts.
From K. E. Smith *et al.*, 1992, "Cloning and Expression of a Glycine Transporter Reveal Colocalization with NMDA Receptors," *Neuron*, 8: 927–935.

Proceedings of a German Society of Biological Chemistry conference on Coordinate Regulation between Transport and Metabolism of Glutamine, Glutamate, and Gamma-Glutamyl Compounds, held May 13–16, 1990, in Titisee, Germany

ISBN 978-1-4899-1163-6 ISBN 978-1-4899-1161-2 (eBook)
DOI 10.1007/978-1-4899-1161-2

Originally published by Plenum Press, New York in 1992.
Softcover reprint of the hardcover 1st edition 1992

FOREWORD

Amino acid transport is a part of each of two larger subjects, amino acid metabolism and the biomembrane transport of various small molecules and ions. Nevertheless in this volume we treat amino acid transport as more than a fragment of either of these two larger subjects. A more comprehensive approach is justified when we remember two historic and ongoing aspects of the title subject.

First, amino acid transport had its beginning and acquired a distinct momentum (even if somewhat interrupted from 1913 until about 1945) as amino acid metabolism with the central and pioneer work of Van Slyke and Meyer in 1913. The reviews in this volume will show that it steadily becomes a larger aspect of amino acid metabolism, broadly perceived. These chapters will show for how many organelles, cells, tissues, organs and organ systems, the transmembrane compartmentations and flows of amino acids play very large parts in their fundamental biological relations. The authors here are tending collectively to evaluate an understanding of amino acid flows across biomembranes, and the regulation of these flows, as necessary to an ultimate understanding of the full range of development and metabolism. Such an understanding goes far beyond the purely substrate-destabilizing contributions by enzymes, which have often been arbitrarily limited to that conceptual entity, "the cell", and which for so long a splendid time had most of biochemical research attention.

Secondly, the title interest had another beginning, which it embraced tightly for at least two decades and to which it steadily returns. This beginning encompassed the full range of the phenomenon of the membrane transport of small molecules and ions, whether organic or inorganic. The historic selection of the amino acids for that attack arose not so much from their unique biological interest as from the advantages they offer for such studies through their wide ranges of structure, shape, configuration, and distribution of charge and polarity. (An attendant tragedy is the circumstance that probably half of the biological world blinds itself to the real electrolytic nature of the amino acids by seeing them as RCH(NH2)COOH rather than RCH(NH3+)COO- greatly handicaps exploitation of this advantage). These features undoubtedly underlie their biological importance too, but they have made pertinent, to a wide range of other organic molecules, the biological means by which their transmembrane flows are determined and expressed. This range includes various other metabolites, hormones, small peptides and their analogs, drugs, and other environmental factors. The relationship of their transmembrane flows to linked movements of inorganic ions speaks also to the fundamental breadth of our subject.

This series of reviews will be found to progress through this succession of central interests. At least some of these chapters will be seen quite appropriately to reflect directly on the phenomenon of membrane transport for its own sake, a subject that still needs finishing while we reach forward to try to understand the whole pattern of the flows of metabolism. Furthermore, as the possibility emerges of identifying and characterizing the proteins catalyzing and regulating these events, the importance of accurately characterizing each participating membrane transport system increases. But for this aspect also, the biology of membrane transport has moved far beyond the naked concept of permeability and its interruption, once supposed to occur solely by enzyme action, or to be mediated by lipids, or to be energized in isolation from other cellular energy transduction. Instead it has grown to be tightly associated with almost all other metabolic events.

One should not introduce our title subject without attempting to construct a bridge crossing the intervening chapters to the concluding essay entitled Interorgan Amino Acid Nutrition, which I have the privilege also to offer. It is an admitted bias to regard that subject as our principal and ultimate target: how are the flows of the amino acids and their metabolites, along with the linked flows of inorganic ions, all ultimately to be integrated into their places, to understand the overall flows of biological function? The concept of that overall flow of metabolism led Peter Mitchell to a striking, if perhaps slightly exaggerated dictum: "it is easier to understand biochemistry in terms of transport, than it is to understand transport in terms of biochemistry." Recent work suggests that transport receptor sites may have proved attractive for secondary exploitations as cell surface receptors for viruses. Thus, perhaps allowing viral infections to contribute as they do to nutritional problems, and transport abnormalities to influence viral pathogenesis (Kim et al., 1991; Wang et al., 1991). Will the discrimination between transporters per se from associated transport regulators present further problems? How far will the emerging tendency of transport mediators to form protein super families go? (Gibson et al., 1991). As James H. Schwartz remarks, "the amount of functional information the can be derived from protein structure is astonishing."

The participating authors have sought to review the subject broadly, rather than to limit attention to studies made in a particular laboratory. We depend, however, on each authorship for sensitivity to the personal knowledge growing out of that area of participation, and we are also aware that some of the subjects are so new that the corresponding chapters arise for the moment from rather few labs.

REFERENCES

Gibson, A.L., Wagner, L.M., Collins, F.S. and Oxender, D.L., 1991, Science 254:109.
Kim, J.W., Closs, E.I., Albritton, L.M. and Cunningham, J.M., 1991, Nature 352:725.
Van Slyke, D.D. and Meyer, G.M., 1913, J. Biol. Chem. 16:197.
Wang, H., Kavanaugh, M.P., North, R.A. and Kabat, D., 1991, Nature 352:728.

H.N. Christensen

PREFACE

Transport of amino acids or their derivatives across biological membranes is a highly controlled process. Nature has evolved more or less specific transport systems which allow amino acid transfer between different subcellular compartments on the one hand and between the cellular interior and its extracellular environment on the other. However, it would be an oversimplification to view the role of amino acid transport only as a substrate translocation from one side of a membrane to the other. For example, flux through many amino acid-metabolizing pathways is controlled by transmembrane and interorgan amino acid transport, rendering transport as an important site for control of amino acid metabolism. More recently, it has become clear that energy-dependent amino acid transporters can build up osmotic gradients across the plasma membrane, resulting in alterations of cellular volume and cellular function, such as protein turnover and cell growth. Thus, amino acid transport also must receive attention in terms of inter-and intracellular signalling.

Recent advances in knowledge are presented here in an indepth treatment by experts in this active research area from the viewpoint of physiology, biochemistry, molecular biology and medicine. This includes the methodological aspects for assessing amino acid transport systems, their energetics, the specific features of amino acid transport in different subcellular organelles and tissues, the functional consequences and physiological importance, ontogenetic aspects, as well as intercellular and interorgan transport. Thus, the book is not only of interest to biochemists, physiologists, and cell biologists, but also for the interested student and physician.

Finally, it is a pleasure to thank the many colleagues who have contributed to this project by giving their advice and critique. We are especially thankful to Dr. H.N. Christensen for overtaking the task to summarize the recent advances in physiological interorgan flows of the amino acids. His professional life has been dedicated to the subject of amino acid transport and his contributions to the field are innumerable. We also are indebted to Ms. Maria Chan for preparing the layout of the book and for help in compiling the entries to the subject index, and to our secretaries Ms. Vanessa Garrison and Ms. Petra Ochsenfarth for their patient efforts on our behalf. We hope that this extensive and integrating account of present knowledge will provide a useful platform for further research in this field.

February 1992

Dieter Häussinger
Michael Kilberg

CONTENTS

PART A: GENERAL ASPECTS

PART B: TISSUE SPECIFIC TRANSPORT AND FUNCTIONAL IMPLICATIONS

PART A

GENERAL ASPECTS

AMINO ACID TRANSPORTERS:

SYSTEMATIC APPROACH AND PRINCIPLES OF CONTROL

Guido G. Guidotti and Gian C. Gazzola

Istituto di Patologia Generale
Università di Parma
Via Gramsci, 14 - 43100 Parma, Italy

INTRODUCTION

All living cells, from autonomous protists to constrained components of the complex multicellular mammalian organisms, require amino acids for replacing their structure, synthesizing essential proteins and supplying sources of metabolic energy. Transport of these molecules across the membranes that fix the boundaries of the cell and generate intracellular discontinuities is one of the determinants of cell survival.

Explicit studies on amino acid transport across membranes began in the fifties (Christensen and Riggs, 1952; Heinz and Walsh, 1958) and the formal identification of distinct transport systems for neutral and cationic amino acids was reached within the sixties (Oxender and Christensen, 1963; Christensen, 1964; Christensen *et al.*, 1967). The characterization of systems that mediate the transport of anionic amino acids and the description of a number of tissue-specific agencies for discrete groups of these molecules are more recent achievements (Bannai and Kitamura, 1980; Kilberg *et al.*, 1980; Gazzola *et al.*, 1981*a*; Stevens *et al.*, 1984; Ellory, 1987; Lerner, 1987; Yorek *et al.*, 1987; Silbernagl, 1988; Van Winkle, 1988). Meanwhile, other studies have attempted to define functionally the concept of 'transporter' by elucidating kinetic behaviors, operational properties, mode of energization and transport mechanisms of some of these systems (Christensen, 1973; Christensen, 1979; Christensen, 1984a; Lerner, 1985; Bussolati *et al.*, 1987, 1992; Collarini and Oxender, 1987; Dall'Asta *et al.*, 1991). The physiologic significance of amino acid transport has been approached by investigating the mechanisms of control of the process at the cellular and supracellular level (Guidotti *et al.*, 1978; Christensen, 1982; Shotwell *et al.*, 1983; Kilberg *et al.*, 1985; Christensen *et al.*, 1986) together with the effects of perturbations of the surroundings from or to which amino acid transport normally occur (Gazzola *et al.*, 1972, 1981, 1991).

In this chapter we will attempt to give a coherent view of the process of amino acid transport across membranes as functionally effected by the major sets of transporters (transport systems) and worked upon by known forms of control. In developing the subject, we have been somewhat restrictive in the choice of topics and the selection of the examples provided reflects to some extent the personal investigative efforts of the authors.

AMINO ACIDS

Chemistry

Amino acids are molecules that must possess at least one carboxyl group and one amino group usually bonded to the same carbon atom (α- or 2-carbon). The other ligands

Mammalian Amino Acid Transport, Edited by M.S. Kilberg and
D. Häussinger, Plenum Press, New York, 1992

of the α-carbon are -H and a R-group of different size, shape, and chemical properties (side chain). With some exceptions (e.g., glycine where the R-chain is -H) the tetrahedral array of the four different ligands about the α-carbon confers optical activity on amino acids giving rise to two stereoisomers, the L and D form.

In aqueous solutions at neutral pH, amino acids are predominantly dipolar ions (zwitterions): the α-amino group is protonated and the α-carboxyl group is dissociated. In this form the molecule is lipid-insoluble, and cannot diffuse across lipid bilayers. Most amino acids lack additional ionizable groups in the side chain and, at physiologic pH, are "neutral". Amino acids that possess a proton-accepting group in the side chain (e.g. arginine, lysine) are defined as "cationic" or "basic"; those that carry one or more extra carboxyl groups (e.g., aspartic acid, glutamic acid, γ-carboxyglutamic acid) are called "anionic" or "acidic" amino acids.

The majority of the amino acids are primary amines ($-NH_3^+$); in some cases alkyl groups substitute for one or more -H of the amino group (e.g. methylamines). Proline and derivatives are secondary amines ($=NH_2^+$). Some biologically important amino acids carry the amino group at the β- (e.g., β-alanine) or at the γ-carbon (e.g., γ-aminobutyric acid or GABA).

Twenty different amino acids are used to build up proteins in all species from bacteria to humans (Table 1). Additional amino acids can be found in some specialized proteins as products of co- or post-translational modification of the original amino acid residues (hydroxyproline and hydroxylysine in collagen, γ-carboxyglutamic acid in some clotting factors, mono- and di-iodotyrosine in thyreoglobulin, etc.), and free in the cytoplasm as intermediates of relevant metabolic pathways (ornithine, citrulline, homocysteine, homoserine, sarcosine, etc.).

Table 1. Repertoire of amino acids used to build up proteins.

Chemical classification	Amino acid	Three-letter abbreviation	One-letter symbol
"Neutral" (zwitterionic) amino acids			
Linear aliphatic	Glycine	Gly	G
	Alanine	Ala	A
Branched aliphatic	Valine	Val	V
	Leucine	Leu	L
	Isoleucine	Ile	I
Hydroxy-containing aliphatic	Serine	Ser	S
	Threonine	Thr	T
ω-Amides	Asparagine	Asn	N
	Glutamine	Gln	Q
Sulfur containing	Cysteine	Cys	C
	Methionine	Met	M
Aromatic	Phenylalanine	Phe	F
	Tyrosine	Tyr	Y
Other heterocyclic	Histidine	His	H
	Tryptophan	Try	W
Secondary ("imino" acids)	Proline	Pro	P
"Anionic" amino acids	Aspartate	Asp	D
	Glutamate	Glu	E
"Cationic" amino acids	Lysine	Lys	K
	Arginine	Arg	R

In animal proteins, constituents are L-amino acids. The D stereoisomers present in mammalian cells are usually inert and are rapidly converted to ketoacids. With the relevant exception of D-aspartate (Gazzola *et al.*, 1981*a*), D amino acids are not as readily accepted by amino acid transport systems as their L counterparts.

Physiology

Amino acids play a crucial role in a variety of biologic processes. Beyond being biosynthetic precursors of proteins, amino acids are used as fuels, as carriers/donors of carbon and nitrogen atoms, as neurotransmitters, as components of buffer systems, and as compatible osmolytes.

Nutritional considerations tend to cluster the amino acids into two groups: those that must be supplied with the diet are called "essential", whereas the others are called "nonessential". This division refers to the needs of an organism as a whole and under particular circumstances. Cell populations growing *in vitro* must be provided with a larger set of "indispensable" amino acids (Eagle, 1955, 1959) that they are unable to synthesize or that they synthesize in amounts insufficient to sustain a normal amino acid utilization and whose composition varies for different cell types. *In vivo*, the interorgan cooperativity in amino acid metabolism contributes at least in part to supply amino acids to the cells by ensuring a distribution of these molecules among organs and tissues (Christensen *et al.*, 1986).

In each cell, the requirement for amino acids must be met by a combination of the biosynthesis of same of the amino acids and by the transport of amino acids into the cell. Transport across the plasma membrane is considered a step in amino acid metabolism and may be subject to regulation to maintain an internal amino pool adequate to support the activities of the cell and to assist interorgan nutrition. Fluxes and transport through intracellular membranes are required for the function of organelles and may be additional control sites for the operation of the intracellular machinery.

THE TRANSPORTERS

The term *transport* covers the events essential to the mediated transfer of a solute across a membrane. Operationally it consists of a three-step process: binding of the solute at one side of the membrane, translocation across the membrane, and release of the solute at the other surface. Kinetically, as the concentration of the solute at one side of the membrane is progressively enhanced, the rate of transfer does not continue to rise in a linear fashion, and the rate eventually reaches a maximum that cannot be exceeded by a further increase of solute concentration. This behavior, described by a rectangular hyperbola, defines the mediated transport.

The transfer of hydrophilic solutes (as the amino acids) through the lipid phase of a membrane requires proper mediating structures that have been termed carriers or *transporters* to imply mobility of the substrate, without implications as to how large a part of the structure actually changes conformation or undergoes reorientation. These structures, sequentially but not simultaneously accessible to the substrate from the two sides of a membrane, are characterized by receptor sites (reactive sites) that recognize the substrate, so that its passage through the membrane is permitted or impelled in one direction or the other as a result of that recognition. Recognition properties define the reactivity to substrates of a transporter, and the whole of transporters endowed with the same recognition properties forms a *transport system.*

A transport system is characterized by several attributes beyond substrate recognition: they include operational characteristics, kinetic behaviors, mode of energization, susceptibility to boundary conditions, and response to control mechanisms. Numerous systems mediate the transport of amino acids across the membranes. Different membranes possess a distinct set of transporters and heterogeneity of their distribution pattern in a given membrane exists among tissues and species (see below).

Characterization and discrimination of the amino acid transport systems have been primarily carried out on the basis of kinetic and physiologic studies, taking advantage of the recognition and operation properties of each transporter set. These studies have provided a great deal of information and the systematic approach followed in this chapter is essentially based on it. The isolation and purification of the components of amino acid transport systems, until recently hindered by the difficulty of dealing with integral membrane proteins, by an apparently low expression of these proteins and by the lack of specific probes for their labelling, begin now to be pursued by methodologies of molecular biology (see D.L. Oxender in this book) and this approach will soon add new insights and perspectives in the field.

Substrate Recognition

Each system of transporters acts on a discrete group of substrate amino acids and for each system recognition of substrates is based on their net charge, on the steric conformation of the molecule, on the size, structure and polarity of the side chain, on the presence of alkyl substitutions in the amino group and on the position of the amino group in the carbon chain. For coupled systems (obligatory exchange or cotransport, see later) the reactivity toward the amino acid depends also on the conformational change of the transporter induced by the binding of the co- or counter-transported substrate (Krupka, 1990).

In mammalian cells, substrate recognition by the receptor site of a transporter is often incomplete and the structural requirements for acceptance of an amino acid may allow considerable latitude. Two main consequences follow:

(1) a transport system usually reacts with more than one amino acid. This degeneracy may have permitted to limit the number of the systems required for the transmembrane transport of the entire range of amino acids, but at the cost of a loss of specificity of the vectorial process. For amino acids channeled to metabolism, however, the high chemical specificity of the enzymes subsequently acting on them provides a further screening and excludes potentially dangerous compounds that may have been accepted by the transporters (Berlin, 1970);

(2) different transport systems exhibit some degree of overlapping substrate reactivity and various amino acid species may be transported by more than one apparently independent agency. Consequently, the study of the transport of a given amino acid across a membrane requires a rigorous definition of the different pathways involved and a precise evaluation of their relative contribution in either direction.

Kinetic Characteristics

Mediated transport can be kinetically dealt with the same kinetic approach developed for enzyme catalysis (Christensen, 1975). Transport-producing systems are made by transporters engaged in accepting and transferring substrates across a membrane and the transmembrane flux is defined by the two familiar parameters V_{max} and K_m. In transport kinetics V_{max} (or J_{max}) is the maximal rate of transport through the system attained at infinite substrate concentration (and is a measure of the system capacity); K_m is the substrate concentration sufficient to produce a half-maximal transport velocity. Moreover, because a transporter can be engaged by a substrate or by a competing analog, the simultaneous saturating effect of the inhibitor provides the related parameter K_i (inhibitor concentration sufficient to reduce the substrate transport rate to half of its maximum), often utilized in competition experiments.

The various transport systems that mediate the fluxes of amino acids through the cell membranes differ widely in their capacity and the constituents transporters possess different reactivity toward the individual substrates. For separate systems that translocate the same amino acid, the substrate reactivity to the transporter may vary greatly.

The transporter-mediated transfer of a substrate is a reversible process. A directional asymmetry of the transport rate can be generated by a vectorial input of energy and by different kinetic properties of the reactive site of the transporters at the two surfaces of the membrane, in turn attributable to changes in conformation or to the physical location as in exchange phenomena.

Exchange phenomena (*cis*-inhibition and *trans*-effects) are associated with the operation of one or another transport system and their occurrence cannot be separated from those systems. *Cis*-inhibition defines the asymmetric interference of a competing analogue at one side of a membrane with the two opposite fluxes of a substrate. *Trans*-inhibition and *trans*-stimulation (*trans*-effects) are exchange phenomena by which a substrate at one side of the membrane causes a deficiency or an excess of transporter reactive sites at the other side of the membrane, thus inhibiting or stimulating the translocation of another identical (homoexchange) or analogous substrate (heteroexchange).

Energy Coupling

The mediated transport of amino acids can be passive (facilitated diffusion) or active. The former case is encountered when the driving force for the flux of the amino acid depends on only the thermodynamic potentials of that species at the two sides of the membrane. With electroneutral species or charged species in the absence of an electric field, the appropriate driving force is the *chemical potential.* If the chemical species is charged, as in the case of anionic and cationic amino acids, and there is an electric potential difference across the membrane, the driving force is the *electrochemical potential.* Facilitated diffusion of neutral amino acids can have either of two consequences: *i)* for the transmembrane transfer of an amino acid whose synthesis and degradation is negligible, the process will proceed to equilibrium by dissipating the chemical potential difference between the two sides of the barrier; *ii)* for the transfer of an amino acid synthesized or consumed, a nonequilibrium steady state will be achieved when the flux across the membrane equals the rate of synthesis or consumption.

A transport process that moves a solute uphill against a concentration difference (or, for charged species, against an electrochemical potential) requires an input of energy. This endergonic flux is defined active transport and the transporter involved can be regarded as a transducer that converts energy obtained directly or indirectly from metabolic reactions into transport work.

When the energy-delivering reaction (usually an exergonic dephosphorylation) is directly coupled to the flux of the transported solute, the process is referred to as *primary active transport.* A few ion transport systems in cell and organelle membranes operate in this way. They generate a situation in which the electrochemical potential of the transported ion is higher on the side of the membrane toward which the metabolic pump is directed. Consequently, a driving force exists for downhill backflux of the ion. This ion flux down its gradient can in turn be coupled to the flux of a second species, driving it uphill against its own thermodynamic potential. Such a coupled process, is called *secondary active transport.* Some amino acid transport systems operates as secondary active ion-coupled cotransports. They involve one or more ions and can generate strong uphill gradients of the amino acid species across cell membranes. The concentration gradient of an amino acid accumulated by a secondary active transport can supply energy for the accumulation of another amino acid species through countertransport. The mechanism driving the uphill flux of the second amino acid has been termed *tertiary active transport.*

Fig. 1 provides models of energization mechanisms for transport systems acting at the cell membrane.

In unicellular organisms living in a very diluted environment, capture and accumulation of solutes by energy-requiring processes must have been critical for the operation of the intracellular metabolic pathways. In cells of more complex organisms, bathed by a *milieu intérieur* that provides a quasi-constant and abundant supply of nutrients (including amino acids), the need of a concentrative transport to feed internal metabolism is less obvious. A passive, downhill facilitated transport may indeed be sufficient to replace the molecules consumed by metabolism. In these organisms,

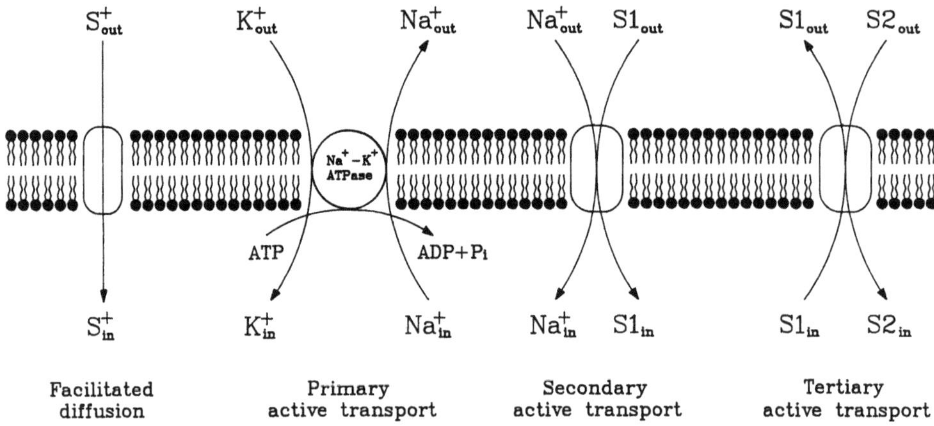

Fig. 1. ATPase-dependent energization of vectorial processes across cell membranes.

however, the generation of concentration gradients are essential for other functions. In nonpolarized cells an active transport of solutes allows the storage of nutrients and contributes to the maintenance of cell volume. Moreover, active transport is needed for polarized cell engaged in absorption, resorption or secretion.

Discrimination and Functional Characterization of Transport Systems

Fundamentals for discriminating individual amino acid transport systems have been set forth since 1966 by H.N. Christensen.

The first step of discrimination among systems is usually represented by a preliminary gross characterization of the various agencies. Time-course, concentration-dependent rate and competition experiments with naturally occurring amino acids and susceptibility of the transport process to boundary conditions (ion composition and concentration of the surroundings, pH changes) can be used to identify the occurrence of saturable transport systems and to define their substrate array. Additional connotations of individual systems can be obtained by extending the study of the transport process to include the associated exchange phenomena (*trans*-inhibition, *trans*-stimulation). In this phase, the identification of transport specific or preferential substrates and the generation of model transport analogs (Christensen, 1988*a*) can lead to a significant improvement in system discrimination. In particular, transport specific analogs provide information on the recognition properties of a system and can be used to obstruct the system for which they are substrates.

Further refinements in system discrimination require a strategy based on kinetic, thermodynamic, physiologic and genetic approaches.

Kinetic experiments in which the transport initial velocity of naturally occurring or nonmetabolizable model amino acids is measured over an extended range of substrate concentrations allow to identify the nonsaturable and saturable components contributing to the transport process within the selected conditions. The analysis, made by graphic transformations or by computerized mathematical procedures (Guidotti *et al.*, 1971), yields kinetic parameters that define the fluxes through each system or nonsaturable route involved (Dall'Asta *et al.*, 1983). Kinetic parameters can then be used to estimate the correct concentrations of substrate and competing analog (either a model or a natural molecule) in inhibition analysis and to evaluate the functional relevance of a transport system in a given biologic preparation (Franchi-Gazzola *et al.*, 1982).

Kinetic proof of shared transport between two competing molecules (substrate and analog) by inhibition analysis requires that the saturating activity of the substrate and that of the analog be essentially the same, i.e., that the K_i for the analog inhibiting the transport of the substrate be close to the K_m value of the substrate, and vice versa

(Christensen, 1985). And, since these parameters may carry large uncertainties under real conditions, a more complex test involving more than two competing molecules may be generated (Christensen, 1975). If in these experiments the competing molecules are transported by systems operating as ion-coupled cotransports, a serious error may be introduced by the analog transfer-dependent membrane depolarization that may simulate (or alter) inhibition of the substrate transport.

The kinetic parameters defining the transport of an amino acid are also relevant to the functional characterization of the systems that translocate it across a membrane: they provide an estimate of the relative contribution of each system to the total transport of that amino acid at different substrate concentrations including those occurring *in vivo* under physiologic and pathologic conditions (Franchi-Gazzola *et al.*, 1982; Dall'Asta *et al.*, 1990).

Thermodynamic approaches help system discrimination as follows. Ion-coupled amino acid cotransports are easily differentiated from ion-insensitive countertransports or facilitated diffusion: ion cotransport-driven systems stop working in the absence of a transmembrane ion gradient, whereas the countertransport-driven systems are still operative in these conditions (as long as the *trans*-accumulated amino acid species can supply energy for the process) and the same is true for any ion-independent passive transport (Gazzola *et al.*, 1980). Moreover, changes in the transmembrane ion gradients may affect differently the operation of separate ion-coupled cotransport systems depending of their coupling efficiency and stoichiometry (ion-to-amino acid flow ratio) (Gazzola *et al.*, 1984*b*). The sensitivity to the electric potential difference across the membrane can also contribute to discriminate among transport systems (Bussolati *et al.*, 1986; Rotoli *et al.*, 1991). For some systems that mediate the transfer of charged amino acid species the electric potential is the primary driving force (Bussolati *et al.*, 1987, 1989; Rotoli *et al.*, 1989); for systems that operate as ion-coupled cotransports the electric potential is a component of the driving force (electrochemical potential, Dall'Asta *et al.*, 1991); other systems mediating countertransport of electroneutral species are insensitive to changes of the electric potential (Bussolati *et al.*, 1986).

Discrimination among transport systems takes also advantage of physiologic approaches. A variety of cellular responses to altered environmental conditions (nutrient deprivation, alteration in the external ionic concentration and composition) and functional modifications associated with cell growth and density or promoted by chemical signals (local mediators, neurotransmitters and modulators, hormones) involve changes in the operation of amino acid transport systems. Some of these changes are specific for individual systems and can be identified on the basis of their time course and molecular mechanism. The discriminative usefulness of the relations between perturbation of cell physiology and changes in transport activity and capacity of systems has been reported (Franchi-Gazzola *et al.*, 1982; Dall'Asta *et al.*, 1990). Its importance is even greater when associated with information provided by the connected intracellular metabolic pathways and cell functions (Burger *et al.*, 1989).

Genetic approaches to the discrimination of transport systems usually combine the methodologies of formal genetics and molecular biology (whose ultimate purpose is to isolate and characterize transporters) with those of cell physiology, including rigorous measurements of transport (kinetics, inhibition analysis, regulatory properties) that identify them. Discriminative procedures based on genetic approaches will be described by D.L. Oxender in this book.

Cellular and Supracellular Functions of Transport Systems

All living cells require a continuous supply of amino acids. Scalar (amino acid biosynthesis) and vectorial processes (transmembrane transport) contribute to generate and maintain an internal pool of these molecules that are used for protein synthesis, cytoplasmic metabolism and transport to organelles.

The vectorial processes through the cell membrane are mediated by several transport systems. A common function of these systems is to take up amino acids from the extracellular fluids. The concentration of amino acids in the intracellular compartment is usually higher (and for some amino acid in some cell type much higher) than in the extracellular compartment and the generation of this in/out gradient by concentrative transport systems has an energetic cost. A high energy expenditure for inward amino acid transport might have little justification if the vectorial processes were merely intended as suppliers of substrates for the intracellular metabolism since a continuous influx of amino acids down their gradient from the external compartment would be adequate in most cases for this purpose. In this context, a high intracellular concentration of amino acids (or of some of them) would be required only to force a resistant step of metabolic pathway(s) providing essential molecules for cell survival. However, the occurrence of energy-consuming systems for the transmembrane transport of amino acids becomes clear when one considers other physiologic functions of these molecules:

(a) Nutritional homeostasis. A transient accumulation of some amino acid species into the cells of particular tissues is the concentrative step of the mechanisms that, by regulating amino acid release into the circulation, contribute to interorgan nutrition and insure organ cooperativity in amino acid metabolism (Christensen, 1982; Christensen *et al.*, 1986).

(b) Transcellular transport. Inward transport of amino acids against a concentration difference is required in cells engaged in absorption and resorption of these molecules from dilute fluids (as intestinal and renal cells, Silbernagl, 1988; Stevens *et al.*, 1984) or in cells that, for specific purposes, selectively extract and concentrate amino acids from the external milieu (Kilberg *et al.*, 1980; Johnson and Smith, 1988; Burger *et al.*, 1989).

(c) Cell volume control. Mechanisms that counteract perturbations of the cell volume involve changes in transmembrane fluxes of ions and organic osmolytes (Hoffmann, 1987; Macknight, 1988); amino acids are compatible organic osmolytes (Yancey, 1982) and their contribution to the volume regulatory mechanisms requires a vigorous uphill accumulation by concentrative, regulatable transport systems (Gazzola *et al.* 1991) and a fast outward release from a highly concentrated internal pool through the reverse operation of the same (or other) systems (Hoffmann and Hendil, 1976; Hoffmann and Lambert, 1983).

(d) Organelle function. Metabolism and activities of intracellular organelles require a continuous flow of amino acids from the cytoplasmic pool to the organelle matrix. Some fluxes are facilitated diffusions energized by the chemical potential generated by the concentration difference of the amino acid(s) across the organelle membrane. Therefore, conditions exist in which amino acids need to be concentrated intracellularly to maintain a sufficiently high rate of transfer from cytoplasm to organelles (Lenzen *et al.*, 1987; Pisoni *et al.*, 1987; Christensen, 1988*b*).

This survey of cellular and supracellular functions is by no means complete. Many other activities are restricted to particular cell types and will be described in more detail in other chapters of the book. What can be inferred here is that some of the amino acid transport systems serve housekeeping functions, essential for cell survival, and others promote specific processes, more or less critical for the survival of the entire organism. The former systems have ubiquitous distribution among cells (*public* systems, Christensen, 1984*b*), the latter are operative in a limited number of cell types (*private* systems, Karasov, 1983). These functional differences reflect a large heterogeneity (qualitative and quantitative) in the expression of the transporter sets among cells and tissues. One can say that the complement of transport systems at the plasma membrane of a cell is adjusted to match at best its own requirements and, at the same time, to cover the functions that the cell must accomplish as a component of the organism.

THE CELLS

Heterogeneity in Amino Acid Transport

The histologist recognizes some two hundred differentiated cell types in primates and a wide cell diversification is encountered in all mammals. Most nucleated cells express, with minor variations, a few constitutive transport systems, while other transporter sets are differentially expressed in individual cell types. As a consequence, each cell type usually owns a unique complement of amino acid transport systems. This heterogeneity holds for unrelated cell types as well as for such closely related cells as

Table 2. Amino acid transport systems in selected cell types of mammals.

	Some reactive substrates	Fibroblasts (human)[1]	Hepatocytes (rat)[2]	Exocrine pancreas cells (rat)[3]	Blastocyst (mouse)[2]	Placenta microvillous (human)[4] membrane vescicles		Enterocytes (rodent)[2] membrane vescicles	
						maternal–face	fetal–face	brush border	basolateral
"Neutral"	Pro, Gly, Ala, Ser Met, Gln	**A**	**A**	**A**	**(A)**	**A**	**A**		**A**
	Ala, Ser, Cys, Thr, Gln	**ASC**	**ASC**	**ASC**			**ASC**		**ASC**
	Small zwitterionic aa			**asc**		**asc**			
	Leu, Phe, Met, Cys, Gln	**L**	L_1, L_2	**L**		**L**	**L**	**L**	**L**
	Gln, Asn, His		**N**					**N**	
	Gly, Sarcosine		*Gly*	*Gly*	*(Gly)*				
	Pro, MeAIB							*IMINO*	
	Most zwitterionic aa							*NBB*	
	Phe, Met							*PHE*	
"Anionic"	Asp, Glu	X^-_{AG}	X^-_{AG}			X^-_{AG}	X^-_{AG}	X^-_{AG}	
	Glu, Cystine	x^-_C	x^-_C						
"Cationic"	Lys, Arg	y^+	y^+	y^+				y^+	
	Cationic aa				b^+_1, b^+_2				
"Cationic and neutral aa"	Most cationic and zwitterionic aa				$B^{0,+}$				
	Bulky cationic and α–β–unbranched zwitterionic aa				$b^{0,+}$				

[1] See Tables 3 and 4

[2] For details see Kilberg and Häussinger (hepatocytes), Van Winkle (blastocyst), and Wright and Stevens (enterocytes) in this book.

[3] Mann and Peran, 1986; Mann *et al.*, 1986; Munoz *et al*, 1988; Von Blankenfeld and Jauch, 1989

[4] Johnson and Smith, 1988; Hoeltzli and Smith, 1989; Moe and Smith, 1989; Hoeltzli *el al*, 1990

fibroblasts and endothelial cells (Gazzola *et al.*, 1980; Mann *et al.*, 1989; Dall'Asta *et al.*, 1990; Herskowitz *et al.*, 1991), lymphocytes and macrophages (Borghetti *et al.*, 1981; Sato *et al.*, 1987), and even in cells of the same type in populations devoted to perform different functions (periportal and perivenous hepatocytes, Haussinger and Gerok, 1983; Taylor and Rennie, 1987; Burger *et al.*, 1989). In polarized cells, a further heterogeneity arises from a nonrandom distribution of the transporter sets within the cell membrane (Fass *et al.*, 1977; Hammerman and Sacktor, 1977; Stevens *et al.*, 1984; Silbernagl, 1988). This intracellular heterogeneity is outstanding in cells engaged in transcellular transport processes and will be described in detail in other chapters of this book.

The heterogeneity in the distribution pattern of transport systems among selected cell types of mammals is exemplified in Table 2. System acronyms reported in boldface agree with the provisional designation rules adopted by a group of students of the field in a statement intended to minimize complexity in the matter (Bannai *et al.*, 1984).

The analysis of the Table shows that three major systems, denominated systems A, ASC and L, serve for neutral, zwitterionic amino acids; two, systems X^{-}_{AG} and x^{-}_{C}, serve for anionic amino acids; and one, system y^{+} mediates the transport of cationic amino acids. These systems have been described in some detail in several biologic preparations. An accurate description of them is available for rat hepatocytes (Kilberg, 1982), where system L is possibly split into two sub-systems (Weissbach *et al.*, 1982) and an additional Na^{+}-dependent agency, known as system N (for nitrogen), mediates the transport of the amidic amino acids (glutamine, asparagine) and histidine (Kilberg *et al.*, 1980).

The amino acid transport systems have been extensively characterized in cultured human fibroblasts. Kinetic behaviour, energization, transport mechanism and physiologic relevance have been defined for each system together with its regulation by local and systemic factors (Gazzola *et al.*, 1980, 1981*b*, 1991; Franchi-Gazzola *et al.*, 1982; Dall'Asta *et al.*, 1983, 1990, 1991; Longo *et al.*, 1985; Bussolati *et al.*, 1987, 1989). This biologic preparation has therefore been selected as a model for the analysis of the transport systems that follows in the next paragraph.

The Fibroblast Complement of Amino Acid Transport Systems

(*a*) <u>Transport systems for neutral amino acids</u> (Table 3)

System A is an adaptive agency that accepts amino acids with unbranched side chain of various lengths as glycine, L-alanine, L-methionine, L-glutamine and the imino acid L-proline (perhaps the most specific natural substrate of System A in human fibroblasts), it has an incomplete stereoselectivity and tolerates the alkylation of the α-amino group. Methylamines as sarcosine and betaine are substrates of this system for which 2-(methylamino)isobutyric acid (MeAIB) has been developed as a nonmetabolizable model transport-specific analog. The operation of System A requires sodium, is sensitive to changes in electric membrane potential ($\Delta\psi$), is inhibited by substrates in the *trans*-compartment (*trans*-inhibition) and has an apparent affinity constant for transported amino acids in the millimolar region. The proposed transport mechanism for System A is an electrogenic sodium symport (secondary active transport, Fig. 2), with a stoichiometry amino acid/sodium of 1:1 (Fig. 3), energized by the gradient of electrochemical potential of sodium ($\Delta\tilde{\mu}_{Na}$, Fig. 4).

System ASC is a constitutive agency that accepts amino acids with linear side chain of various length such as L-alanine and L-glutamine, with some preference for polar substrates such as L-serine and L-cysteine. L-Threonine is the best natural site ASC-reactive substrate in human fibroblasts (Bussolati *et al.*, 1991). System ASC has a stringent stereoselectivity and rejects analogs with alkyl substitutions at the α-amino group. Its apparent affinity constant for substrates ranges between 0.05 and 0.15 mM. System ASC is sodium dependent and sensitive to $\Delta\psi$, but not electrogenic. Its operation is stimulated by substrates in the *trans*-compartment (*trans*-stimulation). A possible transport mechanism for system ASC is an antiport of substrate amino acids obligatorily associated with a movement of sodium in both directions (Bussolati *et al.*, 1992).

Table 3. Transport systems for neutral (zwitterionic) amino acids in human fibroblasts.

Designation	A	ASC	L
Relevant Substrates	Pro,Gly,Ala, Met,Gln,MeAIB	Ala,Ser,Cys Thr,Gln	Leu,Phe,Cys BCH
Recognition:			
-Net charge (pH 7.4)	nil	nil	nil
-Side chain	aliphatic, unbranched	short, polar preferred	long, also branched or ring-shaped
-Stereoselectivity	incomplete	strict	incomplete
-Alkyl substitution at α-N	accepted	rejected	rejected
Operation:			
-Na^+-dependence	yes	yes	no
-$\Delta\psi$-dependence	yes	yes	no
-*trans*-substrate-driven movements	inhibition	stimulation	stimulation
-K_m for substrates	~1 mM	50–150 μM	~10 μM
Transport mechanism:	Na^+-symport (aa:Na = 1:1)	Na^+-symport (aa:Na = 1:1)	Substrate-coupled antiport
Driving force:	$\Delta\tilde{\mu}_{Na}$	*trans*-concentration of Na^+ and substrate aa	*trans*-concentration of substrate aa

System L has a broad specificity, accepting neutral amino acids with linear or branched side chains of various lengths and aromatic bulky structures. 2-Aminobicyclo(2,2,1)heptane-2-carboxylic acid (BCH) has the requisites for a model transport site L-reactive analog. System L does not tolerate alkyl substitutions for hydrogen at the α-nitrogen and has an incomplete stereoselectivity. Its operation is independent of sodium and of electric membrane potential, being strongly stimulated by substrates present in the *trans*-compartment. System L has a very low apparent affinity constant for substrate amino acids (K_m of 10 μM for leucine) and is unable to generate high uphill gradients for its substrates. The proposed transport mechanism for this agency is a substrate-coupled antiport (tertiary active transport, Fig. 5), whose energization is provided by the chemical gradient of transmembrane site L-reactive substrates. The inward transport of amino acids through system L takes advantage of the transmembrane cycle of such amino acids as L-methionine and L-cysteine (site L-reactive

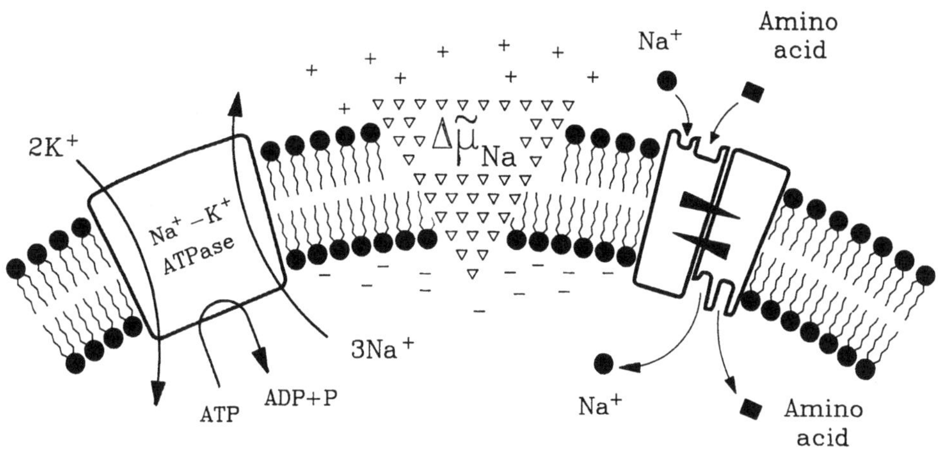

Fig. 2. Proposed transport mechanism for System A. The primary active transport system (Na^+-K^+ ATPase) moves sodium ions from the inside of the cell to the outside (and potassium ions in the opposite direction), generating an electrochemical gradient of sodium ($\Delta\tilde{\mu}_{Na}$). System A is a secondary active transport system that uses the inward backflux of sodium to accumulate neutral amino acids.

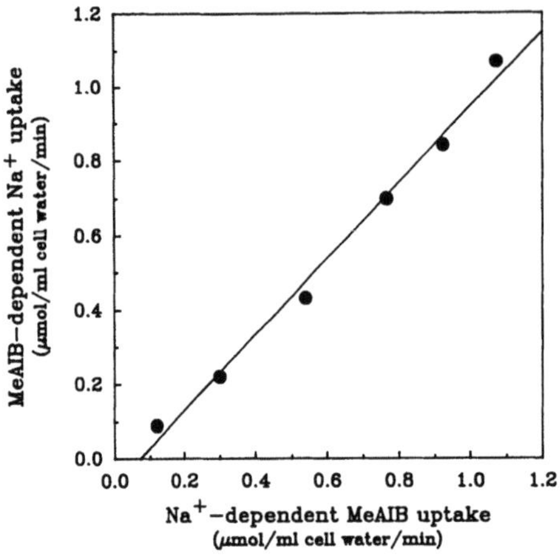

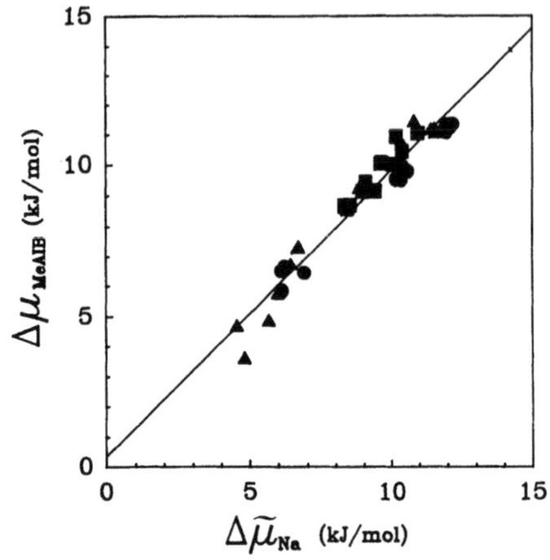

Fig. 3. *(left)* Relationship between the MeAIB-dependent component of Na^+ influx and the Na^+-dependent component of MeAIB influx in cultured human fibroblasts. The linear regression analysis of the data yields a slope of 1.02 ± 0.05, indicative of a 1:1 sodium-MeAIB coupling ratio (from Dall'Asta *et al.*, 1991, with permission).

Fig. 4. *(right)* Accumulation of MeAIB as a function of the Na^+ electrochemical potential. The value of $\Delta\tilde{\mu}_{Na}$ is varied by changing the transmembrane gradient of the Na^+ chemical potential (●), the electric potential difference across the membrane (■) or both components (▲). The linear regression analysis of the data yields a slope of 0.95 (r=0.97), indicative of a constant ratio close to 1 between the molar energy stored in the gradient of MeAIB accumulated intracellularly and the energy made available by the Na^+ electrochemical gradient (from Dall'Asta *et al.*, 1991, with permission).

substrates) that are concentrated intracellularly by uphill-generating agencies with overlapping reactivity toward them. Methionine and cysteine are substrates of systems A and ASC, respectively. Their intracellular preaccumulation *trans*-stimulates the activity of system L and, by exchange, promotes the inward movement of other amino acid substrates of this agency.

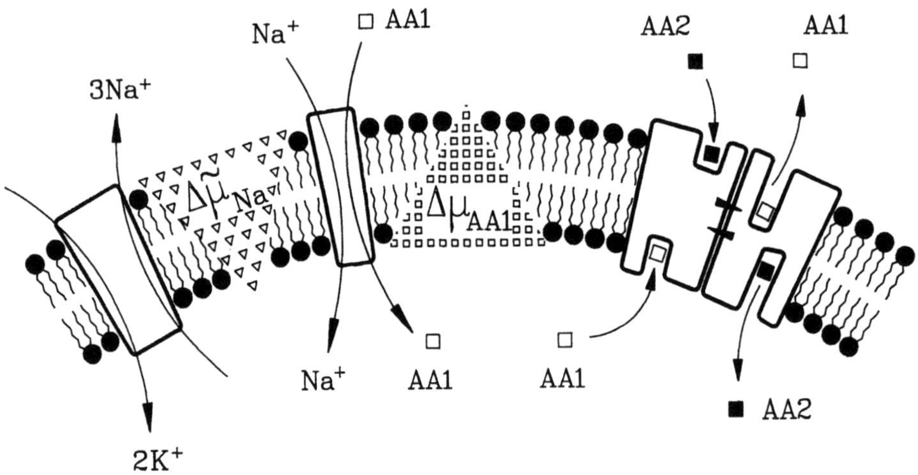

Fig. 5. Proposed transport mechanism for systems L and x^-_C. Secondary transport systems driven by $\Delta\tilde{\mu}_{Na}$ accumulate amino acids (AA1) inside the cell generating a transmembrane chemical gradient of these molecular species (that have overlapping reactivity to systems L or x^-_C). Systems L and x^-_C are tertiary active transport systems that use downhill efflux of AA1 to move inwardly, by countertransport, different amino acid substrates (AA2).

Table 4. Transport systems for anionic and cationic amino acids in human fibroblasts.

Designation	X^-_{AG}	x^-_C	y^+
Relevant Substrates	Asp,Glu	Glu,Cystine	Lys,Arg
Recognition:			
-Net charge (pH 7.4)	negative	negative	positive
-Side chain	linear (2, 3 C atoms)	3 or more atoms (C or S) between negative charges	3, 4 C atom-side chain preferred
-Stereoselectivity	anomalous	incomplete	strict
-Alkyl substitution at α-N	rejected	(?)	rejected
Operation:			
-Na^+-dependence	yes	no	no
-$\Delta\psi$-dependence	yes	no	yes
-*trans*-substrate-driven movements	(?)	stimulation	stimulation
-Km for substrates	~20 μM	100-200 μM	50-100 μM
Transport mechanism:	Na^+-symport (Na:aa ≥ 2)	Substrate-coupled antiport	Facilitated diffusion
Driving force:	$\Delta\tilde{\mu}_{Na}$	*trans*-concentration of substrate aa	$\Delta\psi$

(*b*) Transport systems for anionic and cationic amino acids (Table 4)

System X^-_{AG} has aspartic acid and glutamic acid as primary natural substrates. Its stereoselectivity is anomalous, accepting the L stereoisomer of both amino acids and D-aspartate, but not D-glutamate (Gazzola *et al.*, 1981*a*). System X^-_{AG} is sodium dependent and sensitive to $\Delta\psi$. The apparent affinity constant for both substrates is rather low (20 μM). The proposed transport mechanism is a sodium cotransport energized by $\Delta\tilde{\mu}_{Na}$, but neither the stoichiometry amino acid/sodium (1:2 or 1:>2), nor the requirement and role of K^+ for system operation have been firmly established.

System x^-_C transports L-glutamate and L-cystine (as tripolar ion, a species present in a definite amount in fluids at neutral pH). Its operation does not require sodium, is insensitive to changes in electric membrane potential and is stimulated by substrates present in the *trans*-compartment. System x^-_C has an apparent affinity constant for substrates of 0.1-0.2 mM and is unable to generate high transmembrane amino acid gradients. As for System L, the proposed transport mechanism for system x^-_C is a substrate-coupled antiport (Fig. 5) energized by the chemical gradient of transmembrane substrates of this agency. The glutamine-cystine cycle (Bannai-Ishii, 1988) illustrates how system x^-_C can operate as a tertiary active transport for cystine uptake by human fibroblasts. L-Glutamine enters the cell through systems ASC and A (Dall'Asta *et al.*, 1990) and a significant part of it is converted intracellularly to glutamate. This conversion and the direct accumulation of L-glutamate by system X^-_{AG} generates a high in/out gradient for this amino acid, whose efflux down its gradient through system x^-_C represents the driving force for cystine entry by countertransport.

System y^+ accepts cationic amino acids (as L-lysine and L-arginine) with strict stereoselectivity. Its operation is independent of sodium, strongly dependent of electric membrane potential and stimulated by substrates present in the *trans*-compartment. System y^+ has an apparent affinity constant for substrates of 0.05-0.1 mM. For this system, the transport mechanism is a facilitated diffusion driven by the transmembrane electric potential difference (Fig. 6). One of the major substrates, L-arginine, is concentrated into fibroblasts according to its electrochemical potential and the amino acid distribution ratio at equilibrium exhibits a Nerstian dependence on the electric membrane potential of the cell (Bussolati *et al.*, 1987).

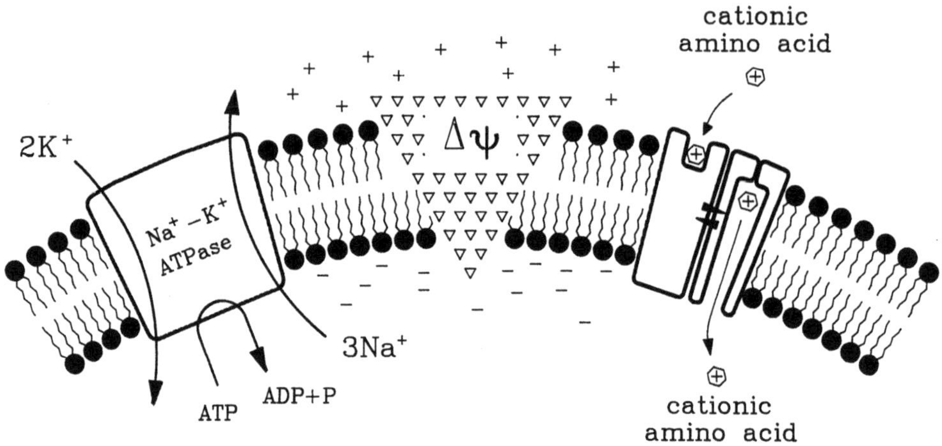

Fig. 6. Proposed transport mechanism for System y^+. Positively charged (cationic) amino acids enter the cell by a mechanism of facilitated diffusion driven by the transmembrane electric potential difference ($\Delta\psi$).

REGULATION OF AMINO ACID TRANSPORT

Control Mechanisms

In the preceding paragraphs we have visualized the biologic transport of amino acids as a process defined by the binding of these molecules to a reactive site at one surface of a membrane, and their release in an unmodified form at the other surface. This cell function has developed early in evolution as a means for capturing amino acids from a dilute environment, since the cell requires these molecules for survival. With the appearance of transport systems endowed with specific recognition properties toward substrates, the cell must have developed some form of regulation to ensure that their activity were adapted to changes in the environmental conditions and, as the cells became organized into a multicellular organism with diversification of their metabolic functions, a more complex regulatory network had to be superimposed to establish mutual controls among the various cells, tissues and organs. Now we know that in multicellular organisms, including mammals, the activity and capacity of amino acid transport systems are modulated or affected by a number of control signals and regulatory mechanisms active at different levels of organization.

Levels of Organization: Cell, Cell Population, Organism

The regulatory network of amino acid transport systems consists of control mechanisms that act at the cell level, at the level of cell populations and at the organismic level with effects that extend to include several tissues and organs (Table 5).

Table 5. Regulation of amino acid transport, level of organization.

Level of organization	Regulatory process	Involved systems
Cell	Adaptive regulation	A - x_C^- - N - L
Cell population	Modulations by cell proliferation and density (cultures)	A - ASC - L - x_C^- - X_{AG}^-
Organism	Controls by chemical signals (hormones, etc.)	A - x_C^- - (L)

Table 6. Regulatory mechanisms of amino acid transport, main features.

Regulatory process	Involved systems	Regulatory signal		Major effects
		Source	Target	
Adaptive regulation	A - x^-_C	external substrate concentration	nucleus plasma membrane	changes in system capacity
Cell density (cultures)	A - ASC x^-_C - X^-_{AG}	plasma membrane (?)	plasma membrane	changes in operation rate
Insulin regulation	A - x^-_C	hormone–receptor binding	nucleus plasma membrane	changes in system capacity

Examples of control mechanisms are adaptive regulation of system A, x^-_C, N and L (cell level, Guidotti *et al.*, 1978; Kilberg *et al.*, 1980; Handlogten *et al.*, 1982; Shotwell *et al.*, 1982, 1983; Dall'Asta *et al.*, 1983), regulations of systems A, ASC, X^-_{AG} and x^-_C associated with cell growth and density (cell populations in culture and tissues, Borghetti *et al.*, 1980; Petronini *et al.*, 1982; Piedimonte *et al.*, 1982; Dall'Asta *et al.*, 1983; Gazzola *et al.*, 1984*a*), and systemic actions of chemical signals as hormones and neurotransmitters on a variety of transport systems (Guidotti *et al.*, 1978; Shotwell *et al.*, 1983; Lerner, 1985). Some of these modulations occur in minutes inducing rapid (and usually transient) changes in the activity of a transport system (Edmondson and Lumeng, 1980; Moule and McGivan, 1987; Woodlock *et al.*, 1989). More commonly they require several hours to develop, affect primarily the capacity of a system and promote a transport change that is maintained over an extended period of time. Adaptive regulation, cell density control and regulation by insulin have been selected as typical examples of slow-acting mechanisms found operative in cultured human fibroblasts and will be described in some detail. Their features are sketched in Table 6. Remarkably, the activity of systems A and x^-_C is modulated by all these mechanisms.

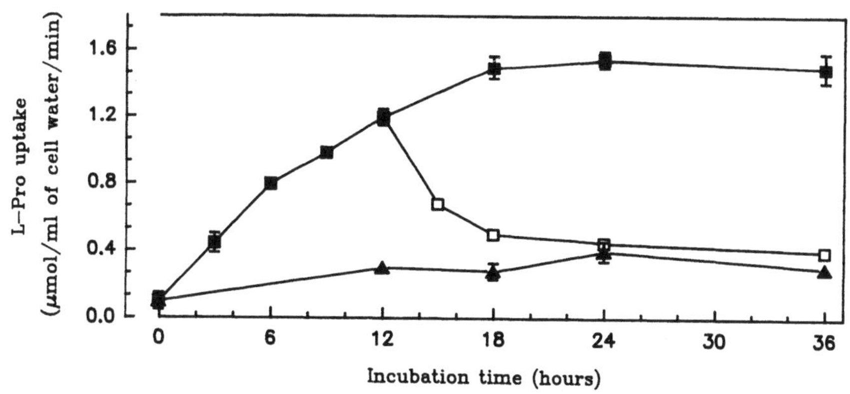

Fig. 7. Starvation-induced enhancement of L-proline uptake by cultured human fibroblasts and its reversal by exposure of the cells to a full complement of naturally occurring amino acids. Fibroblasts grown to confluence in an amino acid-containing culture medium were either maintained in the same medium (repressed controls, ▲) or incubated in an amino acid-free medium (starvation-induced derepression, ■) and, after a 12-h derepression, reinstated to the medium containing a full complement of amino acids (adaptive repression, □). Transport measurements are initial rates of proline entry.

(*a*) Adaptive regulation

Adaptive regulation of system A has been formerly identified as a mechanism of cell response to nutritional stress (amino acid starvation versus replenishment, Gazzola *et al.*, 1972, 1973; Dall'Asta *et al.*, 1978) widely distributed in nature (Guidotti *et al.*, 1975, 1978; Logan *et al.*, 1982; Shotwell *et al.*, 1983; Boerner and Saier, 1985; Kilberg *et al.*, 1985). It consists of a starvation-induced derepression of transport activity and of its reversal by amino acid refeeding. Fig. 7 illustrates the changes in the initial rate of inward transport of L-proline, a naturally occurring site A-reactive substrate, when cultured human fibroblasts are transferred from an amino acid containing- to an amino acid free culture medium (starvation-dependent derepression phase) and subsequently reinstated to the medium containing a full complement of amino acids (repression phase). A similar trend can be observed with substrates of system x^-_C.

The increase in transport activity of system A during the derepression phase reflects a progressive increment in transport maximum (V_{max}) in the absence of

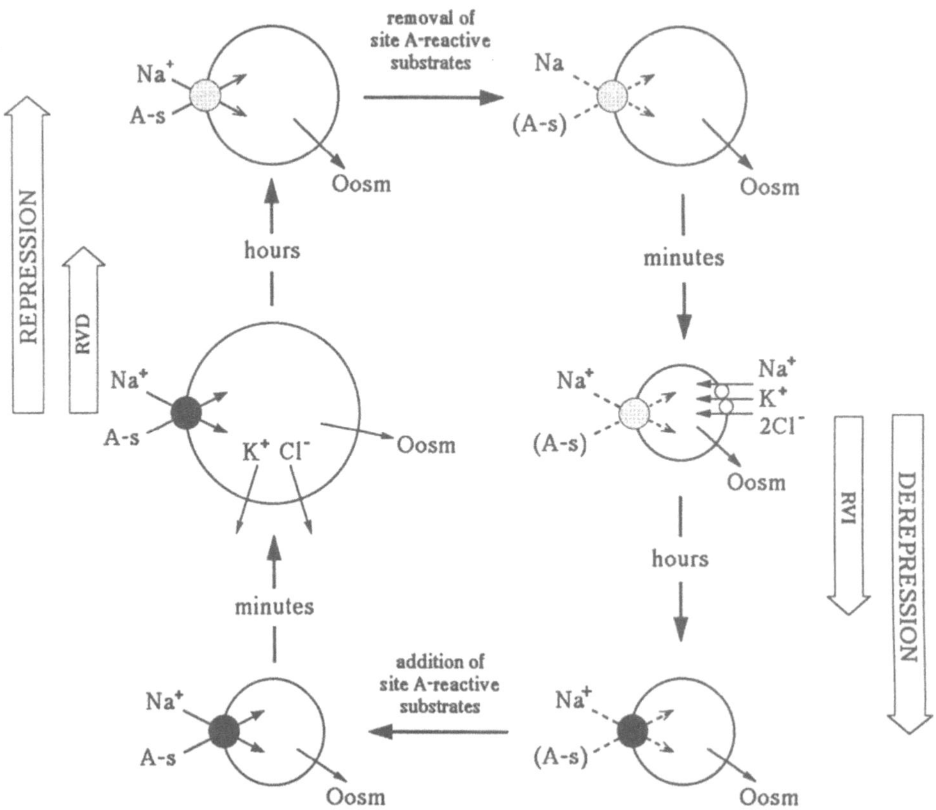

Fig. 8. Role of site A-reactive substrates in the regulation of cell volume and effect of cell volume changes on the activity of transport system A. Regulatory volume increase (RVI) - regulatory volume decrease (RVD) responses complemented by changes in the concentration of intracellular osmolytes induced by the adaptive regulation of transport system A. Model developed for human fibroblasts in culture: A-s, site A-reactive substrates; (A-s), absence of site A-reactive substrates; Oosm, organic osmolytes. Representation of system A: shaded circles, repressed state (low capacity); black circles, derepressed state (high capacity). Arrows through the circles representing system A are continuous when the system is operative in the presence of external substrates and interrupted when the system is nonfunctioning in the absence of external substrates. (from Gazzola *et al.*, 1991, with permission)

appreciable changes in substrate concentration for half-maximal transport (K_m) and requires a *de novo* protein and RNA synthesis (Franchi-Gazzola *et al.*, 1973; Gazzola *et al.*, 1981*b*). An active macromolecular synthesis is also needed for repression, a phase in which transport activity reverts to the initial value. Transport repression is induced by substrate(s) of system A, but not by amino acids that lack site A-reactivity (Gazzola *et al.*, 1973; 1981*b*).

Recently, adaptive regulation of system A has been given an additional role in cell volume control (Gazzola *et al.*, 1991) as a regulator of the transmembrane flux of compatible organic osmolytes (site A-reactive amino acids and methylamines) that protects the cell exposed to conditions of osmotic stress (Fig. 8).

(*b*) Modulation by cell density

In cultured human and murine fibroblasts, the activity of several transport systems is known to vary with cell density. In particular, the inward transport of amino acid substrates of systems A, ASC and x^-_C decreases with the increase of cell density (Borghetti *et al.*, 1980; Dall'Asta *et al.*, 1983), whereas an enhanced cell density of the culture is associated with a rise in transport of site X^-_{AG}-reactive substrates (Dall'Asta *et al.*, 1983). Similar density-dependent decrease in the activity of systems A and ASC has been observed in a Burkitt lymphoma cell line (Piedimonte *et al.*, 1989). The mechanism(s) underlying these modulations are poorly understood. Kinetic studies concerning the inward transport of L-glutamate in cultured human fibroblasts (Dall'Asta *et al.*, 1983) indicate that the cell density-dependent increase or decrease in transport activity of the systems involved (X^-_{AG}, x^-_C, ASC) arise from changes in transport maximum, with no substantial alterations of K_m (Table 7). Whether the changes in V_{max} depend on an altered number of active transporters at the membrane, caused by a change in the rate of their synthesis or by a modulation of their distribution between surface and cell interior, remains to be established.

(*c*) Insulin regulation

Extensive studies in the last three decades have provided abundant information on the effects of hormones and other chemical signals on the transport of amino acids in cells and tissues. The organismic regulation by chemical signalling concerns a variety of metabolic actions, among which transport changes may play a relevant role. Indeed, part of the regulation of metabolism by systemic signals depends upon controlling the access of substrates through membranes. Owing to the interplay between transport and the ensuing reaction steps of metabolism and to the complex interactions among different hormones that concur to elicit a transport response, no attempts will be made to analyze all these aspects in detail. Instead, we report a selected example of regulation of amino acid transport by a systemic signal, insulin. A comprehensive description of the effects of this hormone on the whole complement of amino acid transport systems is available for cultured human fibroblasts (Longo *et al.*, 1985).

In human fibroblasts, insulin markedly enhances the activity of two systems, A and x^-_C, while the four other systems (ASC, L, X^-_{AG} and y^+) found operative in this

Table 7. Kinetic parameters for the uptake of L-glutamic acid by human fibroblasts at different culture densities.

Transport system	Low density		High density	
	K_m	V_{max}	K_m	V_{max}
	mM	$\frac{\mu mol \cdot min^{-1}}{ml\ cell\ water}$	*mM*	$\frac{\mu mol \cdot min^{-1}}{ml\ cell\ water}$
System X^-_{AG}	0.018	0.019	0.015	0.044
System x^-_C	2.46	0.59	2.64	0.27
System ASC	0.33	0.08	0.26	0.03

Table 8. Amino acid transport in cultured human fibroblasts. Effect of insulin*.

Transport system	Substrate	[mM]	v(nmol/min/ml cell water)		Difference (%)
			No insulin	Insulin	
A	L-Proline	0.1	59	380	+ 544
ASC	L-Serine	0.1	380	386	+ 2
X^-_{AG}	L-Aspartate	0.01	58	79	+ 36
x^-_C	L-Cystine	0.05	29	89	+ 207
	L-Glutamate	0.02	4	12	+ 200
L	L-Leucine	0.1	301	171	- 43
y^+	L-Arginine	0.02	62	59	- 5

* Insulin (1 μg/ml) added to the medium (Earle's balanced salt solution) during a 12 h-incubation. Transport measurements are initial entry rates of the indicated amino acid under conditions that restrain its reactivity to a single transport system. (rewritten from Longo *et al.*, 1985, with permission)

biologic preparation are insensitive or barely responsive to the hormone (Table 8). The time-course of insulin action on the transport of L-proline, a site A-reactive substrate, and of L-cystine, a site x^-_C-reactive substrate, indicates that the hormone promotes systems A and x^-_C by a slow-acting mechanism that requires hours to attain its maximal effect (Fig. 9).

The manner in which insulin stimulates the inward transport of substrates appears to be similar in the two responsive systems. In both cases, the half-maximal stimulation of substrate transport is obtained with comparable concentrations of insulin (Fig. 10), at a value (10^{-9} M) that lies within the physiologic concentration range of the hormone in human plasma and approaches that of the affinity constant for insulin-receptor binding in cultured human fibroblasts. In both cases, the stimulation of transport activity by insulin relies on an increase of transport maximum, without substantial changes in the K_m values. The latter result suggests that the hormone enhances the availability of active transporters of system A and system x^-_C at the cell membrane. The slow and long lasting action of the hormone and its suppression by inhibitors of protein and RNA synthesis are compatible with a mechanism of insulin action that involves a *de novo* synthesis of transporters.

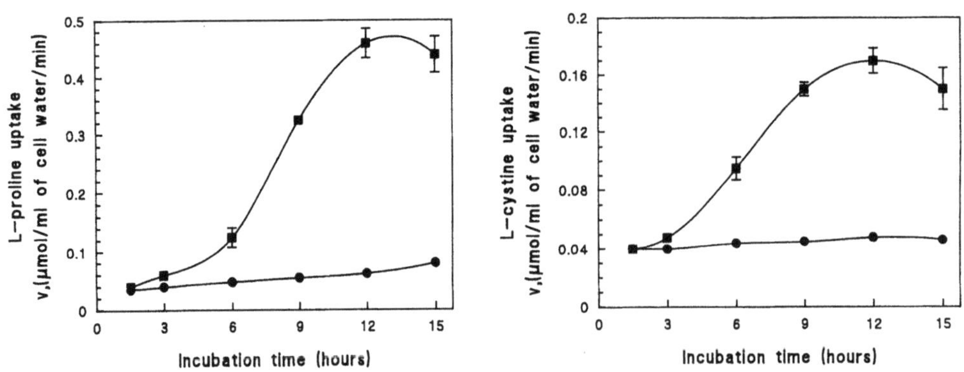

Fig. 9. Time-course of insulin action on the inward transport of L-proline and L-cystine in cultured human fibroblasts. Cells were incubated in Earle's balanced salt solution in the absence (•) and in the presence (■) of insulin (1 μg/ml). Transport measurements are initial entry rates of L-proline through system A (left) and of L-cystine through system x^-_C (right). Each point is the mean (±S.D.) of triplicate determinations. Standard deviations, if not shown, are within the symbol (from Longo *et al.*, 1985, with permission).

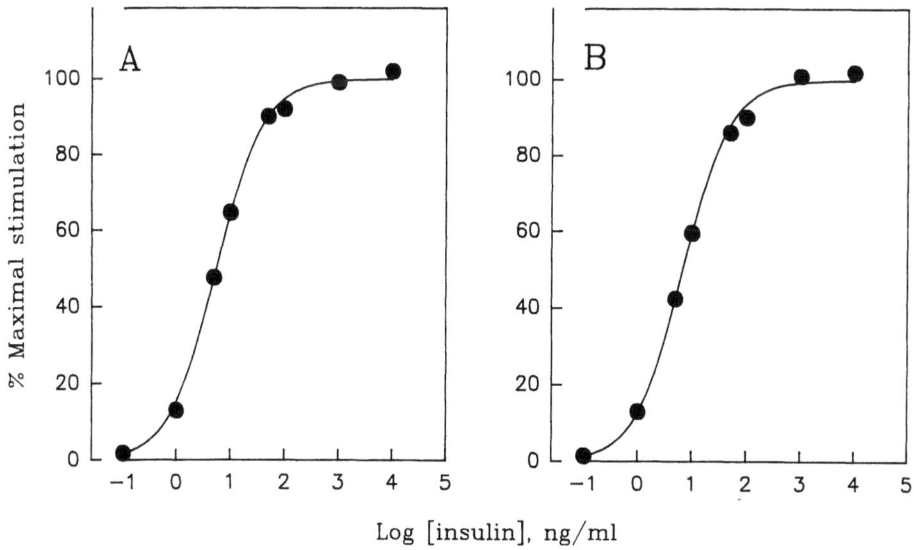

Fig. 10. Dose-response curves for the stimulation of substrate transport through system A (panel A) and system x^-_c (panel B) by insulin. Cultured human fibroblasts were incubated for 6 h in Earle's balanced salt solution containing increasing concentrations of insulin. Transport measurements were initial entry rates of L-proline (system A) and L-cystine (system x^-_c). For each amino acid, insulin-stimulated transport activity corrected for the basal uptake is expressed as percentage of the maximal stimulation. (Panel B from Longo *et al.*, 1985, with permission)

Cooperativity and Coordination of Transport Systems

Cooperation of vectorial and scalar processes in amino acid metabolism occurs at the cell level and among tissues and organs.

The contribution of transmembrane fluxes depends on the activity of multiple amino acid transport systems differently expressed in the various sections of the organism. In cells, system cooperativity tends to adjust the transport of substrates to physiologic changes in their availability and may have a role in promoting useful cell responses under pathologic conditions. At the organismic level, the coordination of amino acid transport systems differently distributed in tissues and organs ensures interorgan nutrition, organ cooperativity in metabolism and, possibly, protective dampening against pathologic derangements.

(*a*) Cooperativity in the cell

Let's consider the uptake of L-alanine, L-serine and L-cysteine by typical mesenchymal cells as human fibroblasts. In these cells, the three amino acids are transported by three overlapping independent routes, systems ASC, A and L (Franchi-Gazzola *et al.*, 1982). The inspection of the apparent affinity constants of systems ASC and A for each substrate (Table 9) indicates that, at extracellular concentrations close to those present in human plasma (Brigham *et al.*, 1960), all the three amino acids enter the cell primarily through system ASC (Fig. 11) and that, under physiologic conditions, their uptake must be dependent on the fluctuations in the extracellular substrate concentration. On the other hand, alanine, serine and cysteine are also substrates of system A, whose activity increases adaptively upon starvation (being inversely correlated with the extracellular concentration of site A-reactive amino acids). Therefore, the contribution of system A to the total uptake of the three amino acids increases in fasting. The results obtained *in vitro* with cultured cells suggest that, *in vivo*, system ASC and A assist each other to maintain and adequate inward flux of alanine, serine and cysteine

Table 9. Kinetic parameters for Na^{+}-dependent uptake of L-alanine, L-serine, and L-cysteine by human fibroblasts.

Amino acid	System ASC		System A	
	K_m	V_{max}	K_m	V_{max}
	mM	*μmol · min⁻¹ / ml cell water*	*mM*	*μmol · min⁻¹ / ml cell water*
L-Alanine	0.10	3.7	0.60	3.6
L-Serine	0.15	3.8	0.73	4.1
L-Cysteine	0.06	2.0	1.4	3.5

into fibroblasts: when the external concentration of these amino acids decreases, the rate of their entry via system ASC slows down, but the enhanced activity of a derepressed system A tends to counteract this slackening; conversely, when the external substrate concentration increases, the enhanced rate of amino acid entry via system ASC tends to buffer the progressive decrease of the activity of system A induced by repression and *trans*-inhibition (Gazzola *et al.*, 1980, 1981*b*).

Another amino acid, L-glutamine, enters human cultured fibroblasts through all the systems that mediate the transport of neutral amino acids in these cells, and the relative contribution of systems A, ASC and L varies as a function of the amino acid concentration in the cell surroundings (Dall'Asta *et al.*, 1990). At glutamine levels comparable to those found in human plasma under physiologic conditions (0.6 mM), system ASC is the leading transport agency but, as the external concentration of the amino acid increases, system A progressively takes over (Fig. 12). Compared with system ASC, system A has a lower affinity but a substantially higher capacity for L-glutamine (in cells that lack the amidic amino acid-specific system N) and these kinetic properties may have significant implications in pathologic conditions associated with enhancements of glutamine in the extracellular fluid. Injury of tissues is associated with the release of large amounts of glutamine from skeletal muscle (Watford, 1989) and damaged muscle cells may increase glutamine outflux (Goldstein, 1986) with enhancement of the amino acid concentration in the local environment. Glutamine promotes DNA synthesis and cell proliferation in cultured fibroblasts (Zetterberg and Engstrom, 1981) and has a central

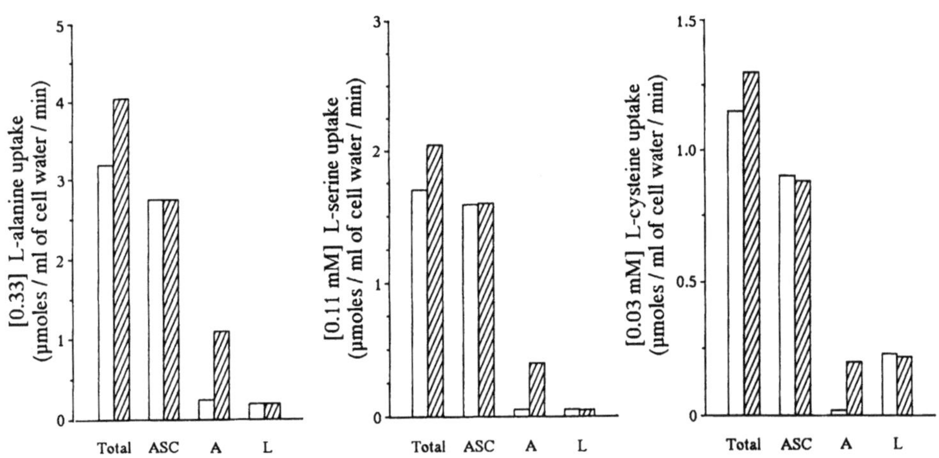

Fig. 11. Pathways of L-alanine, L-serine and L-cysteine uptake by cultured human fibroblasts at amino acid concentrations present in human plasma. Transport measurements in unstarved, repressed (empty bars) and in starved, derepressed cells (shaded bars) are initial entry rates of the indicated amino acid under conditions that restrain its reactivity to a single transport system. (Rewritten from Franchi-Gazzola *et al.*, 1982).

Kinetic parameters for the uptake of L-glutamine by cultured human fibroblasts.

Transport system	K_m	V_{max}
	mM	*$\mu mol \cdot min^{-1}$ / ml cell water*
ASC	0.15	1.67
A	1.66	2.99
L	0.77	1.39

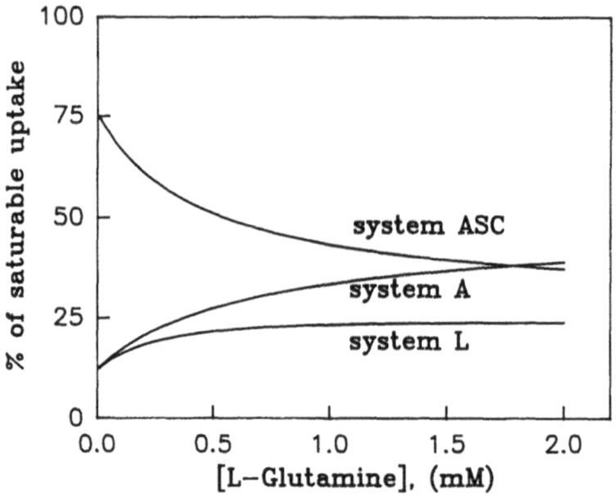

Fig. 12. Expected per cent contribution of systems A, ASC and L to the total saturable uptake of L-glutamine by cultured human fibroblasts as a function of the substrate concentration. The curves are calculated on the basis of the kinetic parameters reported at left. (Rewritten from Dall'Asta *et al.*, 1990).

role in the metabolism of rapidly dividing cells by providing the nitrogen for the synthesis of purines and pyrimidines, and some other nitrogenous compounds needed in the manufacture of DNA, RNA and other macromolecules. The increased availability of glutamine in the environment associated to a site A-mediated faster uptake of this amino acid by the fibroblasts may be of vital importance in wound healing.

(*b*) Interorgan coordination

The potential for interorgan nutrition, interorgan cooperation in amino acid metabolism and dampening actions by tissue reservoirs depends on the inherent distribution among tissues of the enzymes of the relevant metabolic pathways and the transport of amino acids and intermediates of the pathways between organs. This integration of scalar and vectorial processes generates a network of interorgan substrate fluxes that requires control and coordination. The interorgan flows needed to insure an adequate and continuous cell nourishment and to provide substrates for specific cell functions will be dealt by H.N. Christensen later in this book.

TRANSPORT OF AMINO ACIDS IN CELL AGING AND NEOPLASTIC TRANSFORMATION

Cell Aging

Aging at the cell level involves functional changes of membranes, in turn associated with alterations in the transport of nutrients (Polgar *et al.*, 1978; Wheeler, 1981, 1982; Brookes, 1988). Cell aging has been widely studied using cultured human fibroblasts and a few other biologic preparations (Hayflick, 1980; Macieiera-Coelho, 1988), but its effect on the transport of amino acids has received little attention. Even in fibroblasts, information is limited to phenomenologic descriptions of changes that occur along the *in vitro* aging of the cells for systems that mediate the transport of neutral and cationic amino acids.

Experiments in which the effects of aging could be separated from those of cell density of the culture (Gazzola *et al.*, 1984; Bussolati *et al.*, 1990) have evidenced: (*a*) a progressive decline in the activity of system ASC as phase II of the *in vitro* lifespan of human fibroblasts proceeds (Fig. 13); (*b*) a significant decrement in the activity of system A between early- and late-phase II human fibroblasts, dependent on a change in the maximal capacity of the system (Table 10); (*c*) no significant decrease in the activity of the Na^+-independent systems L and y^+ during phase II of human fibroblasts, in spite of some indication of the contrary for system L (Newton *et al.*, 1984) and (*d*) a marked

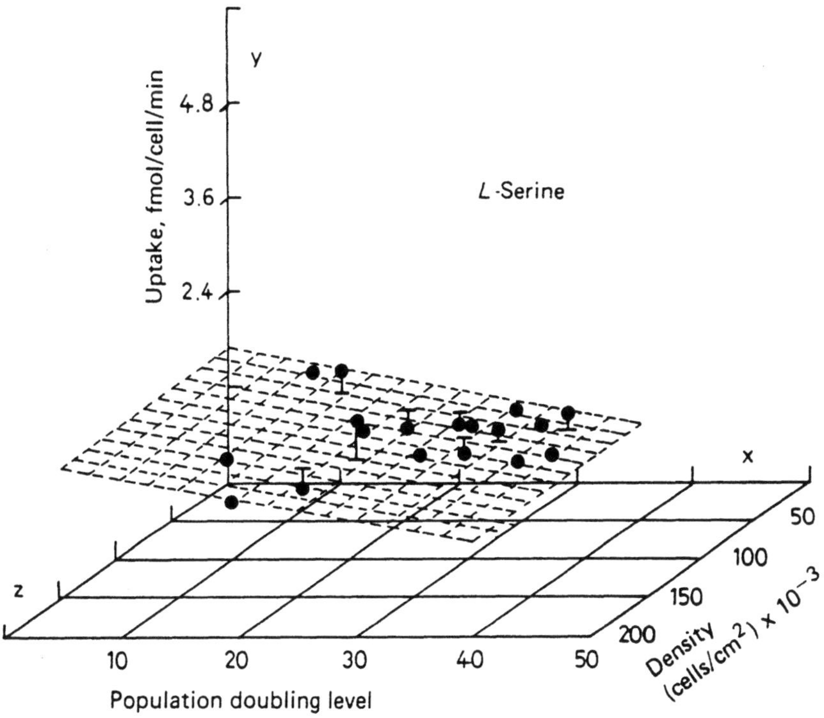

Fig. 13. Effect of *in vitro* cell (aging on transport activity of (system ASC in cultured human fibroblasts. Transport measurements are initial entry rates of L-serine under conditions that restrain the amino acid reactivity to system ASC. The analysis is presented as multiple regression surface of amino acid transport (dependent variable on y axis) versus aging (population doubling level, x axis) and cell density (z axis). Bars are deviations of experimental points from theoretical points (on the plane) predicted by the multiple regression equation (from Gazzola *et al.*, 1984, with permission).

impairment in the activity of all transport systems for neutral amino acids (A, ASC and L) in cultured human fibroblasts during phase III. The latter result suggests that a profound derangement of membrane functions takes place in senescent phase III fibroblasts, i.e., in cells that fail to show a substantial growth upon replating.

As previously shown, a large variety of signals (hormones, growth factors) and external conditions (nutrient availability, osmolarity of the extracellular compartment) converges to affect the activity of system A (Guidotti *et al.*, 1978; Shotwell *et al.*, 1983; Lerner, 1985). The effects of cell aging on the control of the function of this agency, selected as a regulatable model system for amino acid transport in human fibroblasts: (Bussolati *et al.*, 1990), can be summarized as follows. In the passage from early- to late-phase II cultured cells: (*a*) the response to nutritional stress, as assessed by the starvation-induced adaptive enhancement in the activity of system A, decreases; (*b*) the stimulatory

Table 10. Kinetic parameters of L-proline transport in early-phase II and late-phase II cultured human fibroblasts

	V_{max} $\frac{nmol \cdot min^{-1}}{mg\ protein}$	K_m mM	K_d $\frac{nmol \cdot min^{-1} \cdot mM^{-1}}{mg\ protein}$
Early phase II	32.8	1.34	1.40
Late phase II	22.6	1.14	1.78

effect of epidermal growth factor (EGF) on the activity of system A seems to be less efficient; (*c*) the stimulation by insulin of the activity of system A does not change significantly and (*d*) an action of the glucocorticoid dexamethasone develops that depresses the activity of system A. These results suggest that the *in vitro* aging of cultured human fibroblasts is somehow associated with an impairment of the control mechanisms (and/or of their coordination) acting on transport system A. Whether the regulatory derangement of amino acid transport is a direct aging-dependent effect or becomes manifest as a consequence of aging-induced alterations of other cell functions (proliferation rate, post-mitotic changes in gene expression or metabolism) is left to further investigation.

Neoplastic Transformation

An increased rate of amino acid uptake has been considered a consistent feature of tumor cells (Isselbacher, 1972). Results from a number of experiments with murine cells agree with this inference. Enhancements of the transport rate of site A-reactive amino acids are among the early events associated with cell transformation *in vitro* (Perdue, 1980); the uptake of amino acid substrates of the Na^+-dependent systems A and ASC increases markedly in virus- and chemically-transformed cell lines (Borghetti *et al.*, 1980; Boerner and Saier, 1982); rat fibroblasts transfected with *ras* oncogenes exhibit a somewhat faster uptake of MeAIB (a transport-specific substrate of system A) than non-transfected cells (Racker *et al.*, 1985). Other experiments, however, provide different results. In 3T3 mouse fibroblasts transformed with activated *ras* or *neu* oncogenes, the activity of systems A, ASC, L and y^+ is not significantly different from that of untransfected fibroblasts (Longo *et al.*, 1988); instead cell transformation is accompanied by a definite increase in the activity of system x^-_C and by an almost complete suppression of the activity of system X^-_{AG} (Table 11). Similar modulations of transport have been observed when the activities of systems x^-_C and X^-_{AG} of cultured rat astrocytes were compared to those of glioma C-6 cells (Cho and Bannai, 1990). Whether the amino acid transport changes reported in these studies reflect cell transformation or are associated with cell growth remains an open question. Indeed the rate of cell proliferation in the biologic preparations under study may affect the transport process by itself, and different growth rates may introduce disturbing elements of discrepancy.

Little is known on amino acid transport changes in neoplastic cells obtained from spontaneously occurring human tumors. Unchanged activities of the Na^+-dependent systems A and ASC have been found associated with a marked decrease of the activity of the Na^+-independent system L in chronic lymphocytic leukemia (CCL) cells (Segel and Lichtman, 1982; Segel *et al.*, 1984) where V_{max} of the L system transport is one tenth of that of normal blood B lymphocytes (Woodlock *et al.*, 1988).

Table 11. Amino acid transport in normal and transformed mouse 3T3 fibroblasts*.

	Amino acid transport systems					
	A	ASC	L	X^-_{AG}	x^-_C	y^+
	v (*nmol / min / ml cell water*)					
3T3	112 ± 12	118 ± 4	45 ± 1	89 ± 14	26 ± 2	162 ± 18
EJ (*ras*)	86 ± 9	76 ± 7	35 ± 1	3 ± 1	61 ± 4	118 ± 9
XHT (*ras*)	104 ± 2	98 ± 14	44 ± 1	2 ± 1	89 ± 11	106 ± 4
B104 (*neu*)	96 ± 2	80 ± 4	49 ± 10	3 ± 1	177 ± 11	144 ± 7

* Transport measurements are initial entry rates of MeAIB (0.1 mM) for system A, L-serine (0.05 mM) for system ASC, L-leucine (0.01 mM) for system L, L-aspartate (0.01 mM) for system X^-_{AG}, L-glutamate (0.02 mM) for system x^-_C and L-arginine (0.02 mM) for system y^+ under conditions that restrain the amino acid reactivity to the indicated transport system. Data are means (±S.D.) of triplicate determinations. (Rewritten from Longo *et al.*, 1988)

Studies that compared transport activity of human desmoid tumor fibroblasts and normal skin-derived fibroblasts obtained from the same donor (Nucci *et al.*, 1988) did not detect appreciable changes in transport activity of systems A, ASC, L, X^-_{AG} and x^-_C, and showed a consistent decrease in the activity of system y^+ for cationic amino acids. Cultured *in vitro*, fibroblasts from desmoid tumor (a borderline lesion between a stubbornly infiltrative fibroblastic proliferation and a low-grade fibrosarcoma) exhibit growth rates slightly lower than those of normal fibroblasts. Extended to cells of a fibrosarcoma line (HT 1080), studies on the transport of neutral amino acids did not reveal differences in the activity of systems A, ASC and L among HT 1080 cells, desmoid tumor cells and normal fibroblasts (Gazzola *et al.*, 1985). In accord, viral transformation of fetal human fibroblasts did not alter substantially the transport of a model amino acid (AIB) taken up primarily by system A (Patterson *et al.*, 1976). Experiments in progress in our laboratory with IDH4 human fibroblasts (cells immortalized by transfection of the early region of the SV40 genome under the glucocorticoid-sensitive promoter of MMTV, Wright *et al.*, 1989) that can shift from a nonproliferating senescent phenotype to a proliferating immortal phenotype and revert to the quiescent state (depending upon the absence-addition-withdrawal of glucocorticoids) show that the induction of the proliferative state is associated with a definitive increase in the activity of systems A and ASC and with a marked suppression of the activity of system X^-_{AG} (for anionic amino acids). The changes in the activity of systems A and ASC progressively vanish (and transport activities return to the initial value) when fibroblasts revert to the senescent phenotype upon glucocorticoid withdrawal.

CONCLUDING REMARKS

This chapter provides an overview of the systems of transporters that mediate the transfer of amino acids across the cell membrane.

Over three decades of research have indicated that all cell types of mammalian organisms are equipped with multiple amino acid transport systems whose discrimination has been pursued using kinetic, thermodynamic and physiologic approaches. These studies have settled the recognition properties of the systems by defining their individual reactivity to substrate amino acids and established the system operational features (including ion-dependence, sensitivity to electric membrane potential and boundary conditions, associated *trans*-substrate driven movements, predominant vectorial directionality and concentrative capacity). They have identified a number of control mechanisms that adapt the activity of the transport systems to changes in the cell environment and adjust their function to the requirements of tissues and organs within the organism. It has been shown that the regulatory processes of amino acid transport can play a relevant role in metabolism by controlling the access of substrates into the cells.

In this chapter, special emphasis has been given to the amino acid transport in human fibroblasts, a biologic preparation in which the transporter systems have been extensively characterized and assayed for activity under a variety of physiologic and pathologic conditions including *in vitro* cell aging and density, proliferation state and neoplastic transformation.

Recent success in the study of the energization of amino acid transport systems in human fibroblasts has provided insights into the vectorial mechanism underlying the function of some of these agencies and work in progress warrants a rapid increase of our knowledge for most systems active in a number of cell types. These investigations, together with the recognition that the adaptive response of the cell to osmotic changes of the environment requires the contribution of regulatable systems for amino acid transport is likely to add new impetus in the study of cell volume control in mammalian cells.

ACKNOWLEDGMENTS

The authors' work which is quoted in this chapter has been supported by the Consiglio Nazionale delle Ricerche (Progetto Finalizzato Biotecnologie e

Biostrumentazione), by the Ministero della Università e della Ricerca Scientifica e Tecnologica (Progetto Nazionale Biologia e Patologia delle Membrane; Progetto Nazionale Patologia e Fisiopatologia Generale dell'Invecchiamento), and by the Associazione Italiana per la Ricerca sul Cancro.

REFERENCES

Bannai, S., and Ishii, T., 1988, J. Cell. Physiol., 137:360
Bannai, S., and Kitamura, E., 1980, J. Biol. Chem., 255:2372
Bannai, S., Christensen, H.N., Vadgama, J.V., Ellory, J.C., Englesberg, E., Guidotti, G.G., Gazzola, G.C., Kilberg, M.S., Lajtha, A., Sacktor, B., Sepulveda, F.V., Young, J.D., Yudilevich, D., and Mann, G., 1984, Nature, 311:308
Berlin, R.D., 1970, Science, 168:1539
Boerner, P., and Saier, M.H.Jr, 1982, J. Cell. Physiol., 113:240
Boerner, P., and Saier, M.H.Jr, 1985, J. Cell. Physiol., 122:308
Borghetti, A.F., Piedimonte, G., Tramacere, M., Severini, A., Ghiringhelli, P., and Guidotti, G.G., 1980, J. Cell. Physiol., 105:39
Borghetti, A.F., Tramacere, M.R., Ghiringhelli, P., Severini, A., Kay,J.E., 1981, Biochim. Biophys. Acta, 646:218
Brigham, M.P., Stein, W.H., and Moore, S., 1960, J. Clin. Invest., 39:1633
Brookes, N., 1988, J. Neurochem., 51:1913
Burger, H.J., Gebhardt, R., Mayer, C., and Mecke, D., 1989, Hepatology, 9:22
Bussolati, O., Laris, P.C., Longo, N., Dall'Asta, V., Franchi-Gazzola, R., Guidotti, G.G., and Gazzola, G.C., 1986, Biochim. Biophys. Acta, 854:240
Bussolati, O., Laris, P.C., Nucci, F.A., Dall'Asta, V., Longo, N., Guidotti, G.G., and Gazzola, G.C., 1987, Am. J. Physiol., 253:C391
Bussolati, O., Laris, P.C., Nucci, F.A., Dall'Asta, V., Franchi-Gazzola, R., Guidotti, G.G., and Gazzola, G.C., 1989, Am. J. Physiol., 256:C930
Bussolati, O., Longo, N., Dall'Asta, V., Gazzola, G.C., and Guidotti, G.G., 1990, in "Protein Metabolism in Aging", H.L. Segal, M. Rothstein, and E. Bergamini, eds., Wiley-Liss, New York
Bussolati, O., Rotoli, B.M., Laris, P.C., Dall'Asta, V., and Gazzola, G.C., 1991, Biochim. Biophys. Acta, 1070:305
Bussolati, O., Laris, P.C., Rotoli, B.M., Dall'Asta, V., and Gazzola, G.C., 1992, J. Biol. Chem., 267:8330
Cho, Y., and Bannai, S., 1990, J. Neurochem., 55:2091
Christensen, H.N., 1964, Proc. Natl. Acad. Sci. U.S.A., 51:337
Christensen, H.N., 1966, Fed. Proc., 25:850
Christensen, H.N., 1973, Bioenergetics, 4:31
Christensen, H.N., 1975, "Biological Transport", W.A. Benjamin, Reading, Mass.
Christensen, H.N., 1979, Adv. Enzymol., 49:41
Christensen, H.N., 1982, Physiol. Rev., 62:1193
Christensen, H.N., 1984a, Biochim. Biophys. Acta, 779:255
Christensen, H.N., 1984b, Nutr. Rev., 42:237
Christensen, H.N., 1985, J. Membrane Biol., 84:97
Christensen, H.N., 1988a, Trends Biochem. Sci., 13:40
Christensen, H.N., 1988b, Bioscience Reports, 8:121
Christensen, H.N., and Riggs, T.R., 1952, J. Biol. Chem., 194:57.
Christensen, H.N., Liang, M., and Archer, E.G., 1967, J. Biol. Chem., 242:5237.
Christensen, H.N., Curthoys, N.P., Mortimore, G.E., Smith, R.J., Goldstein. L., and Harper, A.E., 1986, Fed. Proc., 45:2165
Collarini, E.J., and Oxender, D.L., 1987, Ann. Rev. Nutr., 7:75
Dall'Asta, V., Gazzola, G.C., and Guidotti, G.G., 1978, Biochim. Biophys. Acta, 507:165
Dall'Asta, V., Gazzola, G.C., Franchi-Gazzola, R., Bussolati, O., Longo, N., and Guidotti, G.G., 1983, J. Biol. Chem., 258:6371
Dall'Asta, V., Rossi, P.A., Bussolati, O., Guidotti, G.G., and Gazzola, G.C., 1990, Biochim. Biophys. Acta, 1052:106
Dall'Asta, V., Bussolati, O., Guidotti, G.G., and Gazzola, G.C., 1991, J. Biol. Chem., 266:1591
Eagle, H., 1955, J. Biol. Chem., 214:839

Eagle, H., 1959, Science, 130:432
Edmondson, J.W., and Lumeng, L., 1980, Biochem. Biophys. Res. Commun., 96:61
Ellory, J.C., 1987, in "Amino Acid Transport in Animal Cells", D.L. Yudilevich and C.A. Boyd, eds., Manchester University Press, Manchester
Fass, S.J., Hammerman, M.R., and Sacktor, B., 1977, J. Biol. Chem., 252:583
Franchi-Gazzola, R., Gazzola, G.C., Ronchi, P., Saibene, V., and Guidotti, G.G., 1973, Biochim. Biophys. Acta, 291:545
Franchi-Gazzola, R., Gazzola, G.C., Dall'Asta, V., and Guidotti, G.G., 1982, J. Biol. Chem., 257:9582
Gazzola, G.C., Franchi, R., Ronchi, P., Saibene, V., and Guidotti, G.G., 1972, Biochim. Biophys. Acta, 266:407
Gazzola, G.C., Franchi-Gazzola, R., Ronchi, P., and Guidotti, G.G., 1973, Biochim. Biophys. Acta, 311:292
Gazzola, G.C., Dall'Asta, V., and Guidotti, G.G., 1980, J. Biol. Chem., 255:929
Gazzola, G.C., Dall'Asta, V., Bussolati, O., Makowske, M., and Christensen, H.N., 1981a, J. Biol. Chem., 256:6054
Gazzola, G.C., Dall'Asta, V., and Guidotti, G.G., 1981b, J. Biol. Chem., 256:3191
Gazzola, G.C., Bussolati, O., Longo, N., Dall'Asta, V., Franchi-Gazzola, R., and Guidotti, G.G., 1984a, Monogr. devl Biol., 17:234
Gazzola, G.C., Dall'Asta, V., Franchi-Gazzola, R., Bussolati, O., Longo, N., and Guidotti, G.G., 1984b, Fed. Proc., 43:1777
Gazzola, G.C., Dall'Asta, V., Franchi-Gazzola, R., Bussolati, O., Longo, N., and Guidotti, G.G., 1985, in "Cell Membranes and Cancer", T. Galeotti, A. Cittadini, G. Neri, S. Papa, and L.A. Smets eds., Elsevier, Amsterdam.
Gazzola, G.C., Dall'Asta, V., Nucci, F.A., Rossi, P.A., Bussolati, O., Hoffmann, E.K., and Guidotti, G.G. 1991, Cell. Physiol. Biochem., 1:131
Goldstein, L., 1986, Fed. Proc., 45:2176
Guidotti, G.G., Borghetti, A.F., Luneburg, B., and Gazzola, G.C., 1971, Biochem. J., 122:409
Guidotti, G.G., Gazzola, G.C., Borghetti, A.F., and Franchi-Gazzola, R., 1975, Biochim. Biophys. Acta, 406:264
Guidotti, G.G., Borghetti, A.F., and Gazzola, G.C., 1978, Biochim. Biophys. Acta, 515:329
Hammerman, M.R., and Sacktor, B., 1977, J. Biol. Chem., 252:591
Handlogten, M.E., Kilberg, M.S., and Christensen, H.N., 1982, J. Biol. Chem., 257:345
Häussinger, D., and Gerok, W., 1983, Eur. J. Biochem., 136:421
Hayflick, L., 1980, Annu. Rev. Gerontol. Geriat., 1:26
Heinz, E.J., and Walsh, P.M., 1958, J. Biol. Chem., 233:1488.
Herskowitz, K., Bode, B.P., Block, E.R., and Souba, W.W., 1991, Am. J. Physiol., 260:L241
Hoeltzli, S.D., and Smith, C.H., 1989, Am. J. Physiol., 256:C630
Hoeltzli, S.D., Kelley, L.K., Moe, A.J., and Smith, C.H., 1990, Am. J. Physiol., 259:C47
Hoffmann, E.K., 1987, Curr. Top. Membr. Transp., 30:125
Hoffmann, E.K., and Hendil, K.B., 1976, J. Comp. Physiol., 108:279
Hoffmann, E.K., and Lambert, I.H., 1983, J. Physiol. (London), 338:613
Isselbacher, K.J., 1972, Proc. Natl. Acad. Sci. U.S.A., 69:585
Johnson, L.W., and Smith, C.H., 1988, Am. J. Physiol., 254:C773
Karasov, W.H., Pond III, R.S., Solberg, D.H., and Diamond, J.M., 1983, Proc. Natl. Acad. Sci. U.S.A., 80:7674
Kilberg, M.S., 1982, J. Membr. Biol., 69:1
Kilberg, M.S., Handlogten, M.E., and Christensen, H.N., 1980, J. Biol. Chem., 255:4011
Kilberg, M.S., Barber, E.F., and Handlogten, M.E., 1985, Curr. Topics Cell. Regul., 25:133
Krupka, R.M., 1990, J. Membrane Biol., 117:69
Lenzen, C., Soboll, S., Sies H., Haussinger, D., 1987, Eur. J. Biochem., 166:483
Lerner, J., 1985, Comp. Biochem. Physiol., 81A:713
Lerner, J., 1987, Comp. Biochem. Physiol., 87B:443
Logan, W.J., Klip, A., and Gagalang, E., 1982, J. Cell. Physiol., 112:229
Longo, N., Franchi-Gazzola, R., Bussolati, O., Dall'Asta, V., Foa, P.P., Guidotti, G.G., and Gazzola, G.C., 1985, Biochim. Biophys. Acta, 844:216
Longo, N., Franchi-Gazzola, R., Bussolati, O., Dall'Asta, V., Nucci, F.A., and Gazzola, G.C., 1988, Ann. N.Y.Acad. Sci., 551:374
Macieiera-Coelho, A., 1988, "Biology of Normal Proliferating Cells *in vitro*: Relevance for *in vivo* Aging", Karger, Basel
Macknight, A.D.C., 1988, Renal Physiol. Biochem., 3-5:114

Mann, G.E., and Peran, S., 1986, Biochim. Biophys. Acta, 858:263
Mann, G.E., Habara, Y., and Peran, S., 1986, Pancreas, 1:239
Mann, G.E., Pearson, J.D., Sheriff, C.J., and Toothill, V.J., 1989, J. Physiol. (London), 410:325
Moe, A.J., and Smith, C.H., 1989, Am. J. Physiol., 257:C1005
Moule, S.K., and McGivan, J.D., 1987, Biochem. J., 247:233
Munoz, M., Emery, P.W., Peran, S., and Mann, G.E., 1988, Biochim. Biophys. Acta, 945:273
Newton, R.B., Sullivan, J.L., and Gib Debusk, A., 1984, Mech. Age. Dev., 27:63
Nucci, F.A., Bussolati, O., Dall'Asta, V., Gazzola, G.C., Giardini, R., and Guidotti, G.G., 1988, in "Chemical Carcinogenesis", F. Feo, P. Pani, A. Columbano, and R. Garcea, eds., Plenum Press, New York.
Oxender, D.L., and Christensen, H.N., 1963, J. Biol. Chem., 238:3686.
Patterson, M.K., Birckbichler, P.J., Conway, E., and Orr G.R., 1976, Cancer Res., 36:394
Perdue, J.F., 1980, in "Virus-transformed Cell Membrane", C. Nicolau, ed., Academic Press, London
Petronini, P.G., Piedimonte, G., and Borghetti, A.F., 1982, Biochim. Biophys. Acta, 693:13
Piedimonte, G., Borghetti, A.F., and Guidotti, G.G., 1982, Cancer Res., 42:4690
Piedimonte, G., Baginski, I., Silvotti, L., Petronini, P.G., and Borghetti, A.F., 1989, Cancer Lett., 47:121
Pisoni, R.L., Flickinger, K.S., Thoene, J.G., and Christensen, H.N., 1987, J. Biol. Chem., 262:6010
Polgar, P., Taylor, L., and Brown, L., 1978, Mech. Age. Dev., 7:251
Racker, E., Resnick, R.J., and Feldman, R., 1985, Proc. Natl. Acad. Sci. U.S.A., 82:3535
Rotoli, B.M., Bussolati, O., Dall'Asta, V., and Gazzola, G.C., 1989, Biochem. Biophys. Res. Comm., 164:1093
Rotoli, B.M., Bussolati, O., Dall'Asta, V., and Gazzola, G.C., 1991, J. Cell. Physiol., 146:417
Sato, H., Watanabe, H., Ishii, T., and Bannai, S., 1987, J. Biol. Chem., 262:1305
Segel, G.B., and Lichtman, M.A., 1982, J. Biol. Chem., 257:9255
Segel, G.B., Simon, W., and Lichtman, M.A., 1984, J. Clin. Invest., 74:17
Shotwell, M.A., Mattes, P.M., Jayme, D.W., and Oxender, D.L., 1982, J. Biol. Chem., 257:2974
Shotwell, M.A., Kilberg, M.S., and Oxender, D.L., 1983, Biochim. Biophys. Acta, 737:267
Silbernagl, S., 1988, Physiol. Rev., 68:911
Stevens, B.R., Kaunitz, J.D., and Wright, E.M., 1984, Annu. Rev. Physiol., 46:417
Taylor, P.M., and Rennie, M.J., 1987, FEBS Lett., 221:370
Van Winkle, L.J., 1988, Biochim. Biophys. Acta, 947:173
Von Blankenfeld, G., and Jauch, P., 1989, Biochim. Biophys. Acta, 980:385
Watford, M., 1989, Trends Biochem. Sci., 14:1
Weissbach, L., Handlogten, M.E., Christensen, H.N., and Kilberg, M.S., 1982, J. Biol. Chem., 257:12006
Wheeler, D.D., 1981, Exp. Gerontol., 16:247
Wheeler, D.D., 1982, Exp. Gerontol., 17:71
Woodlock, T.J., Segel, G.B., and Lichtman, M.A., 1988, J. Clin. Invest., 81:32
Woodlock, T.J., Segel, G.B., and Lichtman, M.A., 1989, J. Cell. Physiol., 141:33
Wright, W.E., Pereira-Smith, O.M., and Shay, J.W., 1989, Mol. Cell. Biol., 9:3088
Yancey, P.H., Clark, M.E., Hand, S.C., Bowlus, R.D., and Somero, G.N., 1982, Science, 217:1214
Yorek, M.A., Dunlap, J.A., and Ginsberg B.H., 1987, J. Biol. Chem., 262:10986
Zetterberg, A., and Engstrom, W., 1981, J. Cell. Physiol., 108:365

ENERGETICS OF AMINO ACID TRANSPORT

A.A. Eddy

Department of Biochemistry
and Applied Molecular Biology
UMIST
P. O. Box 88, MANCHESTER M60 1QD, UK

While leucine absorption by human erythrocytes proceeds by facilitated diffusion, and analogous systems occur on one face of polar epithelial cells, the vertebrate systems of amino acid transport that have been studied most widely concentrate amino acids up to about 50-fold corresponding to a free energy increase of the order of 10 kJ mol^{-1}. Much early work had as its main goal the characterisation of the number and specificity of the carrier systems involved, with the object of understanding both how solute concentration was effected at the expense of metabolic energy (the so-called energy coupling problem) and, further, how relatively hydrophilic or polar molecules traverse the hydrophobic barrier represented by the plasma membrane. In the broader biological context that includes both prokaryotes and eukaryotes, the systems that utilize ATP directly are to be distinguished from those based on redox reactions, on the one hand, and on the other hand from secondary transport systems that utilize an electrochemical gradient of either sodium ions or protons acting across the plasma membrane to drive a flow of organic solute either in the same direction (symport) or in the opposite direction (antiport). Only some of these diverse processes have as yet been studied in any detail (Eddy, 1990). In mammalian tissues, the Na^{+}-linked and proton-linked co-substrate systems probably provide the main mechanisms of amino acid and peptide accumulation. In particular the so-called A system of amino acid transport of mouse ascites tumour cells, that accepts many common neutral amino acids, including alanine, glycine and the metabolically inert analogues 2-aminoisobutyrate and 2-N-methylaminoisobutyrate, has been extensively investigated from the standpoint of its energetic and kinetic characteristics. Furthermore, the glycine and sarcosine carrier of the avian erythrocyte was used by Vidaver in the first formal demonstration of the ability of the sodium gradient in the absence of ATP to drive amino acid transport.

The Glutamyltranspeptidase Cycle

Meister has postulated that the membrane-bound enzyme γ-glutamyl-transpeptidase may participate directly in amino acid transport by donating γ-glutamyl residues from intracellular glutathione to incoming amino acids. The resultant γ-glutamylamino acid would break down into the free intracellular amino acid and 5-oxoproline that was subsequently reincorporated into glutathione (Meister & Anderson, 1983). The overall cycle of reactions is a costly one in energetic terms, consuming 3 ATP per mole of amino acid absorbed. While it is known that various tissues contain the relevant enzymes, the ability of the transpeptidase to transfer amino acid residues vectorially in the manner required by the hypothesis has not been demonstrated. Furthermore, a mutant human fibroblast cell line lacking the transpeptidase concentrated various amino acids in a similar fashion to its genetic parent (Pellefigue et al., 1976). Similarly, in the exocrine

Mammalian Amino Acid Transport, Edited by M.S. Kilberg and
D. Häussinger, Plenum Press, New York, 1992

pancreas, marked inhibition of the enzyme by acivicin failed to inhibit uptake of either glutamine or alanine (Sastre et al., 1991). It seems clear that the primary mechanism of amino acid transport in these instances is not based on the γ-glutamylcycle. As Na^+-dependent or proton-dependent symports have now been detected in many mammalian tissues, arguments in favour of a role for the transpeptidase in a second and parallel mechanism of amino acid accumulation need to be based on a careful assessment of the role of ionic gradients in the overall process.

Kinetic and Thermodynamic Models of the Co-substrate System

The possibility that an outflow of K^+ ions might propel the uptake of glycine into mouse ascites tumour cells was first proposed by Christensen and his colleagues (antiport relationship). The mechanism was envisaged as an extension of the principle that membrane carriers, by virtue of the fact that their solute binding sites are alternately exposed either on one side or the other of the cell membrane, can couple an inflow of one solute A to a counterflow of solute B. In effect an initial gradient of A, acting across the cell membrane is converted to a transient gradient of B. Later work by Crane on glucose uptake in the intestine focussed attention on the further possibility that an inflow of Na^+ ions might drive a parallel inflow of solute (symport relationship).

Equilibrium Carrier Models

Let the difference in electrochemical potential of Na^+ on the two sides of the cell membrane be $\Delta\mu^{Na}$, $\Delta\mu^K$ and $\Delta\mu^S$ being the corresponding quantities for K^+ and the solute, S, respectively. In general n Na^+ ions might accompany each molecule of S through the carrier and where such a system comes to thermodynamic equilibrium

$$\Delta\mu^S + n\Delta\mu^{Na} = 0 \qquad (1)$$

Similarly, an antiport relationship with m K^+ ions leads to the equation

$$\Delta\mu^S - m\Delta\mu^K = 0 \qquad (2)$$

These equations can readily be extended to examples where both Na^+ and K^+ or Cl^- participate as co-substrate ions (Table 1). A useful alternative form of Equation 1 describes the amino acid accumulation ratio $[S]_i/[S]_0$ in terms of $[Na^+]_0/[Na^+]_i$ and the membrane potential ($\Delta\psi$)

$$[S]_i/[S]_0 = ([Na^+]_0/[Na^+]_i)^n \, e^{-n\Delta\psi F/RT} \qquad (3)$$

where F, R and T have their conventional meanings.

Table 1. Representative systems with Na^+ and another cosubstrate ion.

Potassium ions. Aspartate uptake, chick intestine (Wingrove & Kimmich, 1988); glial cells (Barbour et al., 1988); renal epithelia (Berteloot, 1986; Kinne et al., 1988).

Chloride and Potassium ions. Serotonin uptake in platelets (Nelson & Rudnick, 1982).

Chloride. Glycine uptake, pigeon erythrocytes (Vidaver et al., 1976); γ-aminoisobutyrate uptake, brain (Keynan & Kanner,1988); the intestinal imino acid carrier and the intestinal ß-alanine carrier (Munck & Munck, 1990).

Leaks. The notion that the amino acid gradient equilibrates with the gradient of Na^+ is a simplification that is unlikely to be strictly true in practice, if only because many mammalian cells contain a Na^+-symport functioning in parallel with at least one leak pathway for the amino acid. An instructive example is the accumulation of glycine and leucine through the A system of mouse ascites tumour cells. The amino acid accumulation ratio is lowered by the presence of a leak pathway that is considerably smaller in the former instance than the latter. In both instances, when a steady state of amino acid accumulation is established, a net current of Na^+ ions is expected to be maintained through the Na^+-symport and has been detected (Hacking & Eddy, 1981a). Further, the ratio $\Delta\mu^S/n\Delta\mu^{Na}$ is expected to be less than the value of -1 predicted by Equation 2. Kimmich has provided strong experimental evidence for that view in a careful study of carbohydrate transport in chicken epithelial cells (Kimmich & Carter-Su, 1978).

Nonequilibrium Carrier Models: Slip or Uncoupling Intrinsic to the Carrier

A more subtle and experimentally inaccessible reason why Equation 1 might not apply concerns the possibility that the carrier system itself is inherently incapable of bringing the substrate and co-substrate into thermodynamic equilibrium. In the conventional model in which 1 Na^+ and 1 molecule of S participate (Fig. 1) passage of the substrate and co-substrate exclusively as the ternary carrier species ENaS leads eventually to the values of $\Delta\mu^{Na}$ and $\Delta\mu^S$ being equal in magnitude and opposite in sign. However, translocation of one or both of the binary species as well lowers the ratio of $\Delta\mu^S/\Delta\mu^{Na}$. The following equation describes the relatively simple situation in which the membrane potential ($\Delta\psi$) is zero and only Na^+ ions move as the binary species as well as in the ternary form (ENaS). The ratio of the intracellular to extracellular concentrations of S ($[S]_i/[S]_0$) is given by

$$\frac{[S]_i}{[S]_0} = \frac{[Na^+]_0}{[Na^+]_i} \cdot \frac{(1 + \alpha[Na^+]_i)}{(1 + \alpha[Na^+]_0)} \qquad (4)$$

where using the assumptions in Eddy (1968a) $\alpha = k^{Na}/k.K$.

The ratio k^{Na}/k is the ratio of the translocation rate constants of ENa and unliganded E, whereas K is the dissociation constant of the species ENa. These equations have the intuitive meaning that the more readily ENa forms and translocates relative to the free carrier E, the less effectively is the flow of Na^+ through the carrier coupled to that of the substrate S. Thus the large ratio $[Na^+]_0/[Na^+]_i$ corresponding to equation 1 is multiplied by a small factor that in the limit, where the system is uncoupled, approaches $[Na^+]_i/[Na^+]_0$.

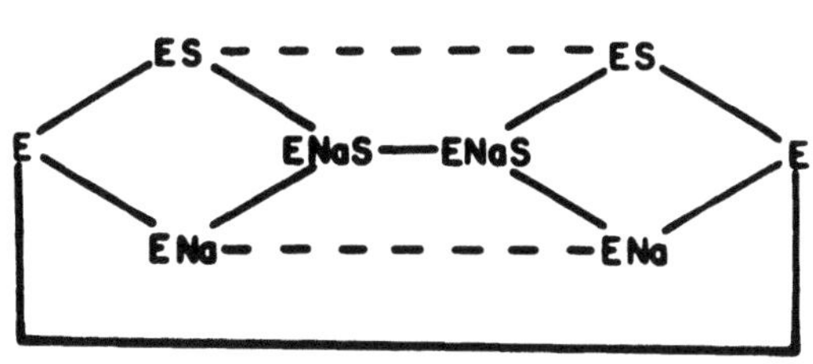

Fig. 1. Kinetic model of the cosubstrate carrier. The carrier E combines randomly with Na^+ and S to give the ternary complex ENaS. Tight coupling between the flow of Na^+ and S requires that translocation of ENaS is preferred to that of the binary species. Translocation of the binary species corresponds to the two slip modes leading to carrier inefficiency (see text).

Analogous considerations apply to the case where the binary species ES translocates. The solute accumulation ratio is then

$$\frac{[S]_i}{[S]_0} = \frac{[Na^+]_0 + \beta}{[Na^+]_i + \beta} \tag{5}$$

where $\beta = K.k^S/k^{NaS}$. The ratio k^S/k^{NaS} is the ratio of the translocation rate constants for ES and ENaS respectively, whereas K is the dissociation constant for the release of Na^+ from the ternary complex. Thus uncoupling of the flow of S from the flow of Na^+ increases with the magnitude of β and corresponds to an increase in the movement of S through ES as opposed to ENaS.

Whereas the above discussion centres on the factors influencing the steady state gradient of solute formed in the presence of a gradient of Na^+, variations in the above parameters α and β will also affect other experimentally accessible parameters, such as the apparent values of K_m with respect to $[Na^+]_0$, $[Na^+]_i$, $[S]_0$ and $[S]_i$ and the corresponding values of J_{max} for both the substrate S and the co-substrate Na^+. In particular translocation of Na^+ through the carrier in the absence of S will lower the amount of extra Na^+ entering in the presence of substrate, below the stoichiometrical ratio of unity corresponding to the composition of the ternary complex.

When either leakage pathways for the substrate are present outside the symport, or when the symport itself only imperfectly couples the flows of substrate and co-substrate, the equations of irreversible thermodynamics apply rather than equation 1. In the latter equation the negative ratio $\Delta\mu^S/\Delta\mu^{Na}$ is equal to the hypothetically fixed stoichiometry *n*. When either slip or leakage pathways exist, however, the stoichiometrical ratio of the flow of Na^+ (J^{Na}) to the flow of S(J^S) is not constant. Instead, the following relations hold

$$J^S = L^S\Delta\mu^S + L\Delta\mu^{Na} \tag{6}$$

$$J^{Na} = L^{Na}\Delta\mu^{Na} + L\Delta\mu^S \tag{7}$$

where the flows of both species are determined by the product of the conjugate force and the magnitude of the respective phenomenological coefficients (L^S, L^{Na} and the cross-coefficient L) in the so-called Onsager domain (Pietrobon & Caplan, 1989). A quantity of interest, corresponding to the equilibrium condition of Equation 1, is the ratio $\Delta\mu^S/\Delta\mu^{Na}$ when J^S is zero. This is the so-called 'static head' condition. In contrast to the equilibrium described by Equation 1, J^{Na} is not, however, then also zero. Thus

$$\Delta\mu^S/\Delta\mu^{Na} = -L/L^S \tag{8}$$

where the right hand ratio corresponds to an apparent stoichiometry of the coupling process. Indeed, it can be shown (Heinz, 1978) that a given value of $\Delta\mu^{Na}$ variations in $\Delta\mu^S$, for instance brought about by varying the amino acid concentration, lead to changes in the flow ratio J^{Na}/J^S whereby

$$(\delta J^{Na}/\delta J^S)_{\Delta\mu^{Na}} = -(\Delta\mu^S/\Delta\mu^{Na})_{J^S=0} \tag{9}$$

An early study of this relationship was used to show that the left hand term was about 0.6 and that the Na^+-gradient hypothesis could not be rejected on energetic grounds (Heinz & Geck, 1974).

Experimental Tests of the Co-substrate Hypothesis

The earlier history of this topic has been reviewed most recently by Kimmich (1990). The main questions are whether (1) cellular preparations or membrane vesicle preparations depleted of ATP in fact concentrate specific solutes when $\Delta\mu^{Na}$ is appropriately manipulated. (2) If so, does the sodium gradient dissipate faster than in the absence of the solute and to an extent consistent stoichiometrically with a carrier mechanism? (3) Can the behaviour of the transport system during energy metabolism be predicted from that of the metabolically depleted preparations? (4) Where a net transfer of Na^+ occurs with the solute S, does the expected membrane depolarization occur, bearing in mind the possible involvement of K^+ or Cl^- in the process?

We shall consider these questions initially in relation to the accumulation of amino acids by mouse ascites tumour cells, a system that has been studied in several laboratories (Johnstone, 1979a; Heinz et al., 1977) and which exhibits a number of distinct amino acid carriers, the best known being the Na^+-dependent A and ASC systems and the Na^+-independent L and y^+ systems (Christensen, 1985).

Na^+ Stoichiometry

The uptake of glycine, 2-aminoisobutyrate and L-methionine through the A carrier system is stimulated markedly by extracellular Na^+, exhibiting first degree saturation kinetics with respect to both the amino acid concentration ([A]) and $[Na^+]$. In general lowering $[Na^+]_0$ raises the K_m with respect to [A] and lowers V^{max} (see, for instance, Inui & Christensen, 1966). At physiological values of $[Na^+]$ each mole of amino acid is accompanied by about 0.9-1.0 equivalent of Na^+ (Table 2). In the absence of Na^+ 2-aminoisobutyrate uptake exhibits saturation kinetics (Jacquez, 1973; Dawson et al., 1983; Martin et al., 1984) but whether the A system is involved is unclear because specific substrates of the L system interfere with the process (Neville et al., 1980). Dawson et al. (1983) argue that protons replace Na^+ as cosubstrate under these conditions, the feasibility of which is supported by a number of parallels between the properties of Na^+ and H_3O^+ (Boyer, 1988). In contrast glycine uptake into pigeon erythrocytes exhibits a stoichiometry (n) nearer 2, like certain other systems described in Table 2. Indeed the values of n deduced by direct assay of the flow of symported ion relative to the flow of substrate tend to cluster around these two integral values. It is usually assumed therefore that the integers simply reflect the relative number of binding sites in the translocated complex EnNaS. On this view the slip components referred to in connection with Equations are small enough, to a first approximation, to be neglected when $[Na^+]$ is relatively large. Table 2 includes the example of isoleucine uptake by the mouse ascites tumour cells, for which n was also inferred to be near 1.

Table 2. Representative estimates of the stoichiometry of Na^+ transfer with a specified amino acid.

Na^+-dependent system	Stoichiometry
Glycine in pigeon erythrocytes (Vidaver et al., 1976)	1.5 - 2.2
Alanine in hepatocytes (Kristensen & Folke, 1983)	1.0
Alanine in rabbit ileum (Paterson et al., 1980)	0.94 ± 0.04
Aspartate in chick intestine (Wingrove & Kimmich, 1988)	2.2 ± 0.1
N-methylaminoisobutyrate in fibroblasts (Dall'Asta et al., 1991)	1.02 ± 0.05
Mouse ascites tumour cells:	
glycine (Eddy, 1968a)	0.90 ± 0.11
2-aminoisobutyrate (Yamamoto & Kawasaki, 1981)	0.8 - 1.0
2-aminoisobutyrate (Schafer & Jacquez, 1967)	0.85 - 1.05
2-aminoisobutyrate (Dawson et al., 1983)	0.90 ± 0.02
methionine (Eddy & Johnson, 1989)	0.96 ± 0.19
isoleucine (Eddy et al., 1988)	1.14 ± 0.12

In fact isoleucine, a typical substrate of system L and a low affinity substrate of system A, was absorbed in the absence of Na^+ by a route that accounted for about 30 per cent of the uptake in the presence of Na^+. Only after correction for this component was the isoleucine stoichiometry of the Na^+ symport found to be similar to that of glycine and methionine (Fig. 2). The correction is based on the assumption that the observed flow of isoleucine in the absence of Na^+ occurs through a second carrier, presumably system L, independently of the Na^+ symport. If instead this flow represents slippage through the symport itself *via* the binary species ES, the flow would be expected to be smaller in

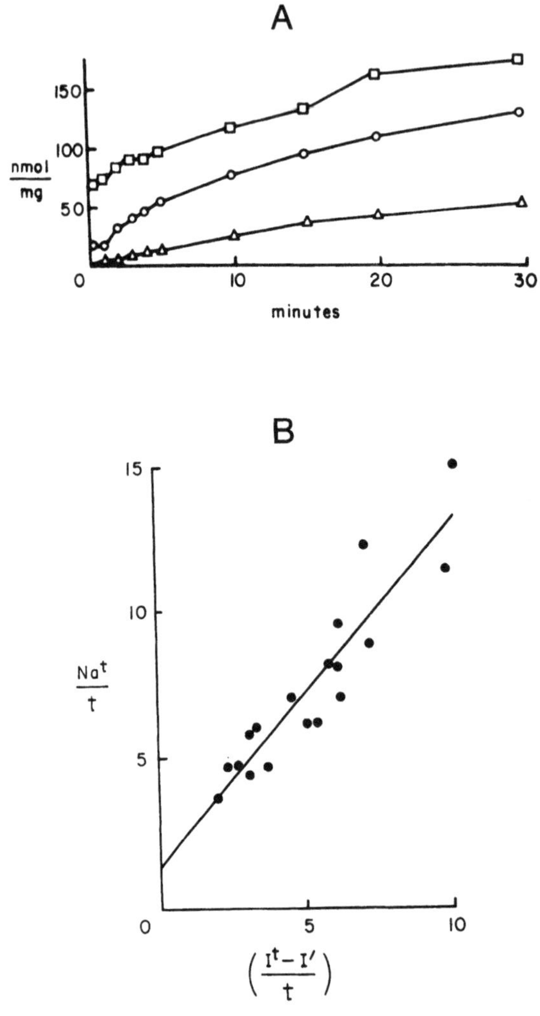

Fig. 2. Assay of the Na^+ stoichiometry of isoleucine uptake by mouse ascites tumour cells (Eddy et al., 1988). A. Uptake of isoleucine from a 40 mM solution, in a Ringer solution containing 32 mM Na^+, 0.3 mM ouabain and valinomycin (10 μg/ml) to clamp $\Delta\psi$. □, Na^{+22} uptake; o, [^{14}C] isoleucine uptake; △, computed basal flow of isoleucine independent of Na^+. B. The slope of the function Na^t/t against $(I^t - I')/t$ defines the Na^+ stoichiometry. Na^t is the Na^+ uptake at time t. $(I^t - I')$ is the total amount of isoleucine absorbed through the symport in time t. Slope = 1.136 ± 0.120.

the presence of Na^+ and a somewhat larger uptake of Na^+ should have occurred than was in fact observed. At present it seems unnecessary, therefore, to invoke the slip hypothesis to explain the behaviour of isoleucine. Similar ambiguities have arisen in attempts to define the Na^+ stoichiometry of alanine and methionine transfer at the intestinal mucosal boundary (Paterson et al., 1980). In the former case two carriers appeared to be involved, whereas in the latter methionine appeared to be absorbed through the same carrier both in the presence and in the absence of Na^+. This issue is of course further complicated by the possibility noted above that H_3O^+ may mimic the actions of Na^+.

Electrical Characteristics

Some of the earliest evidence in favour of the Na^+ co-substrate hypotheses was based on the finding that, for instance, alanine transfer across the gut increases the short circuit current by an amount equal to the net transfer of Na^+ through the tissue (Schultz & Curran, 1970). Membrane depolarization at the mucosal face leads to a complex set of biochemical and electrical changes at the serosal face, consideration of which lies outside the scope of the article (see e.g. Sheppard et al., 1988). Recently whole cell patch clamping has provided important insights into the co-substrate stoichiometry of, for instance, electrogenic glutamate uptake into glial cells (Barbour et al., 1988) and electrogenic alanine uptake, probably *via* the A system, in pancreatic acinar cells (Jauch et al., 1986; Jauch & Lauger, 1986). In the latter instance, the dependence of current on $[Na^+]$ and [alanine] exhibited simple saturation kinetics, with tight coupling corresponding to approximately unit stoichiometry between the flow of substrate and co-substrate. Study of the inwards and outwards currents as a function of $[Na^+]$ and [S] suggested that the carrier was functionally almost symmetrical. Observations with voltage-sensitive dyes indicate that Na^+-dependent amino acid absorption through the A system is electrogenic whereas absorption through the L system is not electrogenic (Laris et al., 1976; Philo & Eddy, 1978a; Hacking & Eddy, 1981a). Important confirmation has been provided by Dawson & Smith (1987) who observed using glass microelectrodes that $\Delta\psi$ initially fell from -22mV to -16mV in the presence of 2-aminoisobutyrate.

Amino Acid Accumulation and $\Delta\mu^{Na}$ in the Absence of ATP

The earliest demonstration that solute accumulation occurred in symport with the Na^+ gradient, in the absence of ATP, was made by Vidaver who used resealed pigeon erythrocyte membrane preparations with glycine as solute (Vidaver et al., 1976). The results were interpreted in terms of a carrier system in which 2 Na^+ and probably 1 Cl^- acted as co-substrates. Gramicidin short-circuited the gradient of Na^+ and virtually abolished glycine accumulation (Terry & Vidaver, 1973). Although full details have yet to be established, such a mechanism would respond to changes in the membrane potential, and to both the chloride gradient ($[Cl^-]_0/[Cl^-]_i$) to the first power and to the square of the Na^+ concentration gradient ($[Na^+]_0/[Na^+]_i$).

In the mouse ascites tumour cells, it early became apparent that preparations thoroughly depleted of ATP were able to accumulate glycine, methionine and various other amino acids without these undergoing significant metabolism, provided the sodium concentration gradient acting across the cell membrane was appropriately manipulated. Amino acid accumulation, represented by the ratio of the intracellular to extracellular concentrations, increased with time up to a maximum value that subsequently diminished. When the ratio of the extracellular to cellular concentration of Na^+ ($[Na^+]_0/[Na^+]_i$) at that peak was deliberately varied, the magnitude of $[A]_i/[A]_0$ itself varied systematically (Eddy, 1968a,b). Reference to Equation 3 shows that if the membrane potential is small enough to be neglected, as studies up to 1970 suggested (cf. Aull, 1967; -11mV), and is indeed 1, then the above two ratios would be equal. In fact $[A]_i/[A]_0$ was consistently larger than $[Na^+]_0/[Na^+]_i$, a circumstance that initially appeared to be due to an antiport function for K^+ but later was shown to depend on the prevailing membrane potential when the latter was systematically varied (Fig. 3). The specific contribution to the energy of the Na^+ gradient ($\Delta\mu^{Na}$) that is represented by $\Delta\psi$ has been verified in a wide range of membrane vesicle preparations, including those from mouse ascites cells, that accumulate amino acids and other organic solutes by symport mechanisms (Bardin & Johnstone, 1978; Johnstone, 1979b; Lever, 1980; Cecchini et al., 1977; Kilberg & Christensen, 1980).

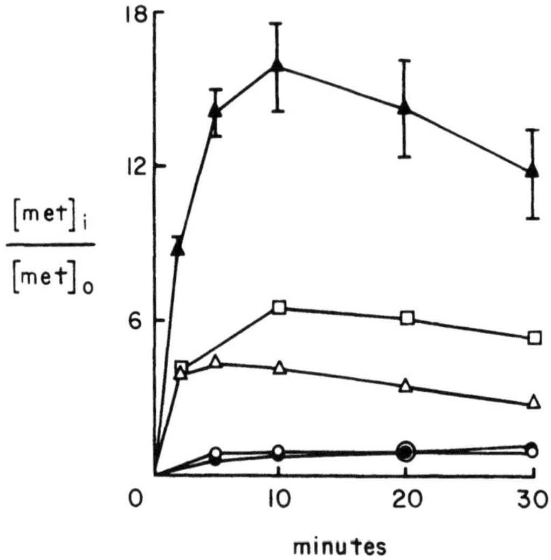

Fig. 3. Methionine accumulation by ATP-depleted mouse ascites tumour cells is larger when valinomycin is used to increase Δψ in the presence of an outwards K^+ gradient (Reid et al., 1974). The tumour cells depleted of ATP were suspended in the presence or absence of valinomycin in Ringer solutions containing different concentrations of Na^+ and K^+. Valinomycin present: (▲) 85mM Na^+, 5mM K^+; (□) 85mM Na^+, 22mM K^+; (●) 45mM Na^+, 115mM K^+; Valinomycin absent: (△) 85mM Na^+, 5mM K^+; (o) 45mM Na^+, 115mM K^+.

Magnitude of Δψ during Energy Metabolism

These earlier studies led to a reassessment of the magnitude of Δψ that the tumour cells maintain in various physiological circumstances. Recent work has clarified some of the controversies in this area (Johnstone et al., 1982; Smith, 1982). Gstrein et al. (1987) implanted capillary microelectrodes into Ehrlich mouse ascites tumour cells, observing stable potentials near -57 mv in a standard Ringer solution, whereas several previous studies had given values in the ranges either -20 mV to -40 mV or below -12 mV (Laris et al., 1976; Lambert et al., 1989). Lambert et al. (1989) suggest that the low values obtained by this technique in different laboratories may be erroneous due to ionic leakage around the implanted microelectrode, but it is also possible that differences in the natural ionic permeabilities of the different cell lines lead to different values of Δψ. In a comprehensive study Lambert et al. (1989) obtained values of -62 mV by means of a voltage-sensitive carbocyanine dye, the system being calibrated by addition of the ionophore gramicidin. Laris et al. (1976) who used 1 μM valinomycin reported a range of values from -18 to -42 mV. Lambert et al. (1989) additionally observed that when 1.5 μM valinomycin was used to calibrate the system, the intrinsic cellular permeability to Cl^- tended to short circuit the K^+ diffusion potential. Thus, at the so-called null point the actual potential would be lower than the equilibrium potential for K^+. Philo & Eddy (1978a) used 2.2 μM valinomycin to calibrate their preparations of another tumour cell line and deduced a value of Δψ of about -55 mV. They also employed metabolic inhibitors to prevent an accumulation of the dye in the mitochondria that would otherwise have been released during the calibration by the ionophore. It is by no means clear that this latter factor is unimportant. Whether the Cl^- factor is of significant magnitude, at the somewhat higher valinomycin concentration, under the above conditions has also to be established. It is interesting that Hoffmann & Lambert (1983) earlier estimated Δψ fluorimetrically as -36 mV with their cell line, using valinomycin and the mitochondrial inhibitors. After correction by the appropriate ionic activity coefficients determined by Dawson & Smith (1986), Δψ would be -21 mV, in agreement with direct determinations using microelectrodes in the same laboratory (see Lambert et al., 1989). The problem of mitochondrial absorption of the cationic dyes was circumvented by Bashford, Pasternak and colleagues (see Bashford et al., 1985) by employing the anionic probe oxonol V with Lettre ascites tumour cells. These workers concluded that $\Delta\psi$ was near -50 mV whether a protonophore or valinomycin was used to calibrate the system. Of considerable interest is their discovery that ouabain depolarizes this strain of tumour cells, $\Delta\psi$ being generated by electrogenic cation pumps rather than by ionic diffusion (Bashford & Pasternak, 1984). The work of Gstrein et al. (1987), Lambert et al. (1989) and others indicates that diffusion of K^+ represents a major component of $\Delta\psi$ in Ehrlich mouse ascites tumour cells. The important issue of the relationship between $\Delta\psi$ and the prevailing Cl^- distribution across the cell membrane has been subject to a penetrating analysis by Geck & Heinz (1986).

Relationship between $\Delta\mu^S$ and $\Delta\mu^{Na}$ during Energy Metabolism: Testing Equation 3

Extracellular amino acid concentration as a variable. The use of ATP-depleted mouse ascites tumour cells showed very clearly that both the terms $[Na^+]_0/[Na^+]_i$ and the magnitude of $\Delta\psi$ contributed to the accumulation of various amino acids (cf Fig. 3). However, the rapidity with which both parameters changed with time made it difficult to assay $\Delta\psi$ and therefore to define the quantitative relationship between $\Delta\mu^S$ and $\Delta\mu^{Na}$. During energy metabolism fairly steady values of all these parameters are maintained and there have been three independent studies made of the effects of systematic changes in $\Delta\mu^{Na}$ on the magnitude of $\Delta\mu^S$ formed in the presence of methionine, 2-aminoisobutyrate or glycine, $\Delta\psi$ being assayed in each instance by means of a voltage-sensitive dye. It has long been known that the magnitude of $[S]_i/[S]_0$ varies markedly with the amino acid concentration. For instance, Jacquez (1975) showed that whereas $[S]_i/[S]_0$ was fairly constant in the range 40-50 when $[S]_0$ was 1 mM or less, the ratio approached 2 when $[S]_0$ was 100 mM. Indeed, in the range of amino acid concentration above about 10 mM the tumour cells maintained an almost constant difference in amino acid concentration of 85 mM across the cell membrane. This was despite a complex pattern of cell swelling and loss of cellular K^+. The studies illustrated in Fig. 4a utilized relatively dilute solutions of the amino acids in conditions where the maximum amino acid concentration ratio would be expected to form. The general trend of the results supported the idea that $\Delta\mu^S$ and $\Delta\mu^{Na}$ were near thermodynamic equilibrium, bearing in mind that the stoichiometrical ratio n was close to 1 (see Table 2). Nevertheless, a consistent deficiency in the energy available from the Na^+ gradient is apparent over the whole range. To explain this Philo & Eddy (1978b) and Henius & Laris (1979) suggested that the effective cytosolic activity of Na^+ might be lowered by segregation of the ion in another cellular compartment. Moreover, if the thermodynamic activity coefficients for cellular and extracellular Na^+, like the corresponding coefficients for K^+, are not in fact equal, both the estimates of $\Delta\psi$ and $[Na^+]_0/[Na^+]_i$ would be changed.

Wastage of energy by amino acid leakage. Philo & Eddy (1978a) and Hacking & Eddy (1981a) attempted to estimate the overall magnitude of the apparent energy deficiency when amino acid leakage outside the Na^+ symport was also taken into account. A mathematical model was used to compute the predicted value of the amino acid gradient at the measured value of $\Delta\mu^{Na}$. This predicted value was then compared with the observed value, which in most instances was larger. The ratio for methionine ranged from 1.4 to 2.4, corresponding to an energy deficit of 0.9 - 2.3 kJ mol^{-1}. Values in the same range were obtained for glycine and 2-aminoisobutyrate. The leak pathways, as defined by amino acid uptake in the absence of Na^+, are especially important when 10 mM solutions of the amino acid are used, the amino acid outflow by leakage resulting in a significant fall in $\Delta\psi$ and in the magnitude of $\Delta\mu^{Na}$. Table 3 shows that while $\Delta\mu^{Na}$ was then larger than $\Delta\mu^S$ by about 2.6 to 4.3 kJ mol^{-1}, a small energy deficit of the order of 1 kJ mol^{-1} was nevertheless apparent with glycine and leucine (Hacking & Eddy, 1981a). This behaviour emphasizes the point that the interpretation of the relative magnitudes of $\Delta\mu^{Na}$ and $\Delta\mu^S$ is problematic unless specific information is available about the relevant rates of amino acid leakage.

Thermodynamic activity coefficients. Estimates of the magnitude of the corrections associated with the apparent activity coefficients for Na^+, K^+ and Cl^- are

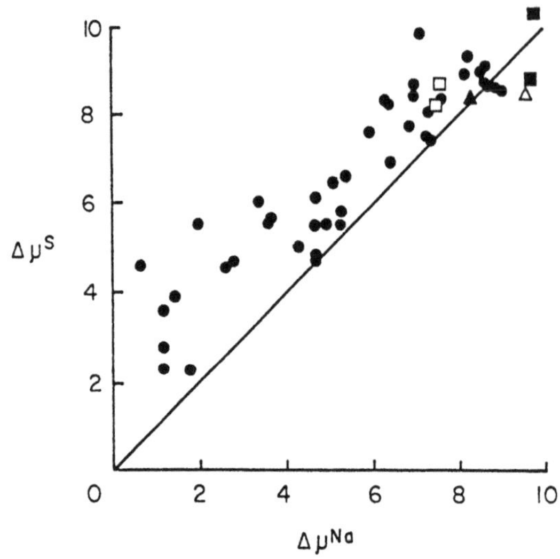

Fig. 4a. Relative magnitudes of $\Delta\mu^S$ and $\Delta\mu^{Na}$ during energy metabolism. The assays made in three laboratories are combined. $\Delta\psi$ was measured fluorimetrically. (●) 1mM methionine (Philo & Eddy, 1978b); (■), 0.1 mM or 1 mM 2-aminoisobutyrate (Hacking & Eddy, 1981a); (□), 0.5 mM 2-aminoisobutyrate (Heinus & Laris, 1979). (△), 1 mM glycine (Hacking & Eddy, 1981a; (▲), 0.01 mM glycine (Hoffman & Lambert, 1983). Gradients are expressed in units of kJ mol^{-1}.

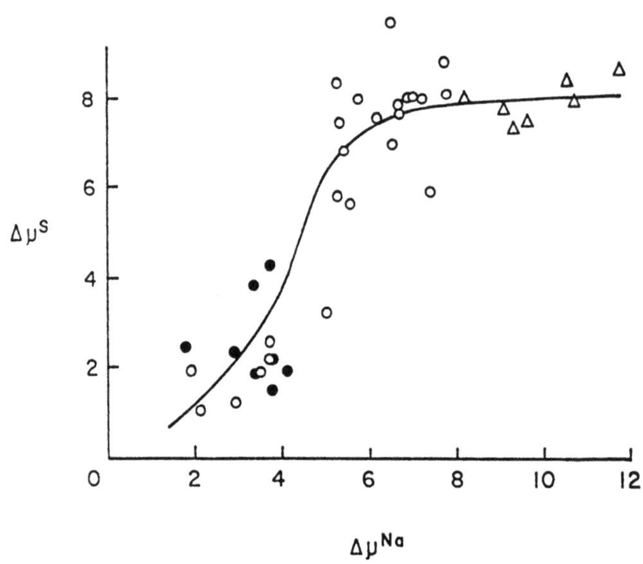

Fig. 4b. Relative magnitudes of $\Delta\mu^S$ and $\Delta\mu^{Na}$ in cell suspensions containing different amounts of nigericin. The assays were made under conditions similar to the 'reversed gradient' experiments illustrated in Fig. 5. The Ringer solution contained 40-50 mM Na^+ and 18 mM K^+ and 18 mM lactate. Thirty two assays were made in the presence of nigericin (o) or nigericin and ouabain (●). Controls (△) contained neither compound (adapted from Eddy et al., (1988)).

Table 3. Observed and predicted values of $\Delta\mu^S$ at values of $\Delta\mu^{Na}$ measured in suspensions of mouse ascites tumour cells in standard Ringer solution.

Amino acid	$\Delta\mu^{Na}$	$\Delta\mu^P$	Energy excess ($\Delta\mu^{Na}$-$\Delta\mu^P$)	$\Delta\mu^S$	Energy deficit ($\Delta\mu^S$-$\Delta\mu^P$)
0.1 mM 2-aminoisobutyrate	9.83	9.00	0.83	10.34	1.34
1 mM 2-aminoisobutyrate	9.71	8.07	1.64	8.77	0.70
10 mM 2-aminoisobutyrate	7.52	4.94	2.58	4.82	-0.12
1 mM methionine	8.45	7.08	1.37	8.79	1.71
1 mM leucine	9.36	5.86	3.50	5.62	-0.24
10 mM leucine	7.95	3.67	4.28	4.37	0.70
1 mM glycine	9.62	7.98	1.64	8.42	0.44
10 mM glycine	8.33	4.57	3.76	5.70	1.13

A mathematical model of the A system was used to predict the size of the amino acid gradient that would be formed at a given value of $\Delta\mu^{Na}$. The difference between the observed $\Delta\mu^S$ and the predicted value $\Delta\mu^S$ is the energy deficit ($\Delta\mu^S$ - $\Delta\mu^P$) in kJ mol^{-1}. The value $\Delta\mu^{Na}$ - $\Delta\mu^P$ is the excess energy expected to be required to drive the system. Adapted from Hacking & Eddy (1981a). When $\Delta\mu^{Na}$ was varied with methionine the energy deficit varied from 0.9-2.2 kJ mol^{-1} (Philo & Eddy, 1978b).

now possible on the basis of the work of Dawson & Smith (1986, 1987), who measured the effective intracellular activities of the three ions by means of ion-selective glass electrodes. We shall assume that their data apply to the different cell line studied by Hacking & Eddy (1981a). The effective activity coefficient for cellular K^+ was about 0.41, whereas that for cellular Na^+ was unexpectedly low, near 0.17 when the tumour cells were maintained in various Ringer solutions. The corresponding extracellular activity coefficients were 0.73 for K^+ and 0.71 for Na^+. When $\Delta\psi$ is assayed fluorimetrically using K^+ as the calibrating ion, the true value is thus about 15 mV less negative than the value based on the assumption that the cellular and extracellular activity coefficients are equal. In contrast $\Delta\mu^{Na}$ is increased by RT ln[(0.71 x 0.41)/(0.17 x 0.73)] or about RT ln 2.34; namely 2.2 kJ mol^{-1}. Thus the evidence presented by Dawson & Smith suggests that previous workers using the fluorimetric technique have overestimated $\Delta\psi$ by about -15mV and underestimated $\Delta\mu^{Na}$ by about 2.2 kJ mol^{-1}. It is therefore very satisfactory that the theoretical estimates for methionine of the magnitude of the overall energy deficit, including the effect of amino acid leakage, were in the range 0.9 to 2.3 kJ mol^{-1}. As envisaged by Philo & Eddy (1978a) and by Henius & Laris (1979) and demonstrated by Dawson & Smith, this deficit is removed when appropriate activity coefficients are used.

Magnitude of the excess energy required to maintain solute accumulation. Dawson & Smith (1987) have studied the relative magnitudes of $\Delta\mu^S$ and $\Delta\mu^{Na}$ for 0.25 mM solutions of 2-aminoisobutyrate, as a function of the concentration of Na^+ in the Ringer solution. The magnitude of $\Delta\psi$ was assayed with implanted glass electrodes, whereas the effective cellular activities of Na^+ were determined by means of Na^+ selective electrodes. The work showed clearly that the quantity $\Delta\mu^S$ - $\Delta\mu^{Na}$ was significantly larger than zero at each of the five values of [Na^+] employed. In the standard Ringer solution, the excess Na^+-gradient was about 1.4 kJ mol^{-1}. These results agree well with the computations in Hacking & Eddy (1981a), based on estimates of the amino acid flow through leakage pathways, and whose conclusions are illustrated in Table 3, the accumulation ratios reported originally having been converted into the equivalent energy units. While the driven gradient and the driving gradient are evidently near equilibrium at the relatively low amino acid concentrations, they are not exactly in equilibrium, if the stoichiometry of the system is 1. Reference to the earlier discussion of Equation 8 shows that the ratio $\Delta\mu^S/\Delta\mu^{Na}$, whether computed or determined experimentally, defines the apparent stoichiometry (n') of the coupling. Thus the energy excess of 1.4 kJ mol^{-1} obtained by

Dawson & Smith (1987) corresponds to a value of 0.87 for n' which fell to 0.82 at the lower concentration of Na^+. These two estimates of n' are generally consistent with the direct determinations of the Na^+ stoichiometry reported in Dawson et al. (1983). These workers observed an increase in the coupling parameter from 0.60 at 10 mM Na^+ to 0.90 at 140 mM Na^+. Dawson et al. (1983) associate the changes in stoichiometry with the substitution of protons for Na^+ as co-substrate, but the leakage pathway has more commonly been viewed as dependent on the L system.

The concentrative uptake of alanine by hepatocytes is another system in which the magnitude of $\Delta\mu^{Na}$ has been shown somewhat to exceed the magnitude of $\Delta\mu^{S}$. Preliminary evidence for the presence of a leak outside the Na^+ symport has been obtained (Kristensen et al., 1983; Kristensen & Folke, 1986)).

In a comprehensive study of the relationship of $\Delta\mu^S$ to $\Delta\mu^{Na}$ in human fibroblasts accumulating 2-methylaminoisobutyrate, Dall'Asta et al. (1991) showed that the ratio $\Delta\mu^S/\Delta\mu^{Na}$ was about 0.95 when $\Delta\mu^S$ increased from 4 kJ mol^{-1} to 12 kJ mol^{-1}, as a result of varying the ionic composition of the culture medium. The Na^+ stoichiometry for this amino acid was very close to 1. These workers used the accumulation of arginine to monitor $\Delta\psi$. Starvation increased the rate of uptake of the amino acid but not the maximum accumulation ratio to a significant extent. All these circumstances suggest that the A system in this cell line brought the sodium gradient and the amino acid gradient very close to thermodynamic equilibrium, if it is correct to suppose that leakage of the amino acid out of the fibroblasts can be neglected. Correction of these results for the effective activity of cellular Na^+, relative to the activity coefficient in the suspension medium, would be partly compensated by the corresponding coefficient for cellular arginine. The magnitude of these corrections is not known.

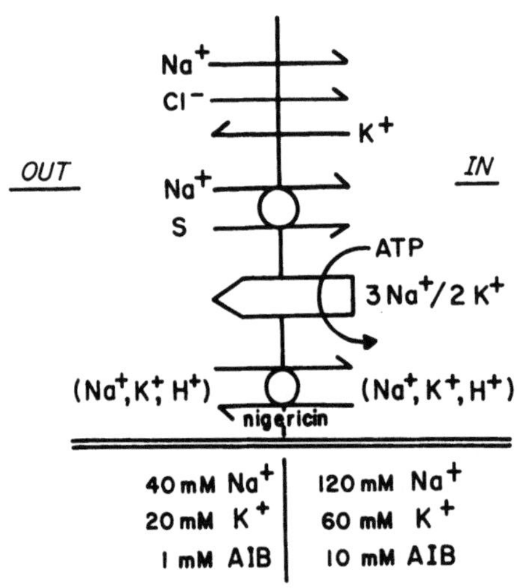

Fig. 5. Amino acid uptake by mouse ascites tumour cells in the presence of nigericin. The conditions are those in the 'reversed' Na^+ gradient experiment of Heinz et al., (1981). Nigericin causes the accumulation of Na^+ in the tumour cells to levels much larger than the 40 mM present outside. Amino acid is nevertheless concentrated intracellularly. The upper part of the diagram illustrates the processes that may contribute to the generation of a relatively large negative membrane potential, as discussed in Eddy (1985). Observations made with the voltage sensitive probe oxonol V indicate that ψ reached 70-80 mV negative under these conditions (Eddy et al., 1988).

The above discussion treats the leakage of amino acids from the tumour cells as occurring through pathways with a constant permeability coefficient. A number of observations indicate that, not only do ion conductive pathways open during amino acid uptake leading to cell volume increase, but that the passive permeability to amino acids increases with cell volume (Hoffmann & Lambert, 1983; Hudson & Schultz, 1988; Lambert et al., 1989). However, the persistent membrane depolarization that large concentrations of amino acid cause was still exhibited when increases in cell volume were prevented by sucrose (Hacking & Eddy, 1981a).

Effect of Nigericin on Amino Acid Accumulation:
A Stringent Test of the Na^+ Gradient Hypothesis

Heinz et al. (1981) devised an ingenious test of the ability of $\Delta\mu^{Na}$ to maintain amino acid accumulation in the seemingly adverse conditions where the mouse ascites tumour cells are kept in the presence of nigericin and lactate or pyruvate (Fig. 5). In these circumstances Na^+ accumulates in the cells at concentrations of 100 mM or more, the ratio $[Na^+]_0/[Na^+]_i$ becoming smaller than 1. Nevertheless 2-aminoisobutyrate or 2-methylaminoisobutyrate were each accumulated extensively, from 12 to about 20 fold (Heinz et al., 1981; Johnson & Eddy, 1985). Heinz et al. (1981) showed that the accumulation still occurred in the presence of 0.1 mM ouabain but later work demonstrated that 2 mM ouabain lowered the accumulation ratio to about 2. Inhibitors of mitochondrial energy metabolism had a similar effect, observations collectively indicating the Na^+ pump was involved in driving the concentration of the amino acids, at the expense of ATP generated by oxidative phosphorylation (Johnson & Eddy, 1985).

Because of the role of intact mitochondria in the system, the use of the cationic carbocyanine dyes to assay $\Delta\psi$ is precluded. Indeed an initial study of the problem by this means suggested that $\Delta\mu^{Na}$ was much too small to maintain the observed amino acid accumulation (Heinz et al., 1981). However, satisfactory assays were later made using the anionic oxonol V as probe (Eddy et al., 1988). The general trend of the results, illustrated in Fig. 4b, showed that there was a rough correlation between the magnitude of $\Delta\mu^S$ and $\Delta\mu^{Na}$, but probably not a linear relationship between the two variables. An energy deficit of up to about 2 kJ mol^{-1} was observed in the presence of nigericin when $\Delta\mu^S$ itself was near 8 kJ mol^{-1}. A detailed knowledge of the cellular activity of Na^+ is required to advance the interpretation of these experiments which, taken as a whole, seem consistent with the Na^+ gradient hypothesis. One aspect that is not understood is that the tumour cells maintained an apparent of $\Delta\psi$ about 70 mV (uncorrected with respect to the activity coefficients of K^+) during lactate metabolism in the low Na^+, low K^+ Ringer solution, with $[Na]_0/[Na]_i$ about 3.6. Nigericin increased $\Delta\psi$ to about 77 mV and the Na^+ ratio to 0.6. Even in the presence of ouabain $\Delta\psi$ considerably exceeded the equilibrium potential for K^+. The possible contribution of lactate and pyruvate transfer to the maintenance of $\Delta\psi$ deserves further study (see Bashford & Pasternak, 1984).

Amino Acid Substrates with Special Properties

In the course of their pioneering work on the specificities of the A, ASC, L and y^+ substrate systems of mammalian cells, Christensen and his colleagues drew attention to the way in which various substitutions in the molecule of 2-aminoisobutyrate affected the maximum accumulation ratio that mouse ascites tumour formed in the presence of the compound (Christensen, 1984). Whereas N-methylation was well tolerated, the introduction of one, two or three fluorine atoms into the alkyl side chain progressively lowered the accumulation ratio, in parallel with a reduced basicity of the α-NH2 group (Christensen & Handlogten, 1975).

It was inferred that the α,α-zwitterion was the preferred substrate of the A-system. Further, whereas the strongly basic arginine cation was not a substrate of the A system, both 2,3-diaminopropionate and 2,4-diaminobutyrate were vigorously accumulated by this Na^+-dependent route (Christensen & Liang, 1966; Christensen, 1973). The diaminobutyrate, in particular, displaces much of the normal content of cellular K^+ and is accumulated together with Cl- in amounts that can lead to cell lysis (Christensen et al., 1952). The kinetic parameters of a series of A system substrates is illustrated in Table

4 which shows that the diaminobutyrate exhibits the largest V_{max} in the series, as well as the largest K_m value. Indeed in terms of the parameter V_{max}/K_m this amino acid is not an especially effective substrate of the system. Christensen (1973) argues that the especially vigorous absorption of the three and four carbon diamino acids, in comparison with the lysine cation, is related to the facility with which they form a zwitterion, about 9 percent of the diaminobutyrate being in this form near pH 7 ($pK_2 = 8.3$). However, the zwitterionic form in free solution appears to be mainly the α,γ species not the α,α species that would be the analogue of 2-aminoisobutyrate or glycine and therefore the most likely form to undergo transport. The substrate thialysine also exhibits large values of K_m and V_{max}, with facile dissociation of the distal protonated amino group. In the scheme Christensen proposes the fast uptake of the α,α-zwitterion is assisted by a hypothetical gradient of chemical potential of protons acting within the membrane but not necessarily in equilibrium with the solutions on either side of the membrane (Christensen & Handlogten, 1975). The effect of this cryptic gradient is to facilitate the removal of the distal proton from the substrate cation on the external membrane face and to reprotonate it at the internal face. This process would raise the chemical potential of the solute at the expense of the inner gradient of protons. On this view the special character of the diamino acids lies in their ability to sense this proposed inner proton gradient, the maintenance of which would require an ancillary proton pump. This proposal was originally envisaged as a mechanism for supplementing the driving force represented by the Na^+ gradient, thereby making good the deficiency of energy supply that that source appeared to represent. Subsequent work comparing the prevailing magnitudes of $\Delta\mu^S$ and $\Delta\mu^{Na}$ has shown that this energy deficit probably does not exist. Nevertheless, the postulate of a cryptic proton gradient had the additional merit of explaining how solutes might be concentrated through the Na^+-independent L system (Christensen, 1973). There is indeed a further category of anomalous unnatural substrates, of which thialysine is an example, that interact with both system A and system L, thus reinforcing the notion that the two might be based on similar mechanisms (Christensen, 1973):

(1) One issue raised by this ingenious hypothesis concerns the nature of the proton pump, which might originate in the redox function of NAD or NADP linked dehydrogenases (Löw & Crane, 1978) or in ATP breakdown (Oshawa et al., 1980). The established ability of ascorbate plus phenazine methosulphate to stimulate amino acid uptake through the A system initially appeared to support the hypothesis, but would now seem to depend on activation of a Ca^{2+}-dependent K^+ channel causing an increase in $\Delta\psi$ (Yamamoto & Kawasaki, 1981; Valdeolmillos et al., 1982). The existence of this proton pump remains problematical.

Table 4. Approximate kinetic parameters for the absorption of various diamino acids and neutral amino acids by mouse ascites tumour cells in standard Ringer solution.

Amino acid	V_{max}	K_m	V_{max}/K_m
2-methylaminoisobutyrate	4	0.3	13
glycine	7.4	4.1	1.8
alanine	6.4	0.4	16
methionine	3.7	0.26	14
2-aminoisobutyrate	7.1	1.1	6.4
N-methylphenylalanine	5.2	10.5	0.5
phenylalanine	7.7	5.5	1.4
2,3-diaminopropionate	17	2.6	6.5
2,4-diaminobutyrate	60	50	1
S-(2-aminoethyl)-cysteine (thialysine)	20	20	1

Uptake in each case is believed to be mainly by the A system. V_{max} is in units of mmoles/kg cell water/min., K_m in mM. The observations were compiled from Inui & Christensen (1966), Christensen & Liang (1966) and Jacquez (1973).

(2) The balance of evidence, including observations made with membrane vesicles and probes of $\Delta\psi$, is that system L is not electrogenic and is not actuated by $\Delta\mu^H$, nor is it affected directly by the presence of proton conductors (Kilberg & Christensen, 1980; Ohsawa et al., 1980; Hacking & Eddy, 1981a). Evidence against the view that concentration of amino acids through the L system is simply due to exchange with other endogenous amino acids has been obtained in two laboratories (Christensen et al., 1973; Hacking & Eddy, 1981b). A hypothesis that does not appear to have been excluded is that the undoubted ability of the mouse ascites tumour cells to concentrate specific L-system substrates in the absence of Na^+ depends on their accumulation in some intracellular compartment, after passage through the L system carrier into the cytosol.

In summary, the hypothetical proton pump envisaged in the model based on localized membrane gradients lacks direct support, nor is it likely that the A and L systems share a common energy source at the membrane level. For instance, marked lowering of $\Delta\mu^{Na}$ in the presence of ouabain had little effect on the accumulation of the norbornane amino acid substrate of the L system (Hacking & Eddy, 1981a). As discussed in Ohsawa et al. (1980), substrates that interact with both systems A and L can behave in a complex way. Thus Hacking & Eddy (1981a) showed that dilute solutions of isoleucine were considerably concentrated in the presence of ouabain and to a far greater extent than would be expected from the magnitude of $\Delta\mu^{Na}$. Because the norbornane amino acid prevented isoleucine accumulation under these conditions, the L system appeared to be responsible (Hacking & Eddy 1981b). In contrast, larger concentrations of isoleucine behaved as though entry occurred mainly with Na^+ ions and exit predominantly by leakage without Na^+ through a separate route. The accumulation of L-methionine, 2-aminoisobutyrate or 2-N-methylisobutyrate in this system produced similar gradients of 8.8 - 10.3 kJ mol^{-1} (Hacking & Eddy 1981b) and was satisfactorily described as a Na^+ symport operating in parallel with a Na^+-independent leak.

Apparent Uncoupling of the Na^+ Gradient from the Amino Acid Gradient

One argument used to support the notion of a cryptic intra-membrane gradient as the primary driving force of amino acid accumulation was that the flow of Na^+ could be 'uncoupled' from that of the amino acid (Christensen et al., 1973). In these experiments the mouse ascites tumour cells first accumulated, for instance, unlabelled 2-aminoisobutyrate or its N-methyl counterpart. The preparation then absorbed the labelled amino acid and the influx of Na^+ was compared with that of the amino acid. Controls were assayed without the initial incubation with unlabelled amino acid. The main conclusion was that there were circumstances, where, in contrast to the behaviour of controls, less than 0.3 equivalents of Na^+ accompanied the extra flow of amino acid the presence of Na^+ caused. The significance of these observations is unclear, partly because of statistical uncertainties but, more fundamentally, because $\Delta\psi$ would have varied in a complex fashion. Loading the tumour cells, with amino acid would hyperpolarize them so that the basal uptake of Na^+ would be faster in a test made in the absence of amino acid. The basal flow of Na^+ would then be lowered when the added labelled amino acid depolarized the tumour cells (Eddy, 1977). Similarly, in the controls the basal uptake of Na^+ would be less in the presence of the amino acid than in its absence. These latter effects (cf Kimmich & Randles, 1980) were examined theoretically by Philo & Eddy (1978b) and experimentally by Dawson et al. (1983). Another experimental strategy for circumventing this problem is considered in Eddy et al. (1988).

The Behaviour of 2,4-diaminobutyrate in Relation to $\Delta\mu^{Na}$

The need to postulate an ancillary proton pump, in connection with the marked accumulation displayed by the two short chain diamino acids, originated in the belief that both interacted as zwitterions with the A system, though not in a manner that led to accumulation of the dissociated proton in the extracellular solution (Christensen et al., 1973). It seems possible that the presence in these cationic substrates of a dissociable proton assists the binding of the molecule at a site where the carrier itself accepts this proton. The need to postulate an ancillary proton pump would then be avoided.

Whereas the normal mode of action of the A system corresponds to a maximum accumulation ratio equivalent to $\Delta\mu^{Na}$ (Equation 3), the accumulation will be enhanced by a further factor of $e^{-\Delta\psi F/RT}$ in the case of a cationic solute, because two equivalents of positive change are transferred. Thus, on the basis that 2-aminoisobutyrate is accumulated about 50 fold, a cationic analogue might be accumulated 125 fold if $\Delta\psi$ is conservatively assumed to be -25 mV. The detailed information required to test this prediction is at present lacking. Indeed the number of equivalents of Na^+ absorbed along with 2,4-diaminobutyrate has not been established and is merely assumed, for the present purpose, to be the same as the value for neutral amino acids.

It seems unlikely that the relatively large values of V_{max} exhibited by the diamino acids (Table 4) can be accounted for simply in terms of the single extra positive charge transferred. This would be expected to increase the rate of transfer by a factor in the order of $e^{-\Delta\psi F/2RT}$, or only 1.6 when $\Delta\psi$ is -25 mV.

The extensive swelling of the mouse ascites tumour cells that takes place in the presence of 30 mM diaminobutyrate is associated with rather small accumulation ratios of about 5-fold (Christensen et al., 1952). The relatively large K_m exhibited by this substrate probably contributes to its ability to accumulate in large amounts intracellularly. Thus early work on the factors influencing the magnitude of $[S]_i$ showed that it tended towards a maximum value as $[S]_0$ was increased (Johnstone & Scholefield, 1965). Moreover, the concentration of $[S]_0$ at which the amino acid pool is half its maximum size is approximately equal to the K_m exhibited when the initial rate of uptake of the amino acid is studied as a function of $[S]_0$. The model of amino acid accumulation based on a pump and leak accounted for this behaviour (Tenenhouse & Quastel, 1960; Johnstone & Scholefield, 1965). While substrates exhibiting a relatively large K_m might accordingly be expected to form relatively large intracellular pools, the magnitude of the leak pathway can also have a decisive influence, as already noted in connection with Table 3. A low rate of leakage outside the Na^+-symport may thus be another significant factor in the behaviour of the diaminobutyrate. Studies with mathematical models indicate that when the leak coefficient is constant, the maximum accumulation ratio ($[S]_i/[S]_0$) forms when $[S]_0$ is near K_m. At higher concentrations the pump tends to saturate on both sides of the cell membrane and net transfer of substrate is retarded (Eddy et al., 1979, 1980).

General and Molecular Considerations

The model of flow coupling based on the notion of a ternary complex ENaS has been scrutinized both theoretically and experimentally for thirty years. It should be emphasized that modern opinion (see Mitchell, 1987) favours a mobile barrier type of mechanism, resembling Patlak's gated binding site with alternating access (Patlak, 1957), rather than the traditional mobile carrier. A rival possibility suggested by Lancaster (1982) that Na^+ and S translocate consecutively in a 'ping-pong' mechanism has proved unsatisfactory for intestinal glucose transport (Kessler & Semenza, 1983) and for alanine transport in acinar cells (Jauch and Lauger, 1986).

Very considerable kinetic complexity is implicit in the ternary complex model as recent reviews emphasize (Kessler & Semenza, 1983; Sanders, 1986; Kimmich, 1990). Such issues as whether ordered or random binding of Na^+ and S occurs and whether a 'sodium well' effect occurs (Kimmich, 1990) have not been resolved for the A system of amino acid transport. However, Jauch & Lauger (1986) concluded that alanine and Na^+ bound to the pancreatic alanine carrier in a random fashion. In contrast, the intestinal imino carrier binding 2 Na^+, behaved as an iso-ordered ter ter system (Stevens & Wright, 1987). Other systems where Na^+ and K^+ are both involved show especially complex kinetics (Berteloot, 1986; Heinz et al., 1988).

The molecular sizes of various Na^+-dependent amino acid transporters, including a proline carrier, in the kidney cortex were in the range 220-270 kDa as determined by the radiation inactivation method (Béliveau et al., 1990). Wright and Peerce (1984) deduced a minimum molecular weight of about 100 kDa for the intestinal proline carrier by means of covalent labelling studies. Béliveau et al. (1990) accordingly suggest that the proline carrier functions as a dimer. Stevens et al. (1990) using radiation inactivation have inferred that the intestinal brush border Na^+/ glucose symport functions as a tetramer based on a 66 kDa subunit. Indeed two binding sites were required to explain the detailed kinetic properties of the renal glucose symports (Koepsell et al., 1990). The discovery of size-fractionated mRNA that stimulates expression of system A amino acid transporter in Xenopus oocytes opens the way towards the cloning of the corresponding cDNA and the determination of the molecular weight of the relevant polypeptides (Tarnuzzer et al., 1990). A large number of symports from prokaryotes and lower eukaryotes, resemble the respective Na^+-dependent and Na^+-independent mammalian glucose carriers in being based on 10-12 putative membrane helices (Eddy, 1990; Henderson, 1990). Whether the functional units of mammalian Na^+-dependent amino acid carriers will also prove to be oligomers of subunits with 12 helices, as the radiation inactivation studies hint, remains to be established.

The need to minimize binary translocations to obtain efficient energy coupling was emphasized in connection with Equations 4 and 5 and was implicit in early studies of these processes (Eddy, 1968a,b; Heinz et al., 1972). Recent work has shown that a number of point mutations in the lactose proton symport of Escherichia coli allow uncoupled transport of protons or galactoside (King & Wilson, 1990; Brooker, 1991). A search for such mutants and for other types of coupling defects (Eddy, 1990) in the mammalian amino acid systems might be rewarding. Studies of the intrinsic tryptophan fluorescence of the Na^+/glucose transporter have identified a Na^+-induced conformational change that results in tighter glucose binding to the carrier. The latter step leads to a further change in tryptophan fluorescence. It is tempting to associate these fluorescence changes with the respective conformations of E, ENa and ENaS (Peerce, 1990). An analogous study has been made using an extrinsic probe covalently bound to the intestinal proline carrier (Wright & Peerce, 1990).

REFERENCES

Aull, F., 1967, J. Cellul. Physiol., 69:21.
Barbour, B., Brew, H. and Attwell, D., 1988, Nature, 335:433.
Bardin, C. and Johnstone, R.M., 1978, J. Biol. Chem., 253:1725.
Bashford, C.L. and Pasternak, C.A., 1984, J. Membr. Biol., 79:275.
Bashford, C.L., Alder, G.M., Gray, M.A., Micklen, K.J., Taylor, C.C., Turek, P.J. and Pasternak, C.A., 1985, J. Cellul. Physiol., 123:326.
Béliveau, R., Demeule, M., Jette, M. and Potier, M., 1990, Biochem. J., 268:195.
Berteloot, A., 1986, Biochim. Biophys. Acta, 861:447.
Boyer, P.D., 1988, Trends Biochem. Sci., 13:5.
Brooker, R.J., 1991, J. Biol. Chem., 266:4131.
Cecchini, G., Payne, G.S. and Oxender, D.L., 1977, J. Supramolec. Struct., 7:481.
Christensen, H.N., 1973, Bioenergetics, 4:31.
Christensen, H.N., 1984, Biochim. Biophys. Acta, 779:255.
Christensen, H.N., 1985, J. Membr. Biol., 84:97.
Christensen, H.N. and Handlogten, M.E., 1975, Proc. Natl. Acad. Sci. U.S.A., 72:23.
Christensen, H.N. and Liang, M., 1966, J. Biol. Chem., 241:5542.
Christensen, H.N., Riggs, T.R., Fischer, H. and Palatine, I.M., 1952, J. Biol. Chem., 198:17.
Christensen, H.N., de Cespedes, C., Handlogten, M.E. and Ronquist, G., 1973, Biochim. Biophys. Acta, 300:487.
Dall Asta, V., Bussolati, O., Guidotti, G.G. and Gazzola, G.C., 1991, J. Biol. Chem., 266:1591.
Dawson, W.D. and Smith, T.C., 1986, Biochim. Biophys. Acta, 860:293.
Dawson, W.D. and Smith, T.C., 1987, Biochim. Biophys. Acta, 897:5.
Dawson, W.D., Robinson, S.C. and Smith, T.C., 1983, J. Cellul. Physiol., 115:131.
Eddy, A.A., 1968a, Biochem. J., 108:195.
Eddy, A.A., 1968b, Biochem. J., 108:489.
Eddy, A.A., 1977, in: Excerpta Medica International Congress Series No. 391., Excerpta Medica, Amsterdam.
Eddy, A.A., 1985, Ann. N.Y. Acad. Sci., 456:51.

Eddy, A.A., 1990, in: Monovalent Cations in Biological Systems, C.A. Pasternak, ed., CRC Press, Inc., Boca Raton.
Eddy, A.A., and Johnson, E., 1989, Methods Enzymol., 173:771.
Eddy, A.A., Seaston, A., Gardner, D. and Hacking, C., 1979, in: Function and Molecular Aspects of Biomembrane Transport, E. Quaglioriello, ed., Elsevier/North Holland, New York.
Eddy, A.A., Season, A., Gardner, D. and Hacking, C., 1980, Ann. N.Y. Acad. Sci., 341:494.
Eddy, A.A., Hopkins, P. and Johnson, E., 1988, Biophys. Chem., 29:119.
Geck, P. and Heinz, E., 1986, J. Membr. Biol., 91:97.
Gstrein, E., Paulmichl, M. and Lang, F., 1987, Eur. J. Physiol., 408:432.
Hacking, C. and Eddy, A.A., 1981a, Biochem. J., 194:415.
Hacking, C. and Eddy, A.A., 1981b, unpublished.
Heinz, E., 1978, Mechanisms and Energetics of Biological Transport, Springer-Verlag, Berlin.
Heinz, E. and Geck, P., 1974, Biochim. Biophys. Acta, 339:426.
Heinz, E., Geck, P. and Willbrandt, W., 1972, Biochim. Biophys. Acta, 255:442.
Heinz, E., Geck, P., Pietrzyk, C., Burckhardt, G. and Pfeiffer, B., 1977, J. Supramolec. Struct., 6:125.
Heinz, A., Jackson, J.W., Richey, B.E., Sachs, G. and Schafer, J.A., 1981, J. Membr. Biol., 62:149.
Henderson, P.J.F., 1990, J. Bioenerget. Biomembr., 22:525.
Henius, G.A. and Laris, P.C., 1979, Biochem. Biophys. Res. Commun., 91:1430.
Hoffman, E.K. and Lambert, I.H., 1983, J. Physiol., 338:613.
Hudson, R.L. and Schultz, S.G., 1988, Proc. Natl. Acad. Sci. U.S.A., 85:279.
Inui, Y. and Christensen, H.N., 1966, J. Gen. Physiol., 50:203.
Jacquez, J.A., 1973, Biochim. Biophys. Acta, 318:411.
Jacquez, J.A., 1975, J. Gen. Physiol., 65:57.
Jauch, P. and Lauger, P., 1986, J. Membr. Biol., 94:117.
Jauch, P., Petersen, O.H. and Lauger, P., 1986, J. Membr. Biol., 94:99.
Johnson, E. and Eddy, A.A., 1985, Biochem. J., 226:773.
Johnstone, R.M., 1979a, Can. J. Physiol. Pharmacol., 57:1.
Johnstone, R.M., 1979b, J. Biol. Chem., 254:12479.
Johnstone, R.M. and Scholefield, 1965, Advan. Cancer Res., 9:143.
Johnstone, R.M., Laris, P.C. and Eddy, A.A., 1982, J. Cellul. Physiol., 112:298.
Kessler, M. and Semenza, G., 1983, J. Membr. Biol., 76:27.
Keynan, S. and Kanner, B.I., 1988, Biochemistry, 27:12.
Kilberg, M.S. and Christensen, H.N., 1980, Membr. Biochem., 3:155.
Kimmich, G.A., 1990, J. Membr. Biol., 114:1.
Kimmich, G.A. and Carter-Su, C., 1978, Am. J. Physiol., 235:C73.
Kimmich, G.A. and Randles, J., 1980, Biochim. Biophys. Acta, 596:439.
King, S.C. and Wilson, T.H., 1990, Molec. Microbiol., 4:1433.
Kinne, R., Sommerfeld, D. and Heinz, E., 1988, Biophys. Chem., 29:105.
Koepsell, H., Fritzsch, G., Korn, K. and Madrala, A., 1990, J. Membr. Biol., 114:113.
Kristensen, L.O. and Folke, M., 1983, Biochem. J., 210:621.
Kristensen, L.O. and Folke, M., 1986, Biochim. Biophys. Acta, 855:49.
Kristensen, L.O., Sestoft, L. and Folke, M., 1983, Am. J. Physiol., 244:G491.
Lambert, I.A., Hoffman, E.K. and Jorgensen, F., 1989, J. Membr. Biol., 111:113.
Laris, P.C., Pershadsingh, H.A. and Johnstone, R.M., 1976, Biochim. Biophys. Acta, 436:475.
Lever, J.E., 1980, CRC Critical Reviews in Biochemistry, p187.
Low, H. and Crane, F.L., 1978, Biochim. Biophys. Acta, 515:141.
Martin, G.M., Matthews, R.H. and Minton, J.P., 1984, Bioelectrochem. Bioenerget., 12:37.
Meister, A. and Anderson, M.E., 1983, Ann. Rev. Biochem., 52:711.
Mitchell, P., 1987, in: Integration and Control of Metabolic Processes: pure and applied aspects, O. L. Kon et al., eds., Cambridge University Press.
Munck, L.K. and Munck, B.G., 1990, Biochim. Biophys. Acta, 1027:17.
Nelson, P.J., and Rudnick, G., 1982, J. Biol. Chem., 257:6151.
Neville, M.C., Lobitz, C.J., Ripoll, E.A. and Tinney, C., 1980, J. Biol. Chem., 255:7311.
Ohsawa, M., Kilberg, M.S., Kimmel, G. and Christensen, H.N., 1980, Biochim. Biophys. Acta, 599:175.
Paterson, J.Y.F., Sepulveda, F.V. and Smith, M.W., 1980, Biochim. Biophys. Acta, 603:288.
Patlak, C. S., 1957, Bull. Math. Biophys., 19:209.

Peerce, B.E., 1990, J. Biol. Chem., 265:1737.
Pellifique, F., Butler, J.D., Spielberg, S.P., Hollenberg, M.D., Goodman, S.I. and Schulman, J.D., 1976, Biochem. Biophys. Res. Commun., 73:997.
Philo, R.D. and Eddy, A.A., 1978a, Biochem. J., 174:801.
Philo, R.D. and Eddy, A.A., 1978b, Biochem. J., 174:811.
Pietrobon, D. and Caplan, S.R., 1989, Methods Enzymol., 171:397.
Reid, M., Gibb, L.E. and Eddy, A.A., 1974, Biochem. J., 140:383.
Sanders, D., 1986, J. Membr. Biol., 90:67.
Sastre, J., Sweiry, J.H., Doolabh, K., Vina, J. and Mann, G.E., 1991, Biochim. Biophys. Acta, 1065:213.
Schaffer, J.A. and Jacquez, J.A., 1967, Biochim. Biophys. Acta, 125:1081.
Schultz, S.G. and Curran, P.F., 1970, Physiol. Rev., 50:637.
Sheppard, D.N., Giraldez, F. and Sepulveda, F.V., 1988, FEBS Lett., 234:446.
Smith, T.C., 1982, J. Cellul. Physiol., 112:302.
Stevens, B.R. and Wright, E.M., 1987, J. Biol. Chem., 262:6546.
Stevens, B.R., Fernandez, A., Hirayama, B., Wright, E.M. and Kempner, E.S., 1990, Proc. Natl. Acad. Sci. U.S.A., 87:1456
Tarnuzzer, R.W., Campa, M.J., Qian, N.-X., Englesberg, E. and Kilberg, M.S., 1990, J. Biol. Chem., 265:13914.
Tenenhouse, A. and Quastel, J.H., 1960, Can. J. Physiol., 38:1311.
Terry, P.M. and Vidaver, G.A., 1973, Biochim. Biophys. Acta, 323:441.
Valdeolmillos, M., Garcia-Sancho, J. and Herreros, B., 1982, Biochim. Biophys. Acta, 689:177.
Vidaver, G.A., Shepherd, S.L., Lagow, J.B. and Wiechelman, K.J., 1976, Biochim. Biophys. Acta, 443:494.
Wingrove, T.G. and Kimmich, G.A., 1988, Am. J. Physiol., 255:C737.
Wright, E.M. and Peerce, B.E., 1984, J. Biol. Chem., 259:14993.
Yamamoto, S. and Kawasaki, T., 1981, Biochim. Biophys. Acta, 644:192.

TECHNIQUES USED IN THE STUDY OF PLASMA MEMBRANE AMINO ACID TRANSPORT

J.D. McGivan

Department of Biochemistry
University of Bristol
Bristol BS8 1TD, U.K.

INTRODUCTION

Techniques for studying amino acid transport in various cells and organelles have been well established for a number of years. Detailed protocols for the measurement of amino acid transport in various systems have recently appeared in a series of articles in Methods in Enzymology (Kilberg, 1989; Eddy and Johnson, 1989; Harvey and Ellory, 1989; Murer et al., 1989; Hopfer, 1989; McGivan, 1989). The purpose of this article is not to recapitulate this information in detail but rather to focus on points of critical importance in various procedures. The discussion will concentrate on aspects of the measurement of plasma membrane transport. Some details of the measurement of transport in subcellular organelles will be found elsewhere in this volume. There is at present increasing interest in the reconstitution of plasma membrane transport systems into artificial phospholipid membrane vesicles. The methodology used for this has not been reviewed previously, and some account of these techniques will also be given here.

Information about plasma membrane transport has been derived from three types of preparation: intact tissues, isolated cells in suspension or culture and plasma membrane vesicles. Aspects of the measurement of transport in each of these systems will be considered.

INTACT TISSUE MEASUREMENTS

While intact tissue preparations may approximate most closely to the physiological situation, such preparations are not particularly well suited to detailed studies of transport mechanism and kinetics, and inferences from intact tissue measurements must be confirmed using simpler systems such as isolated cells or plasma membrane vesicles.

The optimum method for the measurement of amino acid uptake into whole tissues is the paired isotope dilution technique (Yudilevich and Mann, 1982). A tissue or organ is perfused with a medium containing the ^{3}H-labelled amino acid under investigation together with a compound such as ^{3}H-mannitol which is not taken up across the plasma membrane and acts as a marker of the extracellular space. From a knowledge of the isotope content of the medium before and after pass though the tissue, the amino acid concentration and the perfusion rate, the rate of uptake of amino acid by the tissue can be calculated. The system is limited by the fact that a flow rate compatible with adequate oxygenation must be maintained in order to maintain tissue ATP levels and cell membrane potentials and the composition of the perfusion medium can be varied only

Mammalian Amino Acid Transport, Edited by M.S. Kilberg and
D. Häussinger, Plenum Press, New York, 1992

within strict limits. In many systems it is not possible to exclude Na^+ completely from the medium, and the presence of bicarbonate may also be obligatory.

Using this system, accurate rates of amino acid uptake into different tissues under various physiologically important conditions have been obtained. It does not however follow that such results directly reflect the properties and kinetics of the membrane transport systems involved. Tissue preparations may often contain more than one type of cell, and it is difficult to allow for factors such as intracellular or extracellular substrate binding and in particular substrate metabolism, which can be extensive in such experiments.

A recent example of the use of this technique is the discovery of an amino acid transport system for glutamine in skeletal muscle which has a specificity resembling System N in liver (Hundal et al, 1987). The existence of this system was deduced from uptake experiments in perfused muscle, and the properties of the system were later characterised in skeletal muscle sarcolemmal vesicles (Ahmed et al., 1990). The kinetic parameters of transport as measured in vesicles were not similar to those deduced from whole tissue experiments, and the Li^+ tolerance of glutamine transport was only discovered in vesicle experiments since the appropriate experiments could not be performed on the perfused system.

MEASUREMENT OF AMINO ACID TRANSPORT IN ISOLATED CELLS

The transport of amino acids into isolated cells has been studied extensively in cells in suspension, particularly hepatocytes, red blood cells and Ascites Tumour Cells, and also in cells such as fibroblasts and hepatocytes in monolayer culture. Since the intracellular volume is small relative to that of the suspension medium relatively low absolute amounts of substrate are taken up and amino acid uptake is normally measured using radioactively labelled substrates. A number of important experimental criteria have to be satisfied in order to obtain valid results.

Choice of Medium

For transport studies, cells must be suspended in an isoosmotic medium of defined composition, well buffered at the required pH. For many applications it is appropriate to suspend cells in a physiological salt solution such as Krebs-bicarbonate. Glucose, or another energy substrate can be added if necessary to ensure maintenance of the cell ATP level and ion gradients. The substrate amino acid is then added as a very small volume of a concentrated solution and the uptake of radioactivity after a given time is measured. If this system is to be used, the cell suspension should be equilibrated with a gas phase of $95\%O_2/5\%CO_2$ to maintain the pH at 7.4. If this is not done, CO_2 is lost from the medium and the pH increases. To overcome this problem, transport measurements are frequently performed in media where the bicarbonate is replaced with another appropriate buffer such as HEPES or MOPS, the Na^+ concentration being kept constant. These anions do not appear to have any deleterious effect on cell viability. However, effects of ion replacement on the cell membrane potential must also be considered. For example, in isolated hepatocytes, replacement of bicarbonate in the medium causes partial depolarisation of the membrane (Fitz et al., 1989), and greatly reduces the Vmax of Na^+-dependent transport of amino acids (McGivan, 1979) which is a function of the cell membrane potential (Moule and McGivan, 1990). This factor is particularly important in experimental approaches aimed at exploring the relationship between amino acid transport and metabolism. For measurement of Na^+-independent transport the Na^+ in the medium is replaced with an equal concentration of choline or N-methylglucamine.

Interference from Substrate Metabolism

Simple calculations of substrate transport from measurements of radioactivity found inside cells are based on the assumption that the substrate is not significantly metabolised during the uptake period and the radioactivity measured represents only unchanged substrate. If significant metabolism occurs, this can interfere in two different ways. If the cell membrane is impermeable to the products of metabolism, these will accumulate and simple measurements of radioactivity will overestimate the concentration of substrate in the cell. A striking example of this is the uptake of ornithine by isolated mitochondria, where the apparent large accumulation of radioactive ornithine observed (Gamble and Lehninger, 1973) was in fact due to accumulation of radioactive products of the ornithine aminotransferase reaction (McGivan et al., 1977(a)). Conversely, if the products of metabolism can leave the cell then cell radioactivity does not accurately represent the total substrate transported.

In practical terms, this factor may or may not be of great significance for initial rate measurements in isolated cells since little metabolism may occur at the substrate concentrations used over the appropriate time intervals. Nevertheless, in any particular system, it should always be verified that the radioactivity inside the cells represents only the substrate; this can often be done by separating the contents of the cytoplasm by ion-exchange chromatography since the immediate metabolic products of zwitterionic or positively charged amino acids usually have a net negative charge.

The problem of substrate metabolism is serious in experiments where cells are to be loaded with a radioactively labelled amino acid and exchange against an externally added amino acid is to be measured. Here substrate metabolism can greatly affect the results since the identity of the exiting radioactive species is unknown. Similar considerations apply to experiments where the steady-state uptake of various substrates is to be measured. Three procedures have been used to overcome this problem. Firstly, the metabolism of some amino acids can be prevented by the addition of 0.5mM aminooxyacetate (Edmondson et al., 1977). This compound is a general inhibitor of pyridoxal phosphate requiring enzymes such as transaminases, and serine and threonine dehydratases. Aminooxyacetate does not cause depletion of cell ATP over short time intervals, and is of particular use in the measurement of the transport of amino acids such as alanine whose metabolic pathways involve an initial transamination step. Figure 1 shows the importance of adding aminooxyacetate when measuring threonine transport into isolated hepatocytes. In the absence of this compound, radioactive products escape from the cells and the uptake is underestimated. This situation arises particularly when the rate of transport of a substrate is similar to the rate of its metabolism.

Secondly, non-metabolised amino acid analogues can be used as substrates. AIB Methyl-AIB, cycloleucine and BCH have frequently been used for this purpose. The use of these compounds overcomes the problems of possible metabolism and has been important for many purposes but the rate of transport of such analogues does not necessarily even approximate to the rate of transport of natural amino acids.

For example, the half-maximal velocity of Na^+-dependent AIB transport in hepatocytes occurs at 25mM AIB (Le Cam and Freychet, 1977), while the K_m for overall Na^+-dependent alanine transport is about 4mM (Joseph et al., 1978). Since the V_{max} values for the transport of alanine and AIB are similar, the rate of transport of AIB is less than 20% of the rate of transport of alanine at concentrations of each below 1mM, although the same transport systems are involved in each case. Thirdly, use can be made of membrane vesicles in which no metabolism can occur, and this is discussed in a later section.

Separation of Cells from the Medium

This step is critical for the time resolution of transport experiments. The conditions to be used depend largely in the rapidity of amino acid transport. Cells in suspension are separated by rapid centrifugation. In some cases, the extracellular medium can be removed by aspiration and the cells can be resuspended and washed with complete removal of the extracellular radioactivity and virtually no loss of intracellular substrate. In the case of hepatocytes, resuspension and washing leads to loss of intracellular radioactivity. Cells are therefore incubated with a radioactive marker of the extracellular space such as ^{3}H-inulin when ^{14}C-labelled amino acids are used or ^{14}C-sucrose when ^{3}H-amino acids are employed. The supernatant is removed without washing, and the contamination of the cells by the residual extracellular medium is

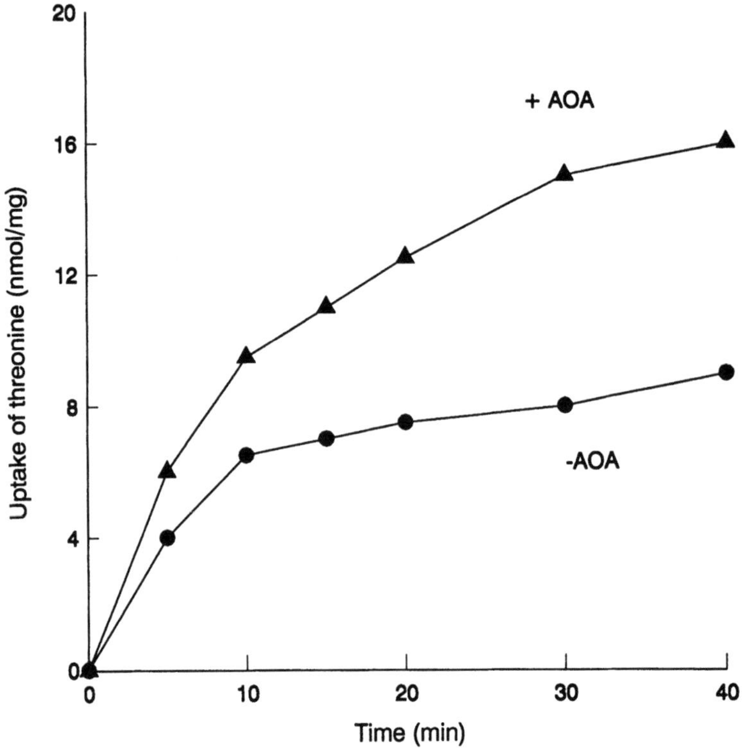

Fig. 1. Effect of 0.5mM aminooxyacetate on the uptake of 1mM ^{14}C-threonine in isolated hepatocytes as measured by isotope distribution. M.R. Hayes and J.D. McGivan, unpublished observations.

quantified using accurate dual-channel scintillation counting. To improve the time resolution and to minimise substrate metabolism, hepatocytes are usually centrifuged through a layer of a mixture of silicone oil and dibutylphthalate into perchloric acid (see e.g. McGivan et al., 1977b). Using this procedure, the time resolution is of the order of a few seconds. For very rapid processes such as the equilibration of leucine in hepatocytes which is 90% complete in 30 seconds even this method does not allow a direct measurement of the initial rate. The uptake data must be linearised using a first order rate plot, and the initial rate calculated from this (see Fig. 2).

In contrast, the uptake of amino acids into monolayer cultures of cells occurs over a period of several minutes. Transport can be terminated simply by removing the extracellular medium and washing rapidly several times. This removes extracellular radioactivity with very little loss of intracellular substrate.

Measurement of Internal Volume

The intracellular volume must be known in order to calculate intracellular concentrations and accumulation ratios of transported substrates. This requires a non-metabolised hydrophilic substance that will equilibrate across the membrane with no significant binding. For cell suspensions, $^{3}H_{2}0$ fulfils this function, and the intracellular volume in the cell pellet is measured from the difference between the pellet content of ^{14}C-sucrose and $^{3}H_{2}O$ in dual labelling experiments. H_2O crosses cell membranes rapidly and is not a suitable choice for methods where cell washing is involved, since $^{3}H_{2}O$ will leak out during the washing process. For cultured cells ^{14}C-urea or ^{14}C-3-O-methyl glucose equilibrate across the cell membrane over periods of minutes, and internal volume can be calculated from the equilibrium uptake of one of these compounds, which are retained during the rapid washing procedure.

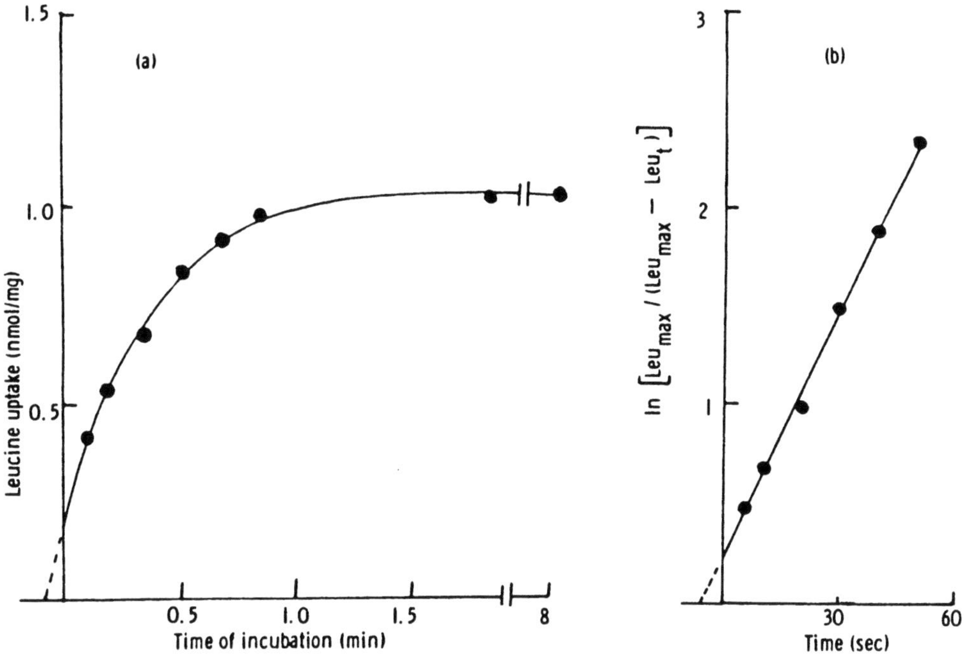

Fig. 2. (a) Time course of leucine (0.5mM) transport into isolated hepatocytes. From McGivan et al. (1977b)

Transport in Polarised Cells

Epithelial cell lines can be cultured on plastic dishes to form confluent monolayers. On achieving confluency the cells polarise such that the apical membranes are in contact with the medium while the basolateral membranes are attached to the plastic. Amino acid transport in such monolayers has been measured simply by the uptake of radioactive amino acids into the cells; this uptake represents mainly transport across the apical membrane. This system does not allow localisation of transport systems to basolateral or apical membranes however, since some basolateral membrane functions, which are essential for cell survival such as Na^+/K^+ ATPase activity, are expressed also on the apical membrane in these circumstances.

To measure localisation of transport systems, cells are grown on permeable membrane supports so that each side of the confluent monolayer is in contact with medium. Amino acids are added from one side of the monolayer or from the other and uptake into the cells is measured. Alternatively the net rate of transport across the monolayer can be determined. When this technique has been used with various renal epithelial cells lines it has been found that Na^+-dependent amino acid transport is located mainly on the basolateral membrane (Rabito and Karish, 1982; Sepulveda and Pearson,1984; Boerner et al., 1986). This result is paradoxical and has not yet been adequately explained. In an interesting extension of this technique, it has been shown by that Na^+-dependent depolarisation of cell monolayers occurs on adding a range of amino acids to the medium on one side of the monolayer, indicating electrogenic cotransport of these amino acids with Na^+ (Schwegler et al., 1989).

THE USE OF MEMBRANE VESICLES TO STUDY PLASMA MEMBRANE TRANSPORT

Plasma membrane vesicles from various tissues have provided much important information about amino acid transport mechanisms and the use of such vesicles has been extensively reviewed (see e.g. Hopfer, 1989; Murer et al., 1989). Vesicles have three major advantages. Firstly, they contain no enzymes, so that problems of substrate metabolism are obviated. Secondly, the medium on each side of the membrane is defined so that the exact ionic requirements for transport and the effects of membrane potential can be unambiguously determined. Thirdly, in polarised cells vesicles can be prepared from specific areas of the plasma membrane so that the localisation of various transport systems can be studied. Vesicle experiments have been particularly important in studying transport in renal and intestinal brush border and basolateral membranes and in liver basolateral and canalicular membranes.

The general protocol for vesicle experiments has been discussed in detail in the above references, but certain points should be emphasised. Brush border membrane vesicles from polarised cells are prepared in quantity by the $MgCl_2$ precipitation method which involves only homogenisation and simple centrifugation (see e.g. Biber et al., 1981). Vesicles from all other membranes are prepared by protocols involving sucrose gradient or Percoll fractionation, and can only be prepared in much lower yield.

The purity of the vesicle preparation and the degree of contamination by endoplasmic reticulum and subcellular organelles must be established by the assay of specific marker enzymes for various membranes. In polarised cells marker enzymes for both apical and basolateral membranes must be included. Na^+/K^+ ATPase is universally used as a marker of basolateral membranes while alkaline phosphatase, glutamyltransferase or 5'-nucleotidase are useful markers for brush border membranes. Mitochondrial contamination is assessed from the distribution of cytochrome oxidase or succinate dehydrogenase, while glucose-6-phosphatase is a marker of the endoplasmic reticulum. Both the specific activity and the recovery of marker enzymes should be recorded to allow for artifactual results due to enzyme activation or inactivation during the vesicle preparation. The sidedness of the vesicles is normally determined by assaying marker enzymes in the presence and absence of a detergent to remove permeability barrier. While this is of importance in studying asymmetric processes such as ATP-driven Ca^{2+} transport or ion transport via the Na^+/K^+ ATPase, it is not of great significance in amino acid transport studies since during the transport process transport proteins must reversibly orientate so that the active site can face either side of the membrane. The direction of net transport depends only on imposed gradients of substrate and Na^+.

In order to study amino acid uptake into vesicles, vesicles are incubated in a medium containing radioactive amino acids plus a high concentration of a Na^+ or K^+ salt (100mM) dissolved in the medium in which the vesicles were equilibrated. The hypertonicity of the external medium prevents osmotic swelling of vesicles during Na^+-substrate cotransport. Swelling must be avoided, since this may lead to rupture of the vesicles. The transport process is terminated by the addition of ice cold medium containing no amino acid and the vesicle suspension is filtered within 15 seconds on a 0.45μm nitrocellulose filter under defined negative pressure. The filter is subsequently washed with more ice-cold medium which removes contamination by extravesicular medium with little or no loss of intravesicular radioactivity if performed sufficiently rapidly. The time resolution of the procedure for kinetic studies is approximately 2 seconds.

In the case of electrogenic Na^+-amino acid cotransport, maximum rates are obtained only in the presence of an imposed membrane potential (negative inside). This potential is created most easily by the presence of the Na^+ salt of a permeant anion such as CNS^- or nitrate. Although Cl^- is often used as a permeant anion, not all membrane vesicles have a high Cl^- permeability and much higher rates of transport may in some cases be obtained in the presence of CNS^-. In order to assess electrogenicity, transport in the presence of NaCNS is compared with that in the presence of the Na^+ salt of a poorly penetrant anion such as cyclamate (Fig. 3).

Alternatively, if the vesicles were equilibrated in a K^+-containing solution and then suspended in a medium containing a lower K^+ concentration, a membrane potential (negative inside) is generated by the addition of the specific K^+ ionophore valinomycin. If the addition of valinomycin does not affect the rate of transport, it should be ensured that an adequate K^+ gradient is present before concluding that transport is not

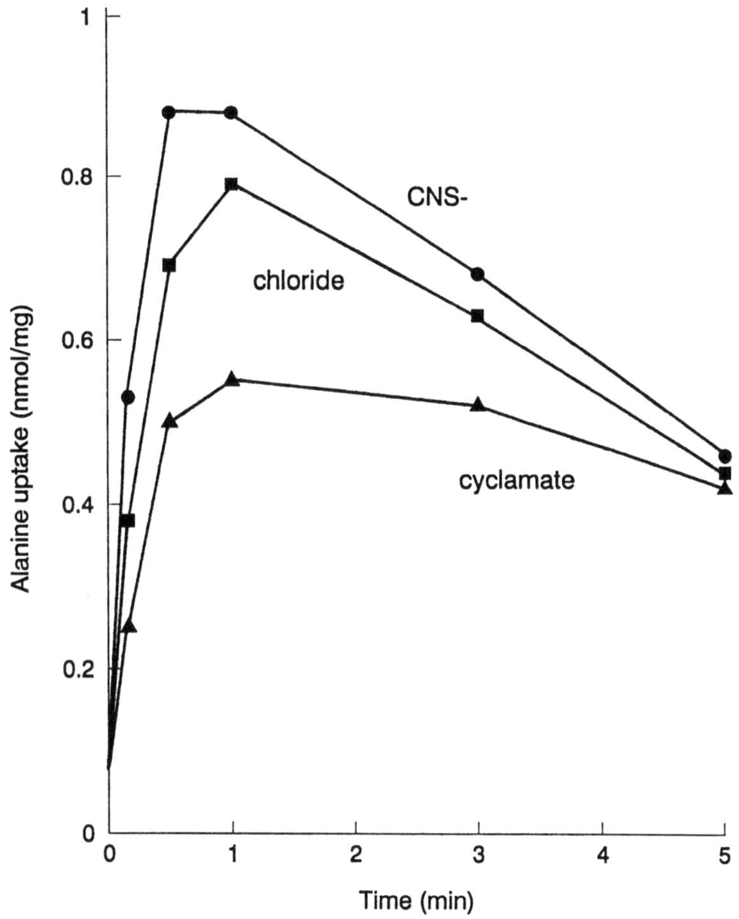

Fig. 3. Anion dependence of the uptake of alanine (0.1mM) into bovine renal brush border membrane vesicles in the presence of a Na^+ gradient. Data from Lynch and McGivan (1977a).

electrogenic. This is most easily done by observing the effect of valinomycin in the same vesicles on a process such as Na^+-glucose cotransport which is known to be electrogenic.

In experiments on vesicles it is important to discriminate carefully between transport into the intravesicular space and binding to the membrane. Binding can be assessed in two ways. Equilibrium uptake of substrate should be determined at a range of osmolarities of the external medium. With increasing osmolarity, the intravesicular space will shrink and the amount of substrate taken up at equilibrium will therefore decrease. A graph of equilibrium uptake versus the reciprocal of osmolarity is constructed. This is extrapolated to infinite osmolarity, and the apparent uptake at infinite osmolarity represents binding, which is usually Na^+-independent. Alternatively, for most amino acids binding to the membrane is much more rapid than transport and is complete within the shortest time that transport can be measured. If the subsequent linear initial rate of transport is extrapolated back to zero time, the apparent uptake at zero time represents binding. It should be verified that estimates of binding from both procedures are consistent. Substantial corrections for binding are required for transport measurements of hydrophobic and aromatic amino acids in vesicles from most cell types and for all amino acids in reconstituted phospholipid vesicles.

The maximal initial rate of Na^+-cotransport into membrane vesicles is a function of three variables - the activity of the transport system, the magnitude of the imposed membrane potential (see above) and the magnitude of the Na^+-gradient. In particular, if the vesicles have a significant permeability to Na^+, the imposed Na^+ gradient will rapidly collapse and the initial rate of transport will be sustained for only a very short time. When amino acid transport in vesicles isolated from similar tissues treated differently (e.g. with hormones) is assessed and differences in the initial rates are found, it is important to confirm that the different rates of uptake are not a consequence of differential Na^+ permeability of the vesicles due to hormone treatment. Similarly, various compounds such as sulphydryl reagents or lysine-modifying reagents have been used as semi-selective inhibitors of amino acid transport in vesicles. In such experiments, the effects of the inhibitors on Na^+-permeability of the vesicles should be assessed to determine the extent to which this factor may be responsible for inhibition of transport.

INTERPRETATION OF KINETIC RESULTS

One of the major aims of the study of amino acid transport is to obtain the kinetic parameters of the transport of various amino acids. This is frequently done by measuring the initial rate of transport at various substrate concentrations and fitting the results to the Michaelis-Menton relationship for a saturable system. A number of essential criteria for the accurate interpretation of such results are often overlooked.

Initial rates of transport are usually determined from uptakes measured at a single time point. An appropriate time point is identified from time courses of uptake usually performed at low substrate concentrations, and is taken as the longest interval over which uptake is proportional to time. It is important to verify that this time point is appropriate to give measurements of initial rate also at the highest substrate concentration used. If at high substrate concentrations equilibration occurs or the time course is non-linear at the time point chosen then the true initial rate at high substrate concentrations is seriously underestimated. A second important point is the distinction between saturable and non-saturable components of uptake. Data should be analysed by regression analysis both to the equation $v = Vmax \cdot S/(K_m + S)$ where K_m is the substrate concentration for half maximal velocity and also to the equation $v = Vmax \cdot S/(Km + S) + K \cdot S$ where K is a constant. If a better fit to the second equation is obtained, then a significant component of the apparent uptake is due to a non-saturable process such as diffusion and this component should be subtracted from the measured velocities to give the transport kinetic parameters.

In practice, non-saturable uptake is minimal for most hydrophilic amino acids such as alanine in both cell suspensions and cells in culture. However, very significant non-saturable components are observed in both systems for the uptake of hydrophobic amino acids such as branched-chain and aromatic amino acids. For example the uptake of leucine in isolated hepatocytes is clearly resolvable into non-saturable and saturable components (McGivan et al., 1977b) and at 10mM leucine the non-saturable component represents 50% of the initial rate of uptake (Fig. 4). In general, K_m values for amino acid transport in mammalian cells are less than 10mM. Apparent higher values can often be attributed to non-saturable components.

The K_m values derived for the transport of amino acids into cells do not necessarily represent the properties of a single transport system, since in many cases amino acids enter on more than one system simultaneously. It is not possible to resolve transport data into parameters for two transport systems adequately unless the K_m values for the systems are widely different.

One very common error is to assume that the kinetic parameters of transport can be derived from simple unweighted linear regression to Eadie-Hofstee or Lineweaver-Burke plots derived from data on substrate concentration and initial velocity. This is not the case since non-systematic errors are introduced. For example, in the Lineweaver-Burk transformation, very large errors in V at high substrate concentrations would produce relatively small errors in 1/V and have little effect on the linear regression line. Conversely, even small errors in V at low concentrations would produce large errors in 1/V and hence greatly affect the best straight line fit. Estimation of kinetic

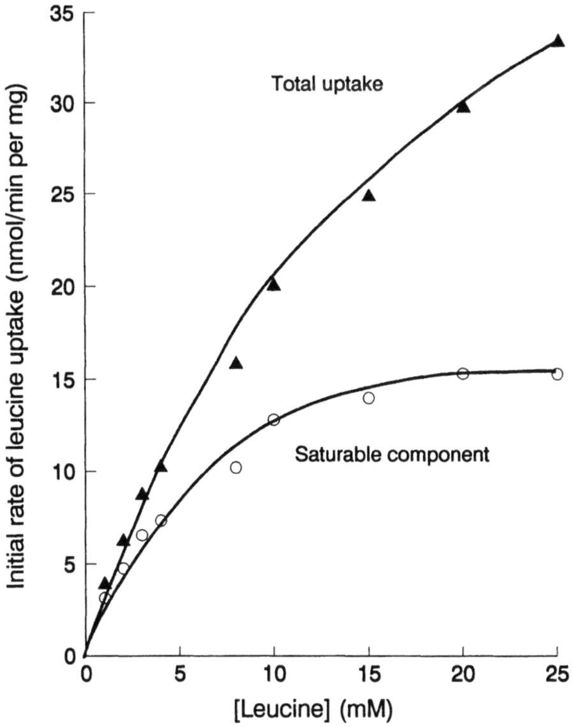

Fig. 4. Concentration of leucine uptake into isolated hepatocytes. Redrawn from McGivan et al. (1977b)

parameters by this method can lead to very large errors, in particular in underestimating K_m values. Such parameters must be derived from non-linear regression analysis of the data to the actual Michaelis Menton equation.

INHIBITION ANALYSIS

Amino acids are transported into mammalian cells via a number of transport systems with widely overlapping specificity. The extent to which a particular transport system is involved in the uptake of a particular amino acid is determined by an analysis of the pattern of inhibition of the uptake of the substrate amino acid by other amino acids. The strategy for the characterisation of particular transport systems in a cell type is discussed very comprehensively by Christensen (1989) and will be considered only briefly here.

The principle used in defining the specificity of transport systems is that if two amino acids are transported on the same system the initial rate of transport of one amino acid should be competitively inhibited by all other amino acids which are transported on the same system. Conversely, if two amino acids are transported on two separate systems with no overlap, either there should be no mutual inhibition of transport or any inhibition should be non-competitive. Thus while lack of inhibition of transport of one amino acid by excess of another is an adequate criterion for lack of transport on the same carrier system, inhibition of transport of one amino acid by another does not necessarily imply that the same carrier system is involved.

In order for inhibition analysis of transport to be valid therefore, the experiments should be performed using an adequate concentration range of both substrate and inhibitor amino acids. It is not uncommon in the literature to see experiments where carrier free radioactive amino acids (at a concentration of 0.5-10μM) are used as substrates and inhibitor amino acids are added at very much higher concentrations, for example in 5000-fold excess. This experimental protocol has a number of limitations. If the inhibitor added contains only 0.5% of an impurity which affects transport, this impurity would be present at concentrations considerably in excess of the substrate. Further, with this discrepancy in concentration between substrate and inhibitor, inhibition is detected with extremely weak inhibitors whereas at more physiological concentrations of substrate no inhibition would occur. Interpretation of experiments designed in this way is difficult. Indeed such experiments are useful only when there is a clear lack of inhibition at high ratios of inhibitor to substrate, establishing that the transport of the two amino acids does not occur on the same system.

In cases where inhibition of the transport of one amino acid by another occurs, it must be determined whether the inhibition is competitive or non-competitive. Non-competitive inhibition of the transport of one amino acid by another was first demonstrated by Joseph et al., (1978) who showed that the uptake of alanine into isolated hepatocytes was non-competitively inhibited by glutamine. It was inferred that alanine and glutamine were transported on different carrier systems. This observation led to the characterisation of System N in hepatocytes by Kilberg et al.(1980); these authors also showed that inhibition of methyl AIB transport by glutamine was non-competitive. Non-competitive inhibition of the transport of one amino acid by another is not an uncommon phenomenon.

From a practical stand-point, inhibition analysis is best performed using Dixon plots (Dixon, 1953). Initial rate determinations are performed with a range of inhibitor concentrations at two or preferably three concentrations of substrate. Graphs are plotted of the reciprocal of the initial rate against the inhibitor concentration. In the case of competitive inhibition, the lines meet behind the y axis and above the x-axis. For non-competitive inhibition the lines intersect behind the y axis but on the x-axis. The value of 1/Ki is given by the point of intersection. Fig. 5 shows that alanine transport in renal brush border membrane vesicles is competitively inhibited by phenylalanine but is non-competitively inhibited by glycine which is transported on a different transport system.

According to Heinz (1972), if two amino acids are transported on a common carrier and this is the only major route of transport, then the transport of each amino acid should be competitively inhibited by the other and in each case the K_i for inhibition of transport should equal the K_t value for the transport of the inhibitory amino acid. Lynch and McGivan (1987a) have shown that these criteria are fully satisfied for the Na^+-dependent transport of alanine, glutamine, phenylalanine and leucine in bovine renal brush border membrane vesicles, and therefore these amino acids are transported by a single common transport system.

The importance of analysis of the competitive or non-competitive nature of inhibition of transport of one amino acid by another has not always been adequately appreciated in the literature particularly when the existence of novel or multiple transport systems for substrates has been proposed. Careful kinetic analysis has been essential in the characterisation of more recently discovered transport systems such as System N in liver (Kilberg et al., 1982), the broad specificity system ($B^{o,+}$) in blastocysts (van Winkle et al., 1985) and the somewhat similar broad specificity transport system in renal brush border membranes (Lynch and McGivan, 1987).

RECONSTITUTION OF AMINO ACID TRANSPORT SYSTEMS

The identification of the proteins responsible for amino acid transport in mammalian cells is a major unsolved problem. The lack of specific tight-binding inhibitors for these transport systems has not assisted attempts to use specific labelling techniques such as those used to identify glucose transport systems. Two approaches are currently being employed. In one approach, mRNA from various tissues is injected into Xenopus oocytes and the transport properties of the oocytes are monitored. Using this procedure, Na^+-independent amino acid transport has been expressed in oocytes after

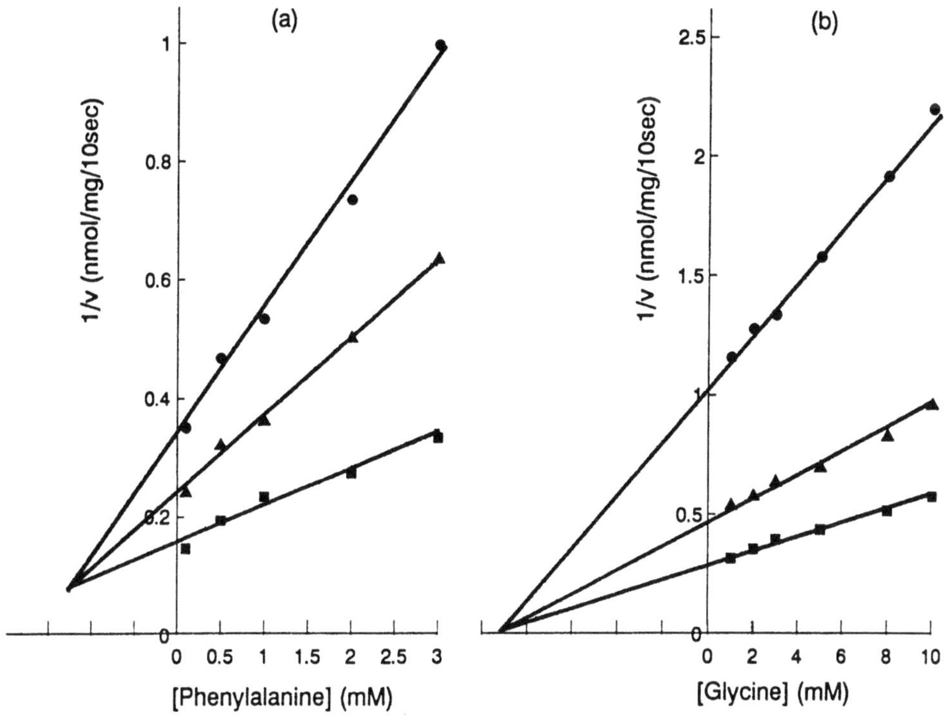

Fig. 5. Dixon plot analysis of the inhibition of alanine transport by phenylalanine and glycine in renal brush border membrane vesicles. Alanine concentrations were ● 0.1mM; ▲ 0.25mM; ■ 0.5mM. A.M. Lynch and J.D. McGivan, unpublished.

injection of mRNA from rat kidney (Tate et al., 1989) and both Na^+-dependent and Na^+-independent amino acid transport has been observed after injection of mRNA from rabbit kidney cortex (Coady et al., 1990). Similarly, injection of mRNA fractions from rat liver has been shown to lead to expression of System A activity (Tarnuzzer et al., 1990; Palacin et al., 1990). Subfractionation of this mRNA may eventually lead to the isolation of cDNA clone for the transport protein, as was the case for the Na^+-glucose transporter from intestine (Hediger et al., 1987).

A complementary approach is to attempt to identify the transport proteins directly using protein fractions reconstituted into artificial phospholipid membranes as an assay system for transport activity. A primary requisite for this approach is the development of a reliable method for the reconstitution of transport activity.

An important criterion for the successful reconstitution of Na^+-dependent amino acid transport activity is that the reconstituted vesicles should show much higher rates of transport in the presence of a Na^+ gradient than in the presence of a K^+ gradient. Koepsell et al. (1984) first described a method for the reconstitution of a number of Na^+-dependent transport activities from pig kidney cortex brush border membrane vesicles. Membranes were dissolved in deoxycholate and were dialysed overnight. Multilamellar liposomes were added, and the suspension was sonicated and then frozen and thawed. After differential centrifugation, a proteolipid fraction was obtained which showed clear Na^+-dependent transport of alanine, proline, glycine, lactate and dicarboxylates. The active fractions contained relatively few protein bands on SDS gel electrophoresis. The vesicles obtained were very large (internal volume approximately 50µl/mg) and impermeable to cations. Alanine uptake continued for 2 hours in this system.

McCormick et al. (1984) introduced a system for the reconstitution of aminoisobutyrate transport from Ascites cells. In this procedure, cell membranes were dissolved in cholate plus urea and dialysed overnight. The dialysed protein was mixed with sonicated asolectin in a K^+-containing medium, and after freezing and thawing, vesicles were collected by centrifugation. The vesicles obtained catalysed Na^+-dependent AIB transport. Bracy et al. (1987) used a modification of this method and were the first to reconstitute System A activity from rat liver plasma membranes. Vesicles obtained by this procedure had an internal volume of 10-20µl/mg protein and amino acid uptake was complete in approximately 10 minutes.

All these methods required overnight dialysis, which is disadvantageous in methods of carrier purification involving protein fractionation. Lynch and McGivan (1987b) reconstituted Na^+-dependent amino acid transport from bovine renal brush border membranes by a method which did not require such a dialysis step. The detergent decanoyl-N-methylglucamide (MEGA-10) employed in this study is readily removed from proteins by passage through Sephadex G-50. Use of this detergent enabled reconstitution of Na^+-dependent amino acid transport to be achieved in 2-3 hours, although the vesicles obtained had a low internal volume in the range 5-10µl/mg. A similar method subsequently achieved the reconstitution of System A and System N activity from rat liver (Quesada and McGivan, 1988). Fafarnoux et al. (1989) devised a rapid method for the reconstitution of System A activity in which rat liver membranes were dissolved in cholate and the protein was precipitated with polyethylene glycol and redissolved in low detergent concentration, thus avoiding a lengthy dialysis step. Tamarappoo and Kilberg (1991) achieved reconstitution and a 20-fold enrichment of System N activity from rat liver using a similar procedure.

These reconstitution methods should eventually be useful in approaches to carrier identification involving protein fractionation. Quesada and McGivan (1988) showed that when solubilised in MEGA-10 the protein responsible for System A activity in liver binds to the mannose-specific lectin concanavalin A while glutamine transport activity does not bind to the same lectin. McCormick and Johnstone (1988) showed that fractionation of cholate-solubilised Ascites cell membranes by various successive procedures led to recovery of a protein fraction with highly enhanced transport activity. The major component of this fraction was a protein of Mr 120-130kD. A polyclonal antibody to this protein immunoprecipitated System A transport activity from solubilised membranes. Based on this evidence, it was proposed that the 120-130kD protein was an essential component of the transport system. McCormick et al. (1991) have recently shown that radiation inactivation of System A transport activity in Ascites cell membranes can be correlated with fragmentation of the 120-130Kd protein in these membranes, which is consistent with this proposal. However, the protein has not yet been isolated, and no reports of the presence of this protein in membranes of other cells which catalyse System A transport activity have appeared to date.

The development of fractionation/reconstitution procedures for membrane transport systems is likely to be of great importance in the identification of these proteins. Ultimately, unambiguous identification of amino acid transport systems will require both the expression of amino acid transport activity in oocytes following injection of a single mRNA species and isolation of the transport protein and demonstration of its transport activity in reconstituted systems. At present, it appears likely that identification of the mRNA for these transport proteins may precede the isolation of the proteins themselves.

REFERENCES

Ahmed, A., Taylor, P. M. and Rennie, M. J., 1990, Am. J. Physiol., 259: E284-291.
Biber, J., Steiger, Haase, W. and Murer, H., 1981, Biochim. Biophys. Acta, 647: 169-176.
Biber, J., Stieger, B., Haase, W. and Murer, H., 1981, Biochim. Biophys. Acta, 647: 169-176.
Boerner, P., Evans-Layng, M., U, H. S. and Saier, M. H., 1986, J. Biol. Chem., 261: 13957-13962
Bracy, D. S., Schenerman, M. A. and Kilberg, M. S., 1987, Biochim. Biophys. Acta, 899: 51-58.
Le Cam, A. and Freychet, P., 1977, J. Biol. Chem., 252: 148-156.
Christensen, H. N., 1989, Methods in Enzymology, 173: 564-575.

Coady, M. J., Pajor, A. M., Toloza, E. M. and Wright, E. M., 1990, Arch. Biochem.Biophys., 283: 130-134.
Dixon, M., 1953, Biochem.J., 55: 170.
Eddy, A. A. and Johnson, E. R., 1989, Methods in Enzymology, 173: 771-776.
Edmondson, J. W., Lumeng, L. and Ti, T-K, 1977, Biochem. Biophys. Res. Commun., 76: 751-757.
Fafarnoux, P., Dudenhausen, E. E. and Kilberg, M. S., 1989, J. Biol. Chem., 264: 4805-4811.
Fitz, J. G., Perisco, M. and Scharschmidt, B. F., 1989, Am.J.Physiol., 255: G491-G500.
Gamble, J. G. and Lehninger, A. L., 1973, J. Biol. Chem., 248: 610-618.
Harvey, C. M. and Ellory, J. C., 1989, Methods in Enzymology, 173: 122-159.
Hediger, M. A., Coady, M. J., Ikeda, T. S. and Wright, E. M., 1987, Nature, 330: 379-381.
Heinz, E., 1972, in: Metabolic Pathways: Metabolic Transport 3rd Edition (Hokin, L.E. ed.,) pp.455-401 Academic Press, New York.,
Hopfer, U., 1989, Methods in Enzymology, 172: 313-330.
Hundal, H. S., Rennie, M. J. and Watt, P. W., 1987, J. Physiol., 408: 93-114.
Joseph, S. K., Bradford, N. M. and McGivan, J. D., 1978, Biochem. J., 176: 827-836.
Kilberg, M. S., 1989, Methods in Enzymology, 173: 564-575.
Kilberg, M. S., Handlogten, M. E. and Christensen, H. N., 1980, J. Biol. Chem., 255: 4011-4019.
Koepsell, H., Korn, K., Ferguson, D., Menhur, H., Ollig, D. and Haase, W., 1984, J. Biol. Chem., 259: 6548-6558.
Lynch, A.M. and McGivan, J. D.,1987(a) Biochim. Biophys. Acta, 899: 176-184.
Lynch, A. M. and McGivan, J. D., 1987(b), Biochem.J., 244: 503-508.
McCormick, J. I., Jette, M., Potier, M., Beliveau, R. and Johnstone, R. M., 1991, Biochemistry, 30: 3704-3709.
McCormick, J. I., Johnstone and R.M., 1988, Proc. Nat. Acad. Sci., 85: 7877-7881.
McCormick, J. I., Tsang, D. and Johnstone, R. M., 1984, Arch. Biophys. Biochem., 231: 355-365.
McGivan, J. D., 1979, Biochem.J., 182: 697-705.
McGivan, J. D., 1989, Methods in Enzymology, 174: 31-38.
McGivan, J. D., Bradford, N. M. and Beavis, A. D., 1977(a) Biochem. J., 162: 147-156.
McGivan, J. D., Bradford, N. and Mendes-Mourao, J., 1977(b) FEBS Letters, 80: 380-394.
Moule, S. K. and McGivan, J. D., 1990, Biochim. Biophys. Acta, 383-397.
Murer, H., Gmaj, P., Stieger, B. and Hagenbuch, P., 1989, Methods in Enzymology, 172: 346-365.
Palacin, M., Werner, A., Dittmer, J., Murer, H. and Biber, J., 1990, Biochem.J., 270: 189-195.
Quesada, A. R. and McGivan, J. D., 1988, Biochem.J., 255: 963-969.
Rabito, C. A. and Karish, M. V., 1982, J. Biol. Chem., 257: 6802-6808.
Schwegler, J. S., Heuner, A. and Silbernagl, S., 1989, PflugersArchiv., 414: 543-550.
Sepulveda, F. V. and Pearson, J. D., 1984, J. Cell. Physiol., 118: 211-217.
Tamarappoo, B. K. and Kilberg, M. S., 1991, Biochem.J., 255: 963-969.
Tarnuzzer, R. W., Campa, M. J., Qian, N. X., Engelsberg, E. and Kilberg, M. S., 1990, J. Biol. Chem., 265: 13914-13917.
Tate, S. S., Urade, R., Getchell, T. V. and Udenfriend, S., 1989, Arch. Biochem. Biophys., 275: 591-596.
van Winkle, L. J., Christensen, H. N. and Campione, A. L., 1985, J. Biol. Chem., 260: 12118-12123.
Yudilevich, D.L. and Mann, G.E., 1982, Fed. Proc., 41: 3045-3053.

MOLECULAR BIOLOGICAL APPROACHES FOR AMINO ACID TRANSPORT

Rufus M. Williamson, Ti Zhi Su and Dale L. Oxender

Department of Biotechnology
Parke-Davis Pharmaceutical Research Division
Warner-Lambert Company
Ann Arbor, MI 48106-1047

INTRODUCTION

While enormous progress has been made in the elucidation of the molecular basis of amino acid transport in bacteria and lower eukaryotes, an understanding of the molecular principles that underlie mammalian amino acid transport is in its infancy. Investigations into the nature of amino acid transport in mammals has yielded a great wealth of knowledge on the substrate specificities and kinetic properties of amino acid transport systems. However, studies which seek to understand their molecular details have been complicated, to a large extent, by the relative low abundance of transport proteins in membranes, and the existence of different transport systems with overlapping specificities for individual amino acids. Several distinct transport systems mediate uptake of amino acids in mammalian cells. The major components for neutral amino acid uptake are systems A, ASC, and L which were first identified in Ehrlich ascites cells (Oxender and Christensen, 1963; Christensen et al., 1967). System A is sodium-dependent and serves predominantly for uptake of amino acids with short, polar or linear side chains. System L is sodium-independent, and shows reactivity toward branched-chain and aromatic amino acids. System ASC is sodium-dependent and shows a strong preference for alanine, serine, and cysteine. In this chapter we will outline and summarize the various approaches that have been used to gain insights into the molecular basis of the mammalian transport process. These approaches naturally divide into two broad areas of endeavor: (i) the identification and characterization of transport proteins, and (ii) the identification of transport genes and their mode(s) of regulating expression. Aspects of some of the approaches described in this chapter are also discussed in another chapter on "Techniques Used in Amino Acid Transport Studies".

APPLICATIONS OF LABELLING REAGENTS FOR MODIFICATION OF AMINO ACID RESIDUES IN AMINO ACID TRANSPORT PROTEINS

Reagents that modify with high reactivity amino acid residues at or near ligand binding sites have been useful in labelling membrane transport proteins for identification, isolation, and characterization. Useful modification reagents form covalent linkages with amino acid residues at or near the binding site. Modification of functional groups by labelling reagents are either reduced or enhanced by the presence of substrates or inhibitors. Protection from modification of functional groups by substrates or competitive inhibitors provide a general indication of the proximity of a modified group relative to the substrate binding site. Thus, modification reagents have proven useful in detecting and correlating specific characteristics of the side chains of amino acid residues in putative transport carriers with functional roles in transport. Two

Mammalian Amino Acid Transport, Edited by M.S. Kilberg and
D. Häussinger, Plenum Press, New York, 1992

experimental problems, however, are inherent in the study of transport proteins using modification reagents. One involves the removal of unreacted reagent prior to examining transport activity (McCormick and Johnstone, 1990). The second concerns the potential side effects of the modifying reagent on membrane permeability and integrity (Will and Hopfer, 1979). An assessment of protein-modifying reagents is described in the chapter on "Techniques Used in Amino Acid Transport Studies". In this section, only selected modification reagents and their applications are described which have proven useful in the identification and isolation of proteins involved in amino acid transport.

p-chloromercuribenzenesulfonate

In 1967, Hare demonstrated that uptake of L-phenylalanine was sensitive to the sulfhydryl-reacting organomercurial p-chloromercuribenzenesulfonate (pCMBS), and suggested that the permeation pathways for amino acids could be distinguished on the basis of their reactivity with a chemical probe. Subsequent studies by Hare (1975) revealed that system L amino acid transport was more sensitive to inhibition caused by pCMBS than system A transport in cultured mouse cells. His observations suggested that system L transport components have sulfhydryl groups lying near the exofacial membrane surface whereas the sulfhydryl reactant-sensitive sites of system A transport components lie near the endofacial surface or in the hydrophobic regions of the membrane. Indeed organomercurials such as pCMBS, which normally react with cysteinyl residues of peptides and proteins (Kenyon and Bruice, 1977), have been reported to be more reactive with sulfhydryl groups in a hydrophobic environment (Means and Feeney, 1971). Several groups of investigators have since shown that amino acid transport systems are inhibited by pCMBS. Chiles et al. (1988) examined the effects of pCMBS on amino acid transport in rat hepatocytes and hepatoma cells, and found that the sulfhydryl-reacting compound strongly inhibited systems A, ASC, and L. Evidence for an essential sulfhydryl group at the substrate binding site of the system A transporter of Ehrlich cell plasma membrane (McCormick and Johnstone, 1990), chick embryo fibroblast (Smith-Johannsen et al., 1977) as well as the system L transporter from Chinese hamster ovary cells (Campbell et al., 1992) was provided from studies employing pCMBS. In each case, an excess of the appropriate amino acid substrate confers partial protection of amino acid uptake from inhibition caused by pCMBS.

N-ethylmaleimide

The alkylating agent N-ethylmaleimide (NEM) also binds covalently to sulfhydryl groups, and preferentially reacts with sulfhydryl groups on the surface of proteins (Means and Feeney, 1971). NEM has been shown to inhibit alanine uptake into hepatocytes and plasma membrane vesicles from rat liver (Kilberg et al., 1980; Sips and vanDam, 1981; Hayes and McGivan, 1983). An excess of L-alanine confers partial protection of alanine uptake from inhibition caused by NEM. Other alkylating agents such as mersalyl and iodoacetate were also found to exhibit inhibition of alanine uptake with protection from inactivation conferred by an excess of L-alanine and sodium (Hayes and McGivan, 1983). Higher concentrations, however, were generally required for half-maximal inhibition than observed with use of NEM (Hayes and McGivan, 1983).

Sensitivity of pCMBS and NEM

One must note that the sensitivity of amino acid transport to inactivation by the reagents pCMBS and NEM has been reported to vary between normal and transformed cell types. For example, inherent differences have been observed in the system A transporters from normal and transformed liver tissue. In some rat hepatoma cell lines, amino acid transport by system A has been shown to be sensitive only to inhibition by pCMBS and unaffected by NEM, whereas in normal rat hepatocytes both sulfhydryl modifying reagents were found to be effective inhibitors of transport (Chiles and Kilberg, 1987; Dudeck et al., 1987; Chiles et al., 1988). Similarly, sodium-dependent transport of a-aminoisobutyric acid, a nonmetalizable substrate of system A from Ehrlich cells, was shown to be inhibited by pCMBS, but is largely insensitive to NEM (McCormick and Jonstone, 1988). These observations are consistent with an earlier report by Weber et al. (1976) indicating that transformed cell lines display alterations in amino acid transport.

Diethyl pyrocarbonate

The acylating reagent diethyl pyrocarbonate (DEPC) which displays reactivity against histidine residues has been shown to inactivate sodium-dependent alanine transport in plasma membranes derived from rat liver (Bertran et al., 1991). The presence of sodium and L-alanine during exposure of membranes to DEPC conferred protection from inactivation of alanine transport.

Azaserine

Segel et al. (1989) reported use of the amino acid derivative, azaserine (O-diazoacetyl-L-serine) as a potentially useful label for identification of system L in T-lymphocytes. Azaserine is transported by system L, and is competitively inhibited 2-amino-2-carboxybicycloheptane. The amino acid derivative forms a highly reactive carbene intermediate and free N_2 upon photolysis with 365 nm incident light which is capable of rapid covalent binding to the adjacent protein.

Fluorescein isothiocyanate

Site-specific reagents coupled to fluorescent probes provide a useful tool for the study of conformational changes resulting from ligand binding. This approach has been used successfully to probe conformational changes in several transport proteins (Carilli et al., 1982; Karlish, 1980; Pick, 1981; Pick and Bassilian, 1981; Jackson et al., 1981; Nigg et al., 1979). However, its application in the study of amino acid transport proteins has been limited. Wright and Peerce (1984) reported use of fluorescein isothiocyanate (FITC), an irreversible group-specific reagent for amino groups, in the identification and study of conformational changes in the proline transporter in brush-border membranes. Binding of FITC to the brush-border proline transporter was determined after reacting nonspecific binding sites with the nonfluorescent analog phenyl isothiocyanate (PITC) in the presence of L-proline and sodium. Proline transport was found to be inhibited irreversibly by PITC and FITC, and transport activity was protected by the simultaneous presence of sodium and L-proline during exposure of membranes to the isothiocyanates. The latter finding enabled a determination of specific binding of FITC to the proline transporter. Addition of sodium to the treated membrane preparations resulted in quenching of FITC fluorescence.

USE OF RECONSTITUTED MEMBRANE VESICLES IN THE CHARACTERIZATION OF PUTATIVE AMINO ACID TRANSPORT PROTEINS

That transported substrates are often metabolized following interaction with membrane-associated carriers poses an obstacle in understanding the molecular basis of amino acid transport in living cells. The study of transport in isolation can potentially overcome the problem. Colombini and Johnstone (1974) were the first to demonstrate that amino acid transport can be reconstituted in mammalian plasma membrane vesicles. These studies demonstrated both sodium-dependent and independent amino acid transport. Reconstitution of both system L (Cecchini et al., 1977) and system A (Johnstone and Bardin, 1976; Hamilton and Nilsen-Hamilton, 1976) transport has been achieved in partially purified membrane vesicles.

Membrane vesicles in conjunction with the use of labelling reagents have been useful in the identification and isolation of proteins believed to be involved in amino acid transport. For example, vesicles derived from rabbit intestinal brush-border membranes were employed to isolate a protein involved in proline transport (Wright and Peerce, 1984). The covalent binding of FITC to the proline transporter was utilized in the identification of the transport polypeptide by SDS/polyacrylamide gel electrophoresis. A single polypeptide band corresponding in size to 100 kDa was found to bind FITC after pretreatment of membranes with PITC. The inclusion of proline and sodium during FITC treatment reduced binding of FITC at the 100 kDa polypeptide band.

Membrane vesicles derived from rat liver have been used to identify a putative carrier protein thought to be involved in alanine transport. Hayes and McGivan (1983) incubated [^{3}H]-NEM in the presence of plasma membrane vesicles derived from rat liver. Membrane proteins were dissolved in detergent and separated by SDS/polyacrylamide gel electrophoresis. A highly labelled protein band with a molecular mass of 20,000 kDa was identified. Titration of NEM binding to the protein band was shown to parallel the inhibition of alanine transport caused by NEM. Furthermore, preincubation of membrane vesicles in the presence of excess L-alanine resulted in reduced binding of [^{3}H]-NEM to the 20,000 kDa protein.

In another example, McCormick and Johnstone (1988) developed an assay to measure sodium-dependent amino acid transport activity from Ehrlich cell plasma membrane vesicles. On the basis of immunoprecipitation and affinity labelling studies of plasma membrane that displayed reconstituted transport activity, a polypeptide component present in purified membranes of Ehrlich cells with a molecular mass of 120-130 kDa was tentatively identified as a major component of the sodium-dependent amino acid transporter.

ESTIMATION OF MOLECULAR SIZES OF AMINO ACID TRANSPORTERS BY THE RADIATION-INACTIVATION METHOD

Insights into the molecular sizes of transporters has also been gained by using the more indirect method of radiation inactivation. This method relies on measurement of the loss of biological activity as a function of increasing radiation exposure. The physical theory for size determination of proteins by radiation inactivation has been reviewed by Kempner and Schlegel (1979). Beliveau et al. (1990) used the radiation inactivation method to study structural differences between the various transport systems present in brush-border membrane vesicles derived from the rat kidney cortex. Initial rates of influx of amino acids into the membrane vesicles were estimated for the substrates L-glutamate, L-alanine, ß-alanine, L-proline, L-lysine, and L-leucine to establish the molecular sizes of carriers. The respective values for L-glutamate, L-proline, L-leucine, and L-alanine carriers were 250, 224, 293, and 274 kDa. The radiation inactivation size of ß-alanine and L-lysine were 127 and 90 kDa respectively.

McCormick et al. (1991) also reported use of the radiation inactivation method to estimate the molecular size of a sodium-dependent amino acid transporter in Ehrlich cell plasma membrane. A value of approximatelty 348 kDa was obtained for the sodium-dependent carrier. This value for the size of the sodium-dependent transporter was, however, 3 times higher than the value obtained from other studies which revealed that a 120-130 kDa polypeptide could be purified and shown to play a role in sodium-dependent transport in the Ehrlich cells (McCormick et al., 1988). In an unique set of studies, the rate of disappearance of the 120-130 kDa polypeptide following radiation exposure was found to be correlated with the loss of sodium-dependent transport activity in irradiated membranes as determined from sodium-dependent α-aminoisobutyric acid uptake, and quantitation of peptide loss by Western blot analysis. These studies suggested that: (i), the 120-130 kDa polypeptide was an integral component of the system A transporter and (ii), the transporter may be an oligomer of size 348 kDa consisting of three subunits of size 129-130 kDa. Alternatively, the 120-130 kDa polypeptide may be part of a heteropolymer associated with other membrane proteins to give the 348 kDa oligomer.

FLUORESCENCE SPECTROSCOPIC STUDY OF TRANSPORT PROTEINS

A potentially valuable approach for future studies into the static and dynamic aspects of mammalian transport protein structure is the design of membrane transport proteins for fluorescence spectroscopic analysis. This approach is being developed by Kaback and his associates (Menezes et al., 1990). In a preliminary report of this method, these investigators prepared the lactose transporter from Escherichia coli for tryptophan fluorescence spectroscopy by replacing tryptophan residues with phenylalanine by using oligonucleotide-directed site-specific mutageneisis. Replacement of tryptophan for phenylalanine had little effect on the activity of the mutagenized protein. However, the fluoresence emission spectrum of the purified wild-type lactose transporter differed from the emission spectrum of the mutagenized transporter. This approach may prove useful in the analysis of structure-function relationships in transport proteins.

GENETIC APPROACHES FOR THE STUDY OF AMINO ACID TRANSPORT GENES AND PROTEINS

Our previous discussion involved approaches which focused on the identification and characterization of transport systems at the membrane level with the aim of isolating putative transport proteins. Several genetic approaches have been advanced in efforts to identify and study amino acid transport genes and proteins. Genetic methods are often an invaluable aid in obtaining cells with altered transport characteristics which can facilitate identification, isolation, and characterization of genes and their proteins involved in amino acid transport.

Gene Amplification Technique

Conditions that favor the amplification of genes has been exploited as a method in initial attempts to clone and characterize mammalian transport genes. The mechanisms which underlie the process of gene amplification has been the subject of several excellent reviews (Hamlin et al., 1884; Schimke, 1984; Stark and Wahl, 1984). Gene amplification by stepwise selection is characteristically observed when cells are subjected to severe, but not absolute, growth limiting conditions (Schimke, 1988). Englesberg and coworkers used this approach to provide evidence that two membrane-bound proteins were associated with system A transport. (Moffett et al., 1988). These investigators took advantage of the fact that growth of a mutant Chinese hamster ovary cell line, auxotrophic for proline, was inhibited by L-alanine. L-alanine inhibits growth of the proline auxotroph by preventing proline transport. Proline auxotrophs with growth resistance to L-alanine were generated by using hydroxurea to increase the frequency of gene amplification in a stepwise fashion. By selecting mutants resistant to increasingly higher concentrations of L-alanine, cell lines with increased proline transport were obtained. Two proteins with molecular weights of approximately 62-66 kDa and 29 kDa, respectively, were identified and shown to exist in increased concentrations by SDS/polyacrylamide gel electrophoresis of proteins from membrane vesicles and endoplasmic reticulum of cells resistant to growth in the presence of 125 mM L-alanine.

Tritium Suicide Selection

E. A. Adelberg and his colleagues exploited tritium suicide selection to obtain mutant mouse lymphocytic cells with decreased system A transport activity (Finkelstein et al., 1977; Dantzig et al., 1979; Dantzig et al., 1982). Mutagenized cells were briefly exposed to ^{3}H-labelled aminoisobutyric acid at extremely high specific activity and immediately frozen. Normal cells, having accumulated a high intracellular concentration of [^{3}H]-aminoisobutyric acid were preferentially killed by radioactive decay, while transport mutants having failed to accumulate the substrate selectively survive. A similar approach was used to obtain mutants with decreased system L activity in Chinese hamster ovary cells (Dantzig et al., 1984; Collarini et al., 1989).

Temperature Sensitive Mutants

Our laboratory has been developing genetic approaches to isolate genes for system L using Chinese hamster ovary cell lines harboring temperature-sensitive mutations in leucyl-tRNA synthetase. The activity of system L has been shown to increase two to three fold in temperature sensitive mutant cell lines when leucyl-tRNA synthetase is inactivated by growth of cells at a restrictive temperature. (Moore et al., 1977; Shotwell et al., 1982). By selecting for temperature resistance, a mutant cell line was obtained which displayed constituitively elevated system L activity (Shotwell et al., 1983).

Genetic Techniques in Regulatory Studies

The availability of mutant cell lines with altered amino acid transport activities have provided an avenue to study expression of amino acid transport. For example, evidence for coordinate regulation of system A transport and the mRNA for the α subunit of the Na^+,K^+-ATPase gene was provided from studies of Chinese hamster ovary cells harboring a constitutive mutation of system A (Qian et al., 1989). Additionally, studies of Chinese hamster ovary cell lines, with altered activity of system L, provided evidence for involvement of a trans-acting regulatory element in its regulation (Shotwell et al., 1983; Cambpell, 1989).

METHODS FOR CLONING AMINO ACID TRANSPORT GENES

Several different approaches have been utilized to clone and sequence genes involved in mammalian amino acid transport. The approaches reflect the relative difficulty investigators have encountered in the isolation of transport genes. In this section, we review three approaches that have been developed in attempts to clone genes involved in amino acid transport.

Expression Cloning of Genes for Transport Proteins

The Xenopus oocyte has been exploited as an in vivo system for study of mRNA and its translation because of their high efficiency for translation of purified exogenous mRNA. Since its introduction as an in vivo system in 1971 by Gurdon et al., a number of transport, membrane and secretory proteins have been expressed in oocytes (Colman, 1984; Snutch, 1988; Aoshima et al., 1988; Sarthy, 1986; Blakely et al., 1980; Sigel et al, 1988; Hediger et al, 1987; Masu et al, 1987; Lubbert et al., 1987; Julins et al., 1988; Takumi et al., 1989). The identification of an unknown mRNA for a known protein or protein activity can be accomplished by introducing mRNA into oocytes via microinjection techniques, and assaying for the synthesis of the protein. Using this approach, Aoshima et al. (1988) microinjected oocytes with mRNA prepared from rat small intestine, and measured their electrical responses to amino acids by electrophysiological methods. Microinjected oocytes displayed depolarization responses upon exposure to several neutral and basic amino acids. The reponses to amino acids were found to increase with incubation time after injection of mRNA, and followed Michaelis-Menten type kinetics. Furthermore, the responses to amino acids were dependent on both sodium concentration and membrane potential, and were inactivated by the sulfhydryl reagent, 5,5-dithiobis (2-nitrobenzoate). These results were interpreted as the expression of sodium/amino acid cotransporters in oocytes injected with mRNA from rat small intestine.

Estimates for the size of mRNA for an amino acid transport protein or protein activity have been obtained by separating total RNA according to size into different fractions, introducing each fraction into oocytes, and assaying for the synthesis of a transport protein or protein activity. Tate et al. (1989) isolated polyadenylated RNA from total RNA from rat kidney and human lymphoid cells by affinity chromatography on oligo(dT) cellulose, and fractionated the polyadenylated RNA according to size by sucrose density centrifugation. The size-fractionated polyadenylated RNA when microinjected into oocytes induced expression of sodium-independent leucine and phenylalanine transport. The highest expression was observed following microinjection of fractions containing mRNAs ranging in size from 3 to 4 kb. Palacin et al. (1990), using microinjection of size-fractionated polyadenylated RNA derived from rat liver reported that a mRNA of 1.9 to 2.5 kb in length induced high expression of sodium-dependent L-alanine transport in Xenopus oocytes. Su et al. (1991) observed that a mRNA of 2 to 3 kb in length from Chinese hamster ovary cells induced expression of L-leucine transport in oocytes following microinjection of size-fractionated polyadenylated RNA.

Genomic and cDNA Libraries

A paucity of attempts to clone genes for amino acid transport proteins from genomic and cDNA libraries have been reported. El-Gewely and Oxender (1985) reported efforts to clone genes responsible for system L activity by utilizing a cosmid library containing human gene sequences (Lau and Kan, 1983). Cosmids are bacterial plasmids to which the cohesive ends of bacteriophage lambda have been attached to allow packaging with extracts of lysogenic bacteria (Collins and Hohn, 1978). A maximum of 45 kilobases of DNA can be inserted into cosmid vectors which increases the likelihood of isolating intact genes from genomic DNA. The strategy employed was to transform a mutant Chinese hamster ovary cell line which displayed conditionally defective system L transport activity with a cosmid library containing human DNA sequences. From previous studies, the locus for system L was found to be located on human chromosome 20 (Lobaton et al., 1984). These investigators reasoned that sequences from the human cosmid library may complement the the defective system L activity of the mutant Chinese hamster ovary cell line. Cosmids which complemented defective system L acivity, and thus restored functional transport in the mutant, could be recovered and inserts containing human DNA could be rescued using an in vitro packaging system for the bacteriophage lambda. While transformants were found with restored system L transport activity, attempts to recover the cosmids and rescue the human DNA sequences have so far been unsuccessful.

Kilberg and coworkers used differential hybridization and a subtracted cDNA library to clone a mRNA species whose accumulation increased during amino acid starvation in rat hepatoma cells (Shay et al., 1990). Levels of mRNAs for amino acid transport proteins were assumed to be increased under the conditions of growth. In their approach, a lambda gt11 library was screened with isotopically labelled cDNA probes derived from rat hepatoma cells that had been cultured in either the presence or absence of amino acids. Positive clones were further subjected to northern hybridization analysis of size-fractionated RNA derived from rat hepatoma cells. A 409 bp cDNA was identified which reproducibly yielded increased hybridization to RNA from amino-acid starved cells. Nucleotide sequence analysis suggested that the cDNA corresponded to the 3' end of a mRNA.

Polymerase Chain Reaction Technique

The Polymerase Chain Reaction (PCR) (Mullis and Faloona, 1987) represents an exciting new approach for the amplification and cloning of mammalian genes. PCR is an in vitro method in which a particular segment of DNA can be specifically replicated and amplified. Two oligonucleotide primers that flank a DNA fragment of interest are used as templates for amplification by repeated cycles of primer extension. These primers hybridize to opposite strands of the target sequence such that DNA synthesis by DNA polymerase proceeds across the region between primers. The high specificity of PCR makes it an ideal method for the isolation of particular nucleotide sequences. Preliminary studies have been initiated to design oligonucleotide template primers for amplification of mammalian transport genes (Su, Williamson, and Oxender, unpublished results). In this approach, the design of degenerate oligonucleotide primers is based upon observed sequence conservation among a number amino acid transport genes which have been identified in yeast, and the assumption that transport proteins have structural and functional similarities. DNA sequences amplified by PCR may be used as probes to isolate the corresponding full-length mRNA which can be subsequently reversed transcribed by reverse transcriptase to obtain a full-length cDNA of the respective gene. Full-length cDNAs can be screened for the ability to restore transport activity in a mutant cell line deficient in the transport activity.

CONCLUDING REMARKS

The approaches outlined in this chapter have provided some insights into the molecular aspects of amino acid transport. These approaches have shown that critical histidine and cysteinyl residues in transport proteins appear to be essential for substrate binding. Additionally, these approaches have allowed for estimates of the sizes of amino acid transport proteins and their corresponding mRNAs as well as some aspects on the mode of gene expression. To date none of the genes for mammalian amino acid transport proteins have been reported in the literature with the exception of the recent identification of the mouse ecotropic retrovirus receptor as a transporter of cationic amino acids (Kim et al., 1991). The continued development of these approaches should lead to a wealth of information on the molecular basis of amino acid transport.

REFERENCES

Aoshima, H., Tomita, K., and Sugio, S., 1988, Arch. Biochem. Biophys., 265:73.

Beliveau, R., Demeule, M., Jette, M., and Potier, M., 1990, Biochem. J., 268:195.

Bertran, J., Roca, A., Pola, E., Testar, X., Zorzano, A., and Palacin, M., 1991, J. Biol. Chem., 266:798.

Blakely, R. D., Robinson, M. B., and Amara, S. G., 1980, Proc. Natl. Acad. Sci. U.S.A., 85:9846.

Cambpell, G. S., Yu, J. H. and Oxender, D. L., 1992, J. Biol. Chem. (in press).

Carilli, C. T., Farley, R. A., Perlman, D. M., and Cantley, L. C., 1982, J. Biol. Chem., 257:5601.

Cecchini, G., Payne, G. S., and Oxender, D. L., 1977, J. Supramolec. Struct., 7:481.

Chiles, T.C. and Kilberg, M. S., 1987, J. Cell Physiol., 129:321.

Chiles, T.C., Dudeck-Collart, K. L., and Kilberg, M. S., 1988, Am. J. Physiol., 255:C340.

Christensen, H. N., Liang, M., and Archer, E. G., 1967, J. Biol. Chem., 242:5237.

Collarini, E. J., Campbell, G. S., and Oxender, D. L., 1989, J. Biol. Chem., 264:15856.

Collins, J. and Hohn, B., 1978, Proc. Natl. Acad. Sci. U.S.A., 75:4242.

Colombini, M. and Johnstone, R. M., 1974, J. Membr. Biol., 15:261-276.

Colman, A., 1984, in "Transcription and Translation: A Practical Approach," Hames, B. D., and Higgins, S. J., Eds., pp.271-300, IRL Press, Washington, D. C.

Dantzig, A. H., Slayman, C. W., and Adelberg, E. A., 1982, Somat. Cell Genet., 8:509.

Dantzig, A. H., Fairgrieve, M., Slayman, C. W., and Adelberg, E. A., 1984, Somat. Cell Mol. Genet., 10:113.

Dudeck, K. L., Dudenhausen, E. E., Chiles, T. C., Fafournoux, P., and Kilberg, M. S., 1987, J. Biol. Chem., 262:12565.

El-Gewely, M. R. and Oxender, D. L., 1985, Ann. New York Acad. Sci., 456:417.

Finkelstein, M. C., Slayman, C. W., and Adelberg, E. A., 1977, Proc. Natl. Acad. Sci. U.S.A., 74:4549.

Gurdon, J. B., Lane, C. D., Woodland, H. R., and Marbaix, G., 1971, Nature (London), 233:177.

Hamilton, R. T. and Nilsen-Hamilton, M.,1976, Proc. Natl. Acad. Sci. U.S.A., 73:1907.

Hamlin, J. L., Milbrandt, J. D., Heintz, N. H., and Azizkhan, J. C., 1984, Int. Rev. Cytol., 90:31.

Hare, J. D., 1967, Cancer Res., 27:2357.

Hare, J. D. , 1975, Arch. Biochem. Biophys., 170:347.

Hayes, M. R. and McGivan, J. D., 1983, Biochem. J., 214:489.

Hediger, M. A., Ikeda, T., Coady, M., Gunderson, C. B., and Wright, E. M., 1987, Proc. Natl. Acad. Sci. U.S.A., 84:2634.

Jackson, R. J., Mendlein, J., and Sachs, G., 1983, Biochim. Biophys. Acta, 731:9.

Johnstone, R. M. and Bardin, C., 1976, J. Cell Physiol., 89:801.

Julins, D., MacDermott, A. B., Axel, R., and Jessell, T. M., 1988, Science, 241:558.

Kahlenberg, A.,1977, J. Supramolec. Struc., 7:37.

Karlish, S. J. D., 1980, J. Bioenerg. Biomembr., 12:111.

Kempner, E. S. and Schlegel, W., 1979, Anal. Biochem., 92:2.

Kenyon, G. and Bruice, T., 1977, Methods Enzymol., 47:407.

Kilberg, M. S., Handlogten, M. E., and Christensen, H. N., 1980, J. Biol. Chem., 255:4011.

Kim, J. W., Closs, E. I., Albritton, L. M. and Cunningham, J. M., 1991, Nature (London) 352:725.

Lau, Y. F. and Kan, Y. W., 1983, Proc. Natl. Acad. Sci. U.S.A., 80:5225.

Lobaton, C. D., Moreno, A., and Oxender, D. L., 1984, Mol. Cell Biol., 4:475.

Lubbert, H., Hoffman, B. J., Snutch, T. P., Van Dyke, T., Levine, A. J., Hartig, P. R., Lester, H. A., and Davidson, N., 1987, Proc. Natl. Acad. Sci. U.S.A., 84:4332.

Masu, Y., Nakayama, K., Tamaki, H., Harada, Y., Kuno, M., and Nakanishi, S., 1987, Nature (London), 329:836.
Means, G. E. and Feeny, R. E., 1971, in "Chemical Modifications of Proteins," Holden-Day Inc. San Franscico, CA, pp.194-200 and 219.
McCormick, J. and Johnstone, R. M., 1988, Proc. Natl. Acad. Sci. U.S.A., 85:7877.
McCormick, J. and Johnstone, R. M., 1990, Biochem. Cell Biol., 68:512.
McCormick, J. I., Jette, M., Potier, M., Beliveau, R., and Johnstone, R. M., 1991, Biochemistry, 30:3704.
Menezes, M. E., Roepe, P. D., and Kaback, H. R., 1990, Proc. Natl. Acad. Sci. U.S.A., 87:1638.
Moffett, J., Mendiaz, E., Jones, M., and Englesberg, E., 1988, Somat. Cell Genet., 14:1.
Moore, P. A., Jayme, D. W., and Oxender, D. L., 1977, J. Biol. Chem., 252:7427.
Mullis, K. B. and Faloona, F. A., 1987, Methods Enzymol., 155:335.
Nigg, E. A. and Cherry, R. J., 1979, Biochemistry, 18:3457.
Oxender, D. L. and Christensen, H. N., 1963, J. Biol. Chem., 238:3686.
Palacin, M., Werner, A., Dittmer, J., Murer, H., and Biber, J., 1990, Biochem. J., 270:189.
Pick, U., 1981, FEBS Lett., 123:131.
Pick, U. and Bassilian, S., 1981, FEBS Lett., 123:127.
Qian, N.-X., Jones, M., McDonough, A., and Englesberg, E., 1989, Proc. Natl. Acad. Sci. U.S.A., 86:7984.
Sarthy, V., 1986, Mol. Brain Res., 1:97.
Schimke, R. T., 1984, Cell, 37:705.
Schimke, R. T., 1988, J. Biol. Chem., 263:5989.
Shotwell, M. A., Mattes, P. M., Jayme, D. W., and Oxender, D. L., 1982, J. Biol. Chem., 257:2974.
Shotwell, M. A., Collarini, E. J., Mansukhani, A., Hampel, A. E., and Oxender, D. L., 1983, J. Biol. Chem., 258:8183.
Segel, G. B., Woodlock, T. J., Murant, F. G., and Lichtman, M. A., 1989, J. Biol. Chem., 264:16399.
Shay, N. F. , Nick, H. S., and Kilberg, M. S., 1990, J. Biol. Chem., 265:17844.
Sigel, E., Baur, R., Prozig, H., and Reuter, H., 1988, J. Biol. Chem., 263:14614.
Smith-Johannsen, H., Perdue, J. F., Ramjeesingh, M. and Sips, H. J., and Van Dam, K., 1981, J. Membr. Biol., 62:231.
Su, T. Z., Logsdon, C. D., and Oxender, D. L., 1991, FASEB J., 5:4740.
Snutch, T. P., 1988, Trends Neurosci., 11:250.
Stark, G. R. and Wahl, G. M., 1984, Annu. Rev. Biochem., 53:447.
Takumi, T. Ohkubo, H., and Nakanishi, S., 1988, Science, 242:1042.
Tate, S. S., Urade, R., Getchell, T. V., and Udenfriend, S., 1989, Arch. Biochem. Biophys., 275:591.
Weber, M. J., Hale, A. H., Yau, T. M., Buckman, T., Johnson, M., Brady, T. M., and LaRossa, D. D., 1976, J. Cell Physiol., 89:711.
Wright, E. M. and Peerce, B. E., 1984, J. Biol. Chem., 259:14993.

AMINO ACID TRANSPORT DURING EMBRYOGENESIS

Lon J. Van Winkle

Department of Biochemistry
Chicago College of Osteopathic Medicine
Downers Grove, IL 60515

INTRODUCTION

Amino acid transport during mammalian embryogenesis has been characterized best in preimplantation conceptuses and chorioallantoic placentas. Most of the amino acid transport systems in the placental trophoblast appear to be similar to the well-known systems that are present in many other types of mammalian cells (e.g., Yudilevich and Sweiry, 1985; Ganapathy et al., 1986; Kudo et al., 1987; Johnson and Smith, 1988; Wheeler and Yudilevich, 1989; Moe and Smith, 1989; Karl et al., 1989; Hoeltzli and Smith, 1989; Hoeltzli et al., 1990; Kudo and Boyd, 1990). Although the placental systems probably contribute to net transfer of most amino acids from mother to fetus, the mechanisms of these net transfers remain to be determined. Similarly, the developmental etiologies and fates of the placental systems and their controls are not yet known. For these reasons, we will focus mainly on developmental changes in the activities of several novel amino acid transport systems in preimplantation mouse conceptuses. Since embryogenesis is preceded by oogenesis, we will also discuss available information concerning amino acid transport in mouse oocytes and their associated follicular cells. In addition, we will consider the endogenous amino acid transport systems in *Xenopus* oocytes since these cells may prove to be useful as a means of helping to isolate and characterize mammalian amino acid transporters and, eventually, their genes.

Many of the amino acid transport systems in preimplantation mouse conceptuses are different from the systems in other mammalian cells. The systems in other cells, referred to here as "better-known systems", select anionic, cationic or zwitterionic amino acids as substrates, and they are either Na^+-dependent or Na^+-independent (Table 1). These criteria of Na^+ dependence and substrate selectivity define six classes of transport systems (e.g., Bannai et al., 1984; Christensen, 1984). This conventional classification scheme becomes inadequate, however, as illustrated by the designations for the conspicuous embryonal systems, $B^{o,+}$ and $b^{o,+}$. Both of the latter systems accept some cationic and some zwitterionic amino acids about equally well as substrates (Table 1). In fact, the substrate selectivities of five of the ten known amino acid transport systems in preimplantation mouse conceptuses do not correspond well to the selectivities of the heretofore better-known systems (Van Winkle et al., 1985a, 1988a, 1990c; Van Winkle and Campione, 1990). A similar conclusion is not yet warranted for the systems in mouse oocytes and follicular cells, perhaps because characterization of the systems in these cells began only recently.

Mammalian Amino Acid Transport, Edited by M.S. Kilberg and
D. Häussinger, Plenum Press, New York, 1992

Table 1. Conventional classification of α-amino acid transport systems[1]

Category of systems	For zwitterionic amino acids	For cationic amino acids	For anionic amino acids
Na^+-Dependent			
Better-known systems	A,Gly,ASC,N	None known	X_{AG}^-
Systems in mouse conceptuses	B,Gly,$B^{o,+}$	$B^{o,+}$	X_{AG}^-
Na^+-Independent			
Better-known systems	L,asc_1,T	y^+	x_C^-,x_G^-
Systems in mouse conceptuses	L,T,$b^{o,+}$	$b^{o,+}$,b^+_1,b^+_2	x_C^-

[1]References: e.g.; Bannai et al, 1984; Christensen, 1984, 1989; Van Winkle et al., 1985a, 1988a,b, 1990a,c-e; Van Winkle, 1988; Van Winkle and Campione, 1990; Mann et al., 1991 and our unpublished data.

AMINO ACID TRANSPORT IN OOCYTES AND FOLLICLES

Oocyte-follicular cell-complexes (i.e., follicles) appear to contain the better-known amino acid transport systems, Gly and L, and free mouse oocytes also contain system Gly (Haghighat and Van Winkle, 1990). On the other hand, Na^+-dependent L-alanine transport in follicles (unpublished data) is not affected by substances that inhibit system A (i.e., 2-(methylamino)isobutyrate; MeAIB) or system ASC (e.g., L-penicillamine or L-arginine) (Kilberg et al., 1981; Christensen, 1984, 1989). Therefore, a novel Na^+-dependent system may transport alanine in follicles. Moreover, mediated Na^+-independent glutamate transport appears to be induced in follicles simply by incubating them *in vitro* for 3 h (unpublished data). For these reasons, the amino acid transport systems in follicles warrant further characterization as do the mechanisms that control their expression *in vitro* and *in vivo*. Nevertheless, some of the functions of the amino acid transport processes in follicles are already obvious.

Oocytes probably need to take up amino acids in order to grow and develop normally (Buccione et al., 1990), and this uptake appears to be enhanced when oocytes are associated with follicular cells (Collona and Mangia, 1983; Haghighat and Van Winkle, 1990). The enhancement of uptake in oocytes by follicular cells is blocked reversibly by the inhibitor of gap junctional communication, 18-α-glycyrrhetinic acid (Haghighat and Van Winkle, 1990). Therefore, follicular cell-enhanced amino acid uptake into oocytes appears to require the gap junctional communication known to exist between these two types of cells. Amino acids are probably taken up by follicular cells and transferred to oocytes via gap junctions (e.g., Eppig, 1979; Moor et al., 1980; Bachvarova et al., 1980; Heller et al., 1981; Collona and Mangia, 1983), although further studies are needed to prove the latter hypothesis (Haghighat and Van Winkle, 1990). Enhancement of uptake of an amino acid into oocytes by follicular cells also is influenced by the presence of other amino acids. Amino acids can directly influence uptake of one another through competition for the same transport processes. Such competition also may indirectly increase or decrease the ability of follicular cells to enhance uptake by shifting most of the uptake of an amino acid from one system or process to another (Haghighat and Van Winkle, 1990). Experiments concerning the interactions among amino acids and between follicular cells and oocytes can be conducted more easily with follicles *in vitro*. Ultimately, however, these interactions must be studied *in vivo* since the interactions may change when the follicles are removed from the ovarian environment and cultured *in vitro*.

In this regard, mediated Na^+-independent glutamate uptake appears in follicles when they are incubated for 3 h in Waymouth medium under an atmosphere of 5% CO_2 in air. This otherwise conspicuous transport activity is virtually absent from freshly isolated follicles (unpublished data). Induction of Na^+-independent glutamate transport via system x_C^- in other mammalian cells seems to help them resist oxidative stress when they are cultured *in vitro* (Takada and Bannai, 1984; Hishinuma et al., 1986; Watanabe and Bannai, 1987; Bannai and Ishii, 1988; Bannai et al., 1989). In the latter case, glutamine is taken up, converted to glutamate and exchanged for cystine via system x_C^-. The cystine accumulated in this manner is reduced and converted to glutathione which can help cells withstand oxidative insult. Follicles appear to be sensitive to oxidative stress since viable oocytes can be grown in association with follicular cells in an atmosphere of 5% O_2 (Eppig and Schroeder, 1989; Daniel et al., 1989), but follicles eventually lose their organization and integrity and they come apart when they are cultured in an atmosphere of 20% O_2 (Eppig, personal communication). Therefore, appearance of system x_C^- in follicles *in vitro* could help them resist oxidative stress. The ease with which system x_C^- activity can be induced in follicles, and its apparent involvement in resisting oxidative stress *in vitro* open the question of whether such a mechanism also operates *in vivo*. In fact, it needs to be confirmed that the characteristics and mechanisms of control of most amino acid transport systems are the same *in vivo* and *in vitro*. The characteristics of the systems *in vivo* obviously must be determined before we can fully understand their physiological functions. The more tractable problems of determining which systems represent different transport proteins and elucidation of the molecular mechanisms that control their activities in cells *in vitro* undoubtedly will receive the most immediate attention. Nevertheless, solution of the latter problems may ultimately make *in vivo* studies easier to perform.

Techniques to isolate and characterize most amino acid transporter proteins and their genes may be available soon. For example, injection of poly(A^+) mRNA from several types of mammalian cells into *Xenopus* oocytes apparently leads to expression of the mammalian systems A, ASC and L in the oocytes (Tate et al., 1989; Palacin et al., 1990; Tarnuzzer et al., 1990). Injection of some mRNA fractions results in greater amino acid uptake in oocytes than does injection of other fractions. Moreover, injection of mRNA from cells that express more system A activity leads to more 2-aminoisobutyrate (AIB) uptake by oocytes than injection of mRNA from cells that express less system A activity (Tarnuzzer et al., 1990). The results of these studies must be interpreted cautiously, however, because the studies were performed with oocytes that had been treated with collagenase to help isolate them from their follicular cells. The use of collagenase during isolation of oocytes can lead to loss of endogenous amino acid transport activity (Campa and Kilberg, 1989). The lost endogenous activity is not normally regained upon incubation of oocytes *in vitro* (Campa and Kilberg, 1989), so it is conceivable that injection of mammalian poly(A^+) mRNA helps restore endogenous transporter activities lost as a result of collagenase treatment. For example, collagenase treatment might damage other membrane components that are needed for full expression of one or more amino acid transport system. Injection of mammalian mRNA might replenish these other membrane components and thus restore endogenous amino acid transport. In this regard, the amounts of alanine, AIB, MeAIB and possibly leucine transport activities expressed in mRNA-injected oocytes (Tate et al., 1989; Palacin et al., 1990; Tarnuzzer et al., 1990) does *not* exceed the amounts of these activities that are expressed in manually isolated oocytes that have never been exposed experimentally to collagenase (Campa and Kilberg, 1989). Alanine, AIB and MeAIB uptake were utilized to measure the activities that are attributed to systems ASC and A in oocytes injected with mammalian poly(A^+) mRNA (Palacin et al, 1990; Tarnuzzer, et al., 1990), and leucine uptake was used to measure presumptive system L activity (Tate et al., 1989). Moreover, tail-bud stage *Xenopus* embryos appear to express considerable system A and system L activities (Mlot et al., 1978), so these systems conceivably can be induced in nucleated *Xenopus* cells. For these reasons, the transport activities detected in collagenase-treated *Xenopus* oocytes after they

Table 2. Expression of system $B^{o,+}$ in oocytes and early conceptuses of mice, frogs and sea urchins

Stage of Development	Mouse	*Xenopus*	Sea Urchin
Immature oocyte	-[1]	++	N.D.
Unfertilized egg	-	±	-
Fertilized egg	-	N.D.	++
Cleavage stages	?	N.D.	N.D.
Blastula or blastocyst	++	N.D.	N.D.

[1]Key: -, not detected; +, detected; ++, relatively conspicuous activity detected; ?, insufficient or ambiguous data; N.D., not determined.

are injected with mammalian mRNA could result from induction or restoration of endogenous transporter activities rather than from incorporation of mammalian amino acid transporters into the oocytes' plasma membranes.

Although manually isolated *Xenopus* oocytes may express transporter activities that resemble better-known systems (Campa and Kilberg, 1989), much of the endogenous transport activity expressed by *Xenopus* oocytes resembles the mouse blastocyst's system $B^{o,+}$ (Van Winkle, 1988; Campa and Kilberg, 1989; Taylor et al., 1989). System $B^{o,+}$ activity appears to be relatively high in fully grown *Xenopus* oocytes, but it is reduced by 98% before mature oocytes are deposited in pond water (Table 2) (Jung and Richter, 1983, 1984; Van Winkle, 1988). Since even highly concentrative amino acid transport systems operate for exodus as well as for uptake (White and Christensen, 1983), we suggested that this decrease in amino acid transport activity is needed to prevent Xenopus eggs from losing excessive amounts of amino acids to pond water (Van Winkle, 1988). In contrast, an apparent variant of system $B^{o,+}$ (Van Winkle et al., 1985a; Van Winkle, 1988) increases dramatically in activity in sea urchin eggs when they are fertilized (Epel, 1972; Allemand et al., 1984, 1985). In the latter case, however, the Na^+ in sea water could help to drive net uptake of amino acids even at the relatively low concentrations of amino acids that are present in the sea (Manahan et al., 1983; Van Winkle, 1988). In preimplantation mouse conceptuses, system $B^{o,+}$ becomes fully expressed in blastocysts where its activity appears to become susceptible to uterine regulation near the time of implantation (Van Winkle and Campione, 1987; Van Winkle, 1988; Van Winkle et al., 1990c,f and see below). We suggested that the chronology of expression of system $B^{o,+}$ in oocytes and early conceptuses may have changed during evolution of frogs, mice and sea urchins from a common ancestor (Table 2) in order better to support development of these animals in different environments (Van Winkle, 1988). A phyletic change in the developmental chronology of expression of a characteristic, such as system $B^{o,+}$, is termed heterochrony, and heterochrony is believed to be a mechanism by which species evolve (Gould, 1977). When the developmental changes in several amino acid transport systems have been studied in the oocytes and early conceptuses of a variety of species, a complex array of individual instances of apparent heterochrony might be discovered. Although each apparent instance of heterochrony would, by definition, have direction either to earlier or later times during development, the timing of expression of each transport system might not have changed in the same direction. Such a complex array of individual instances of apparent heterochrony might be explained better by developmental reprogramming of expression of the systems in different species (Thomson, 1988). In either case, study of the mechanisms and control of amino acid transport in mammalian oocytes and preimplantation conceptuses may have significance at several levels of biological inquiry including the biochemical, developmental, physiological and evolutionary levels. In this regard, it may soon be possible to study control of expression of system $B^{o,+}$. For example, the system $B^{o,+}$ amino acid receptor site in *Xenopus* oocytes appears to bind the glutamine analog, 6-diazo-5-oxo-L-norleucine, covalently and in a Na^+-dependent manner (Taylor et al., 1990). The latter studies could lead to isolation and characterization not only of the system $B^{o,+}$ transporter protein and

Table 3. Summary of α-amino acid transport systems in mouse oocytes and preimplantation conceptuses[1]

Transport System[2]	Stage of development				
	Immature oocyte	Egg[3]	2-cell	8-cell[4]	Blastocyst
Na⁺-Dependent Systems					
$B^{o,+}$	-[5]	-	?	+(?)	++
B	N.D.	-	?	+(?)	+
X_{AG}^-	N.D.	-	-	+	++
Gly	++	++	++	++	-
Na⁺-Independent Systems					
$b^{o,+}$	N.D.	+	+	+	++
b^+_2	N.D.	-	-	?	+
T	N.D.	++	++	N.D.	++
x_C^-	N.D.	++	+	+	±
L	N.D.	++	++	++	++
b^+_1	N.D.	+	+	?	-

[1]References: Epstein and Smith, 1973; Borland and Tasca, 1974, 1975; Van Winkle, 1977, 1981, 1988; Van Winkle and Dabich, 1977; DiZio and Tasca, 1977; Holmberg and Johnson, 1979; Lindqvist et al., 1981; Kaye et al., 1982; Keefer and Tasca, 1984; Van Winkle et al., 1985a,b, 1988a,b, 1990a-f; Hobbs and Kaye, 1985, 1990; Zuzack et al., 1985; Van Winkle and Campione, 1987, 1990; Haghighat and Van Winkle, 1990; Mann et al., 1991 and our unpublished data.
[2]With the possible exception of system B (Van Winkle et al., 1990c), systems provisionally designated in part by an upper or lower case "B" were first identified in preimplantation mouse conceptuses.
[3]The same systems have been detected in ovulated eggs before and after fertilization.
[4]Mixtures of compacted and non-compacted conceptuses have usually been studied at this stage.
[5]Key: -, not detected; +, detected; ++, relatively conspicuous activity detected; ?, insufficient or ambiguous data; N.D., not determined.

gene expressed in *Xenopus* oocytes, but also the ones expressed in preimplantation mouse conceptuses, depending upon the number of other membrane proteins, such as γ-glutamyl transferase, that also react with this and similar amino acid analogs (e.g., Inoue et al., 1977).

TRANSPORT IN PREIMPLANTATION CONCEPTUSES

Systematic assessment of the ability of various amino acids and amino acid analogs to inhibit uptake of particular cationic, zwitterionic or anionic amino acid-substrates has led to delineation of five novel and a total of ten mediated α-amino acid transport processes in preimplantation mouse conceptuses (first column in Table 3). Four of these transport systems are Na^+-dependent and six are Na^+-independent, although, conversely, cations strongly *inhibit* at least one of the "Na^+-independent" systems (Van Winkle et al., 1990a,b). The activity of one of the Na^+-dependent systems is sigmoidally related to the exogenous Na^+ concentration, and the system interacts selectively with glycine and sarcosine, is stimulated by Cl^-, and accepts neither Li^+ nor choline as a substitute for Na^+ in stimulating transport (Hobbs and Kay, 1985, 1990; Van Winkle, et al., 1988b). By these criteria, the latter system appears to be the well-known system Gly (Table 1). Another of the Na^+-dependent transport processes in conceptuses appears to be system X_{AG}^- because it interacts strongly with smaller anionic amino acids, such as L-aspartate, D-aspartate, L-glutamate, L-cysteinesulfinate and L-cysteate, but not with larger ones, such as L-homocysteate, L-aminopimelate and L-aminoadipate (Gazzola et al., 1981; Mann et al., 1991 and our unpublished data). Moreover, the K_m values for L-aspartate and L-glutamate uptake are nearly equal to their K_i values for inhibition of uptake of each other ("AB" part of the "ABC" test; reviewed by Van Winkle, 1988; Christensen, 1989), and such is also the case for inhibition of L-aspartate uptake by D-aspartate. Therefore, these three amino acids probably compete for transport via system X_{AG}^- in mouse blastocysts.

One of the other two Na^+-dependent transport systems in preimplantation mouse conceptuses (Table 3) was identified for the first time in mouse blastocysts (Van Winkle et al., 1985a), and the fourth system may also be novel (Van Winkle, et al., 1990c,d). The Na^+-dependent system $B^{o,+}$ receives a broad scope of cationic, zwitterionic and bicyclic amino acids as substrates (Van Winkle et al., 1985a, 1988b, 1990c,d). Thus, it is unlike any of the better-known amino acid transport systems in mammalian cells (Table 1). Standard "ABC" testing has shown that L-alanine, 3-amino*endo*bicyclo[3.2.1]octane-3-carboxylic acid (BCO), L-lysine, L-leucine, glycine, L-tryptophan and probably L-arginine, L-valine and 2-amino*endo*bicyclo[2.2.1]heptane-2-carboxylic acid (BCH) compete directly for transport via system $B^{o,+}$. Numerous other amino acids are strong inhibitors of system $B^{o,+}$, so they are probably also good substrates of it, whereas N-methyl, anionic and β-amino acids are largely or completely excluded by the system. It has been suggested that system $B^{o,+}$ may be the same as broad-scope systems in renal and intestinal epithelia (Van Winkle et al., 1985a; Christensen, 1989). These renal and intestinal systems resemble the fourth Na^+-dependent system in conceptuses better than they resemble system $B^{o,+}$, however (Van Winkle et al., 1988b, 1990c,d). This fourth system, provisionally designated B, is much less conspicuous than system $B^{o,+}$, so ABC testing has not been used to study it. Nevertheless, inhibition analyses are consistent with the conclusion that this system interacts strongly with and transports glycine, L-alanine, L-tryptophan and probably most other ordinary zwitterionic α-amino acids. The latter broad scope of system B along with its rejection of MeAIB and L-penicillamine as inhibitors distinguish it from systems A and ASC (Van Winkle et al., 1990c,d). Moreover, weak interaction of system ASC with cationic amino acids has been used to help deduce the structure of its amino acid receptor site (Thomas and Christensen, 1970), but system B does not interact even weakly with cationic amino acids (Van Winkle et al., 1988b, 1990c,d). Thus, system B, but *not* system $B^{o,+}$, resembles systems in the brush borders of bovine kidney (Lynch and McGivan, 1987) and rabbit intestine (Stevens et al., 1982) because each of these transporters interacts relatively strongly with a broad scope of zwitterionic but *not* cationic amino acids. Somewhat surprisingly, lysine- and arginine-sensitive Na^+-dependent alanine uptake has been detected in the apical membranes of confluent cultures of the Madin-Darby canine kidney (MDCK) epithelial cell line (Boerner et al., 1986). The strength of the latter inhibition and whether cationic amino acids are also substrates of the system in confluent MDCK cells could be determined utilizing "ABC" testing. Cationic amino acids are substrates of system $B^{o,+}$ (Van Winkle et al., 1985a), whereas they are competitive inhibitors but not substrates of system ASC (Thomas and Christensen, 1970). Therefore, more complete characterization of the system in confluent MDCK cells would reveal whether its substrate selectivity resembles that of system ASC, system $B^{o,+}$ or neither of these systems.

While it is possible, as discussed above, to draw some parallels between the novel *Na^+-dependent* systems in preimplantation conceptuses and some of the transport activities in other mammalian cells, it is not as easy to draw such parallels for the novel *Na^+-independent* systems in conceptuses. System $b^{o,+}$ preferentially interacts with bulky cationic and zwitterionic α-amino acids whose carbon skeletons do not branch at the α or β positions (Van Winkle et al., 1988a). Thus, the substrate selectivity of system $b^{o,+}$ is clearly different *even* from the selectivity of system $B^{o,+}$ which accepts the small and branched bulky amino acids that system $b^{o,+}$ rejects (Van Winkle et al., 1985a, 1988b). Moreover, uptake of cationic amino acids by this "Na^+-independent" system is inhibited relatively strongly and competitively by most organic and inorganic cations including Na^+ (Van Winkle et al., 1990a,b,f). We know of no published report of another amino acid transport system with these unusual characteristics, although transport of cationic amino acids by system y^+ is inhibited weakly by Na^+ (Christensen et al., 1969). The latter effect of Na^+ is stimulated by zwitterionic amino acids which are cotransported with Na^+ by system y^+. In contrast, transport of zwitterionic amino acids by system $b^{o,+}$ does not require Na^+ or other cations (Van Winkle et al., 1990b). In this respect, system $b^{o,+}$ resembles system asc_1. Yet system asc_1 differs from system $b^{o,+}$ because Na^+ does not appear to inhibit transport of cationic amino acids by system asc_1 (Young et al., 1988). Nevertheless, there are enough similarities among the characteristics of systems asc_1, $b^{o,+}$, y^+ and ASC to draw a limited parallel among their substrate receptor sites (Thomas and Christensen, 1970; Young et al., 1988; Van Winkle et al., 1990b), and this parallel may reflect an evolutionary relationship among the systems (Van Winkle et al., 1990b). As for system $b^{o,+}$, the relative strengths of the interactions of systems b^+_1 and b^+_2 in conceptuses with various cationic and zwitterionic amino acids also easily distinguish these systems from known transporters (Van Winkle and Campione, 1990), although future studies with other mammalian tissues could of course reveal amino acid transport systems that more closely resemble systems b^+_1, b^+_2 or $b^{o,+}$.

Preimplantation mouse conceptuses also appear to contain the better-known Na^+-independent systems x_C^-, L and T (Van Winkle et al., 1988a, 1990a,d, and unpublished data). System x_C^- selects for larger anionic amino acids including L-glutamate, and it also transports cystine (e.g., Makowske and Christensen, 1982). System L selects for bulky, zwitterionic amino acids, whereas system T prefers benzenoid substrates. We suggested that system T might, nevertheless, be a member of an extended system L family (Van Winkle et al., 1988a). Contrary to the latter hypothesis, the affinity of system T for its substrates is not diminished when the substrates are N-methylated (Lopez-Burillo et al., 1985; Van Winkle et al., 1990d). Since system L rejects N-methyl amino acids, system T probably should *not* be viewed simply as a form of system L with a somewhat narrower substrate specificity (Van Winkle et al., 1990d). Similarly, it is unlikely that system T is the product of a dispensable duplicate of a gene for system L, although such could conceivably be the case for multiple forms of a system that are expressed simultaneously in the same cell.

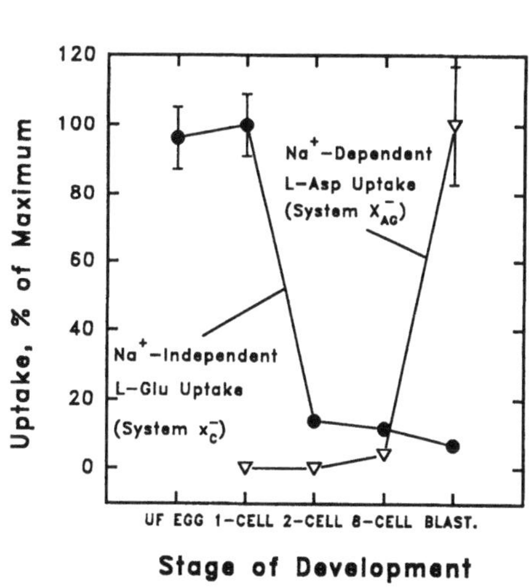

Fig. 1. Developmental changes in Na^+-dependent L-aspartate uptake and Na^+-independent L-glutamate uptake by preimplantation mouse conceptuses via systems X_{AG}^- and x_C^-, respectively. The mean ± S.E. uptake was determined as described elsewhere for anionic (Van Winkle et al., 1990e; Mann et al., 1991) and other (e.g., Van Winkle et al., 1985a) amino acids, and these values were normalized assuming 100% uptake by conceptuses at the stage of maximum transport. When error bars are not visible they were within the symbols. UF EGG, unfertilized egg; BLAST., blastocyst.

DEVELOPMENTAL REGULATION AND POSSIBLE FUNCTIONS OF SYSTEMS

The activities of eight of the ten known amino acid transport systems in preimplantation mouse conceptuses undergo dramatic developmental changes relative to the activities of systems L and T (Van Winkle et al., 1988b, 1990a,c-e; Van Winkle and Campione, 1990). The activities of systems L and T per conceptus or per ng of protein decrease somewhat during preimplantation development. In contrast, much larger changes in the activities of several of the other systems occur in association with blastocyst formation from 8-cell conceptuses. During the latter time, the activities of systems $B^{o,+}$, $b^{o,+}$, X_{AG}^- and possibly B increase several fold, whereas system Gly becomes undetectable (Table 3). These large changes probably occur in association with plasma membrane synthesis or remodeling that may help transform the undifferentiated 8-cell conceptus into an epithelial trophoblast which encloses the blastocyst cavity and the inner cell mass. Small amounts of systems $b^{o,+}$ activity can be detected in conceptuses at all stages of development prior to blastocyst formation, whereas systems $B^{o,+}$ and B have not been detected until at least the 2-cell stage, and system X_{AG}^- becomes detectable in 8-cell conceptuses (Van Winkle et al., 1988b, 1990a,c-e; Mann et al., 1991 and our unpublished data). Therefore, the activities of the latter systems do not all appear to be controlled by the same single mechanism. In contrast to these other systems, the most dramatic change in system x_C^- activity occurs in conceptuses between the 1- and 2-cell stages (Table 3, Fig. 1). The latter decrease in system x_C^- activity is a reflection of a more than 90% reduction in the V_{max} value for L-glutamate uptake (unpublished data). Moreover, it occurs in association with a large decrease in the amounts of maternal total RNA and poly(A^+) mRNA and an increase in the rate of RNA synthesis in conceptuses (e.g., Piko and Clegg, 1982). Therefore, expression of system x_C^- may depend on translation of maternal mRNA. Conversely, disappearance of this transport activity may require transcription of the conceptus' genome.

In contrast to the quantitatively large and relatively abrupt developmental changes in the activities of systems x_C^-, X_{AG}^-, $B^{o,+}$, $b^{o,+}$, B and Gly, the changes in the activities of systems b^+_1 and b^+_2 in conceptuses appear to represent a gradual qualitative transition from system b^+_1 to system b^+_2 between the 2-cell and blastocyst stages of development (Van Winkle and Campione, 1990). We think that an intermediate of these two forms of system b^+, rather than either or both of systems b^+_1 and b^+_2, is expressed in 8-cell conceptuses. The system in both uncompacted and compacted 8-cell conceptuses interacts more strongly with L-arginine than with L-homoarginine. By this criterion, the system in 8-cell conceptuses resembles our definition of system b^+_2 but not system b^+_1 (Van Winkle and Campione, 1990). The system in 8-cell conceptuses also interacts more strongly with L-histidine than with L-tryptophan, however, so it resembles system b^+_1 but not system b^+_2 by the latter criterion. It is intellectually satisfying to suppose that these different forms of system b^+ result from, say, operation of the system in different biochemical environments in cells or from post-translational modification of the system. On the other hand, discontinuities among systems that are present simultaneously in cells could result from expression of separate genes to yield their transporter proteins. These hypotheses will become testable when most amino acid transport proteins and their genes can be isolated. Moreover, such technical advances will allow us to study the transcriptional and post-transcriptional mechanisms by which the activities of amino acid transport systems are controlled during oogenesis and preimplantation development. It is, of course, possible that some of the novel systems in conceptuses are produced from the same genes as better-known systems, or even that some of the different, better-known systems are produced from the same gene. It seems more likely, however, that greater rather than less heterogeneity among amino acid transporter proteins and their genes will be discovered after they are isolated and characterized. For example, several isoforms of what had been thought to be one glucose transporter protein were discovered after one of their genes was identified (e.g., Fukumoto et al., 1988; Koivisto et al., 1991), and such could become the case for plasma-membrane transporters of other circulating forms of metabolizable energy.

From a physiological standpoint, it is largely unclear why preimplantation conceptuses have so many developmentally regulated amino acid transport systems, especially since many of the systems have overlapping substrate specificities.

Preimplantation mouse conceptuses do not increase their protein content until implantation is imminent (Brinster, 1967; Weitlauf, 1973; Schiffner and Spielmann, 1976; Sellens et al., 1981), and 2-cell mouse conceptuses can develop into blastocysts in the absence of exogenously supplied amino acids (e.g., Brinster, 1963; Borland and Tasca, 1974, 1975; Van Winkle et al., 1990e). Hence, it might be argued that early mouse conceptuses have little need to transport amino acids. Conceptuses do not develop in the absence of amino acids *in vivo*, however, and an imbalance of exogenous amino acids can inhibit development. For example, 10 mM L-lysine or L-arginine almost completely inhibit development of blastocysts from 2-cell conceptuses *in vitro*, and this inhibition can be reversed by 10 mM L-leucine (unpublished data). Ten mM leucine alone has no obvious morphological effect on development. For these and other reasons, the contributions of transport systems to amino acid homeostasis should be viewed broadly as allowing conceptuses to interact with their environments via amino acid exodus and exchange as well as uptake. For example, exodus of glutamate via the relatively high level of system x_C^- in 1-cell mouse conceptuses (Fig. 1) could be involved in the mechanism by which exogenously supplied glutamine improves development of 1-cell conceptuses *in vitro* (Chatot et al., 1989). After glutamine is taken up, exodus of glutamine's metabolite, glutamate, via system x_C^- may be beneficial to conceptuses. Unlike other mammalian cells in culture, however, the mechanism for the latter possible beneficial effect is *not* to increase cystine uptake (e.g., Bannai et al., 1989 and see above). Cystine is not present in culture media for preimplantation mouse conceptuses (e.g., Brinster, 1963; Chatot et al., 1989) although its inclusion in media along with glutamine might be another way to improve development of conceptuses especially at the 1-cell stage (Fig. 1).

Preimplantation conceptuses may also need several overlapping amino acid transport systems because different cells in conceptuses or even different regions in the same cell may differ biochemically and morphologically. These different cells or regions of the plasma membrane in the same cell could contain different proportions of various amino acid transport systems. Even if the systems are distributed evenly in the plasma membrane of the same cell, however, the different systems could conceivably serve preferentially different pools in the cell for protein synthesis, protein degradation, amino acid catabolism, and conversion of amino acids to other substances (Van Winkle, 1988). Furthermore, if particular substrates are present at high concentrations in cells or in secretions of the reproductive tract, then systems with partially overlapping substrate specificities may be needed to maintain transport of all amino acids. For example, if the glycine concentration in uterine secretions of the mouse is as high as it is in the rabbit (e.g., Miller and Schultz, 1987), then system $B^{o,+}$ would serve primarily to transport this amino acid in blastocysts *in situ*. Other systems, such as $b^{o,+}$, that do not interact strongly with glycine, might then be needed to transport other amino acids. We have found that the ability of preimplantation mouse conceptuses to take up glycine appears to allow them to counteract the detrimental effects of the high inorganic ion concentrations in the secretions of the reproductive tract (Van Winkle et al., 1990e). Some systems may also have other functions, such as transport of hormones or other substances, in addition to or instead of transport of ordinary amino acids (Zhou et al., 1990; Van Winkle et al., 1990d).

In this regard, system $B^{o,+}$ activity increases about three-fold within 20 min after mouse and rat blastocysts are disturbed by removing them from the uterus or by gently massaging the uterus containing blastocysts with a blunt instrument within about 5 h prior to implantation (Van Winkle and Campione, 1987; Van Winkle et al., 1990c,f). Such increases do *not* occur, however, for other systems in the latter blastocysts nor for system $B^{o,+}$ in blastocysts disturbed 24 h prior to implantation. Similarly, aspects of metabolism with pleiotropic effects, such as adenyl cyclase activity, change rapidly in the rat uterus when it is physically stimulated, but only while it is receptive to implantation (Bekairi et al., 1984a,b). We suggested that development of the ability to rapidly change these and other metabolic processes may be needed in the blastocyst and in the uterus in order for them to send and respond to chemical or physical signals from each other (Van Winkle and Campione 1987; Van Winkle, 1988; Van Winkle et al., 1990c,f). Development of this ability to send and respond to signals could conceivably help to coordinate implantation or other interactions between the uterus and conceptus.

SUMMARY AND FUTURE DIRECTIONS

Mouse oocytes and preimplantation conceptuses contain both novel and better known amino acid transport systems, and the activities of these systems are developmentally regulated (Table 3). Elucidation of the mechanisms of these developmental regulations probably will require isolation and characterization of the amino acid transporters and their genes. To fully understand this regulation we will also need to learn how the activities and the substrate selectivities of the amino acid transport systems are influenced by developmental changes in their chemical and physical environments and by post-translational modifications.

Developmental changes in the amino acids that are actually transported by the systems in preimplantation conceptuses also depend on the amino acid concentrations in secretions of the reproductive tract. Although these concentrations have not been determined for most amino acids in the mouse, they have been shown to change as preimplantation development proceeds in the rabbit (Miller and Schultz, 1987). Moreover, the concentrations of amino acids in secretions of the reproductive tract may be influenced by the concentrations of these substances in blood plasma. We have shown that the concentrations of several amino acids decrease in mouse blood plasma during the preimplantation period of development (unpublished data), whereas the concentrations of both free and bound tryptophan increase in rat serum within one day after eggs are fertilized (Badawy, 1988). The increase in circulating tryptophan is correlated with inhibition of rat liver tryptophan 2,3-dioxygenase activity, so the increase may result from decreased tryptophan catabolism via the kynurenine-anthranilate pathway. Nevertheless, some catabolism of tryptophan via this pathway appears to be beneficial to early development. The quantitatively minor metabolite of tryptophan, picolinate, improves development of preimplantation mouse conceptuses when it is consumed by mice deprived of dietary tryptophan, niacin and vitamin B_6 during the first three days of pregnancy (Van Winkle et al., 1983). Elucidation of the mechanisms that produce changes in the amino acid concentrations in blood plasma during *pre*implantation development and the effects of these changes on the reproductive process should help inform and be informed by related studies on interorgan amino acid flows (Christensen, 1990 and this volume). In this regard, a more or less continuous flow of amino acids to the conceptus is probably needed for it to grow *after* implantation. Little is known, however, about development and regulation of amino acid transport in either the yolk-sac or the chorioallantoic placenta. Interestingly, free amino acids in blood plasma may not be the major source of these nutrients for early *post*-implantation conceptuses. Rather, the visceral yolk sac appears to take up and digest serum proteins in lysosomes and to release the resultant amino acids to the early post-implantation rat embryo, at least in conceptuses developing *in vitro* (e.g., Beckman et al., 1990, 1991). Therefore, amino acid transport across lysosomal as well as plasma membranes of the visceral yolk sac may be needed in order adequately to nourish the embryo before the chorioallantoic placenta has begun to function. After the chorioallantoic circulation is established, abnormal development or function of amino acid transport systems in the chorioallantoic placenta could lead or contribute to fetal abnormalities including intrauterine growth retardation. Finally, different developmental programs for the expression of amino acid transport systems may have helped different species evolve different reproductive strategies (Van Winkle, 1988). Expression of systems at different stages of oogenesis or embryogenesis (e.g., Table 2) may reflect a need in conceptuses of different species to interact differently with their different environments in order to maintain intracellular amino acid homeostasis. Therefore, description of the developmental programs for the expression of amino acid transport systems in conceptuses of several representative species could help us learn how amino acid transport, metabolism and homeostasis are involved in various developmental processes.

ACKNOWLEDGEMENTS

The author thanks Drs. Halvor N. Christensen, Helen R. Dickinson and David F. Mann for critically reading the manuscript and for useful discussions. The help of Barbara Le Breton and Howard Wasserlauf in preparing the manuscript is gratefully acknowledged. Work in the author's laboratory was supported by USPHS grant HD21801.

REFERENCES

Allemand, D., DeRenzis, G., Ciapa, B., Girard, J.-P., and Payan, P., 1984, Biochim. Biophys. Acta, 772:337.
Allemand, D., DeRenzis, G., Maistre, C., Girard, J.-P., and Payan, P., 1985, J. Membr. Biol. 87:217.
Bachvarova, R., Baran, M.W., and Tejblum, A., 1980, J. Exp. Zool., 211:159.
Badawy, A.A.-B., 1988, Biochem J., 255:369.
Bannai, S., and Ishii, T., 1988, J. Cell. Physiol., 137:360.
Bannai, S., Christensen, H.N., Vadgama, J.V., Ellory, J.C., Englesberg, E., Guidotti, G.G., Gazzola, G.C., Kilberg, M.S., Lajtha, A., Sacktor, N., Sepulveda, F.V., Young, J.D., Yudilevich, D., and Mann, G., 1984, Nature, 311:308.
Bannai, S., Sato, H., Ishii, T., and Sugita, Y., 1989, J. Biol. Chem., 264:18480.
Beckman, D.A., Pugarelli, J.E., Jensen, M., Koszalka, T.R., Brent, R.L., and Lloyd, J.B., 1990, Placenta, 11:109.
Beckman, D.A., Pugarelli, J.E., Koszalka, T.R., Brent, R.L., and Lloyd, J.B., 1991, Placenta, 12:37.
Bekairi, A.M., Sanders, R.B., and Yochim, J.M., 1984a, Biol. Reprod., 31:752.
Bekairi, A.M., Sanders, R.B., Abulaban, F.S., and Yochim, J.M., 1984b, Biol. Reprod., 31:752.
Boerner, P., Evans-Layng, M., U, H.S., and Saier, M.H., 1986, J. Biol. Chem., 261:13957.
Borland, R.M., and Tasca, R.J., 1974, Dev. Biol., 36:169.
Borland, R.M., and Tasca, R.J., 1975, Dev. Biol., 46:192.
Brinster, R.L., 1963, Exp. Cell Res., 32:205.
Brinster, R.L., 1967, J. Reprod. Fert., 13:413.
Buccione, R., Schroeder, A.C., and Eppig, J.J., 1990, Biol. Reprod., 43:543.
Campa, M.J., and Kilberg, M.S., 1989, J. Cell. Physiol., 141:645.
Chatot, C.L., Ziomek, C.A., Bavister, B.D., Lewis, J.L., and Torres, I., 1989, J. Reprod. Fert., 86:679.
Christensen, H.N., 1984, Biochim. Biophys. Acta, 779:255.
Christensen, H.N., 1989, Methods Enzymol., 173:576.
Christensen, H.N., 1990, Physiol. Rev., 70:43.
Christensen, H.N., Handlogten, M.E., and Thomas, E.L., 1969, Proc. Natl. Acad. Sci. U.S.A., 63:948.
Collona, R., and Mangia, F., 1983, Biol. Reprod., 28:797.
Daniel, S.A.J., Armstrong, D.T., and Gore-Langton, R.E., 1989, Gamete Res., 24:109.
DiZio, S.M., and Tasca, R.J., 1977, Dev. Biol., 59:198.
Epel, D., 1972, Exp. Cell Res., 72:74.
Eppig, J.J., 1979, J. Exp. Zool., 209:345.
Eppig, J.J., and Schroeder, A.C., 1989, Biol. Reprod., 41:268.
Epstein, C.J., and Smith, S.A., 1973, Dev. Biol., 33:171.
Fukumoto, H., Seino, S., Imura, H., Seino, Y., Eddy, R.L., Fukushima, Y., Byers, M.G., Shows, T.B., and Bell, G.I., 1988, Proc. Natl. Acad. Sci. U.S.A., 85:5434.
Ganapathy, M.E., Leibach, F.H., Mahesh, V.B., Howard, J.C., DeVoe, L.D., and Ganapathy, V., 1986, Biochem. J., 238:201.
Gazzola, G.C., Dall'Asta, V., Bussolati, O., Makowske, M., and Christensen, H.N., 1981, J. Biol. Chem., 256:6054.
Gould, S.J., 1977, "Ontogeny and Phylogeny," The Belknap Press of Harvard University Press, Cambridge, Massachusetts.
Haghighat, N., and Van Winkle, L.J., 1990, J. Exp. Zool., 253:71.
Heller, D.T., Cahill, D.M., and Schultz, R.M., 1981, Dev. Biol., 84:455.
Hishinuma, I., Ishii, T., Watanabe, H., and Bannai, S., 1986, In Vitro, 22:127.
Hobbs, J.G., and Kaye, P.L., 1985, J. Reprod. Fert., 74:77.
Hobbs, J.G., and Kaye, P.L., 1990 Reprod. Fertil. Dev., 2:651.
Hoeltzli, S.D., and Smith, C.H., 1989, Am. J. Physiol., 256:C630.
Hoeltzli, S.D., Kelley, L.K., Moe, A.J., and Smith, C.H., 1990, Am. J. Physiol., 259:C47.
Holmberg, S.R.M., and Johnson, M.H., 1979, J. Reprod. Fert., 56:223.
Inoue, M., Horiuchi, S., Morino, Y., 1977, Eur. J. Biochem., 73:335.
Johnson, L.W., and Smith, C.H., 1988, Am. J. Physiol., 254:C773.
Jung, D., and Richter, H.-P., 1983, Cell Biol. Int. Rep., 7:697.
Jung, D., and Richter, H.-P., 1984, Cell Differ., 14:99.

Kaye, P.L., Schultz, G.A., Johnson, M.H., Pratt, H.P.M., and Church, R.B., 1982, J. Reprod. Fert., 65:367.
Karl, P.I., Tkaczevski, H., and Fisher, S.E., 1989, Ped. Res., 25:19.
Keefer, C.L., and Tasca, R.J., 1984, J. Reprod. Fert., 70:399.
Kilberg, M.S., Handlogten, M.E., and Christensen, H.N., 1981, J. Biol. Chem., 256:3304.
Koivisto, U.-M., Martinez-Valdez, H., Bilan, P.J., Burdett, E., Ramlal, T., and Klip, A., 1991, J. Biol. Chem., 266:2615.
Kudo, Y., Yamada, K., Fujiwara, A., and Kawasaki, T., 1987, Biochim. Biophys. Acta, 904:309.
Kudo, Y., and Boyd, C.A.R., 1990, J. Physiol., 426:381.
Linqvist, I., Nilsson, B.O., and Ronquist, G., 1981, Acta Physiol. Scand., 111:35.
Lopez-Burillo, S., Garcia-Sancho, J., and Herreros, B., 1985, Biochim. Biophys. Acta, 820:85.
Lynch, A.M., and McGiven, J.D., 1987, Biochim. Biophys. Acta, 899:176.
Makowske, M., and Christensen, H.N., 1982, J. Biol. Chem., 257:5663.
Manahan, D.T., Davis, J.P., and Stephens, G.C., 1983, Science, 220:204.
Mann, D.F., Weimer, B.D., and Van Winkle, L.J., 1991, FASEB J., 5:A1188.
Miller, J.G.O., and Schultz, G.A., 1987, Biol. of Reprod., 36:125.
Mlot, C., Prahlad, K.V., and Hampel, A., 1978, Dev. Biol., 67:65.
Moe, A.J., and Smith, C.H., 1989, Am. J. Physiol., 257:C1005.
Moor, R.M., Smith, M.W., and Dawson, R.M.C., 1980, Exp. Cell Res., 126:15.
Palacin, M., Werner, A., Dittmer, J., Murer, H., and Biber, J., 1990, Biochem. J., 270:189.
Piko, L., and Clegg, K.B., 1982, Dev. Biol., 89:362.
Schiffner, J., and Spielmann, H., 1976, J. Reprod. Fert., 47:145.
Sellens, M.H., Stein, S., and Sherman, M.I., 1981, J. Reprod. Fert., 61:307.
Stevens, B.R., Ross, H.J., and Wright, E.M., 1982, J. Membr. Biol., 66:213.
Takada, A., and Bannai, S., 1984, J. Biol. Chem., 259:2441.
Tarnuzzer, R.W., Campa, M.J., Qian, N.-X., Englesberg, E., and Kilberg, M.S., 1990, J. Biol. Chem., 265:13914.
Tate, S.S., Urade, R., Getchell, T.V., and Udenfriend, S., 1989, Arch. Biochem. Biophys., 275:591.
Taylor, P.M., Hundal, H.S., and Rennie, M.J., 1989, J. Membr. Biol., 112:149.
Taylor, P.M., Mackenzie, B., Hundal, H.S., and Rennie, M.J., 1990, Biol. Chem. Hoppe-Seyler, 371:304.
Thomas, E.L., and Christensen, H.N., 1970, Biochem. Biophys. Res. Comm., 40:277.
Thomson, K.S., 1988, "Morphogenesis and Evolution," Oxford University Press, Oxford.
Van Winkle, L.J., 1977, J. Exp. Zool., 202:275.
Van Winkle, L.J., 1981, J. Exp. Zool., 218:239.
Van Winkle, L.J., 1988, Biochim. Biophys. Acta, 947:173.
Van Winkle, L.J., and Dabich, 1977, Biochem. Biophys. Res. Comm., 78:357.
Van Winkle, L.J., and Campione, A.L., 1987, Biochim. Biophys. Acta, 925:164.
Van Winkle, L.J., and Campione, A.L., 1990, Biochim. Biophys. Acta, 1028:165.
Van Winkle, L.J., Doman, D.R., and Campione, A.L., 1983, J. Nutr. Sci. Vitaminol., 29:701.
Van Winkle, L.J., Christensen, H.N., and Campione, A.L., 1985a, J. Biol. Chem., 260:12118.
Van Winkle, L.J., Campione, A.L., and Kester, S.E., 1985b, J. Exp. Zool., 235:141.
Van Winkle, L.J., Campione, A.L., and Gorman, J.M., 1988a, J. Biol. Chem., 263:3150.
Van Winkle, L.J., Haghighat, N., Campione, A.L., and Gorman, J.M., 1988b, Biochim. Biophys. Acta, 941:241.
Van Winkle, L.J., Campione, A.L., Gorman, J.M., and Weimer, B.D., 1990a, Biochim. Biophys. Acta, 1021:77.
Van Winkle, L.J., Campione, A.L., and Gorman, J.M., 1990b, Biochim. Biophys. Acta, 1025:215.
Van Winkle, L.J., Campione, A.L., and Farrington, B.H., 1990c, Biochim. Biophys. Acta, 1025:225.
Van Winkle, L.J., Mann, D.F., Campione, A.L., and Farrington, B.H., 1990d, Biochim. Biophys. Acta, 1027:268.
Van Winkle, L.J., Haghighat, N., and Campione, A.L., 1990e, J. Exp. Zool., 253:215.
Van Winkle, L.J., Iannaccone, P.M., Campione, A.L., and Garton, R.L., 1990f, Dev. Biol., 142:184.
Watanabe, H., and Bannai, S., 1987, J. Exp. Med., 165:628.
Weitlauf, H.M., 1973, Anat. Rec., 176:121.

Wheeler, C.P.D., and Yudilevich, D.L., 1989, Biochim. Biophys. Acta, 978:257.
White, M.F., and Christensen, H.N., 1983, J. Biol. Chem., 258:8028.
Young, J.D., Mason, D.K., and Fincham, D.A., 1988, J. Biol. Chem., 263:140.
Yudilevich, D.L., and Sweiry, J.H., 1985, Biochim. Biophys. Acta., 822:169.
Zhou, Y., Samson, M., Osty, J., Francon, J., and Blondeau, J.-P., 1990, J. Biol. Chem., 265:17000.
Zuzack, J.S., Tasca, R.J., and DiZio, S.M., 1985, J. Cell. Physiol., 122:379.

AMINO ACID TRANSPORT BY LYSOSOMES

Ronald L. Pisoni[1] and Jerry A. Schneider[2]

[1]Department of Pediatrics
and Communicable Diseases
University of Michigan
Ann Arbor, Michigan

[2]Department of Pediatrics
University of California, San Diego
La Jolla, California

INTRODUCTION

Lysosomes are a major intracellular site for the degradation of a large variety of macromolecules including proteins, nucleic acids, complex carbohydrates, and lipids. These macromolecules are delivered to lysosomes by the processes of pinocytosis, receptor-mediated endocytosis, phagocytosis, or autophagy. In the lysosome these large molecules are degraded by hydrolases, many of which function optimally at the acidic pH of the intralysosomal environment. The resulting degradation products often require specific transport systems in order to exit the lysosomal compartment. This review focuses on the amino acid transport systems of mammalian lysosomes which allow the release of amino acids generated within the lysosome following proteolysis. As in the plasma membrane, the specificity of lysosomal amino acid transport systems is based upon the net charge and dimensions of the amino acid. Thus, there are specific transport routes for the cationic amino acids, for the anionic amino acids, and multiple routes for the neutral amino acids. In addition, three highly specific lysosomal transport routes have been characterized for cystine, cysteine, and proline. All of the lysosomal transport systems described thus far are Na^+-independent.

Lysosomal amino acid transport systems have been characterized very recently. Consequently, the properties of many of these systems have been determined in only one, or at best a few, cell types. Some differences have already been noted in the properties of lysosomal transport systems in various cell types. This may be especially important when lysosomal carriers play a critical role in supporting the differentiated functions of a particular cell type. We hope the lysosomal amino acid transport systems described here will provide a solid framework for future investigations of the processes regulating translocation of amino acids across the lysosomal membrane.

LYSOSOMAL CYSTINE TRANSPORT: ROLE OF THE INHERITED DISEASE CYSTINOSIS IN ITS DELINEATION

The era of characterizing lysosomal transport systems began as a consequence of efforts to understand the basis of the genetic disease, nephropathic cystinosis. The earliest clinical descriptions of this disease appeared in the 1930's by Fanconi in

Mammalian Amino Acid Transport, Edited by M.S. Kilberg and
D. Häussinger, Plenum Press, New York, 1992

Switzerland, de Toni in Italy and Debre in France. All three of these pediatricians described children who had stunted growth, rickets and glycosuria. Many years later it was recognized that the internal organs of these children contained large accumulations of the disulfide amino acid, cystine. Since all attempts to delineate an enzyme defect in cystine metabolism in tissues from these children were unsuccessful, the possibility of a defect in cystine transport was suggested (Segal, 1965; Schneider et al., 1968). However, early studies of cystine transport in cystinotic cells were difficult to interpret because the cystine was rapidly metabolized upon entering the cell and the cystinotic cells were already "preloaded" with about 100 times more free (non-protein) cystine than normal control cells (Schneider et al., 1968).

Since cystine is toxic to so many cellular enzyme systems, it was difficult to understand how cystine storage within the cells of these patients could be compatible with life. This suggested that the cystine might be compartmentalized within the cells and not available as a toxic material to various intracellular enzyme systems. Differential centrifugation of extracts from both peripheral leucocytes and cultured skin fibroblasts demonstrated that cystine is indeed compartmentalized within cystinotic cells (Schneider et al., 1967a and 1967b). Schulman et al. (1969) showed that cystine co-sedimented with lysosomal enzymes during sucrose gradient centrifugation of cystinotic cells. This finding led to a flurry of activity attempting to identify lysosomal enzyme systems that could metabolize cystine, but none were found.

Since cystine is localized in lysosomes of cystinotic cells, lysosomal cystine transport seemed a likely candidate for the inherited defect in this disorder. However, at that time there was no evidence for the transport of any amino acid by normal lysosomal membranes. Investigators sought a method to load normal lysosomes with cystine and the fact that cystine is so insoluble made this an especially difficult problem. One solution to this problem was the use of cysteine-glutathione mixed disulfide. When normal cultured skin fibroblasts were placed in 30 mM cysteine-glutathione mixed disulfide for 24 hours the cystine content of the cells increased 6-7 fold (Jonas et al., 1982a). This increased cystine was located in the lysosomes of the cells and rapidly disappeared from the cells when they were placed in a cystine-free medium ($t_{1/2}$ = 20 min). The cystine content of cystinotic cells remained stable under these conditions whether or not they were first incubated in the mixed disulfide.

A breakthrough in these studies came with the realization that lysosomes could be loaded with any amino acid if incubated in the methyl ester of that amino acid (Goldman and Kaplan, 1973; Reeves, 1979). Amino acid dimethyl esters permeate the lysosomal membrane because of their lysosomotropic properties and are specifically broken down within the lysosome by acid hydrolases yielding methanol and the corresponding free amino acid. Conditions could be established which led to a much greater rate of methyl ester hydrolysis than the rate of amino acid egress from the lysosome. This resulted in lysosomes which were "loaded" with the amino acid under investigation. Using cystine dimethyl ester to load lysosomes, investigators found that cystine egress was defective in cystinotic lysosomes from peripheral leukocytes (Gahl et al., 1982a and 1982b), Epstein Barr Virus transformed lymphoblasts (Jonas et al., 1982b; Jonas et al., 1983), and cultured skin fibroblasts (Pisoni et al., 1985; Smith et al., 1987).

Peripheral leucocytes are an excellent system for demonstrating the defect in cystinotic lysosomes. In the study demonstrated in Fig. 1, Gahl et al. (1982a) found that the initial velocity of cystine egress from normal peripheral leucocyte lysosomes increased as a function of cystine loading until reaching saturation. In contrast, cystinotic lysosomes demonstrated virtually no cystine egress regardless of load level. Cystinosis is an autosomal recessive disease and heterozygotes should have one normal and one defective gene for the lysosomal cystine transporter. In this study, heterozygous lysosomes demonstrated approximately half the normal maximal velocity for cystine egress. Counter-transport proved to be an excellent technique to study lysosomal transport (Gahl et al., 1983). These investigators used this technique to show that the system was stereospecific for the L-isomer and that cystathionine, cystamine and cysteamine-cysteine mixed disulphide share the same system, but at lower activity than cystine.

Lysosomal cystine egress was also observed in Epstein Barr virus transformed lymphoblasts, but only after the special requirement for ATP was understood (Jonas et al., 1982b; Jonas et al., 1983). Lysosomes contain an ATPase which can be distinguished from mitochondria F_1-ATPase on the basis of responses to various inhibitors. This ATPase activity is necessary for acidification of lysosomes in the presence of MgATP and is present at normal concentrations in cystinotic lysosomes (Jonas et al., 1983). The pattern of response of this ATPase to inhibitors is identical in cystinotic and normal control lysosomes. In addition, inhibitors of this ATPase also inhibit cystine egress from normal lysosomes. The proton translocator carbonylcyanide-m-chlorophenylhydrazone (CCCP) does not inhibit ATPase activity but does inhibit lysosomal cystine egress. A detailed study of cystine egress from human skin fibroblast lysosomes showed effects of both the transmembrane pH gradient and the membrane potential on this process (Smith et al., 1987).

Some investigators have found lysosomal cystine egress to be stimulated by MgATP while others have not. These variable findings appear to be due to differences in the requirement for ATP which is less at very high load levels of cystine, and the fact that N-ethylmaleimide, which is required in some systems for studying lysosomal cystine egress, inhibits lysosomal ATPase (Greene et al., 1987). Cystine transport has also been studied in lysosomes prepared from rat liver (Jonas, 1986). In this system cystine egress is also stimulated by MgATP,as well as by divalent cations which may alter the lysosomal membrane potential. Lysosomal cystine transport has recently been characterized in mouse L-929 fibroblasts (Greene et al., 1990). The Michaelis constant for infinite-trans influx was 0.27 mM. Significant requirements for recognition by the transporter were: the presence of amino groups, L configuration, and a chain length not greater than 8 atoms. A net positive or negative charge was not required.

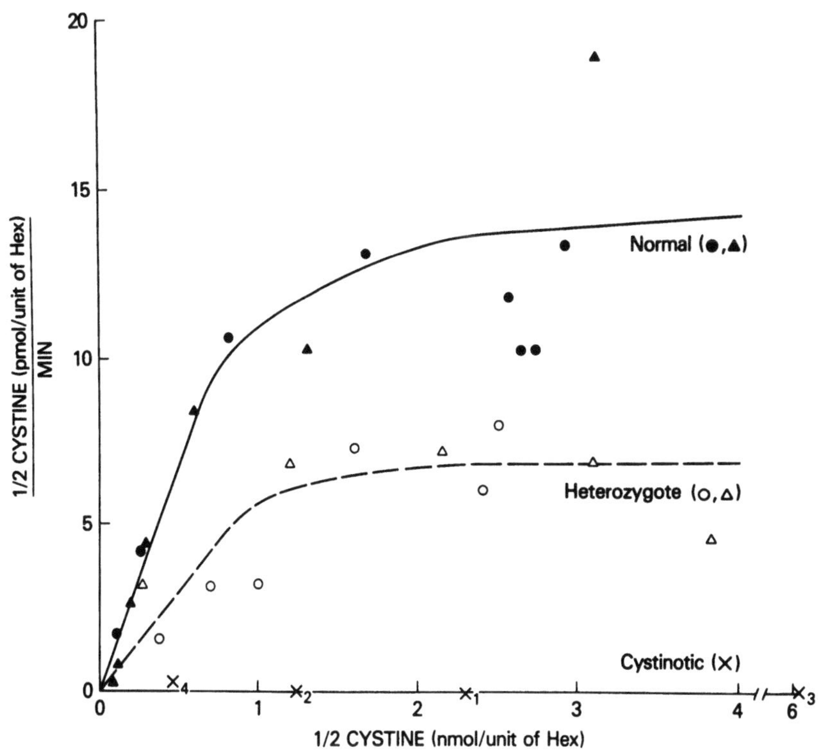

Fig. 1. Initial velocity of cystine transport as a function of lysosomal cystine loading in human leukocytes. (From Gahl et al. [1982a]. Copyright 1982 by the American Association for the Advancement of Science.)

CATIONIC AMINO ACID TRANSPORT: LYSOSOMAL SYSTEM c

Cationic amino acids have been shown to be transported across the lysosomal membrane of human fibroblasts by a specific route designated as lysosomal system c. Many of the properties of this transport system have been revealed by lysosomal uptake studies employing L-[^{14}C]-arginine as a model substrate (Pisoni et al., 1987b). Arginine uptake by human fibroblast lysosomes is saturable, displaying a Km of 0.32 mM at pH 7.0 and 37°C, which is 8 times greater than the Km for arginine transport across the human fibroblast plasma membrane by system y^+ (White et al., 1982). Specificity of this transport route for cationic amino acids was shown by analog inhibition analysis demonstrating that L-[^{14}C]-arginine uptake is inhibited by the cationic amino acids L-2,4-diaminobutyrate, L-ornithine, S-2-aminoethyl-L-cysteine, L-lysine, L-arginine, and L-homoarginine. In contrast, neutral and anionic amino acids have little effect on lysosomal arginine uptake. An unmodified α-carboxyl group appears to be important in recognition by lysosomal system c as suggested by the inability of lysine methyl ester to inhibit lysosomal arginine uptake.

Competitive inhibition constants indicate that lysosomal system c only moderately discriminates among stereoisomers, recognizing L-arginine with 4 fold greater affinity than D-arginine. In addition, recognition of L-arginine is decreased only slightly by methylating its α-amino group, and trimethylation of lysine's ε-amino group decreases recognition of lysine 4-fold. These characteristics suggest that lysosomal system c has a broader substrate specificity than system y^+ which displays little affinity for methylated analogs of cationic amino acids and strongly discriminates between D- and L-stereoisomers (White et al., 1982). Lysosomal system c and system y^+ also differ significantly in two other respects: (1) lysosomal arginine influx by lysosomal system c increases 10-fold as the extralysosomal pH is raised from pH 5 to 7 whereas arginine transport across the plasma membrane by system y^+ is insensitive to pH changes in this range; (2) certain neutral amino acids such as L-homoserine and L-ε-hydroxy-α-aminocaproate, in the presence of Na^+, are accepted as surrogate substrates by system y^+ (Christensen et al., 1969; Thomas et al., 1971) but are not accepted by lysosomal system c.

Additional characteristics of lysosomal system c have been revealed by an examination of L-[^{14}C]-lysine exodus from human fibroblast lysosomes (Pisoni et al., 1985). Most important of these was the demonstration of a trans-stimulation property associated with lysine exodus. Exodus of L-[^{14}C]-lysine from fibroblast lysosomes occurs with a half-time of 24 min at 25°C and pH 6.5. In the presence of external 2 mM L-lysine, however, L-[^{14}C]-lysine exodus is accelerated 2-3 fold. This trans-stimulation of lysosomal lysine efflux is dependent on the external concentration of L-lysine and displays preference for the L-stereoisomer. In addition to lysine, the cationic amino acids arginine, ornithine, diaminobutyrate, histidine, and the mixed disulfide of cysteamine and cysteine all trans-accelerate lysine exodus from human fibroblast lysosomes whereas neutral and anionic amino acids have no effect. Similar to lysosomal arginine uptake, lysosomal lysine efflux increases over the pH range from pH 5 to 7.5, but only to the extent of ~1.3 fold. MgATP, however, has opposite effects on efflux and influx, accelerating lysine efflux 2-3 fold, but retarding lysosomal arginine uptake 3-fold. As shown for some other lysosomal transport systems, these effects by MgATP on lysosomal system c activity could be due to changes in the lysosomal membrane potential and/or lysosomal ion or proton gradients (Smith et al., 1987).

LYSOSOMAL TRANSPORT OF DICARBOXYLIC AMINO ACIDS: SYSTEM d

An investigation of lysosomal glutamate and aspartate uptake has demonstrated the existence of carrier-mediated passage of anionic amino acids across the human fibroblast lysosomal membrane (Collarini et al., 1989). This route, designated as lysosomal system d, has a similar and somewhat broader substrate specificity than the plasma membrane system, X^-_{AG} (Gazzola et al., 1981; Dall'Asta et al., 1983). The uptake of glutamate and aspartate by human fibroblast lysosomes is saturable, displaying Km values of 4-12 μM at pH 7.0 and 37°C. Lineweaver-Burk and V vs V/S plots are linear suggesting that only one route is involved in lysosomal uptake of glutamate and aspartate. Furthermore, double reciprocal plots of aspartate inhibition of lysosomal glutamate uptake or glutamate inhibition of lysosomal aspartate uptake indicate that glutamate and aspartate competitively inhibit each other's uptake with Ki values of 5-7 μM. A small trans effect was observed to be associated with glutamate uptake.

An analysis of the ability of different amino acid analogs to inhibit the lysosomal uptake of 10 μM L-[^{3}H]-glutamate demonstrates the specificity of this transport route for anionic amino acids. Aspartate, aspartate-β-hydroxamate, cysteate, cysteinesulfinate, homocysteate and α-aminoadipate strongly inhibit glutamate uptake, whereas the longer anionic amino acid, α-aminopimelate, is only weakly inhibitory. In addition, leucine, serine, MeAIB, and cystine display no significant effect on lysosomal glutamate uptake. Lysosomal system d moderately discriminates between D- and L-stereoisomers of aspartate and glutamate, but as the chain length increases to 6 carbons, in the case of α-aminoadipate, strong preference is given to the L-stereoisomer. N-Methylation of L-aspartate and L-glutamate greatly decreases their recognition by lysosomal system d, whereas N-methylation of D-aspartate slightly enhances its recognition by lysosomal system d. Finally, the β-amino acid, 3-aminoglutarate, and the cyclic glutamate analog, kainate, are very weak inhibitors of lysosomal glutamate uptake.

The pH profile of lysosomal aspartate uptake reveals a linear 2-3 fold increase in aspartate uptake over the range from pH 5-8. Since the charge of aspartate's amino and carboxyl groups does not significantly change over most of this pH range, other factors such as changes in transmembrane ion or proton gradients, membrane potential, or titration of an amino acid(s) on the transport protein may be responsible for this pH dependency. An examination of lysosomal glutamate exodus demonstrates the ability of lysosomal system d to rapidly translocate anionic amino acids across the lysosomal membrane, exhibiting a half-time of 5-7 min for glutamate exodus at pH 7.0 and 37°C.

LYSOSOMAL TRANSPORT OF SMALL NEUTRAL AMINO ACIDS: SYSTEMS e, f, AND p

Similar to the plasma membrane, the passage of small neutral amino acids across the lysosomal membrane is mediated by several different carrier-mediated transport routes. The existence of these multiple transport routes, designated as systems e, f, and p, has been described in an investigation of the uptake of proline, alanine, serine, and threonine by human fibroblast lysosomes (Pisoni et al., 1987a). The majority of information concerning lysosomal transport of small neutral amino acids has come from investigating lysosomal proline uptake. Proline uptake by human fibroblast lysosomes displays saturation at [Pro] ≥ 0.4 mM and exhibits a bell-shaped pH activity curve with maximal uptake over the range from pH 6.25-7.0. Lysosomal proline uptake is highly specific for the L-stereoisomer of proline and is inhibited ≥ 90% by L-proline and 3,4-dehydro-L-proline. In contrast, cationic and anionic amino acids, or large neutral amino acids such as leucine, methionine, and phenylalanine, have no effect on lysosomal proline uptake. Small neutral amino acids having side chains 1 or 2 carbons in length, however, inhibit lysosomal proline uptake by 50-75%. Alanine and α-amino-n-butyrate, which have unbranched side chains 1 and 2 carbons in length, respectively, inhibit 0.03 mM L-[^{14}C]-proline uptake by 70-75% at relatively low concentrations. In contrast, norvaline, which contains an unbranched side chain 3 carbons long, is a poor inhibitor of lysosomal proline uptake indicating that the transport protein receptor site is unable to accommodate amino acids in which the side chain is greater than 2 carbons long. Glycine, which lacks a side chain, and valine and threonine, which have branched side chains, inhibit proline uptake but with a substantially lower affinity than does alanine. This difficulty in accommodating small neutral amino acids having branched side chains is especially evident with 2-aminoisobutyrate, 2-(methylamino)isobutyrate, and trans and cis 4-hydroxy-L-proline, all of which are very weak inhibitors of lysosomal proline uptake. Two other characteristics regarding the ability of small neutral amino acids to inhibit lysosomal proline uptake were revealed: (1) the β-amino acids, 3-aminoisobutyrate and β-alanine, are recognized nearly as well as the corresponding α-amino acids, and (2) mono-N-methylation of the α-amino group is tolerated as shown by the ability of sarcosine and N-methyl-L-alanine to inhibit lysosomal proline uptake by 50-70%; however, further methylation of the α-amino group is not acceptable as indicated by the complete failure of N,N-dimethyl glycine and betaine to inhibit lysosomal proline uptake.

Thus, carrier-mediation of proline uptake by human fibroblast lysosomes is shared with small neutral amino acids having side chains not exceeding 2 carbons in length and which are not highly branched. Biphasic Lineweaver-Burk and V/S vs V plots of lysosomal proline uptake indicate that at least two different transport routes mediate proline uptake by human fibroblast lysosomes: a high affinity system with an apparent Km of 0.01 mM accounts for ~60-70% of the total proline uptake and a lower affinity route displaying an apparent Km of 0.07 mM serves for the remainder of lysosomal proline uptake. Additional support for the existence of at least two routes mediating lysosomal proline uptake was provided by the demonstration that 75% of total proline uptake is sensitive to inhibition by alanine whereas the remaining 25% of proline uptake is not sensitive to inhibition by alanine. The high affinity route mediating the major part of lysosomal proline uptake which recognizes alanine and presumably other small neutral amino acids was designated, lysosomal system f (Fig. 2). L-Proline and 3,4-dehydro-L-proline inhibit L-[^{14}C]-proline uptake ≥ 90% suggesting that these two amino acids are recognized not only by the alanine-sensitive route (system f) but also by the second route of lower affinity which was named system p for proline-specific route (Fig. 2).

Analog inhibition analysis of the uptake of [^{14}C]-alanine, [^{14}C]-serine, and [^{14}C]-threonine by human fibroblast lysosomes has demonstrated that in addition to lysosomal systems f and p, at least one other carrier mediates the passage of small neutral amino acids across the lysosomal membrane (Pisoni et al., 1987a). This transport system, designated lysosomal system e, is the major route for the lysosomal uptake of alanine, serine, and threonine, and excludes amino acids in which the α-amino group is a secondary amine as occurs in proline, sarcosine, and N-methyl-L-alanine. Thus, alanine, serine, and threonine strongly inhibit the lysosomal uptakes of one another, whereas, N-methyl-L-alanine and L-proline inhibit the lysosomal uptake of alanine, serine, and threonine by only 10-20%. In addition, leucine which has no effect on lysosomal proline uptake, inhibits the lysosomal uptake of alanine, serine, and threonine 40-70%. The lysosomal transport systems for small neutral amino acids are shown in Fig. 2, and are analogous with regard to general substrate specificity to the Na^+-dependent transport systems, ASC and A, of the human fibroblast plasma membrane (Gazzola et al., 1980; Franchi-Gazzola et al., 1982). System ASC, similar to lysosomal system e, mediates the major part of alanine and serine transport across the human fibroblast plasma membrane, but excludes amino acids in which the α-amino group is a secondary amine. System A, somewhat similar to lysosomal system f, is a minor route of small neutral amino acid transport which recognizes small neutral amino acids including those in which the α-amino group is a secondary amine.

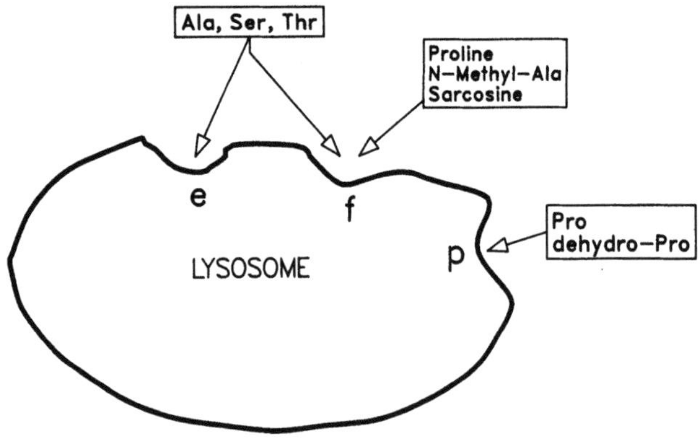

Fig. 2. Schematic diagram of the transport systems for small neutral amino acids in the lysosome of the human fibroblast. (From Pisoni et al. [1987a]. Copyright 1987 by the American Society of Biological Chemists, Inc.)

CYSTEINE-SPECIFIC LYSOSOMAL TRANSPORT SYSTEM

Although most of the lysosomal amino acid transport systems are believed to serve for the net removal of amino acids from the lysosomal compartment, a cysteine-specific lysosomal transport system has been described recently which appears to be a major route for the delivery of thiol into human fibroblast lysosomes (Pisoni et al., 1990). The existence of this transport route was first indicated when examining the fate of L-[^{35}S]-cystine following its uptake by intact human fibroblasts. When human fibroblasts are incubated for 2-5 min with 20 μM L-[^{35}S]-cystine in Krebs-Ringer phosphate solution at pH 7.4, a minimum of 50-60% of the total radioactivity taken up by the fibroblasts is found sequestered within lysosomes in the form of cysteine. This degree of sequestration of cysteine in such a short time interval is exceptional upon considering that lysosomes comprise only 4% of the intracellular volume of a normal human fibroblast (Oude Elferink et al., 1983). The possibility that cysteine sequestration by fibroblast lysosomes was mediated by the lysosomal neutral amino acid transport systems e or f, was excluded by demonstrating that when intact human fibroblasts were incubated with L-[^{14}C]-alanine, which is a substrate for systems e and f (Pisoni et al., 1987a), no accumulation of [^{14}C]-alanine within lysosomes was observed. These results indicate that the pathway mediating lysosomal accumulation of cysteine does not recognize alanine and is distinct from lysosomal transport systems e and f.

An examination of the ability of isolated human fibroblast lysosomes to take up L-[^{35}S]-cysteine demonstrated the existence of a lysosomal transport system highly specific for cysteine which exhibits characteristics favorable for the delivery of cysteine into lysosomes. Lysosomal cysteine uptake by this route displays saturable Michaelis-Menten kinetics with a Km of 53 μM at pH 7.0 and 37°C. Cysteine transport is observed in lysosomes of both normal and cystinotic human fibroblasts and is highly temperature dependent with a Q_{10} of 3.2 which is the highest activation energy found for any of the known lysosomal transport systems. The highly selective substrate specificity of this lysosomal cysteine-specific transport system is indicted by the inability of 5 mM concentrations of other neutral, anionic, or cationic amino acids to inhibit the lysosomal uptake of 35 μM L-[^{35}S]-cysteine. Analogs very similar in structure to cysteine, such as homocysteine, penicillamine, and serine, are, at best, very weak inhibitors of lysosomal cysteine uptake. D-Cysteine and the deaminated cysteine analog, β-mercaptopropionate, also are poor inhibitors of lysosomal cysteine uptake indicating that both the L-stereoisomer and α-amino group of cysteine are important structural determinants for recognition by the cysteine-specific transporter. In contrast, cysteamine which is the decarboxylated analog of cysteine, strongly inhibits lysosomal cysteine uptake suggesting that the α-carboxyl group of cysteine is not important for recognition. If this feature implies cysteamine transport it is highly exceptional among amino acid transport systems.

The pH activity curve of the lysosomal cysteine-specific transport system favors the delivery of cysteine into lysosomes. Very little cysteine transport is observed in the acidic range between pH 5 and 6, but then transport activity increases 7-10 fold between pH 6 and 7.5. If the pH dependence of this transport system is similar on both sides of the lysosomal membrane, cysteine influx into fibroblast lysosomes originating at the cytosolic pH of 7.0 would be greatly favored over lysosomal cysteine efflux originating at the intralysosomal pH of 5.3, thereby favoring net cysteine accumulation within lysosomes.

The delivery of cysteine into lysosomes by the cysteine-specific transport system is thought to support lysosomal proteolysis in two ways. Some of the lysosomal proteases, such as cathepsins B,H, and L are thiol dependent enzymes (Kooistra et al., 1982). Thus, the delivery of cysteine into the intralysosomal space provides these proteases with thiol to maintain maximal levels of hydrolytic activity. Secondly, several studies have indicated that lysosomal degradation of disulfide-rich proteins is accelerated by the presence of cysteine and other thiols even when the lysosomal proteinases which are involved do not require thiols for enzymatic activity (Kooistra et al., 1982; Griffiths and Lloyd, 1979; Mego, 1984). This stimulatory effect of thiols is postulated to be due to thiols reacting with disulfide bridges of the protein substrate causing dissociation of disulfide bonds thereby permitting the proteins to unfold and allow proteinases greater access to susceptible linkages within the protein substrate (Kooistra et al., 1982).

Two different cycles involving the interchange between cystine and cysteine appear to be interconnected in serving the needs for cellular metabolism within the cytosolic and lysosomal compartments of the human fibroblast (Fig. 3). Cystine is transported by system X^-_{CG} across the plasma membrane into the cytosol where it is rapidly reduced to cysteine by GSH (Oshima et al., 1976; Bannai and Kitamura, 1980; Bannai and Tateishi, 1986). Cysteine, thus formed, can be transported from the cytosol back into the extracellular space where it is re-oxidized to cystine. Alternatively, cytosolic cysteine can be used for the synthesis of proteins and glutathione, metabolized for the production of other cellular constituents, or transported into lysosomes. Within the lysosomal compartment, cysteine is postulated to react with protein disulfide bridges by sulfhydryl/disulfide exchange reactions yielding protein-bound cysteine and protein-bound cystine at the site of each original disulfide bridge, with these products eventually released as free cystine and cysteine within lysosomes after proteolysis (Lloyd, 1986). The lysosomal cystine carrier then transports cystine from the lysosomal compartment into the cytosol where cystine is reduced by GSH to cysteine. Cysteine formed within lysosomes can also egress from this compartment, although the extent to which lysosomal

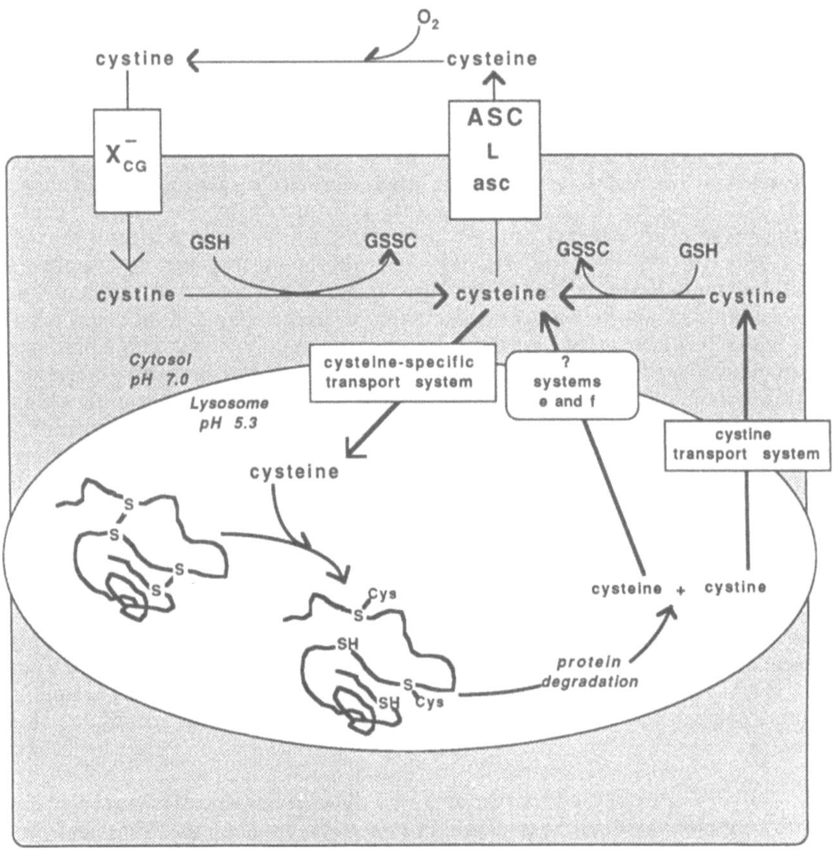

Fig. 3. Schematic diagram showing (a) the interconversions between cystine and cysteine in different cellular compartments, (b) the role of transport systems in maintaining this cycle, and (c) the role of cysteine in supporting lysosomal proteolysis. Shaded area represents the cytosol whereas the clear area represents the lysosomal compartment; ASC, L, asc, and X^-_{CG} are plasma membrane amino acid transport systems. (From Pisoni et al. [1990]. Copyright 1990 by the Rockefeller University Press.)

cysteine egress is mediated by lysosomal systems e and f, the lysosomal cysteine-specific transport route, or by some other possible route remains to be determined.

Especially striking is how the two lysosomal transport systems in the above cyclic model have adapted their properties to utilize the reducing environment of the cytosol and the pH gradient across the lysosomal membrane to direct their flow of substrate. The pH dependence of the lysosomal cysteine-specific transport system appears to take advantage of the pH gradient across the lysosomal membrane to direct transport of cysteine into the lysosomal compartment. In contrast, the lysosomal cystine carrier appears to operate solely for lysosomal exodus of cystine, a need arising from the rapid reduction of cystine by GSH when cystine enters the cytosol.

LYSOSOMAL SYSTEM h: TRANSPORT OF LARGE, NEUTRAL HYDROPHOBIC AMINO ACIDS AND ITS HORMONAL REGULATION BY THYROTROPIN IN FRTL-5 CELLS

A transport system mediating the lysosomal passage of large neutral amino acids, and designated as lysosomal system h, has been characterized in the Fischer rat thyroid epithelial cell line, FRTL-5. The existence of this transport route was first shown by demonstrating a 5-fold trans-stimulation of lysosomal [^{3}H]-tyrosine uptake when FRTL-5 lysosomes were pre-loaded to a high level with unlabeled tyrosine (Bernar et al., 1986). The degree of [^{3}H]-tyrosine countertransport was directly proportional to the intralysosomal concentration of unlabeled tyrosine until saturation was observed at intralysosomal [Tyr] ≥ 1.5 nmol Tyr/unit hexosaminidase. Tyrosine uptake by FRTL-5 lysosomes displays a temperature coefficient (Q_{10}) of 1.9 and exhibits a Km of 20 μM at pH 7.0 and 37°C (Andersson et al., 1990; Bernar et al., 1986).

Analog inhibition analysis has demonstrated that this transport route mediating lysosomal tyrosine uptake in FRTL-5 cells is specific for the large neutral amino acids. Leucine, isoleucine, phenylalanine, tyrosine, and tryptophan at concentrations of 0.5 mM strongly inhibit the lysosomal uptake of 8 μM [^{3}H]-tyrosine (Bernar et al., 1986). Valine and methionine inhibit lysosomal tyrosine uptake to an intermediate level whereas cystine, lysine, glutamate, and the small neutral amino acids, alanine, proline, and glycine, have no effect on tyrosine uptake by FRTL-5 lysosomes. In addition, lysosomal system h is highly selective for the L-stereoisomer as indicated by the poor ability of D-tyrosine to inhibit [^{3}H]-L-tyrosine uptake.

Recently, monoiodotyrosine (MIT) was shown to be transported by lysosomal system h (Tietze et al., 1989; Andersson et al., 1990). When FRTL-5 lysosomes are pre-loaded to a high level with unlabeled monoiodotyrosine, the lysosomal uptakes of [^{125}I]-MIT, [^{3}H]-Phe, [^{3}H]-Tyr, and [^{3}H]-Leu are trans-stimulated several-fold. Furthermore, lysosomal MIT uptake is strongly inhibited by large neutral amino acids but is not inhibited by proline or cystine. Lysosomal system h recognizes monoiodotyrosine with 10-12 times greater affinity than tyrosine, displaying a Km of 1.5-1.8 μM for monoiodotyrosine at pH 7.0 and 37°C (Andersson et al., 1990). Enhanced recognition due to the presence of a ring halogen is also observed with other halogenated derivatives of phenylalanine irrespective of the mass of the halogen or its position on the phenyl ring. Phenylalanine derivatives containing aromatic ring substituents which decrease the hydrophobicity of phenylalanine (e.g. p-nitro-L-phenylalanine), however, are weaker than phenylalanine as competitors of lysosomal MIT uptake. Several other characteristics of lysosomal system h were revealed from the studies of monoiodotyrosine uptake. Maximal lysosomal uptake of monoiodotyrosine is observed in the range from pH 7.0-7.5 but decreases sharply at pH values alkaline to this range. Neither 3-(p-hydroxyphenyl)-propionic acid, L-tyrosine methyl ester, or tyramine inhibit lysosomal MIT uptake indicating that both the α-amino group and α-carboxyl group are necessary for recognition by lysosomal system h. Leucine displays very similar competitive inhibition constants (Ki's) of 3.5 μM and 3.2 μM, respectively, for the inhibition of lysosomal tyrosine and lysosomal MIT uptakes, indicating that leucine is recognized by lysosomal system h with greater affinity than tyrosine but with less affinity than monoiodotyrosine.

The proteolytic activity of lysosomes is particularly relevant to the specialized function of thyroid epithelial cells in which lysosomal degradation of thyroglobulin is responsible for producing the thyroid hormones, thyroxine and triiodothyronine (Dumont et al., 1989). In addition, significant amounts of mono- and diiodotyrosine are formed during lysosomal proteolysis of thyroglobulin with these degradation products passing from lysosomes into the cytosol where they are deiodinized and re-incorporated into newly synthesized thyroglobulin molecules (Ekholm, 1981). Lysosomal system h plays a central role in this salvage pathway by mediating the passage of monoiodotyrosine, and possibly diiodotyrosine, from lysosomes into the cytosol (Tietze et al., 1989; Andersson et al., 1990). Harper et al. (1988) have demonstrated that lysosomal system h transport activity is stimulated 4-7 fold in FRTL-5 cells when cells are cultured in media containing the hormone thyrotropin (TSH). Stimulation by TSH requires a minimum incubation period of 48 hrs, and is prevented by the presence of 1 nM actinomycin D or 20 μM cycloheximide indicating that RNA synthesis and protein synthesis are necessary for stimulation of lysosomal system h transport activity. In the absence of TSH, cholera toxin and dibutyryl cAMP are each able to mimic TSH stimulation of lysosomal system h transport activity indicating that TSH stimulation is mediated via a cAMP signal. The net effect of TSH stimulation of lysosomal system h activity in FRTL-5 cells may be to increase the ability of thyroid cells to salvage monoiodotyrosine from lysosomes for the synthesis of new thyroglobulin molecules, thereby supporting the ability of thyroid cells to maintain a high rate of thyroglobulin synthesis and thyroxine release. This is the first description of hormonal regulation of a lysosomal transport system and encourages the search to understand how lysosomal transport systems may be regulated in other tissues, especially those in which salvage of lysosomal metabolites is important for operation of specialized cellular functions.

The extent to which lysosomal system h is present in other cell types, in addition to FRTL-5 cells, remains to be determined. Significant differences have been observed between FRTL-5 cells and several other cell types in the transport of large neutral amino acids. In contrast to the trans effect observed for lysosomal system h in FRTL-5 cells, trans effects have not been seen for phenylalanine exodus from human fibroblast lysosomes (Pisoni et al., 1985), tyrosine exodus from human leucocyte lysosomes (Bernar et al., 1986), and leucine exodus from rat liver lysosomes (Reeves, 1979). The rate of tyrosine exodus from FRTL-5 lysosomes (Bernar et al., 1986) and phenylalanine exodus from human fibroblast lysosomes (Pisoni, R. and Christensen, H.N., unpublished data) is rapid at 37°C displaying half-times of < 5 min, whereas half-times of 17-26 min have been measured for efflux of methionine and tyrosine from leucocyte lysosomes (Gahl et al., 1982b). Recently, Stewart et al. (1989) have found a low level of saturable uptake of phenylalanine, leucine, and tryptophan by human fibroblast lysosomes with Km values ranging from 5-30 μM at pH 7.0 and 37°C. Patterns of analog inhibition suggest, however, that carrier-mediation of large neutral amino acids across the fibroblast lysosomal membrane may occur by more than one route. In addition, the L-stereoisomer of inhibitory analogs was only slightly more effective than D-stereoisomers in striking contrast to lysosomal system h of FRTL-5 cells which is highly selective for the L-stereoisomer.

BIBLIOGRAPHY

Andersson, H. C., Kohn, L. D., Bernardini, I., Blom, H. J., Tietze, F. and Gahl, W. A., 1990, J. Biol. Chem., 265:10950.

Bannai, S. and Kitamura, E., 1980, J. Biol. Chem., 255 (6):2372.

Bannai, S. and Tateishi, N., 1986, J. Membr. Biol., 89:1.

Bernar, J., Tietze, F., Kohn, L. D., Bernardini, I., Harper, G. S., Grollman, E. F. and Gahl, W. A., 1986, J. Biol. Chem., 261:17107.

Christensen, H. N., Handlogten, M. E. and Thomas, E. L., 1969, Proc. Natl. Acad. Sci. U. S. A., 63:948.

Collarini, E. J., Pisoni, R. L. and Christensen, H. N., 1989, Biochim. Biophys. Acta, 987:139.

Dall'Asta, V., Gazzola, G. C., Franchi-Gazzola, R., Bussolati, O., Longo, N. and Guidotti, G. G., 1983, J. Biol. Chem., 258:6371.

Dumont, J. E., Vassart, G. and Refetoff, S., 1989, in: " The Metabolic Basis of Inherited Disease," C. R. Scriver, A. L. Beaudet, W. S. Sly and D. Valle, eds. McGraw Hill, New York.
Ekholm, R., 1981, Mol. Cell. Endocrinology, 24:141.
Franchi-Gazzola, R., Gazzola, G. C., Dall'Asta, V. and Guidotti, G. G., 1982, J. Biol. Chem., 257:9582.
Gahl, W. A., Bashan, N., Tietze, F., Bernardini, I. and Schulman, J. D., 1982a, Science, 217:1263.
Gahl, W. A., Tietze, F., Bashan, N., Steinherz, R. and Schulman, J. D., 1982b, J. Biol. Chem., 257:9570.
Gahl, W. A., Tietze, F., Bashan, N., Bernardini, I., Raiford, D. and Schulman, J. D., 1983, Biochem. J., 216:393.
Gazzola, G. C., Dall'Asta, V. and Guidotti, G. G., 1980, J. Biol. Chem., 255:929.
Gazzola, G. C., Dall'Asta, V., Bussolati, O., Makowske, M. and Christensen, H. N., 1981, J. Biol. Chem., 256:6054.
Goldman, R. and Kaplan, A., 1973, Biochim. Biophys. Acta, 318:205.
Greene, A. A., Clark, K. F., Smith, M. L. and Schneider, J. A., 1987, Biochem. J., 246:547.
Greene, A. A., Marcusson, E. G., Pintos Morell, G. and Schneider, J. A., 1990, J. Biol. Chem., 265:9888.
Griffiths, P. A. and Lloyd, J. B., 1979, Biochem. Biophys. Res. Commun., 89:428.
Harper, G. S., Kohn, L. D., Bernardini, I., Bernar, J., Tietze, F., Andersson, H. C. and Gahl, W. A., 1988, J. Biol. Chem., 263:9320.
Jonas, A. J., Greene, A. A., Smith, M. L. and Schneider, J. A., 1982a, Proc. Natl. Acad. Sci. USA., 79:4442.
Jonas, A. J., Smith, M. L. and Schneider, J. A., 1982b, J. Biol. Chem., 257:13185.
Jonas, A. J., Smith, M. L., Allison, W. S., Laikind, P. K., Greene, A. A. and Schneider, J. A., 1983, J. Biol. Chem., 258:11727.
Jonas, A. J., 1986, Biochem. J., 236:671.
Kooistra, T., Millard, P. C. and Lloyd, J. B., 1982, Biochem. J., 204:471.
Lloyd, J. B., 1986, Biochem. J., 237:271.
Mego, J. L., 1984, Biochem. J., 218:775.
Oshima, R. G., Rhead, W. J., Thoene, J. G. and Schneider, J. A., 1976, J. Biol. Chem., 251:4287.
Oude Elferink, R. P., Harms, E., Strijland, A. and Tager, J. M., 1983, Biochem. Biophys. Res. Commun., 116:154.
Pisoni, R. L., Thoene, J. G. and Christensen, H. N., 1985, J. Biol. Chem., 260:4791.
Pisoni, R. L., Flickinger, K. S., Thoene, J. G. and Christensen, H. N., 1987a, J. Biol. Chem., 262:6010.
Pisoni, R. L., Thoene, J. G., Lemons, R. M. and Christensen, H. N., 1987b, J. Biol. Chem., 262:15011.
Pisoni, R. L., Acker, T. L., Lisowski, K. M., Lemons, R. M. and Thoene, J. G., 1990, J. Cell Biol., 110:327.
Reeves, J. P., 1979, J. Biol. Chem., 254:8914.
Schneider, J. A., Bradley, K. and Seegmiller, J. E., 1967a, Science, 157:1321.
Schneider, J. A., Rosenbloom, F. M., Bradley, K. and Seegmiller, J. E., 1967b, Biochem. Biophys. Res. Commun., 29:527.
Schneider, J. A., Bradley, K. H. and Seegmiller, J. E., 1968, Pediatr. Res., 2:441.
Schulman, J. D., Bradley, K. H. and Seegmiller, J. E., 1969, Science, 166:1152.
Segal, S., 1965, Ann. Intern. Med., 62:847.
Smith, M. L., Greene, A. A., Potashnik, R., Mendoza, S. A. and Schneider, J. A., 1987, J. Biol. Chem., 262:1244.
Stewart, B. H., Collarini, E. J., Pisoni, R. and Christensen, H. N., 1989, Biochim. Biophys. Acta, 987:145.
Thomas, E. L., Shao, T. C. and Christensen, H. N., 1971, J. Biol. Chem., 246:1677.
Tietze, F., Kohn, L. D., Kohn, A. D., Bernardini, I., Andersson, H. C., Adamson, M. D., Harper, G. S. and Gahl, W. A., 1989, J. Biol. Chem., 264:4762.
White, M. F., Gazzola, G. C. and Christensen, H. N., 1982, J. Biol. Chem., 257:4443.

AMINO ACID TRANSPORT IN MITOCHONDRIA

J.D. McGivan

Department of Biochemistry
University of Bristol
Bristol BS8 1TD, U.K.

INTRODUCTION

The transport of metabolites across the inner mitochondrial membrane was initially studied in the late 1960's and was intensively investigated over the next fifteen years. By about 1980, the major transport systems had all been characterised and their mechanisms and kinetic parameters had been established. This work preceded much of the study of transport in many mammalian cell systems and in bacteria. More recent work has concentrated on the isolation and characterisation of particular transport proteins. Reviews on the development of the field are available (Meijer and Van Dam, 1974; LaNoue and Schoolwerth, 1979; Kovacevic and McGivan, 1983; Schoolwerth and LaNoue 1985). Meijer et al. (1990) include aspects of amino acid transport in a recent review on urea synthesis. In now reviewing this area in the context of amino acid transport, it is not proposed to give a detailed account of earlier controversies which are discussed in the above literature. It is rather intended to illustrate the similarities and differences in the mechanisms of amino acid transport in the mitochondrial inner membrane and cell plasma membrane in the context of cellular metabolism, and to focus on questions in this area which are still unresolved.

AMINO ACID METABOLISM AND ITS INTRACELLULAR COMPARTMENTATION

The intracellular compartmentation of amino acid metabolism has been studied mostly in liver and kidney. In liver, many important metabolic pathways involve enzymes both in the cytosol and in the mitochondrial matrix (see Meijer et al. (1975), Meijer and Hensgens (1982), Meijer et al.,1990). In urea synthesis, the enzymes carbamoyl phosphate synthase and ornithine transcarbamylase are found only in the mitochondrial matrix, while argininsuccinate synthetase, argininosuccinase and arginase are cytosolic. The important enzymes phosphate-dependent glutaminase, glutamate dehydrogenase and N-acetylglutamate synthase are also exclusively mitochondrial while isozymes of aspartate amino transferase and alanine amino transferase are located in both matrix and cytoplasm. The initial enzymic reactions in the degradation of glycine, proline and ornithine are also localised in the mitochondrial matrix.

Transport of amino acids across the inner mitochondrial membrane is thus an essential process in amino acid metabolism. In urea synthesis, ornithine formed from arginine in the cytosol must enter the mitochondria and citrulline must leave. The transport of glutamine, glycine, proline and possibly serine into the mitochondria is

Mammalian Amino Acid Transport, Edited by M.S. Kilberg and
D. Häussinger, Plenum Press, New York, 1992

necessary for their catabolism, while glutamate, as a product of cytosolic transamination reactions, must enter the mitochondria to be further metabolised. Aspartate transport is involved both in the urea cycle and in the transfer of reducing equivalents across the mitochondrial membrane.

In tissues which do not synthesise urea, mitochondrial amino acid transport is relatively less important. Nevertheless, the intramitochondrial location of phosphate-dependent glutaminase means that mitochondrial glutamine transport is essential in all tissues which metabolise glutamine, especially kidney, intestine and brain and cells of the immune system. Aspartate transport is involved in cytosolic NADH oxidation in many tissues and glutamate transport systems are also widely distributed.

PERMEABILITY PROPERTIES OF THE INNER MITOCHONDRIAL MEMBRANE

The development of knowledge of mitochondrial transport coincided with the development of the chemiosmotic theory of oxidative phosphorylation (Mitchell, 1966) and any consideration of transport must be placed in this context. It is established that the inner mitochondrial membrane has a relatively very low intrinsic electrogenic permeability to ions. Oxidation of NADH via the respiratory chain leads to the ejection of protons and the generation of a protonmotive force of the order of -220 mV which in isolated mitochondria is manifested as a membrane potential (negative inside) of some -180 mV together with a pH gradient of approximately 0.5 pH units. Return of H^+ through the mitochondrial ATPase results in ATP synthesis driven by this proton-motive force.

Mitochondrial transport systems have the primary function of allowing communication between cytosolic and matrix enzymes in metabolic pathways. Some of the mechanisms involved in metabolite transport across the mitochondrial membrane are necessarily different from those involved in plasma membrane transport and different proteins are therefore involved. It was predicted in the earliest versions of the chemiosmotic theory that for negatively charged metabolites such as phosphate to enter the matrix, specific transport systems must exist and must catalyse electroneutral transport so that anion entry should not be influenced by the membrane potential.

The first report of a mitochondrial transport system was that by Pfaff et al. (1965) who demonstrated the existence of the adenine nucleotide carrier. The presence of a number of metabolite transport systems in the inner mitochondrial membrane was reported by Chappell et al. (1968). Over the next ten years, the mechanism of a number of transport systems was elucidated in terms of the ionic species involved (see LaNoue and Schoolwerth, 1979). The transport systems now known to exist in liver mitochondria are summarised in Table 1. In general, metabolite transport is electroneutral in nature. In cases such as the citrate/malate exchanger where the two substrates have different charges, the carrier transports a proton together with the citrate so that electroneutrality is preserved. The proton-linked electroneutral transport of substrates leads to only a small accumulation of metabolites in mitochondria, since only the pH component of the proton motive force is involved. It should be noted that anion/OH^- exchange and anion/H^+ symport are formally equivalent. Although in early work the term anion/OH^- exchange was commonly used, it is now usual to refer to anion/H^+ symport by analogy with plasma membrane and bacterial transport systems.

Two transport systems catalyse electrogenic transport. The adenine nucleotide carrier transports ADP^{3-} in exchange for ATP^{4-}, and the glutamate/aspartate antiporter transports negatively charged glutamate plus 1 H^+ in exchange for negatively charged aspartate. Here the membrane potential influences both the rate and equilibrium position of the transport process. Implications of the electrogenicity of the adenine nucelotide carrier for ATP synthesis have been extensively discussed elsewhere (see e.g. LaNoue and Schoolwerth, 1979). The glutamate/aspartate exchanger is considered in a later section.

Table 1. Metabolite Transport Systems in Liver Mitochondria.

Transporter	Mechanism	Metabolic Function
Phosphate	Pi^-/H^+ symport	ATP synthesis
Dicarboxylate	$malate^{2-}/Pi^{2-}$ exchange	Urea synthesis Gluconeogenesis
2-oxoglutarate	$OG^{2-}/malate^{2-}$ exchange	NADH oxidation
Citrate	$(citrate^{3-} + H^+)/malate^{2-}$ exchange	Fatty acid synthesis
Pyruvate	$pyruvate^-/H^+$ symport	Glucose oxidation
Carnitine	carnitine/acylcarnitine exchange	Fatty acid oxidation
Adenine nucleotide	ADP^{3-}/ATP^{4-} exchange	ATP synthesis
Amino acid transport systems		
Neutral amino acid carrier	neutral amino acid uniport	Amino acid degradation
Glutamate	$Glutamate^-/H^+$ symport	Urea synthesis Amino acid degradation
Glutamate/ aspartate	$(Glutamate^- + H^+)/$ $aspartate^-$ exchange	Urea synthesis NADH oxidation
Ornithine	$ornithine^+/H^+$ exchange	Arginine degradation
Ornithine/ citrulline	$ornithine^+/(citrulline + H^+)$ exchange	Urea synthesis

Not all the transport systems shown in Table 1 occur in every tissue. In general, tissues express only those systems required for the metabolic pathways they catalyse. Thus only the pyruvate, phosphate and adenine nucleotide transport systems are ubiquitous. In general, where transport has been studied in mitochondria from tissues other than liver the properties of the transport systems present have been shown to be similar to the liver transport systems. Only transport in liver and kidney will be considered here. Early work on other tissues is covered in the above-mentioned reviews.

MEASUREMENT OF MITOCHONDRIAL TRANSPORT: TECHNIQUES AND LIMITATIONS

Much of the initial work on mitochondrial transport was done using the technique of ammonium swelling. Mitochondria are suspended in an isoosmotic solution of the ammonium salt of the anion under investigation. Since the membrane is permeable to ammonia but impermeable to ammonium ions, net salt entry and hence osmotic swelling occurs only if the anion can enter the mitochondria together with a proton or if its transport is linked to the transport of a second anion together with a proton. The technique of mitochondrial swelling has also been widely used for substrates with no overall net charge such as neutral amino acids. However, it cannot be used to study antiport reactions and the behaviour of mitochondria towards high concentrations of substrates may not reflect their permeability to these substrates at physiological concentrations. Another indirect method involves incubating mitochondria with inhibitors of respiration and observing the rate of intramitochondrial NAD^+ reduction on adding a substrate of an intramitochondrial dehydrogenase. This has been useful where it can been clearly shown that the rate of NAD^+ reduction is limited by transport rather than by enzyme activity.

Direct measurements of transport involve measurement of the uptake of isotopically labelled substrates into mitochondrial suspensions, and in some ways this is more problematical than the methods used for measurement of substrates transport across the cell plasma membrane. A major problem involves prevention of substantial metabolism of the added substrate during the incubation. The intramitochondrial volume is approximately 1 μl/mg protein, and since substrates in general undergo little accumulation, the amount of substrate transported is relatively small. The time course of transport is also rapid and equilibration is often complete within seconds.

The method of separation of mitochondria from the medium is critical. Filtration is generally unsatisfactory since little protein can be used and there is metabolism of, or loss of, the substrate during washing. Centrifugation has been generally used, and to avoid metabolism as far as possible, mitochondria are normally sedimented through a layer of silicone oil into perchloric acid with immediate deproteinisation. Even with this technique, extramitochondrial water is carried down with the mitochondria and may represent up to 80% of the pellet water. This must be corrected for by using an non-penetrant marker of the extramitochondrial space such as sucrose together with accurate dual-channel scintillation counting.

Reliable measurements of initial rates are limited by these considerations. In cases where an efficient inhibitor is available, the "inhibitor stop" method is used. This involves termination of the reaction by adding high concentrations of inhibitor followed by immediate centrifugation. This gives a time resolution of 1-2 seconds. Where such an inhibitor is not available, the transport reaction must be terminated by centrifugation and this has much poorer time resolution of 10-15 seconds.

In order to study mechanisms of transport, mitochondria are incubated with respiratory inhibitors, the substrate is added, and a complete balance of ion movements is measured. Membrane potentials in this system can be generated by adding valinomycin in a K^+-free medium and the proton permeability of the membrane can be varied by adding uncoupling reagents. Antiport reactions are studied by artificially loading the mitochondria with one substrate and observing its efflux on addition of the second substrate. A combination of these approaches are necessary to define the mechanism and kinetic parameters of transport processes.

TRANSPORT OF NEUTRAL AMINO ACIDS ACROSS THE INNER MITOCHONDRIAL MEMBRANE

Mitochondrial transport systems which are specific for amino acids are shown in Table 1. The evidence for the existence of these systems and their properties is considered below.

Transport of Glutamine

Glutamine is one of the most important amino acids which act as a primary substrate in tissue metabolism. The major route of glutamine metabolism is hydrolysis in a reaction catalysed by the enzyme phosphate-dependent glutaminase which is present either in the mitochondrial matrix, or associated with the inner surface of the inner mitochondrial membrane. Glutamine transport across the inner mitochondrial membrane is thus an important process in glutamine metabolism.

In isolated liver mitochondria, glutamine equilibrates across the inner membrane within 10 seconds at 30°C (Joseph and Meijer, 1981). This was within the resolution time of the transport assay, so that kinetic parameters could not be resolved. Glutamine transport was inhibited by mersalyl but not by N-ethylmaleimide, and L-glutamine was transported faster than D-glutamine. These properties, plus the fact that glutamine does not accumulate indicate the presence of an electroneutral glutamine uniporter in the inner mitochondrial membrane. However, Häussinger et al (1985) have performed studies in which liver was perfused with glutamine and rapidly frozen and the distribution of metabolites was measured by fractionation in non-aqueous solvents. Under these conditions there was an apparent 3-4 fold accumulation ratio of glutamine between matrix and cytosol.

In a recent reinvestigation of glutamine transport in liver mitochondria (Soboll et al., 1991), glutamine uptake was studied at 10°C in order to obtain initial rate measurements. It was found that the initial rate of glutamine transport increased as the intramitochondrial pH was increased from 7.6 to 8.7, and the distribution of labelled glutamine across the mitochondrial membrane could be correlated with the pH gradient. It was suggested that glutamine transport is linked to the pH difference across the mitochondrial membrane by an electroneutral mechanism. This study did not allow for possible metabolism of the labelled glutamine at alkaline pH in some experiments and further details of this proposed mechanism will be required.

In kidney, glutamine hydrolysis is greatly increased in metabolic acidosis. There has been much controversy about whether glutamine transport across the mitochondrial membrane is a potential regulatory step in kidney glutamine metabolism; this matter has been reviewed (Kovacevic and McGivan, 1983, Schoolwerth and LaNoue 1985) and will be discussed only briefly here. Glutamine transport in kidney mitochondria is catalysed by a mersalyl-sensitive electroneutral mechanism similar to that in liver. A number of early reports indicated that on addition of ^{14}C-glutamine to mitochondria from control and acidotic rats followed by rapid centrifugation, increased radioactivity was found in the matrix of acidotic mitochondria compared with the controls but all the radioactivity was in the form of glutamate. It was concluded that glutamine transport was rate-limiting for glutamine hydrolysis and that the stimulation of glutamine hydrolysis in acidosis was therefore due to stimulation of glutamine transport.

Later work showed that kidney mitochondrial preparations were contaminated with phosphate-independent glutaminase; when this was remedied, glutamine could be measured in the mitochondrial matrix. In principle, accurate transport kinetics cannot be measured in the absence of a specific and effective inhibitor of the active kidney mitochondrial glutaminase. Goldstein and Boylan (1978) nevertheless attempted this using a rapid mixing and filtration technique. Kovacevic and Bajin (1982) loaded mitochondria with glutamine at 0°C in a phosphate-free medium. Mersalyl was added to inhibit loss of glutamine during washing and glutamine efflux was then measured on removal of the mersalyl with an excess of dithiothreitol. These latter two studies indicated that the rate of transport of glutamine may be an order of magnitude faster than the activity of glutaminase, and that mitochondrial glutamine transport probably does not therefore play a regulatory role in glutamine hydrolysis.

In order to avoid complications due to glutamine metabolism, Sastrasinh and Sastrasinh (1989, 1990) have studied glutamine transport in rat kidney submitochondrial particles. This work confirmed that glutamine transport occurred by a mersalyl-sensitive electroneutral uniport mechanism, and showed that there was little difference in glutamine transport characteristics in submitochondrial particles from control and acidotic rats. While the rate of transport doubled on reducing the external pH from 7.5 to 6.5, this was an effect of pH per se and the rate of glutamine transport was not sensitive to the pH difference across the mitochondrial membrane.

Transport of Other Neutral Amino Acids

In liver, the neutral amino acids alanine, glycine and proline are necessarily transported into mitochondria since they are metabolised in reactions catalysed by matrix enzymes. Leucine and isoleucine must also cross the mitochondrial membrane, since these amino acids are observed to activate intramitochondrial glutamate dehydrogenase under some conditions (McGivan et al., 1973).

As discussed above for glutamine, the measurement of neutral amino acid transport into mitochondria is not straightforward. Extensive investigations have been performed in which mitochondria were suspended in isotonic solutions of amino acids and swelling was monitored. Cybulski and Fisher (1977) using this technique observed that rapid swelling occurred in solutions of L-serine, L-alanine, L-valine, L-threonine, L-methionine and L-leucine, while swelling in the D-isomers of these amino acids was relatively much slower in each case. Swelling in all of these L-amino acids was inhibited by diazobenzenesulphonate and by mersalyl but not by N-ethylmaleimide. Inferences about permeability obtained from swelling in isotonic solutions of substrates do not necessarily reflect the situation at physiological substrate concentrations, but where transport of low concentrations of amino acids has been measured, the same general picture is observed i.e. very rapid equilibration of the substrate, little or no accumulation, no energy dependence, inhibition by mersalyl (King and Diwan, 1973; McGivan et al., 1973). A more detailed study of proline transport (Meyer, 1977) indicated a small, unexplained energy dependence, but again transport was rapid and inhibited by mersalyl. Unfortunately since equilibration is so rapid, it is not possible to do adequate competition experiments to determine whether all the amino acids are transported on a common carrier.

All the present available data can be explained by the existence of a single transport system for amino acids of high activity and broad specificity. This transport system is probably the same as that responsible for glutamine transport and is inhibited by mersalyl but not by N-ethylmaleimide. The transport system does not discriminate between the amino acids whose transport is physiologically necessary (alanine, glycine, proline) and a range of other amino acids. Such a system is possibly analogous to System L in cell membranes which catalyses Na^+-independent neutral amino acid uptake.

TRANSPORT OF GLUTAMATE

Glutamate$^-$/H^+ Symporter

The existence of this transport system was first inferred by Azzi et al. (1967) from studies of the swelling of liver mitochondria in ammonium glutamate. The carrier is inhibited by N-ethylmaleimide, but not mersalyl. The mechanism of transport was identified as electroneutal glutamate/OH^- antiport (or glutamate/H^+ symport) by Meijer et al. (1972). The kinetics of the transport system have been reported (Bradford and McGivan, 1973; Meyer and Vigais, 1974; Hoek and Niogu 1976, 1980). The Vmax of the transport system depends strongly on the pH difference across the mitochondrial membrane but has been reported to be adequate to account for rates of glutamate deamination observed in intact liver. In liver, the carrier can work in either direction. The physiological function of the carrier is to transport glutamate formed by cytosolic transamination reactions into the mitochondrial matrix where it is deaminated via the glutamate dehydrogenase reaction. Under experimental conditions where ammonia rather than amino-acid derived glutamate is the major nitrogen source for urea synthesis or where glutamine is generated intramitochondrially e.g. by proline oxidation, the transporter operates in the reverse direction to limit intramitochondrial accumulation of glutamate (Figure 1).

In kidney mitochondria, the glutamate/H^+ carrier has relatively low activity. Schoolwerth et al. (1984) have shown that glutamate efflux via this transport system is strongly inhibited by decreasing external pH and have suggested that this may contribute to increased matrix glutamate concentrations and increased flux through glutamate dehydrogenase in acidosis.

The Electrogenic Glutamate/Aspartate Antiporter

Azzi et al. (1967) first showed that aspartate will enter respiration-inhibited mitochondria only in the presence of glutamate. Evidence for, and elucidation of the

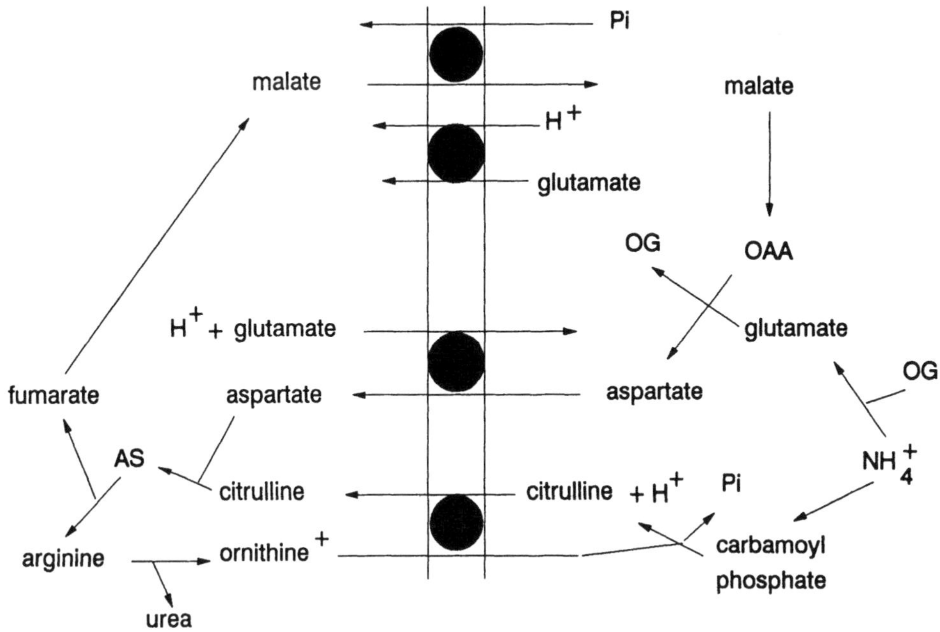

Figure 1. Urea Synthesis from Ammonia (see Meijer et al., 1975).

mechanism of, the glutamate/asparate antiporter has been reviewed in detail elsewhere (LaNoue and Schoolwerth, 1979). The key experiments were performed by LaNoue and Tischler (1974). Using respiration-inhibited mitochondria loaded with aspartate they measured aspartate efflux on addition of glutamate. When a membrane potential was generated by the addition of valinomycin one glutamate molecule plus one H^+ was observed to enter the mitochondria for each molecule of aspartate that left the matrix. These and other considerations established the mechanism of the carrier as an electrogenic (glutamate$^-$ + H^+)/aspartate$^-$ antiporter; the exchange involves the net outward movement of negative charge and is hence driven by the membrane potential. These results explained the earlier findings that aspartate production and efflux from mitochondria oxidising glutamate is energy-dependent. The kinetic properties of this transport system have been extensively studied (LaNoue et al., 1979, Williamson et al., 1979).

The glutamate/aspartate transport system is primarily in oxidation of cytosolic NADH by mitochondria (see LaNoue and Schoolwerth, 1979). Extramitochondrial NADH does not penetrate the inner mitochondrial membrane, and is oxidised via the malate/asparate shuttle. The imbalance of charge in the glutamate/aspartate exchange system implies that the membrane potential pushes the exchange in the direction of aspartate efflux and this accounts in part for the high NADH/NAD^+ ratio observed in the matrix compared with the cytosol. The glutamate/aspartate transport system is also involved in urea synthesis (see Meijer et al. 1975) where intramitochondrial glutamate oxaloacetate transaminase provides aspartate for the cytosolic argininosuccinate synthetase reaction (Fig. 1); here the mitochondrial membrane potential effectively displaces the equilibrium of the intramitochondrial glutamate oxaloacetate transaminase reaction in the direction of aspartate formation and efflux.

TRANSPORT OF ORNITHINE

Ornithine is formed in the liver cytosol by the hydrolysis of arginine. In the urea cycle ornithine reacts with carbamoyl phosphate to form citrulline in a reaction catalysed by the mitochondrial matrix enzyme ornithine transcarbamylase. The transport of ornithine from cytosol to matrix and of citrulline from matrix to cytosol is thus an integral part of the urea cycle (Fig. 1). Ornithine is also a substrate for the intramitochondrial enzyme ornithine aminotransferase which catalyses the formation of glutamate plus glutamate semialdehyde from ornithine and oxoglutarate. Glutamate semialdehyde is subsequently oxidised to glutamate in the matrix. This is the major pathway for degradation of excess arginine in liver.

At physiological pH values, ornithine has a net positive charge. The first systematic study of mitochondrial ornithine and citrulline transport (Gamble and Lehninger, 1973) reported that mitochondria swell in solutions of isotonic ornithine phosphate but only if succinate was present as an energy source. Swelling occurred spontaneously in isotonic solutions of citrulline. From these results and from data on isotope uptake it was concluded that transport of the ornithine$^+$ cation was electrogenic and was influenced by the membrane potential. This study did not however take adequate account of the considerable binding of the ornithine cation to the mitochondrial membrane nor of the extensive metabolism of added radioactive ornithine via the ornithine aminotransferase reaction with endogenous 2-oxoglutarate. Further, it was found that the permeability of mitochondria to ornithine and citrulline at isoosmotic concentrations was not the same as that at more physiological concentrations. Subsequent studies (Bryla et al.,1976, McGivan et al., 1977) which took these factors into consideration showed that ornithine transport is in fact electroneutral and occurs via a specific ornithine$^+$/H$^+$ antiport system. This was confirmed in later work (Hommes et al., 1983) which showed that lysine is also transported on the same system.

The question of citrulline transport has proved more difficult. A simple uniport system does not seem to exist, since although mitochondria swell rapidly in isotonic citrulline (Gamble and Lehninger, 1973), labelled citrulline at physiological concentrations penetrates the mitochondrial membrane very slowly. However, mitochondria can be loaded with citrulline by incubation under hypotonic conditions where the permeability of the membrane is artificially increased. The citrulline is retained when the mitochondria are returned to isotonic medium (Bradford and McGivan, 1980). Addition of extramitochondrial ornithine then caused specific displacement of intramitochondrial citrulline, and lysine had a similar effect. The uptake of ornithine was higher in citrulline-loaded mitochondria than in control mitochondria. On the basis of these and other data it was suggested that ornithine enters the matrix via a second transport system in exchange for citrulline (Bradford and McGivan, 1980). The fact that energisation of the membrane was without effect on ornithine/citrulline exchange suggested that this process was electroneutral and that a H$^+$ ion was transported together with citrulline in exchange for the ornithine cation. Hommes et al. (1983) were unable to show increased ornithine uptake when mitochondria were preincubated with citrulline. However, they provided no evidence that any of the citrulline in their experiments was in fact intramitochondrial.

It has been argued that the postulate of a second ornithine transport system is unnecessary since the matrix enzymes ornithine aminotransferase and ornithine transcarbamylase cannot distinguish between ornithine transported on the two different carriers (Hommes et al., 1983, Inoue et al, 1988). However, the situation is closely analogous to that of the glutamate/Hsymporter and the glutamate/asparate antiporter discussed previously. Ornithine can enter the mitochondria in exchange for citrulline if this is present in the mitochondria. Citrulline can only leave the mitochondria via the ornithine/citrulline exchanger in exchange for an incoming ornithine molecule. Ornithine entry in excess of that required for citrulline formation can only occur via the ornithine/H$^+$ antiport system. The postulate of two ornithine transport systems has the following merits:- (i) it accounts for the fact that ornithine entry in the absence of urea synthesis (i.e. via the ornithine/H$^+$ antiporter) is slower than that necessary to account for urea synthesis when this is occurring (McGivan et al., 1977, LaNoue and Schoolwerth, 1979); (ii) it explains the low unidirectional permeability of the mitochondrial membrane to citrulline (iii) it provides a pathway (ornithine$^+$/(citrulline + H$^+$) antiport) by which the H$^+$ generated in the ornithine transcarbamylase reaction can leave the matrix during urea synthesis. However until specific inhibitors are found to discriminate between these systems, or until they are identified at the molecular level, the evidence for these mechanisms will remain of an indirect nature.

Recent interest in mitochondrial ornithine transport has centred on the hyperornithinaemia, hyperammonemia and homocitrullinuria syndrome (HHH disorder). This has been proposed to be due to a defective mitochondrial ornithine transport system. Inoue et al. (1988) have studied various parameters in mitochondria isolated from liver biopsy samples of patients suffering from this disorder. In these samples, the activities of carbamoyl phosphate synthase and ornithine transcarbamylase the intramitochondrial content of N-acetylglutamate and the integrity of the isolated mitochondria were normal, but the rate of mitochondrial citrulline synthesis from ammonia was impaired. A decreased uptake of ornithine when measured in the absence of citrulline synthesis was also observed. These results are consistent with a decrease in the activity of both ornithine transport systems. The identification and isolation of mitochondrial ornithine transport systems will be of great interest in relation to this inherited disorder.

Recently, Passarrella (1990) has produced evidence that ornithine transport in kidney mitochondria occurs via an ornithine/phosphate antiport system which does not occur in liver. The mechanism and physiological significance of such a system in kidney are obscure at present.

IS MITOCHONDRIAL AMINO ACID TRANSPORT A REGULATORY STEP IN AMINO ACID METABOLISM?

The question of whether mitochondrial amino acid transport can exert a regulatory role in nitrogen metabolism has frequently been discussed. In general, the kinetic parameters measured for transport indicate that the transport systems have excess capacity over that required to account for metabolism. Certainly the rate of transport of glutamine and other neutral amino acids across the mitochondrial membrane in liver is extremely rapid. One exception may be the ornithine/H^+ antiporter, which has relatively low activity (McGivan et al., 1977) and may limit the effective activity of ornithine aminotransferase when this enzyme is highly induced in liver. However, it is difficult to test this hypothesis in the intact liver cell when ornithine is also participating in the urea cycle.

The role of mitochondrial transport in metabolic regulation could be easily investigated if specific inhibitors of these processes were available, but no such inhibitors have been found. Meijer (1989) has shown that during synthesis of urea from increasing concentrations of a mixture of amino acids in isolated hepatocytes the level of intramitochondrial carbamoyl phosphate remained very low. This, together with the finding that carbamoyl phosphate does not exert feedback inhibition on carbamoyl phosphate synthase, led to the conclusion that ureagenesis under physiological conditions cannot be controlled at any step after the carbamoyl phosphate synthase reaction. This eliminates the possibility of regulation at the level of ornithine/citrulline exchange or glutamate/aspartate exchange under these conditions.

The question of the possible involvement of kidney mitochondrial glutamine transport in the regulation of glutamine metabolism in acute acidosis has been considered earlier. It appears that when measured under appropriate conditions, glutamine transport into kidney mitochondria is rapid relative to glutamine deamidation. The increase in glutamine metabolism in acute acidosis is now thought to be due to activation of mitochondrial 2-oxoglutarate dehydrogenase activity as an indirect result of low cytosolic pH (Lowry and Ross, 1980). A reduced level of intramitochondrial 2-oxoglutarate perturbs the glutamate dehydrogenase equilibrium and allows increased flux through glutamate dehydrogenase and hence through glutaminase; no primary change in the activity of the glutamine transport system is involved.

In general therefore, there is now no compelling evidence to suggest that the mitochondrial amino acid transport systems play any more than a permissive role in cellular metabolism.

MOLECULAR ASPECTS OF MITOCHONDRIAL TRANSPORT SYSTEMS

Over the past few years, considerable progress has been made in the identification and isolation of a number of the mitochondrial transport systems. The adenine nucleotide carrier was purified by solubilisation of mitochondria with Triton X-100 followed by chromatography on hydroxyapatite (Klingenberg et al., 1978). It was subsequently found that the protein fraction passing through hydroxyapatite contained not only the adenine nucleotide carrier but also a number of other membrane transport proteins. A number of groups using extensions of this procedure have reported the complete or partial purifications of the phosphate carrier, 2-oxoglutarate carrier, dicarboxylate carrier, citrate carrier monocarboxylate carrier and carnitine carriers (see Kramer and Palmieri, 1989 for a review). The identities of these proteins have been confirmed by functional reconstitution into artificial phospholipid membranes; all these carriers have subunit Mr values in the range 27-35 kDa. The primary amino acid sequences of the adenine nucleotide carrier (Aquila et al., 1982), phosphate carrier (Runswick et al., 1987), brown adipose tissue uncoupling protein (Aquila et al., 1985) and recently the 2-oxoglutarate carrier (Runswick et al., 1990) have been determined; these four proteins have closely related structural features and may have evolved from a common ancestor. Sequence data for the other transport systems which have been identified should soon be forthcoming.

The only amino acid transport system that has been studied at the molecular level is the glutamate/aspartate antiporter. This carrier was isolated as a 68 kDa protein from solubilised beef heart mitochondria after hydroxyapatite chromatography and was reconstituted into phospholipid membranes with a 570-fold enhancement of activity (Kramer et al., 1986). It was shown that high concentrations of pyridoxal phosphate could be used as an inhibitor in this system and, therefore, it was possible to measure transport kinetics accurately. The reconstituted carrier showed different Km values for glutamate and aspartate when measured from the outside or from the inside, and this confirmed that the carrier was itself asymmetric and was also inserted asymmetrically into the membrane (Dierks and Kramer, 1988). This system also allowed characterisation of the mechanism of the exchange process. Recently, it has been shown that when the reconstituted glutamate/aspartate carrier is modified with certain mercurials, it ceases to catalyse exchange and instead catalyses net efflux of anions via a pore-like mechanism; this effect is reversible on removing the mercurial. It is suggested that the carrier contains an intrinsic channel which is essential for amino acid transport during the antiport function of the carrier, but that the inward and outward amino acid movements are uncoupled in the presence of mercurials (Dierks, T. et al., 1990).

With regard to other amino acid transport systems, there is a single report of the reconstitution of ornithine transport activity from solubilised, unfractionated mitochondria (Hommes et al., 1984). There is no information on the identification of the glutamate/H^+ carrier, the ornithine/citrulline exchanger or the neutral amino acid transport system. In view of the fact that it appears that these carriers are not readily isolated by hydroxyapatite chromatography and that no specific inhibitors are available, the identification and isolation of the carriers is likely to be more difficult.

In general the study of mitochondrial transport systems at the molecular level is at present more advanced than that of the plasma membrane amino acid transport systems. It is, however, likely that amino acid sequences and structural information for both mitochondrial and plasma membrane amino acid transport systems will become available within the next few years. It will be of great interest to investigate the structural relationships of mitochondrial amino acid transport systems with those of other mitochondrial transport systems on the one hand and with plasma membrane systems for amino acid transport on the other.

REFERENCES

Aquila, H., Misra, D., Eulitz, M. and Klingenberg, M., 1982, Hoppe Seyler's Z. Physiol. Chem., 363: 345-349.

Aquila, H., Link, T. A. and Klingenberg, M., 1985, EMBO J. 4: 2369-2376.
Azzi, A., Chappell, J. B. and Robinson, B. H., 1967, Biochem Biophys. Res. Commun., 29: 148-152.
Bradford, N. M. and McGivan, J. D., 1973, Biochem. J., 134: 1023-1029.
Bradford, N. M. and McGivan, J. D., 1980, FEBS Lett., 113: 294-298.
Bryla, J. and Harris, E. J., 1976, FEBS Lett, 72: 331-336.
Chappell, J. B., 1968, Brit. Med. Bull., 24: 150-157.
Cybulski, R. L. and Fisher, R. R., 1977, Biochemistry, 16: 5116-5120.
Dierks, R. and Kramer, R., 1988, Biochim. Biophys. Acta, 937: 112-126.
Dierks, T., Salentin, A. and Kramer, R., 1990, Biochim. Biophys. Acta, 1028: 281-288.
Gamble, J. G. and Lehninger, A. L., 1973, J. Biol. Chem., 248: 610-618.
Goldstein, L. and Boylan, J. M., 1978, Am. J. Physiol., 234: F514-521.
Häussinger, D., Soboll, S., Meijer, A. J., Gerok, W., Tager, J. M. and Sies, H., 1985, Eur. J. Biochem., 152: 597-603.
Hoek, J. B. and Niogu, 1976, FEBS Lett, 71: 341-346.
Hoek, J. B. and Niogu, R. M., 1980, J. Biol. Chem., 255: 8711-8718.
Hommes, F. A., Kitchings, L. and Eller, A. G., 1983, Biochem. Med., 30: 313-321.
Hommes, F. A., Eller, A. G., Evans, B. A. and Carter, A. L., 1984, FEBS Lett., 170: 131-134.
Inoue, I., Saheki, T., Kayanuma, K., Uono, M., Nakajima, M., Takeshita, K., Koike, R., Yuasa, T., Miyatake, T. and Sakoda, K., 1988, Biochim. Biophys. Acta, 964: 90-95.
Joseph, S. K. and Meijer, A. J., 1981, Eur. J. Biochem., 119: 523-529.
King, M. J. and Diwan, J. J., 1973, Arch. Biophys. Biochem., 159: 166-173.
Klingenberg, M., Riccio, P. and Aquila, H., 1978, Biochim. Biophys. Acta, 503: 193-210.
Kovacevic, Z. and Bajin, K., 1982, Biochim. Biophys. Acta, 687: 291-295.
Kovacevic, Z. and McGivan, J. D., 1983, Physiological Reviews, 63: 547-605.
Kramer, R. and Palmieri, F., 1989, Biochim. Biophys. Acta, 974: 1-23.
Kramer, R., Kurzinger, G. and Herberger, C., 1986, Arch. Biophys. Biochem., 251: 166-174.
LaNoue, K. F. and Schoolwerth, A. C., 1979, Ann. Rev. Biochem., 48: 871-922.
LaNoue, K. F. and Tischler, M. E., 1974, J. Biol. Chem., 249: 7522-7528.
LaNoue, K. F., Duszynski, J. and Watts, J. A., 1979, Arch. Biochem. Biophys., 195: 578-590.
Lowry, M. and Ross, B. D., 1980, Biochem. J., 190: 771-780.
McGivan, J. D., Bradford, M. N., Crompton, M. and Chappell, J. B., 1973, Biochem. J., 134: 209-215.
McGivan, J. D., Bradford, N. M. and Beavis, A. D., 1977, Biochem. J., 162: 147-156.
Meijer, A. J., 1989, in: Anion Carriers of Mitochondrial Membranes (Azzi, A., Nalecz, K. A., Nalecz, M. J. and Wojczak, L. E., eds) pp. 307-316 Springer-Verlag, Berlin.
Meijer, A. J. and Hensgens, H. E. S. J., 1982, In: Metabolic Compartmentation, (Sies, H., ed.) pp. 259-286 Academic Press, New York.
Meijer, A. J. and van Dam, K., 1974, Biochim. Biophys. Acta, 346: 213-244.
Meijer, A. J., Brouwer, A., Reijngoud, D. J., Hoek, J. B. and Tager, J. M., 1972, Biochim. Biophys. Acta, 283: 421-429.
Meijer, A. J., Gimpel, J. J. A., Deleeuw, G. A., Tager, J. M. and Williamson, J. R., 1975, J. Biol. Chem., 250: 7728-7738.
Meijer, A. J., Lamers, W. H. and Chamuleau, R. A. F. M., 1990, Biochim. Biophys. Acta, 70: 701-747.
Meyer, J., 1977, Arch. Biophys. Biochem., 178: 387-395.
Meyer, J. and Vignais, P. M., 1973, Biochim. Biophys. Acta, 325: 375-384.
Mitchell, P., 1966, Chemiosmotic coupling in Oxidative and Photosynthetic Phosphorylation. Glynn Research Ltd., Bodmin, Cornwall, England.
Passarrella, S., 1990, Eur. J. Biochem., 193: 221-227.
Pfaff, E., Klingenberg, M. and Heldt, H. W., 1965, Biochim. Biophys. Acta, 314-315.
Runswick, M. J., Powell, S. J., Nyren, P. and Walker, J. E., 1987, EMBO J., 6: 1367-1373.
Runswick, M. J., Walker, J. E., Bisaccia, F., Iacobazzi, V. and Palmieri, F., 1990, Biochemistry, 29: 11033-11040.
Sastrasinh, M. and Sastrasinh, S., 1990, Am. J. Physiol., 259: F863-866.
Sastrasinh, S. and Sastrasinh, M., 1989, Am. J. Physiol., 257: F1050-1058.
Schoolwerth, A. C., LaNoue, K. F. and Hoover, W. J., 1984, Am. J. Physiol., 246: F266-271.
Schoolwerth, A. C. and LaNoue, K. F., 1985, Ann. Rev. Physiol. 47: 143-171.
Soboll, S., Lenzen, C., Rettich, D., Grundel, S. and Ziegler, B., 1991, Eur. J. Biochem., 197: 113-117.

Williamson, J. R., Murphy, E., Viale, O. R. and Cole, K. E., 1979, in: Function and Molecular Aspects of Biomembrane Transport (Quagliariello, E., Palmieri, F., Papa, S., Klingenberg, M., eds) pp.355-364 Elsevier, Amsterdam.

AMINO ACID TRANSPORT, CELL VOLUME

AND REGULATION OF CELL GROWTH

Dieter Häussinger[1], Florian Lang[2] and Michael S. Kilberg[3]

[1]Medizinische Universitätsklinik und Poliklinik
Freiburg, Germany
[2]Institut für Physiologie
Universität Innsbruck, Austria
[3]Department of Biochemistry and Molecular Biology
University of Florida, Gainesville, USA

INTRODUCTION

There is a close relationship between cell volume homeostasis and amino acid transport. On the one hand the cumulative, mostly Na^+-driven amino acid uptake into the cells creates an intra/extracellular osmotic gradient, which shifts free water into the cells. This leads to cell swelling and volume regulatory mechanisms are required to defend the constancy of cell volume. On the other hand, some cells accumulate or release amino acids in response to cell volume challenges in order to balance the intracellular osmolarity. Thus, amino acid transport is involved in both, cell volume perturbation and regulation. During the past three years, it became increasingly clear that alterations of cell volume markedly influence a variety of metabolic pathways and modify cell growth and proliferation. Specifically in the liver, metabolic functions, such as protein and carbohydrate metabolism, which result in formation or disappearance of osmotically active small molecules, are extremely sensitive to alterations of cell volume. Furthermore, indirectly or directly, cell volume changes modify a number of metabolic functions with no obvious impact on intracellular osmolarity. It appears that cell swelling and cell shrinkage lead to certain opposite patterns of cellular metabolic function. Apparently, amino acids and also hormones can trigger those patterns simply by altering cell volume. In view of this, the function of amino acid transport systems in the plasma membrane can no longer be identified with amino acid translocation alone; the role of these transporters may rather be seen to constitute an integral part of a transmembrane signalling system, which triggers cellular function by altering cell volume (Häussinger et al. 1991, Häussinger and Lang 1991a,b).

Excellent reviews have appeared in the past few years on the various aspects of cell volume regulation (Ballanyi and Grafe 1988, Beck et al. 1988, Chamberlin and Strange 1989, Gilles 1988, Graf et al. 1988, Grinstein and Foskett 1990, Hoffmann et al. 1988, Lang et al. 1990, Macknight 1988, Reuss 1988, Völkl et al. 1988, Balaban and Burg 1987). These topics will be delt with only briefly here and emphasis will be layed upon the interplay between amino acid transport, cell volume and regulation of cell growth.

AMINO ACID TRANSPORT AND CELL VOLUME

Effect of Amino Acid Transport on Cell Volume

One of the most important challenges for cell volume homeostasis is the cumulative uptake of osmotically active substances, such as amino acids (Kristensen 1980,

Mammalian Amino Acid Transport, Edited by M.S. Kilberg and
D. Häussinger, Plenum Press, New York, 1992

Kristensen and Folke 1984, 1986, Bakker-Grunwald 1983, Wettstein et al. 1990, Häussinger et al. 1990a,b, Baquet et al. 1990, Häussinger and Lang 1990,1991a, Hallbrucker et al. 1991a). In liver, the Na^+-dependent transporters build up intra/extracellular amino acid concentration gradients of up to 20 (Kristensen 1980, Kristensen and Folke 1986, Häussinger et al. 1990a). Na^+ entering the hepatocyte together with the amino acid is extruded in exchange for K^+ by the electrogenic Na^+/K^+ ATPase. The accumulation of amino acids and K^+ into the cells leads to hepatocyte swelling, which in turn triggers volume regulatory K^+ efflux (Kristensen 1980, Kristensen and Folke 1984, 1986, Bakker-Grunwald 1983, Häussinger et al. 1990a, Wettstein et al. 1990). This is illustrated in **Fig. 1.** Following addition of glutamine to isolated perfused rat liver within about 12 min an intra/extracellular glutamine concentration gradient of about 12 is realized. During the first 2 min of glutamine accumulation liver mass increases rapidly, which quantitatively reflects under these conditions cell swelling, as validated by measurements of intracellular water space (vom

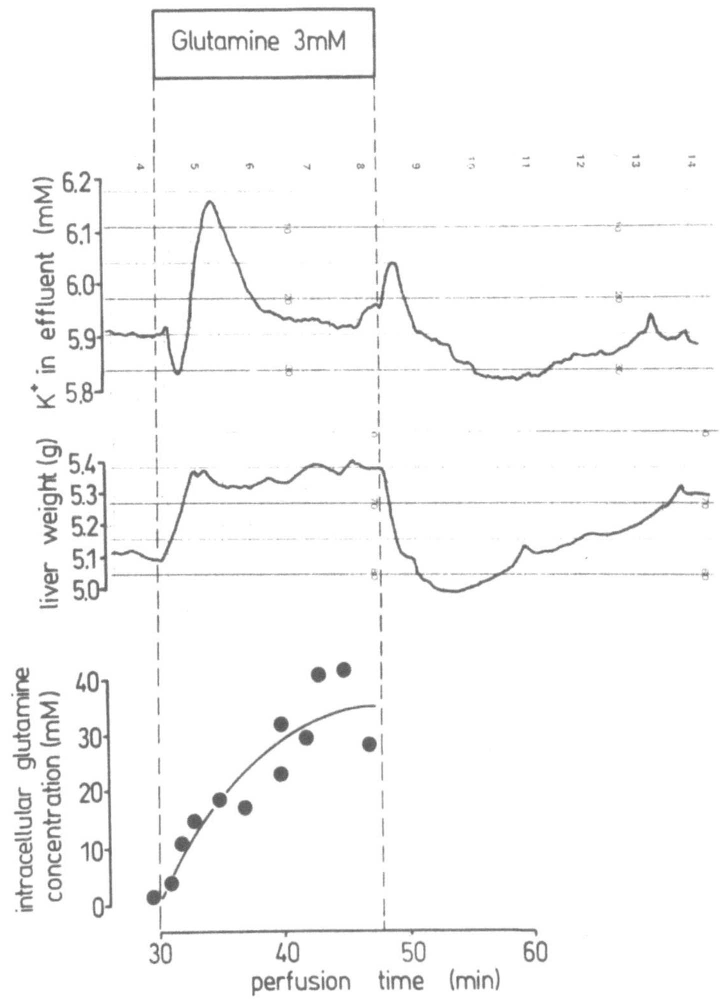

Figure 1. Effect of glutamine (3 mM) addition to influent perfusate of isolated, single-pass perfused rat liver on intracellular glutamine accumulation, liver mass and volume-regulatory K^+ fluxes. Liver mass changes following glutamine addition are exclusively due to alterations of intracellular water space as shown in (vom Dahl et al. 1991c). From (Häussinger et al. 1990).

Dahl et al. 1991c). The simultaneous net K^+ uptake during this phase is explained by extrusion of cotransported Na^+ by Na^+/K^+ ATPase. Thereafter, no further increase of liver cell volume, as reflected by liver mass, is observed, despite continuing accumulation of glutamine inside the cell. This is achieved by a volume-regulatory K^+ efflux from the liver which occurs until the steady state intracellular glutamine concentration of about 35 mM is reached (Fig. 1). During the rapid washout of glutamine following its withdrawal from the influent perfusate, the hepatocytes shrink rapidly as evident from the respective decrease of liver mass. The cell shrinkage leads to a volume-regulatory net K^+ uptake (Fig. 1). Cell swelling and volume-regulatory K^+ efflux were shown to occur upon addition of glutamine, alanine, proline, serine, glycine, 2-aminoisobutyrate, phenylalanine, hydroxyproline and a complete amino acid mixture in physiological concentrations (Bakker-Grunwald 1983, Wettstein et al. 1990, Baquet et al. 1990, Häussinger et al. 1990a,b, Hallbrucker et al. 1991a), but not upon addition of glucose or leucine (Wettstein et al. 1990); i.e., compounds which are not concentrated by liver cells (Baur and Heldt 1976, Kilberg et al. 1985). No cell swelling and volume-regulatory K^+ responses are observed upon addition of glutamate (Wettstein 1990), although this amino acid is taken up by liver cells via Na^+-independent and Na^+-dependent mechanisms. Glutamate-induced volume changes can probably not be detected, because the uptake of this amino acid is restricted to only a very small perivenous hepatocyte population, which comprises less than 10% of all hepatocytes (Häussinger et al. 1989, Stoll et al. 1991).

The mechanisms underlying volume-regulatory K^+ efflux and reuptake, as they occur during amino acid-induced cell swelling and amino acid deprivation-induced cell shrinkage, respectively, are probably identical to those observed during volume regulatory decrease and increase following anisoosmotic cell exposure. In liver, the volume-regulatory decrease following hypoosmotic cell swelling is largely achieved by activation of Ba^{++}- and quinidine-sensitive K^+ channels in parallel to SITS-sensitive anion channels (Graf et al. 1988, Haddad et al. 1989, Haddad and Graf 1989, Lang et al. 1989, Häussinger et al. 1990d). The activation of the K^+ channels leads to the respective hyperpolarization of the cell membrane (Graf et al. 1988, Howard and Wondergem 1987). Hepatocyte swelling apparently opens stretch-activated unspecific ion channels, which allow passage of Ca^{++} into the cell (Baer 1990). The increase of intracellular Ca^{++} may then activate Ca^{++}- sensitive K^+ channels (Baer and Petersen 1987). However, the activation of K^+ channels by increasing intracellular Ca^{++} may not be the only mechanism allowing regulatory cell volume decrease in hepatocytes, because K^+ channels also have been described, which do not require increase of intracellular calcium activity (Hendersen et al. 1989). In proximal renal tubules, entry of Ca^{++} during cell swelling was inhibited by pretreatment with pertussis toxin, suggesting involvement of a G-protein in the activation mechanisms (Suzuki et al. 1990), but in these cells K^+ channels also can be directly activated by cell membrane stretch (Sackin 1987). Theoretically, activation of the K^+ channels could suffice to stimulate cellular loss of potassium and anions, since the hyperpolarization increases the driving force for anion extrusion. It must be kept in mind that activation of K^+ channels may be paralleled or even surpassed by the activation of anion channels and if sufficient K^+ channels are operative even at normal intracellular calcium activity, regulatory cell volume decrease is not really dependent on the activation of K^+ channels by Ca^{++}.

On the other hand, net K^+ uptake by the liver during volume regulatory increase, as it occurs following hyperosmotic exposure of liver cells, is sensitive to inhibition by amiloride and ouabain, but not to loop diuretics (Haddad and Graf 1989, Lang et al. 1989, Häussinger et al. 1990d). Accordingly, regulatory volume increase in liver is apparently accomplished by activation of Na^+/H^+ exchange with subsequent extrusion of sodium in exchange with potassium.

NaCl-KCl-cotransport appears not to appreciably participate in hepatic regulatory volume increase (Häussinger et al. 1991d), even though the carrier exists in the hepatic cell membrane and its activation by insulin is followed by increase of cell volume (Häussinger et al. 1991, Häussinger and Lang 1991b, Hallbrucker et al. 1991b,c). Cell differentiation and cell growth under the influence of growth factors and insulin are coupled to activation of amiloride-sensitive Na^+/H^+ exchange and bumetanide-sensitive NaCl-KCl cotransport, which is accompanied by cell swelling (Häussinger et al. 1991a, vom Dahl et al. 1991a, Hallbrucker et al. 1991b,c). In liver, both, insulin-induced cell swelling and intracellular K^+ accumulation are inhibited by about 60% in presence of bumetanide and insulin-induced cellular K^+ accumulation is completely blocked in presence of bumetanide plus amiloride (Hallbrucker et al. 1991b,c, Häussinger et al. 1991a). Similarly, in mouse 3T3 cells amiloride added together with bumetanide completely blocked the growth factor-induced exit from G_0/G_1-phase and entry into S-phase (Panet et al. 1986a,b). This may suggest a role of cell volume changes in mediating the action of growth factors.

It recently became clear that amino acid-induced cell swelling and volume-regulatory responses occur upon exposure to amino acids in the physiological concentration range (Wettstein et al. 1990, Häussinger et al. 1990a,b, Hallbrucker et al. 1991a,b Bode and Kilberg, 1991). Glutamine-induced cell swelling is half-maximal at a concentration of 0.6-0.8 mM, i.e. the concentration normally found in portal venous blood, and is maximal at 2 mM (Wettstein et al. 1990). Accordingly, physiological fluctuations of the portal amino acid concentration are accompanied by parallel alterations of liver cell volume (Wettstein et al. 1990, Hallbrucker et al. 1991a). The degree of amino acid-induced cell swelling seems largely to be related to the steady state intra/extracellular amino acid concentration gradient. Indeed, Ehrlich ascites tumor cells accumulate 2,4-diaminobutyrate to levels that induce cell swelling sufficient to cause cell lysis (Christensen et al. 1952). Furthermore, the amino acid concentration gradient and accordingly the degree of cell swelling is modified by hormones and the nutritional state in a complex way. Their action not only involves effects on the expression of plasma membrane transport systems and the membrane potential as driving force for Na^+ coupled transport, but also alterations of intracellular amino acid metabolism. For example, insulin, which causes itself liver cell swelling (Häussinger et al. 1991a) due to intracellular K^+ accumulation by activating Na^+/K^+ ATPase, Na^+/H^+ exchange and NaCl/KCl cotransport (Fehlmann and Freychet 1981, Jakubowski and Jakob 1990, Hallbrucker et al. 1991b,c) will also lead to a cycloheximide-sensitive stimulation of Na^+-dependent amino acid transport System A (LeCam and Freychet 1978, Fehlmann et al. 1981, Prentki et al. 1981), which tends to swell the cells further. In line with this, glycine-induced liver cell swelling was found to be enhanced in presence of insulin (Hallbrucker et al. 1991b). Glucagon, like insulin, increases Na^+/K^+ ATPase activity (for reviews see Moule and McGivan 1990), stimulates amino acid uptake via System A (Prentki et al. 1981 Kilberg et al. 1985). Whereas this tends to swell the cell, glucagon in fact shrinks the cells (Häussinger et al. 1991, Hallbrucker et al. 1991b,c, vom Dahl et al. 1991b,c, Häussinger and Lang 1991a,b) due to a glucagon-induced net K^+ release and dissipation of the amino acid concentration gradient across the plasma membrane due to a glucagon-induced stimulation of amino acid breakdown (Joseph and McGivan 1978). Starvation increases the swelling potency of glycine 3 to 4-fold (Hallbrucker et al. 1991a) probably due to induction of the glycine-transporting System A (Kilberg et al. 1985). The osmotic gradient built up during amino acid accumulation, and therefore the degree of cell swelling, is sometimes dissipated by draining amino acids into metabolic pathways, which yield products that equally distribute on both sides of the plasma membrane. Because the rate of alanine transport into hepatocytes controls its metabolism (Sips et al. 1980, Groen et al. 1982), inhibition of alanine metabolism by transaminase inhibitors markedly enhances the alanine-induced volume-regulatory K^+ efflux (Wettstein et al. 1990). Hormones stimulating amino acid metabolism can indirectly affect cell volume by either dissipating the steady state amino acid concentration gradient across the plasma membrane or resulting in the intracellular accumulation of a product derived from the added amino acid. Stimulation of glutaminase following addition of glutamine will, e.g., result in marked intracellular accumulation of (almost impermeable) glutamate as soon as urea synthesis is blocked. As expected, leucine, which is transported via Na^+-independent System L, does itself not lead to cell swelling (Wettstein et al. 1990, Baquet et al. 1990), however, may potentiate cell swelling induced by other amino acids (Baquet et al. 1990). This effect of leucine is due to an inhibition of urea cycle enzymes (for review see Meijer and Hensgens 1982) or to an impaired exit of System A substrates via System L.

Amino acid-induced cell swelling not only occurs in liver but also during concentrative transport in intestine (Schultz, 1981) and the renal proximal tubule (Lang et al. 1986). It is likely that also other Na^+ coupled transport systems, such as those for bile acids and fatty acids (for reviews see (Petzinger et al. 1989)) can contribute to cell swelling in liver under certain conditions.

Effect of Cell Volume Changes on Amino Acid Transport

Whereas the concentrative uptake of amino acids into cells represents a challenge for cell volume homeostasis, and activates volume-regulatory mechanisms, there is some evidence that cell volume changes in turn affect amino acid transport. Whereas in some cell types such alterations of amino acid transport following cell volume changes cannot directly be linked to cell volume homeostasis, in others such alterations are primarily used to balance intra/extracellular osmolarity gradients and therefore to maintain cell volume homeostasis. Here, amino acids serve as intracellular osmolytes, i.e. compounds specifically accumulated or released by the cells, when they undergo anisoosmotic cell volume changes. This "osmolyte strategy" of cell volume regulation is widely distributed among living cells and the general classes of organic osmolytes apart from amino acids also include sugars, polyols, urea and methylamines (for review see (Chamberlain and Strange 1989)). For example, taurine participates in volume regulation of fish red cells and Ehrlich ascites tumor cells (Hoffmann and Lambert 1983). Glutamate, proline, glycine, serine, alanine, β-alanine, proline and the cyclic amino acid ectoine are involved in osmoregulation in bacteria, marine invertebrates and plants (for reviews see (Gilles 1987, Chamberlain and Strange 1989)). An increased intracellular concentration of these amino acids in response to hyperosmotic stress is, depending on the cell type under study, not only achieved by increased intracellular synthesis and protein degradation, but probably also by increased uptake from extracellular fluid. Conversely, amino acids are exported from the cells in response to hypoosmotic environments. In fish red blood cells, during volume regulatory decrease in response to osmotic cell swelling, the NaCl-dependent taurine uptake pathway is inhibited, whereas NaCl-independent taurine efflux is increased (Fincham et al. 1987), resulting in a volume-regulatory net taurine efflux.

Apart from a role of taurine in volume regulation of Ehrlich ascites tumor cells (Hoffmann and Lambert 1983) there are only a few data concerning a role of amino acid transport for cell volume regulation in mammals. In cultured human fibroblasts, the cycloheximide-sensitive adaptive increase of amino acid transport System A activity following amino acid deprivation is potentiated during osmotic cell shrinkage and counteracted during osmotic cell swelling (Nucci et al. 1990, Gazzola et al. 1991). There is good evidence that such an effect of cell volume on the expression of amino acid transport activity contributes to the overall cell volume constancy (Gazzola et al. 1991).

Amino acid transport System N (for reviews see Bode et al. 1990, Kilberg et al. 1985) was shown to be activated by amino acids in a cycloheximide-insensitive way (Weissbach and Kilberg 1984). This stimulation was not explained by trans-stimulation and non-substrates of System N such as alanine, serine, and the non-metabolizable analogue 2-aminoisobutyrate, substrates of the Na^+-dependent System A, were effective. Substrates of Na^+-independent System L, such as leucine were ineffective (Weissbach and Kilberg 1984, Bode and Kilberg 1991). This amino acid-dependent short-term stimulation of amino acid transport is dependent on cell swelling (Bode and Kilberg 1991). The amino acid-dependent induction was blocked by preventing the resulting cell swelling and the activation of the System N transporter could be initiated by incubation of the cells in hypoosmotic (Bode and Kilberg 1991). The latter effect also is observed for intact liver (Haussinger et al. 1990a). Such a stimulation of amino acid transport following cell swelling cannot primarily be viewed as a volume-homeostatic response, because it tends to aggravate the underlying amino acid-induced cell volume perturbation. Hypoosmotic cell swelling increases hepatic alanine uptake in starved rats (Häussinger and Lang 1990), whose intracellular degradation rate was shown to be controlled by transport rather than by metabolism (Sips et al. 1981, Groen et al. 1982). A swelling-induced hyperpolarization of the plasma membrane (Graf et al. 1988) may also explain why sinusoidal glutathione efflux from the liver is increased following hypoosmotic liver cell swelling (Häussinger et al. 1990c), because glutathione release was shown to be under the control of the membrane potential (Fernandez-Checa et al. 1988).

AMINO ACID-INDUCED CELL VOLUME CHANGES AS A TRIGGER OF CELL FUNCTION

Although volume regulatory mechanisms are activated following cumulative amino acid uptake or exposure of cells to hypoosmotic media, a complete recovery of cell volume is never achieved; instead, the cells are left in a slightly swollen state (Fig. 1). Current evidence, obtained in experiments with rat liver, suggests that the extent of this remaining volume deviation somehow influences cell function (Graf et al. 1988, Lang et al. 1989, Häussinger 1990a-c, Baquet et al. 1990, Häussinger and Lang 1990, 1991a,b, Häussinger et al. 1991, Hallbrucker et al. 1991a,b, vom Dahl et al. 1991a,b). Accordingly, it was suggested that cell volume alterations may represent a new principle of metabolic control (Häussinger and Lang 1991a,b) and evidence also has been given that hormones such as insulin and glucagon exert their metabolic effects in part via modulation of liver cell volume (Hallbrucker et al. 1991b,c, vom Dahl et al. 1991b,c, Häussinger and Lang 1991a,b, Häussinger et al. 1991). Several long-known, but mechanistically poorly understood effects of amino acids, which could not be related to their metabolism, such as stimulation of glycogen synthesis (Katz et al. 1976, 1979, Golden et al. 1979, Rognstad 1985, Chen and Lardy 1985, Lavoinne et al. 1987) or inhibition of proteolysis (Schworer and Mortimore 1979, Mortimore and Pösö 1984, 1987, Seglen et al. 1980, Sommercorn and Swick 1981, Pösö et al. 1982a,b, Seglen and Gordon 1984, Hopgood et al. 1977, Leverve et al. 1987) can be mimicked by hypoosmotic swelling of the cells just to the extent as the amino acids do (Häussinger et al. 1990b, Baquet et al. 1990, Hallbrucker et al. 1991a, vom Dahl et al. 1991a, Häussinger et al. 1991). Thus, the above-mentioned amino acid effects are probably mediated by amino acid-induced cell swelling. Because anisoosmotic liver cell swelling was shown to inhibit glycogenolysis, glycolysis (Graf et al. 1988, Lang et al. 1989) and proteolysis (Häussinger et al. 1990b, 1991, Hallbrucker 1991a) and to stimulate glycogen synthesis (Baquet et al. 1990), amino acid uptake (Häussinger et al. 1990a,c, Häussinger and Lang 1990), ureogenesis from amino acids (Häussinger et al. 1990c, Häussinger and Lang 1990) and also protein synthesis (Stoll et al., 1992), it seems that cell swelling acts like an "anabolic signal" (Häussinger and Lang 1991a,b, Watford 1990). Thus, amino acid-induced cell swelling may explain in part the long known anabolic effect of amino acid infusion. Because the extent of amino acid-induced cell swelling is largely dependent upon the activity of concentrative, Na^+-dependent amino acid transport systems in the plasma membrane, amino acid transporters may be seen not only as amino acid-translocating systems, but also as transmembrane signalling systems (Häussinger et al. 1991, Häussinger and Lang 1991a,b). In this respect, it should be noted that cell volume alterations also have been implicated as a "second messenger" of hormone action (Häussinger et al. 1991, Häussinger and Lang 1991a). In both perfused livers and isolated hepatocytes insulin increases and glucagon decreases cellular volume. These changes are dependent upon the activation of the respective K^+ transport systems (Häussinger et al. 1991, Hallbrucker et al. 1991b,c, vom Dahl et al. 1991b,c, Häussinger and Lang 1991b). Insulin activates both loop diuretic-sensitive NaCl-KCl-cotransport and amiloride-sensitive Na^+/H^+ exchange, whereas glucagon stimulates cellular potassium release through barium and quinidine-sensitive K^+ channels (Hallbrucker 1991b,c).

The finding that cell volume changes induced by amino acid accumulation modulate cell function, raises the questions about the signals which couple cell volume changes to the persistent alterations of cellular function. This issue is not yet settled, however, in rat liver hypoosmotic cell swelling was shown to increase intracellular Ca^{++} (Corasanti et al. 1990, Baquet et al. 1991a), to decrease intracellular pH (Gleeson et al. 1990), to hyperpolarize the plasma membrane (Graf et al. 1988), to decrease the cAMP/cGMP ratio (Haussinger et al., unpublished result) and to stimulate the formation of inositoltrisphosphate (vom Dahl et al. 1991a). Recent studies on volume regulation in Ehrlich Ascites tumor cells have indicated an involvement of eicosanoids (Lambert 1987, Lambert et al. 1987, Hoffmann et al. 1988), but no evidence for this was obtained in studies with perfused rat liver (vom Dahl et al. 1991a). Although not demonstrated yet, it is likely that the above-mentioned alterations of intracellular signalling pathways also occur during amino acid-induced cell swelling, but the significance of these alterations for the swelling-induced modulation of hepatic metabolism remains to be established.

Amino Acids, Cell Volume and Proteolysis

Hepatic proteolysis is under the control of amino acids (Schworer and Mortimore 1979, Seglen et al. 1980, Pösö et al. 1982a,b, Mortimore and Pösö 1984, 1987, Sommercorn and Swick 1981, Seglen and Gordon 1984, Hopgood et al. 1977, Leverve et al. 1987) and hormones such as insulin and glucagon (Mortimore and Mondon 1970, Woodside et al. 1974, Hopgood et al. 1980, 1981, Neely et al. 1974, Draznin and Trowbridge 1982), but the underlying mechanisms remained obscure (Mortimore and Pösö 1987). It recently became clear that hypoosmotic cell swelling inhibits proteolysis in liver, whereas hyperosmotic cell shrinkage stimulates protein breakdown under conditions when the proteolytic pathway is not already fully activated (Häussinger et al. 1990b,1991a, Hallbrucker et al. 1991a, vom Dahl et al. 1991a). The known antiproteolytic effect of several amino acids can at least in part be ascribed to an amino acid-induced cell swelling (Häussinger et al. 1990b, Hallbrucker et al. 1991a, Häussinger et al. 1991). A close relationship is observed between the amino acid-induced degree of cell swelling and the inhibition of proteolysis (Häussinger et al. 1990b,1991, Häussinger and Lang 1991b, Hallbrucker et al. 1991a). The antiproteolytic effect of glutamine and glycine can be explained fully by the amino acid-induced cell swelling, because inhibition of proteolysis by these amino acids is quantitatively mimicked by hypoosmotic cell swelling when induced to the same extent as by the above-mentioned amino acids (Häussinger et al. 1990b,1991, Hallbrucker et al. 1991a). Inhibition of proteolysis by glutamine and glycine is additive to the same extent as additivity is observed with respect to cell swelling induced by both amino acids (Hallbrucker et al. 1991a, Häussinger et al. 1991). In livers from fed rats, the antiproteolytic effect of glycine is only about one fourth compared to that found in livers from starved rats. This is explained by a 3-fold greater swelling potency of glycine following starvation, possibly as a consequence of a higher activity of glycine-transporting amino acid transport System A (Kilberg et al. 1985). On the other hand, glutamine-transporting System N is not up-regulated during starvation in vivo (Hayes and McGivan 1982, Kilberg et al. 1985, Bode et al. 1990) and consequently, glutamine-induced cell swelling and proteolysis inhibition are only slightly different in the fed and starved state, respectively (Hallbrucker et al. 1991a). Interestingly, proteolysis control by insulin and glucagon also can be explained quantitatively on the basis of cell volume changes (Häussinger et al. 1991, Häussinger and Lang 1991b, vom Dahl et al. .1991b). As shown in Figure 2, there is a close relationship between the proteolytic activity in perfused liver and the alterations of cell volume; proteolysis is inhibited during cell swelling and stimulated following cell shrinkage. This relationship is maintained regardless of how cell volume is modified (Fig. 2). Apparently, the extent of cell volume change produced by these effectors is the common denominator for their effect on proteolysis.

The antiproteolytic effect of amino acids other than glutamine or glycine can only in part be explained by cell swelling. For example, proteolysis inhibition by alanine, serine or proline is about twice as strong as one would predict from anisotonic swelling experiments producing the same degree of cell swelling as these amino acids (Hallbrucker et al. 1991a). Also, the known antiproteolytic effect of ammonia (in concentrations up to 3 mM) can in part be ascribed to ammonia-induced cell swelling (Haussinger et al.). For other amino acids including phenylalanine, or a complete amino acid mixture, the contribution of cell volume to the antiproteolytic action is less than 50%, even though a linear relationship between cell volume and the extent of proteolysis-inhibition is still maintained (Hallbrucker 1991a). Cell volume changes appear not be involved in proteolysis control by leucine, because leucine does not lead to appreciable cell swelling (Wettstein et al. 1990), but exhibits an antiproteolytic effect (for review see (Mortimore and Pösö 1984)).

Like proteolysis, mRNA breakdown involves the autophagic pathway and is controlled by amino acids and hormones (Lardeux and Mortimore 1987, Lardeux et al. 1987, Balavoine et al. 1990). These findings suggest that amino acid-induced cell swelling also may participate in the control of mRNA degradation.

Amino Acid-induced Cell Volume Changes and Carbohydrate Metabolism, Lipogenesis, Ketogenesis, and Bile Formation

Several amino acids such as glutamine, proline asparagine, and the non-metabolizable amino acid analogue AIB are known to stimulate glycogen synthesis in hepatocytes from fasted rats (Katz et al. 1976, 1979, Golden et al. 1979, Rognstad 1985, Chen and Lardy 1985, Lavoinne et al. 1987). This stimulation was suggested to be primarily due to an activation of glycogen synthase resulting largely from a stimulation of synthase phosphatase (Lavoinne et al. 1987). It now appears that the stimulation of glycogen synthesis by amino acids is a consequence of amino acid-induced cell swelling (Baquet

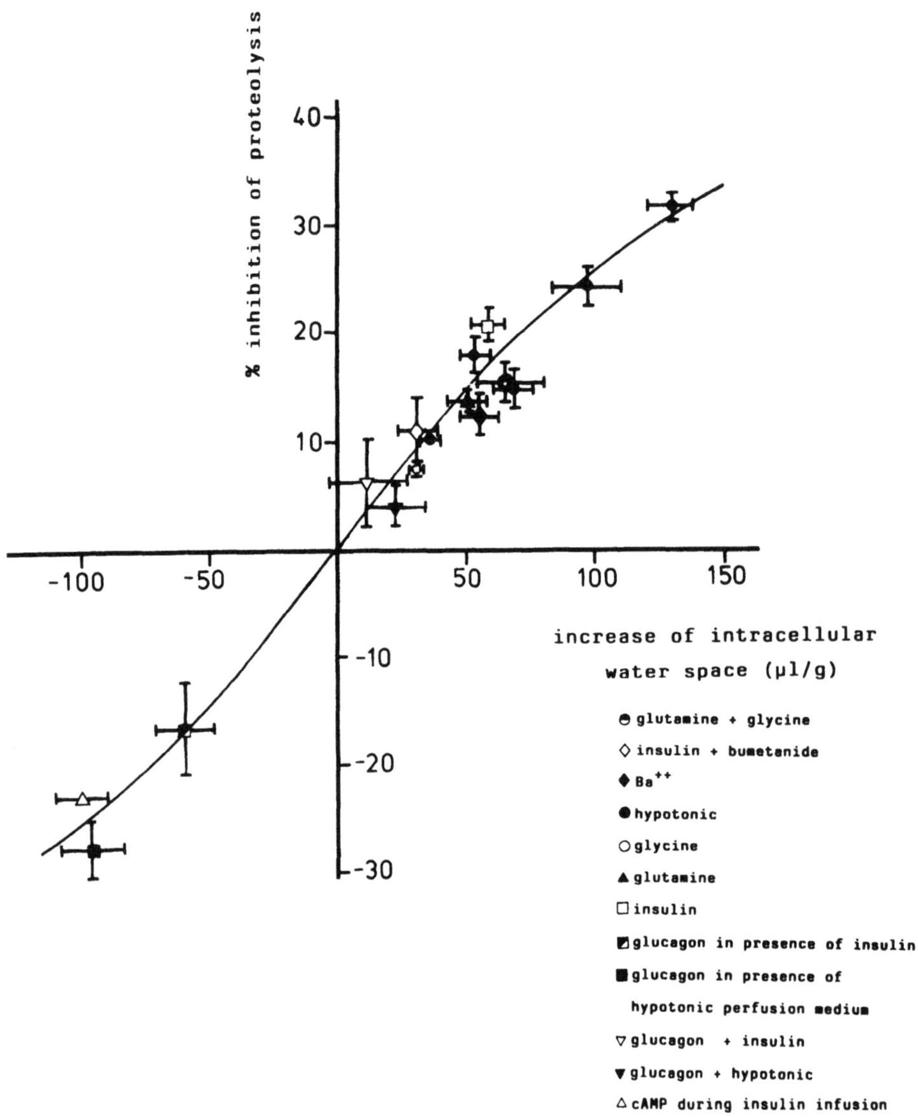

Fig. 2. Relationship between cell volume and proteolysis in liver. Cell volume in perfused liver was determined as intracellular water space (vom Dahl et al. 1991b) and proteolysis was assessed as (^{3}H)leucine release in effluent perfusate from perfused livers from rats, which were prelabelled in vivo by intraperitoneal injection of (^{3}H)leucine 16h prior to the perfusion experiment. Cell shrinkage stimulates proteolysis, whereas cell swelling inhibits. Cell volume changes were induced by insulin, cAMP, glucagon, amino acids, Ba^{++} or anisoosmotic exposure. From (Häussinger et al. 1991).

et al. 1990). The extent of stimulation of glycogen synthesis paralleled the degree of cell swelling elicited by the amino acids and was prevented by hyperosmolar incubation of the hepatocytes (Baquet et al. 1990). The higher rates of glycogen synthesis from amino acids in periportal hepatocytes (for review see (Jungermann and Katz 1989) may be attributed to a higher degree of amino acid-induced cell swelling in periportal compared to perivenous cells (Plomp et al. 1990). Only the proline-stimulation of glycogen synthesis was higher than expected from a similar degree of cell swelling induced by hypoosmotic incubation (Baquet et al. 1990). Interestingly, proline induced inhibition of proteolysis can only in part be ascribed to cell volume alterations ((Hallbrucker et al. 1991a). Hypoosmotic cell swelling in perfused rat liver inhibits glycogenolyis and glycolysis (Graf et al. 1988, Lang et al. 1989) and stimulates net uptake of lactate and pyruvate (vom Dahl et al. 1991). It is likely that such alterations of carbohydrate metabolism also occur upon amino acid-induced cell swelling.

The inhibition of hepatic ketogenesis by glutamine, proline, alanine or asparagine (Nosadini et al. 1980, Sugden et al. 1982, Lavoinne et al. 1987, Baquet et al. 1991b) is probably not due to amino acid-induced cell swelling. Hypoosmotic cell swelling was without effect on ketogenesis from glucose (Baquet et al. 1991b), tyrosine, or octanoate (Häussinger and Lang, 1991c). Furthermore, ketogenesis from ketoisocaproate (Häussinger and Lang, 1991c) and endogenous substrates (Baquet et al. 1991b) was even stimulated by cell swelling. Conversely, no effect on ketogenesis was observed with AIB despite causing cell swelling (Baquet et al. 1991b).

Hypoosmotic incubation of hepatocytes slightly stimulates lipogenesis from glucose (Baquet et al. 1991b). On the other hand, a more pronounced stimulation of lipogenesis occurs upon addition of glutamine, proline, or alanine, but not AIB, histidine, or asparagine (Lavoinne et al. 1987, Baquet et al. 1991b). No correlation was detectable between the potency of these amino acids to swell the hepatocytes on the one hand and their potency to stimulate lipogenesis on the other. Thus, it is unlikely that volume changes play a major role in mediating the lipogenic effect of glutamine or other amino acids.

Amino Acid-induced Cell Swelling and Nitrogen Metabolism

Cell swelling not only affects amino acid transport across the plasma membrane (see above), but also nitrogen metabolism. In rat liver, perfused with a physiological portal glutamine concentration (0.6 mM), exposure to hypoosmotic perfusion fluid switches hepatic glutamine balance from net release to net uptake. This is due to a stimulation of flux through glutaminase in periportal hepatocytes and a simultaneous inhibition of glutamine synthesis in perivenous hepatocytes (Häussinger et al. 1990a). Conversely, hyperosmotic cell shrinkage inhibits flux through glutaminase, whereas flux through glutamine synthetase is stimulated. Cell swelling-induced activation of glutaminase is most likely due to mitochondrial swelling, which is known to activate glutaminase (Lacey et al. 1981, Joseph et al. 1981, McGivan et al. 1985, for review see Halestrap 1989). Also, glycine uptake and oxidation is stimulated in perfused rat liver following hypoosmotic exposure (Haussinger et al., unpublished result). When livers from fed but not from 24h-starved rats, are perfused with an amino acid mixture roughly mimicking physiological portal concentrations in the postabsorptive state, hypoosmotic liver swelling stimulates urea synthesis and ammonia formation from the amino acids (Häussinger et al. 1990c, Häussinger and Lang 1990). Taking into account that amino acids per se induce cell swelling, these findings on hypoosmotic cell swelling suggest that amino acid-induced cell swelling may augment amino acid breakdown. In this respect, the swelling-induced stimulation of glutaminase may be especially important, because the activity of this enzyme is a crucial parameter augmenting urea cycle flux (for review see (Häussinger 1990)).

Asparagine and glutamine, as well as other substrates for the Na^+-dependent amino acid transport System A and N are potent inducers of ornithine decarboxylase in a variety of tumour cell types (Chen and Canellakis 1977, Rinehart et al 1985, Fong and Law 1988). Ornithine decarboxylase is also induced during hypoosmotic cell incubation (Munro et al 1975, Friedmann et al 1977, Perry and Oka 1980, Kapyaho and Janne 1982,

Poulin and Pegg 1990) and by insulin (Manzella et al 1991). Since both, amino acids and insulin lead to cell swelling, one is tempted to attribute the induction of ornithine decarboxylase by amino acids at least in part to amino acid-induced cell swelling.

Clinical Relevance

The finding that hypoosmotic, hormone- and amino acid-induced cell swelling inhibits proteolysis in liver, whereas conversely cell shrinkage stimulates proteolysis and inhibits protein synthesis, may shed a new light on the understanding of catabolic states and the known anabolic effect of insulin and amino acids such as glutamine (Rennie et al. 1986, MacLennan et al. 1987, 1988, Roth et al. 1990). Protein turnover in skeletal muscle is complexely regulated, but glutamine seems to play an important role (for review see (Sugden and Fuller 1991)). In skeletal muscle, glutamine is transported by a Na^+-dependent carrier, which builts up a remarkably high intra/extracellular glutamine concentration gradient. A correlation between the intramuscular glutamine concentration on the one hand and protein synthesis and inhibition of proteolysis on the other was described ((McLennan et al. 1987, 1988), for discussion see (Watford 1989)). Protein catabolic states, such as sepsis are characterized by a marked lowering of intracellular glutamine concentrations in skeletal muscle (for review see (Roth et al. 1990)) and accordingly, a lowering of the intra/extracellular glutamine gradient. Under these conditions cell shrinkage could in part be responsible for the increased protein breakdown.

Amino acid-induced liver cell swelling not only modulates liver cell function, but is probably also a parameter regulating kidney function. Infusion of glutamine or serine, but not of glutamate, in near physiological concentrations into the portal vein leads to a rapid decrease of glomerular filtration rate (GFR) and of urinary flow (Lang et al. 1991a). When the same amounts of glutamine or serine were infused into the jugular vein, renal function remained unaffected. Simultaneous portal administration of glutamine together with glucagon, which reverses the glutamine-induced hepatocyte swelling, prevented the decrease of GFR. Hepatic denervation abolished the effect of portal glutamine on GFR and after unilateral kidney denervation, GFR decreased only in the control kidney. These data suggest the existence of a serotoninergic (Lang et al. 1991b) hepatorenal reflex, which may be triggered by amino acid-induced cell volume changes in the liver (Lang et al. 1991a,b).

Most recent evidence points to altered cell volume in NIH fibroblasts expressing the Ha-ras oncogene. The products of ras oncogenes are GTP binding proteins resistant to inactivation by GTPase activating protein (McGormick 1989). Injection of ras oncogenes into quiescent cells leads to growth factor independent cellular proliferation or differentiation (Bar-Sagi and Feramisco 1985, Doppler et al. 1987, Feramisco et al. 1984, Stacey and Kung 1984). The expression of ras oncogene may lead to intracellular alkalinization (Doppler et al. 1987) and to enhanced furosemide sensitive rubidium flux (Meyer et al. 1991). These observations point to enhanced activity of both, Na^+/H^+ exchanger and $Na^+,K^+,2Cl$-cotransport. The concerted action of these transporters leads to an increase of cell volume (Meyer et al 1991, Lang et al. 1991). Both, ras oncogene expressing cells (+ras) and cells not expressing the ras oncogene (-ras) respond to osmotic cell swelling with regulatory cell volume decrease and to osmotic cell shrinkage with regulatory cell volume increase (Lang et al. 1991). Since both, +ras and -ras display twosided cell volume regulation in anisotonic media, their set points for cell volume regulation must be different. Thus, expression of ras oncogene leads to a shift of the set point for cell volume regulation. Accordingly, the activation of the Na^+/H^+ exchanger and the Na^+,K^+,Cl^- cotransport by expression of the ras oncogene can be viewed as the volume regulatory response of the volume regulatory carriers to the arising discrepancy between the new set point and the actual value of cell volume. At present, the role of enhance cell volume in the regulation of cell proliferation remains elusive. Nevertheless, similar to the insulin stimulated transition of 3T3 cells from G0/G1- to S- phase (Panet et al. 1986), the accelerated growth of ras oncogene expressing cells in serum depleted media is abolished in the presence of dimethylamiloride and bumetanide.

RELATION BETWEEN AMINO ACID TRANSPORT AND CELL PROLIFERATION

Cell Cycle

Sander and Pardee (1972) monitored AIB uptake during the cell cycle in cultured CHO and L cells. AIB uptake was probably reflecting System A transporter activity (Shotwell et al., 1981). Early in the G phase of the cell cycle they detected a two-fold increase in the transport rate. This increase preceded that of DNA synthesis by approximately 5-10 hours and was elevated throughout the remaining portion of the cell cycle. The transport rate returned to basal levels during mitosis. The presence of the protein synthesis inhibitor cycloheximide prevented the increase in AIB transport as well as the progression through the cell cycle. Furthermore, blocking the cells in metaphase with colcemid also prevented the increase in transport activity. Both the cycloheximide and colcemid effects were reversible upon removal of the inhibitors. The induction of transport was characterized kinetically as an increase in V_{max} with no change in K_m; a result consistent with an increase in transporter number rather then an altered affinity for substrate.

That System A transport is altered during the cell cycle in vivo is suggested by monitoring amino acid transport following partial hepatectomy in the rat. A transient increase is AIB uptake by whole tissue (Walker and Whitfield, 1978) or isolated hepatocytes (LeCam et al., 1979) is observed 6-12 h following 70% hepatectomy. Given that the majority of hepatocytes are in a quiescent or growth arrested state in the intact liver, this transient peak in AIB transport is probably the result of the synchronous division of the remaining hepatocytes in the liver remnant. The increase in transport activity may be the result of synthesis of additional transporters (Dolais-Kitabgi et al., 1981). In support of that hypothesis, isolated plasma membrane vesicles from regenerating livers retain the increase in System A transport (Fowler and Mailliard, personal communication). The specificity of the response is illustrated by the fact that Systems ASC and N are not altered in activity at anytime during the five days of liver regeneration. The spike of System A activity during the 12 h following hepatectomy coincides with initiation of cell replication and probably reflects the increased transport observed during the cell cycle (Sander and Pardee, 1972). Presumably, the hepatocytes remaining in the liver remnant have been synchronized by the signals involved in regeneration. A second much smaller peak in System A activity observed following the first 24 h probably reflects the loss of cell synchronization. The return to a low basal level of transport activity rather than the elevated levels typical of rapidly-growing transformed liver tissues (Scott et al., 1972) suggests that the transient increase observed following hepatectomy is not the result of a long-term adaptation to elevated growth rate.

Growth Factors and Mitogens

With regard to agents which can be categorized as "growth factors", the number of reports describing the effects of insulin on amino acid transport are by far the greatest. Insulin either in vivo or in vitro has been shown to induce System A selectively over other transporters in a wide variety of tissues and cell types (Guidotti et al., 1978; Shotwell et al., 1983). In nearly all cases, the resulting effect of insulin was: 1) an increase in V_{max} with little or no change in the K_m for substrate, 2) blocked by inhibitors of either RNA or protein synthesis, and 3) generally ascribed to de novo synthesis of additional transport proteins. Based on mutation analysis of CHO cells, Moffett et al. (1987) have proposed that insulin inactivates a repressor protein of the System A gene.

A number of other growth factors have been shown to modulate amino acid transport. Boerner et al. (1985) demonstrated that epidermal growth factor (EGF) stimulates amino acid transport in rat kidney cells. Experiments designed to test individual systems demonstrated the Systems A and L were enhanced by the growth factor, whereas there was little or no change in System ASC. In hepatocytes, EGF has been reported to cause a rapid, but transient 30-40% stimulation of amino acid transport thought to be the result of membrane hyperpolarization (Moule and McGivan, 1987). In contrast, EGF has been shown to suppress the stimulation of hepatic AIB transport induced by glucagon or cAMP (Morin et al., 1981), but not that of insulin (Auberger et al., 1983). Visciano and Fehlmann (1984) demonstrated that de novo protein synthesis was required for the EGF-dependent inhibition of cAMP stimulation of hepatic transport. Bereta et al. (1990) reported that exposure of cultured hepatocytes to EGF for 2 days increased basal uptake, and further enhanced the elevated transport induced by interleukin-6 (IL-6). It appears that the response of amino acid transport to EGF will be tissue-specific and multi-factoral.

When tested in vivo both interleukin-1 (IL-1) and tumor necrosis factor α (TNFα) increase amino acid accumulation by the liver (Argiles et al., 1989). Isolated hepatocytes in primary culture did not respond to either IL-1 or TNFα, but IL-6 was effective in stimulating AIB uptake by these cells. These results suggest that the induction by IL-1 or TNFα in vivo is mediated by a secondary response. In vitro treatment of hepatocytes with TNFα further enhances the stimulation of basal transport by glucagon (Warren, 1987). Transforming growth factor-ß (TGF-ß) alone inhibits AIB transport in hepatocytes and suppresses the induction by IL-6 (Bereta et al., 1990). On the other hand, TGF-ß stimulates System A activity, and to a lesser extent System L, in rat kidney cells (Boerner et al., 1985). For both of these transport activities, the induction in these cells by EGF and TGF-ß may be additive. Interestingly, it also was noted that the stimulation of System A activity by TGF-ß was greater in confluent cell cultures (Boerner et al., 1985).

A number of reports have demonstrated effects on System A activity by the tumor-promoting compounds termed phorbol esters. For example, Kitagawa et al. (1986) reported stimulation of AIB uptake in 3T3 cells following a 3 h treatment with phorbol esters. Down-regulation of protein kinase C activity by long-term pre-exposure of the cells to phorbol ester abolished the increased transport typically observed in response to phorbol ester, platelet-derived grow factor (PDGF), or the calcium ionophore A23187, but not the stimulation in response to EGF or insulin (Kitagawa et al., 1986). Boerner and Saier (1988) demonstrated that phorbol esters stimulate System A activity in MDCK cells. Interestingly, the tumor promoters had no effect on MDCK cells that already exhibited elevated rates of transport following amino acid starvation or chemical transformation. Other investigators have reported no effect of either phorbol esters or diacylglycerol on amino acid uptake by primary cultures of renal proximal tubular cells (Freidlander and Amiel, 1989). Dawson and Cook (1987) demonstrated a temporal relationship between stimulation of System A transport activity and re-distribution of protein kinase C to the plasma membrane following treatment of LLC-PK_1 cells with phorbol esters. Both the transfer of protein kinase C to the membrane and the activation of System A transport activity was blocked by cytochalasin B, but only the stimulation of System A activity was blocked by the protein synthesis inhibitor cycloheximide. These results suggest that activation of protein kinase C is only one step in a sequence of events by which phorbol esters increase amino acid transport through enhanced transcription of the System A gene.

Cell Growth and Density

A number of specific transport systems have been shown to exhibit altered velocity with changes in cell density. There appear to be at least two different mechanisms by which net transport can be altered, either through a change in the absolute amount of a particular transport protein or as a result of trans-effects due to changes in the intracellular amino acid pool. In either case, the result is modulation of transport by the cell to best suit the metabolic needs as cell density changes.

The activity of the Na^+-independent System L has been monitored in a number of different cell types to correlate changes in activity with cell density. Cecchini et al. (1976) reported that System L activity increased as 3T3 fibroblasts increased in cell density. Further study showed that the increase in System L activity was the result of trans-stimulation as the intracellular amino acid levels increase when cell growth is arrested by either approach to confluency, removal of serum, or by the additional of protein synthesis inhibitors (Oxender et al., 1977). Incubation of quiescent cells in an amino acid free-medium for a time sufficient to deplete the amino acid levels resulted in System L transport activity similar to that in sub-confluent cells. Borghetti et al. (1980) also reported that System L activity remained unchanged with increasing cell density if the intracellular amino acid pools were depleted prior to assay. However, further investigation lead them to conclude that, independent of amino acid depletion, cell density also may modulate System L transport activity by alternate mechanisms in SV40-transformed 3T3 cells (Petronini et al., 1982).

Heterogeneity of Na^+-independent leucine, histidine, and 2-aminobicyclo (2,2,1) heptane-2-carboxylic acid (BCH) transport in primary cultures of rat hepatocytes lead Weissbach et al. (1982) to conclude that there are two components, termed Systems L_1 and L_2, in liver tissue. Despite the fact that hepatocytes in primary culture do not exhibit significant levels of cell division, and therefore do not change in absolute cell density, the authors observed a decrease in Na^+-independent leucine transport and a concurrent increase in Na^+-independent histidine transport with time in culture. Prior to this observation, the Na^+-independent uptake of these two amino acids was thought to be mediated by a single component, namely System L. Both kinetic and competitive inhibitor analysis demonstrated the existence of two Na^+-independent carriers with overlapping substrate specificity. The high affinity component showed a increase in V_{max} with time in culture, whereas the maximum velocity for the low affinity component was reduced over the same time period (Weissbach et al., 1982). Interestly, hepatocytes from both fetal and newborn rats as well as several hepatoma cell lines exhibit the high affinity system only (Kilberg et al., 1983).

The transport activity of System A has been demonstrated to exhibit significant changes in relation to cell density. The large number of reports describing the relation between changes in cell density and System A transport activity precludes description of each within this chapter, but representative studies from a number of different cells types will be discussed. One of the earliest reports was that of Foster and Pardee (1969) who demonstrated that AIB uptake was reduced by approximately 30% when 3T3 cells reached confluence. Uptake of amino acids such as glutamate, glutamine, or arginine, substrates for transport systems other than System A, were unaffected by the growth rate of the cells. The repression of System A transport activity upon cell growth arrest in 3T3 cells has been confirmed by a number of other investigators (Oxender et al., 1977; Borghetti et al., 1980). Growth arrest by removal of serum from sub-confluent cells also causes a reduction in System A transport activity which can be returned to the elevated rate by transferring the cells to serum-containing medium (Oxender et al., 1977). Boerner and Saier (1982) have extensively characterized density effects on amino acid transport in Madin-Darby canine kidney (MDCK) cells. Their results demonstrated that activity of System A decreased with increasing cell density, but that the activity of System ASC was reduced as well. Interestingly, Borghetti et al. (1980) also reported a reduction in System ASC activity in confluent 3T3 cells. One can distinguish cell density and cell growth effects by showing that addition of fresh serum to confluent cells results in an increase in transport.

Transformation

In addition to an effect of cell density on basal System A-mediated amino acid transport, a number of laboratories have reported altered regulation of transporter activity in response to abnormal cell growth. Weber et al. (1984) observed only a slight increase in basal AIB transport activity in Rous sarcoma virus-transformed chicken embryo fibroblasts compared to non-transformed cells. However, induction of transport by amino acid starvation resulted in a large increase in transport activity in the transformed cells, but not in the control cells. They also observed that the stimulation of System A activity by amino acid deprivation was suppressed at high cell density. Boerner and Saier (1985) reported that the induction of System A following starvation of confluent cells was only 40% of the value observed in rapidly growing cells. Likewise, in rat hepatoma cells, the degree of induction by substrate deprivation increased slightly with growth at low cell densities, but declined significantly as cells approached confluence (Kilberg et al., 1985). Collectively, these observations suggest that cellular growth rates as well as cell density impact on the regulatory signals governing the expression of the System A gene in response to substrate availability.

The activity of System A-mediated transport has been linked to chemical transformation of MDCK cells (Boerner and Saier, 1982). The increased transport activity observed following transformation was not the result of increased growth rate given that the parent and transformed MDCK cell lines grew at similar rates (Erlinger and Saier, 1982). Leister et al. (1988) provided evidence that the turnover of the System A transporter in chemically-transformed C3H-10T1/2 cells was significantly slower then that in the parent cell line, an observation that could easily account for elevated transport activity in the transformed cells if the synthesis rates were similar.

Detailed reviews have documented the influence of viral-transformation on System A-mediated transport (Perdue, 1977; Parnes and Isselbacher, 1978; and Koren, 1980). Beginning with the pioneering work of Foster and Pardee (1969) several independent investigations demonstrated that viral-transformation of 3T3 cells resulted in increased Na^+-dependent AIB uptake. In general, stimulation of System A transport activity is observed in the transformed cells compared to the parental cell line, although there are a few reports of transformed cells which exhibit no change (Nakamura and Weber, 1979; Patterson et al., 1976; and Miller et al., 1975). In those investigations reporting a difference, kinetic analysis typically revealed an increase in V_{max} rather then a change in Km values for the substrate; probably the result of de novo carrier synthesis. Additional evidence for this interpretation comes for studies documenting that plasma membrane vesicles isolated from control or virally-transformed cells reflect the increased transport observed in whole cells (Parnes et al., 1976; Lever, 1976).

As mentioned above, Weber et al. (1984) detected no change in basal transport activity following viral transformation of chick embryo fibroblasts, but did observe a difference in the starvation response of System A. Data such as these have been taken to indicate that the increased transport observed in transformed cells is the result of a cellular response to the transformed state rather than a prerequisite to achieve transformation. Borghetti et al. (1980) demonstrated that the increase in System A and ASC activities following SV40 transformation of 3T3 cells remained elevated in SV40-3T3 revertant cells which had regained density-dependent inhibition of growth. These results, as well as other reports, suggest that in some revertant cell lines one can separate the changes in transport activity and cell growth. In contrast, a temperature-sensitive SV40 mutant has been used to allow investigators to shift cells from "normal" to the "transformed" state by incubation at the growth-permissive or growth-restrictive temperature (Chou, 1985). At the permissive temperature of 33°C, SV40-transformed hepatocytes express the transformed phenotype detected as rapid cell division and reduced synthesis of liver-specific proteins. In contrast, at the restrictive-temperature of 40°C, T-antigen is no longer expressed, cell division is considerably reduced, and liver-specific proteins once again are produced. System A transport activity is expressed at high levels in the rapidly growing cells maintained at 33°C, but at low levels in the slowly dividing cells (40°C) (Handlogten and Kilberg, 1988). Transfer of the cells maintained at 40°C to 33°C results in increased transport activity over the following 72 hours. The induction of transport following transfer to the permissive-temperature could be prevented by protein synthesis inhibitors. Furthermore, the elevated transport activity present at 33°C was retained in isolated plasma membrane vesicles as well as reconstituted proteoliposomes prepared following solubilization of the transport activity. Collectively, the data suggest that transfer from 40°C to 33°C results in an increased de novo biosynthesis of the System A transporter.

A number of mammalian cells have been shown to exhibit a transformed phenotype and altered amino acid transport following expression of one or more oncogenes. Racker et al. (1985) demonstrated that transformation of rat fibroblasts with either ras or myc oncogenes did not significantly change the basal System A activity, but the sensitivity of the transformed cells to TGF-ß stimulation of System A activity was enhanced. Longo et al. (1988) transformed 3T3 cells with either ras or neu oncogenes and established that transport by Systems A, ASC, L, or y^+ was not significantly different following transformation, but the activity of a Na^+-independent system for glutamic acid transport (x^-) was increased in the transformed cells that were allowed to reach confluence. The parental 3T3 cell line did not show this density dependent effect, nor was the Na^+-dependent glutamate transport system (X^-) changed in its activity. Obviously, these initial reports describing the effect of oncogene expression on amino acid transport represent the beginning of an interesting and potentially fruitful area of research aimed at understanding the relationship between amino acid transport and regulation of cell proliferation.

REFERENCES

Argiles, J.M., Lopez-Soriano, F.J., Wiggins, D. and Williamson, D.H., 1989, Biochem. J., 261:357.

Auberger, P., Samson, M., and LeCam., 1983, Biochem. J., 210:361.

Bakker-Grunwald, T., 1983, Biochim. Biophys. Acta., 731:239.

Balaban, R.S., and Burg, M.B., 1987, Kidney. Int., 31:562.

Balavoine, S., Feldmann, G. and Lardeux, B., 1990, Eur. J. Biochem., 189:617.

Ballanyi, K., and Grafe, P., 1988, Renal. Physiol., 11:142.

Bar-Sagi, D., and Feramisco, J.R., 1985, Cell., 42:841.

Baquet, A., Hue, L., Meijer, A.J., van Woerkom, G.M. and Plomp, P.J.A.M., 1990, J. Biol. Chem., 265:955.

Baquet, A., Lavoinne, A. and Hue, L., 1991b, Biochem. J., 273:57.

Baquet, A., Meijer, A.J. and Hue, L., 1991a, FEBS Lett., 278:103.

Baur, H., and Heldt, H.W., 1976, In Use of isolated liver cells and kidney tubules in metabolic studies (Tager, J.M., Söling, H.D. and Williamson, J.R., eds.) pp. 29-40, North Holland Elsevier Amsterdam.

Bear, C.E., and Petersen, O.H., 1987, Pflüger's Arch., 410:342.

Bear, C.E., 1990, Am. J. Physiol., 258:C421.

Beck, F.-X., Dörge, A. and Thurau, K., 1988a, Renal Physiol., 11:174.

Bereta, J., Szuba, K., Fiers, W., Gauldie, J., and Koj, A., 1990, FEBS Lett., 266:48.

Bode, B., and Kilberg, M., 1991, J. Biol. Chem., 266:7376.

Bode, B., Tamarappoo, B.K., Mailliard, M. and Kilberg, M.S., 1990, J. Parent. Ent. Nutr., 14:51S.

Boerner, P., Resnick, R.J., and Racker, E., 1985, Proc. Natl. Acad. Sci., USA 82:1350.

Boerner, P., and Saier, M.H., Jr., 1982, J. Cell. Physiol., 113:240.

Boerner, P., and Saier, M.H., Jr., 1988, J. Cell. Physiol., 137:117.

Borghetti, A.F., Piedimonte, G., Tramacere, M., Severini, A., Ghiringhelli, P., and Guidotti, G.G., 1980, J. Cell. Physiol., 105:39.

Cecchini, G., Lee, M., and Oxender, D.L., 1976, J. Supramol. Struct. 4:441.

Chamberlin, M.E., and Strange, K., 1989, Am. J. Physiol., 257:C159.

Chen, K.Y.,and Canellakis, E.S., 1977, Proc. Natl. Acad. Sci., 74:3791.

Chen, K.S., and Lardy, H.A., 1985, J. Biol. Chem., 260:14683.

Chou, J., 1985, Methods Enzymol. 109:385.

Christensen, H.N., Riggs, T.R., Fischer, H., and Palatine, I.M., 1952, J. Biol. Chem., 198:1.

Corasanti, J.G., Gleeson, D., Gautam, A. and Boyer, J.L., 1990, Renal Physiol. Biochem., 13:164.

Dawson, W.D., and Cook, J.S., 1987, J. Cell. Physiol., 132:104.

Dolais-Kitabgi, J., Rey, J.F., Fehlmann, M., Morin, O., and Freychet, P., 1981 Endocrinol., 109:868.

Doppler, W., Jaggi, R., and Groner, B., 1987, Gene., 54:147.

Draznin, B., and Trowbridge, M., 1982, J. Biol. Chem., 257:11988.

Erlinger, S., and Saier, M.H., Jr., 1982, In Vitro, 18:196.

Fehlmann, M., and Freychat, P., 1981, J. Biol. Chem., 256:7449.

Fehlmann, M., Morin, O., Kitabgi, P. and Freychat, P., 1981, Endocrinology, 109:253.

Feramisco, J.R., Gross, M., Kamata, T., Rosenberg, M., and Sweet, R.W., Cell., 38:109.

Fernandez-Chica, J.C., Ren, C., Awe, T.Y., Ookhtens, M. and Kaplowitz, N., 1988, Am. J. Physiol., 255:G403.

Fincham, D.A., Wolowyk, M.W. and Young, J.D., 1987, J. Membr. Biol., 96: 45.

Fong, W.F., and Law, C. L., 1988, Biochem. Biophys. Res. Commun., 155:937.

Foster, D.O., and Pardee, A.B., 1969, J. Biol. Chem., 244:2675.

Freidlander, G., and Amiel, C., 1989, J. Biol. Chem., 264:3935.

Friedman, Y., Park, S., Levasseur, S. and Burke, G., 1977, Biochem. Biophys. Res. Commun., 77:57.

Gazzola, G.C., Dall'Asta, V., Nucci, F.A., Rossi, P.A., Bussolati, O., Hoffmann, E.K. and Guidotti, G.G., 1991, Cell. Phsiol. Biochem., in press
Gilles, R., 1987, In: Cell volume control: fundametnatal and comparative aspects in animal cells (Kleinzeller, A., ed.) pp. 205-247, Academic Press New York.
Gilles, R., 1988, Renal Physiol., 11:277.
Gleeson, D., Corasanti, J.G. and Boyer, J.G., 1990, Am. J. Physiol., 258: G299.
Golden, S., Wals, P.A., Okajima, F. and Katz, J., 1979, Biochem. J., 182: 727.
Graf, J., Haddad, P., Häussinger, D. and Lang, F., 1988, Renal Physiol. Biochem., 11:202.
Grinstein, S., and Foskett, J.K., 1990, Ann. Rev. Physiol., 52:399.
Groen, A.K., Sips, H.J., Vervoorn, R.C. and Tager, J.M., 1982, Eur. J. Biochem., 122:87.
Guidotti, G.G., Borghetti, A.S., Gazzola, G.C., 1978, Biochim. Biophys. Acta, 515:329.
Haddad, P., and Graf, J., 1989, Am. J. Physiol., 257:G357.
Haddad, P., Thalhammer, T. and Graf, J., 1989, Am. J. Physiol., 256: G563.
Handlogten, M.E., and Kilberg, M.S., 1985, Am. J. Physiol., 255:C261.
Häussinger, D., Stoll, B., Stehle, T. and Gerok, W., 1989, Eur. J. Biochem., 185:189.
Häussinger, D., 1990, Biochem. J., 267:281.
Häussinger, D., Hallbrucker, C., vom Dahl, S., Lang, F. and Gerok, W., 1990b, Biochem. J., 272:239.
Häussinger, D., Hallbrucker, C., vom Dahl, S., Decker, S., Schweizer, U., Lang, F. and Gerok, W., 1991, FEBS. Lett., 283:70.
Häussinger, D., and Lang, F., 1990, J. Cell. Biochem., 43:355.
Häussinger, D., and Lang, F., 1991a, Biochem. Cell. Biol., 43:1.
Häussinger, D., and Lang, F., 1991b, Cell. Physiol. Biochem., 1:121.
Häussinger, D., and Lang, F., 1991c, Biochim. Biophys. Acta, 1071: 331.
Häussinger, D., Lang, F., Bauers, K. and Gerok, W., 1990a, Eur. J. Biochem., 188:689.
Häussinger, D., Lang, F., Bauers, K. and Gerok, W., 1990c, Eur. J. Biochem., 193:891.
Häussinger, D., Stehle, T. and Lang, F., 1990d, Hepatology, 11:243.
Halestrap, A.P., 1989, Biochim. Biophys. Acta, 973:355.
Hallbrucker, C., vom Dahl, S., Lang, F. and Häussinger, D., 1991a, Eur. J. Biochem., 197:717.
Hallbrucker, C., vom Dahl, S., Lang, F. Gerok, W. and Häussinger, D., 1991b, Eur. J. Biochem., 199:467.
Hallbrucker, C., vom Dahl, S., Lang, F., Gerok, W. and Häussinger, D., 1991c, Pflügers Arch. Physiol., 418:519.
Hayes, M.R., and McGivan, J.D., 1982, Biochem. J., 204:365.
Henderson, R.M., Graf, J. and Boyer, J.L., 1989, Am. J. Physiol., 256: G1028.
Hoffmann, E.K., and Lambert, I.H., 1983, J. Physiol., 338:613.
Hoffmann, E.K., Lambert, J.H. and Simonsen, L.O., 1988, Renal Physiol. Biochem., 11:221.
Hoffmann, E.K. and Simonsen, L.O., 1989, Physiol. Rev., 69:315.
Hopgood, M.F., Clark, M.G. and Ballard, F.J., 1977, Biochem. J., 164:399.
Hopgood, M.F., Clark, M.G. and Ballard, F.J., 1980, Biochem. J., 186:71.
Hopgood, M.F., Clark, M.G. and Ballard, F.J., 1981, Biochem. J., 196:33.
Howard, L.D., and Wondergem, R., 1987, J. Membr. Biol., 100:53.
Jakubowski, J., and Jakob, A., 1990, Eur. J. Biochem., 193:541.
Joseph, S.K., and McGivan, J.D., 1978, Biochim. Biophys. Acta, 543:16.
Joseph, S.K., McGivan, J.D. and Meijer, A.J., 1981, Biochem. J., 194:35.
Jungermann, K., and Katz, N., 1989, Physiol., Rev. 69:708.
Katz, J., Golden., S. and Wals, P.A., 1976, Proc. Natl. Acad. Sci., 73: 3433.
Katz, J., Golden, S. Wals, P.A., 1979, Biochem. J., 180:389.
Kilberg, M.S., Barber, E.F. and Handlogten, M.E., 1985, Curr. Top. Cell., Reg. 25:113.
Kilberg, M.S., Han, H.P., Barber, E.F., Chiles, T.C., 1985, J. Cell. Physiol., 122:290.
Kilberg, M.S., Weissbach, L., and Barber, E.F., 1983, "Isolation Characterization, and Use of Hepatocytes,"Elsevier Science Publishing, New York.
Kitagawa, K., Nishino, H., and Iwashima, A., 1986, Experientia, 42:1038.
Kophaho, K., and Janne, J., 1982, Biochim. Biophys. Acta, 714:93.
Koren, R., 1980, Inter. Review Cytol., 68:127.
Kristensen, L.O., and Folke, M., 1984, Biochem. J., 221:265.
Kristensen, L.O., and Folke, M., 1986, Biochim. Biophys. Acta., 855:49.
Kristensen, L.O., 1980, J. Biol. Chem., 255:5236.
Lacey, J.H., Bradford, N.M., Joseph, S.K. and McGivan, J.D., 1981, Biochem. J., 194:29.

Lambert, I.H., 1987, J. Membrane Biol., 98:207.
Lambert, I.H., Hoffmann, E.K. and Christensen, P., 1987, J. Membrane Biol., 98:247.
Lang, F., Messner, G. and Rehwald, W., 1986, Am. J. Physiol., 250:F953.
Lang, F., Öttl, I., Freudenschuss K., Honeder, M., Tschernko, E. and Häussinger, D., 1991b, Pflüger's Arch. Physiol., 419:111.
Lang, F., Ritter, E., Woell, Weiss, H., Häussinger, D., Maly, K., Grunicke, H., 1992, Pflüger's Arch. Physiol., 420:424.
Lang, F., Stehle, T. and Häussinger, D., 1989, Pflügers Arch., 413:209.
Lang, F., Tschernko, E., Öttl, I., Schulze, E., Hallbrucker, C. and Häussinger, D., 1991a, Hepatology, 14:590.
Lang, F., Völkl, H. and Häussinger, D., 1990, Comp Physiol, 4:1.
Lardeux, B.R., Heydrick, S.J. and Mortimore, G.E., 1987, J. Biol. Chem., 262:14507.
Lardeux, B.R., and Mortimore, G.E., 1987, J. Biol. Chem., 262:14514.
Lavoinne, A., Baquet, A. and Hue, L., 1987, Biochem. J., 248:429.
Le Cam, A., and Freychat, P., 1978, Diabetologia, 15:117.
Le Cam, A., Rey, J-F., Fehlmann, M., Kitabgi, P., and Freychet, P., 1979, Am. J. Physiol., 236:E594.
Leister, K.J., Schenerman, M.A., and Racker, E., 1988, J. Cell. Physiol., 135: 163.
Leverve, J.E., 1976, J. Cell. Physiol., 89: 779.
Leverve, X.M., Caro, L.H.P., Plomp, P.J.A.M. and Meijer, A.J., 1987, FEBS Lett., 219:455.
Longo, N., Franchi-Gazzola, R., Bussolati, O., Dall'Asta, V., Nucci, F.A., and Gazzola, G.C., 1988 Ann. N.Y. Acad. Sci., 551:374.
Macknight, A.D.C., 1988, Renal Physiol., 11:114.
McGivan, J.D., Vadher, M., Lacey, J.H. and Bradford, N.M., 1985, Eur. J. Biochem., 148:323.
McGormick, F., 1989, Cell 56:5.
MacLennan, P.A., Brown, R.A. and Rennie, M.J., 1987, FEBS Lett., 215:187.
MacLennan, P.A., Smith, K., Weryk, B., Watt, P.W. and Rennie, M.J., 1988, FEBS Lett., 237:133.
Manzella, J.M., Rychlik, W., Rhoads, R.E., Hershey, J.W.B., and Blackshear, P.J., 1991, J. Biol. Chem., 266:2383.
Meijer, A.J., and Hensgens H.E.S.J., 1982, In: Metabolic Compartmentation (Sies, H., ed.) pp 259-286, Academic Press London.
Meyer, M., Maly, K., Uberall, F., Hoflacher, J., and Grunicke H., 1991, J. Biol. Chem. 266, 8230.
Miller, M.S., Kwock, L., and Wallach, D.F.H. 1975, Cancer Res., 35:1826.
Moffett, J., Perier, F., Jones, M., and Englesberg, E., 1987, Proc. Natl. Acad. Sci (USA) 84:8040.
Morin, O., Forest, C., and Fehlmann, M., 1981, FEBS Lett., 127:109.
Mortimore, G.E., and Mondon, C.E. 1970, J. Biol. Chem., 245:2375.
Mortimore, G.E., and Pösö, A.R., 1984, In: Glutamine metabolism in mammalian tissues (Häussinger, D. and Sies, H., eds.) pp. 138-157, Springer Verlag Heidelberg.
Mortimore, G.E., and Pösö, A.R., 1987, Ann. Rev. Nutr., 7:539.
Moule, S.K., and McGivan, J.D., 1987, Biochem. J., 247:233.
Moule, S.K., and McGivan, J.D., 1990, Biochim. Biophys. Acta, 1031:383.
Munro, G.F., Miller, R.A., Bell, C.A., and Verderber, E.L., 1975, Biochim. Biophys. Acta, 411:263.
Nakamura, K.D., and Weber, M.J., 1979, J. Cell. Physiol., 99:15.
Neely, A.N., Nelson, P.B. and Mortimore, G.E., 1974, Biochim. Biophys. Acta, 338:458.
Nosadini, R., Datta, H., Hodson, A. and Alberti, K.G.M.M., 1980, Biochem. J. 190:323.
Nucci, F.A., Rossi, P.A., Hofmann, E.K., Dall'Asta, V., Bussolati, O., Guidotti, G.G. and Gazzola, G.C., 1990, Renal Physiol. Biochem., 13: 172.
Oxender, D.L., Lee, M. and Cecchini, G., 1977, J. Biol. Chem., 252:2680.
Panet, R., Amir, I., and Atlan, H., 1986a, Biochim. Biophys. Acta, 859:117.
Panet, R., Snyder, D., and Atlan, H., 1986b, Biochem. J., 239:745.
Parners, J.R., Garvey, T.Q. III, and Isselbacher, K.J., 1976, J. Cell. Physiol., 89:789.
Parnes, J.R., and Isselbacher, K.J., 1978, Prog. Exp. Tumor Res., 22:79.
Patterson, M.K., Jr., Birckbichler, P.J., Conway, E., and Orr, G.R., 1976, Cancer Res., 36:394.
Perdue, J.F., 1977, "Virus-Transformed Cell Membranes", Academic Press, London.
Perry, J.W., and Oka, T., 1980, Biochim. Biophys. Acta 629:24.

Petronini, P.G., Piedimonte, G., and Borghetti, A.F., 1982, Biochim. Biophys. Acta, 693:13.
Petzinger, E., Kinne, R. and Sies, H., eds. 1989, Hepatic transport of organic substances. Springer Verlag Heidelberg.
Plomp, P.J.A.M., Boon, L., Caro, L.H.P., van Woerkom, G.M. and Meijer, A.J., 1990, Eur. J. Biochem., 191:237.
Pösö, A.R., Schworer, C.M. and Mortimore, G.E., 1982b, Biochem. Biophys. Res. Commun., 107:1433.
Pösö, A.R., Wert, J.J. and Mortimore, G.E., 1982a, J. Biol. Chem., 257:12114.
Poulin, R., and Pegg, A.E., 1990, J. Biol. Chem., 265:4025.
Prentki, M., Crettaz, M. and Jeanrenaud, B., 1981, J. Biol. Chem., 256: 4336.
Rennie, M.J., Hundal, H.S., Babij, P., MacLennan, P., Taylor, P.M., Watt, P.W., Jepson, M.M. and Millward, D.J., 1986, Lancet ii, 1008.
Reuss, L., 1988, Renal Physiol., 11:187.
Rinehart, C.A., Viceps-Madore, D., Fong, W.F., Ortiz, J.G., and Canellakis, E.S., 1985, J. Cell. Physiol. 123:435.
Rognstad, R., 1985, Biochem. Biophys. Res. Commmun., 130:229.
Roth, E., Karner, J. and Ollenschläger, G., 1990, J. Parent. Ent. Nutr., 14: 130S.
Sackin, H., 1987, Am. J. Physiol., 253:F1253.
Sander, G., and Pardee, A.B., 1972, J. Cell. Physiol., 80:267.
Schultz, S.G., 1981, Am. J. Physiol., 241:F579.
Schworer, C.M., and Mortimore, G.E., 1979, Proc. Natl. Acad. Sci., 76:3169.
Scott, D.F., Butcher, F.r., Potter, V.R., Morris, H.P., 1972, Cancer Res., 32: 2127.
Seglen, P.O., and Gordon, P.B., 1984, J. Cell. Biol., 99:435.
Seglen, P.O., Gordon, P.B. and Poli, A., 1980, Biochim. Biophys. Acta, 630: 103.
Shotwell, M.A., Jayme, D.W., Kilberg, M.S., and Oxender, D.L., 1981, J. Biol. Chem., 256:5422.
Shotwell, M.A., Kilberg, M.S., and Oxender, D.L., 1983, Biochim. Biophys. Acta., 737:267.
Sips, H.J., Groen, A.K. and Tager, J.M., 1980, FEBS Lett., 119:271.
Sommercorn, J.M., and Swick, R.W., 1981, J. Biol. Chem., 256:4816.
Stacey, D.W., Hsiang-Fu, 1984, Nature 310:508.
Stoll, B., McNelly, S., Buscher, H.P. and Häussinger, D., 1991, Hepatology, 12: 247.
Stoll, B., Gerok, W., Lang, F., and Häussinger, D., 1992, Biochem. J., in press.
Sugden, M.C., and Watts, D.I., 1982, Biochim. Biophys. Acta, 717:385.
Sugden, P.H., and Fuller, S.J., 1991, Biochem. J., 273:21.
Suzuki, M., Kawahara, K., Ogawa, A., Morita, T., Kawaguchi, Y., Kurihara, S. and Sakai, O., 1990, Am. J., Physiol., 258:F690.
Visciano, G., and Fehlmann, M., 1984, Mol. Cell. Endocrinol., 37:103.
Völkl, H., Paulmichl, M. and Lang, F., 1988, Renal Physiol., 11:1583.
vom Dahl, S., Hallbrucker, C., Lang, F., Gerok, W. and Häussinger, D., 1991b, Biochem. J., 278:771.
vom Dahl, S., Hallbrucker, C., Lang, F. and Häussinger, D., 1991a, Eur. J. Biochem., 198:73.
vom Dahl, S., Hallbrucker, C., Lang, F. and Häussinger, D., 1991c, Biol. Chem., Hoppe-Seyler 372:411.
Walker, P.R., and Whitfield, J.F., 1978, Proc. Natl. Acad. Sci., 75:1394.
Warren, R.S., Donner, D.B., Starners, H.F., Jr., and Brennen, M.F., 1987, Proc. Natl. Acad. Sci. (USA) 84:8619.
Watford, M., 1989, Trends Biochem. Sci,. 14:1.
Watford, M., 1990, Trends Biochem. Sci., 15:329.
Weber, M.J., Evans, P.K., Johnson, McNair, T.F., Nakamura, K.D., and Salter, D.W., 1984, Fed. Proc., 43:107.
Weissbach, L., Handlogten, M.E., Christensen, H.N., and Kilberg, M.S., 1982, J. Biol. Chem., 257:12006.
Weissbach, L., and Kilberg, M.S., 1984, J. Cell. Physiol., 121:133.
Wettstein, M., vom Dahl, S., Lang, F., Gerok, W. and Häussinger, D., 1990, Biol. Chem., Hoppe-Seyler., 371:493.
Woodside, K.H., Ward, W.F. and Mortimore, G.E., 1974, J. Biol. Chem., 249: 5458.

PART B

TISSUE SPECIFIC TRANSPORT AND FUNCTIONAL IMPLICATIONS

AMINO ACID TRANSPORT IN LIVER

Michael S. Kilberg[1] and Dieter Häussinger[2]

[1]Department of Biochemistry and Molecular Biology
University of Florida, Gainesville, USA

[2]Medizinische Universitätsklinik und Poliklinik
Freiburg, Germany

INTRODUCTION

The liver is recognized to play an important role in controlling the levels of free amino acids in the plasma, through both import and export of individual amino acids into or out of the hepatocyte (Schimassek and Gerok 1965, Ishikawa 1975). Much effort has been devoted to elaborate the mechanisms and regulatory properties of amino acid transport in liver and its relevance for overall liver cell function. Regarding amino transport in liver, two phenomena deserve special attention. First, hepatocytes in situ are polar cells exhibiting differences in amino acid transport properties across the basolateral (sinusoidal) and the canalicular membrane. Second, despite similar appearance under the light-microscope, hepatocytes are functionally different and there is marked hepatocyte heterogeneity in amino acid transport across the sinusoidal membrane. This must be kept in mind when transport studies with isolated hepatic plasma membrane vesicles or isolated hepatocytes are considered. In these experimental systems the cellular polarity is lost and cell heterogeneity regarding transport is frequently lost. Several approaches have been used to overcome these problems. They include preparations of plasma membrane vesicles from the sinusoidal and canalicular domains of the cell membrane, attempts to isolate hepatocytes from the periportal or perivenous areas of the acinus, the use of histoautoradiography and the use of radiolabel incorporation studies in the intact perfused liver by following the metabolic fate of an amino acid taken up into pathways which are characteristic for certain hepatocyte subpopulations.

HORMONAL REGULATION OF AMINO ACID TRANSPORT

Rat hepatocytes are a popular model to investigate the hormonal regulation of hepatic amino acid transport because they can be prepared in high yield and viability, and reflect "normal tissue" with the caveat mentioned above concerning cell polarity and heterogeneity. Hepatocytes in suspension or in primary culture can be used soon after isolation to monitor in vivo hormone effects, whereas long-term (24-48h) monolayer cultures minimize the impact of the in vivo status of the rat. The availability of numerous hepatoma cell lines allows investigations of transformation, and fetal hepatocytes permit the study of ontogeny. To compliment studies in whole cells, a number of laboratories have eliminated concerns of metabolism through the use of isolated plasma membrane vesicles. In general, hormones alter the V_{max} of the transport system, rather than the Km for the substrate amino acid (Shotwell et al., 1983), but a

Mammalian Amino Acid Transport, Edited by M.S. Kilberg and
D. Häussinger, Plenum Press, New York, 1992

number of factors must be considered when interpreting transport data based soley on kinetic parameters (Christensen, 1975). An increase in the V_{max} of a transport system is usually regarded to be the result of an increase in the number of active transporters within the membrane. Whether this increase arises from activation of carriers already present or synthesis of new carriers will not be known until the appropriate molecular probes are available. In many instances, inhibitors of either RNA or protein biosynthesis are employed to provide supportive evidence for de novo synthesis of transporters. Transport of an amino acid by multiple systems warrants extra caution in the interpretation of kinetic values. Hormonal regulation of nutrient transfer between the various organs represents an important process by which animals maintain interorgan coordination of metabolism. General reviews of hormonal control of amino acid transport have been published (Shotwell et al., 1983; Kilberg et al., 1985; Cheeseman, 1991), as has a discussion of interorgan transfer of amino acids (Christensen, 1990) and genetic approaches to transport (Englesberg and Moffett, 1986). Although a few of the transport systems for the amino acids show some modulation by hormones, System A activity exhibits far more extensive regulation then any other. Given that it is the best characterized process and to provide a more in depth description of hormone action, much of the following summary will focus on the glucagon-dependent induction of hepatic System A activity. Regulation of hepatic amino acid transport by numerous growth factors is discussed in the chapter of this monograph by Häussinger, Lang, and Kilberg. The effects of other selected hormones will be summarized briefly.

Glucagon

Hormone-mediated induction of hepatic System A occurs in response to a broad spectrum of hormone classes (Shotwell et al., 1983). Some of these hormones, such as polypeptide hormones, act through plasma membrane receptors, whereas others like glucocorticoids interact with nuclear receptors. Clearly, any hypothesis aimed at the mechanism(s) for hormonal control of hepatic System A must account for action by the wide range of hormone types. Glucagon-induced transport is typical and will be used as an example.

The relation between glucagon binding and stimulation of Na^+-dependent AIB uptake has been studied for isolated hepatocytes in suspension (Morin et al., 1982; Canivet et al., 1980; Fehlmann et al., 1981) or in monolayer culture (Morin et al., 1982). The half-maximal stimulation occurs at 0.5 to 1.5 nM glucagon (Morin et al., 1982; LeCam and Freychet, 1976), whereas the maximal stimulation is observed at about 100 nM (LeCam and Freychet, 1976). Interestingly, maximal induction of AIB uptake occurs when only 50-70% of the hormone receptors were occupied (Freychet and LeCam, 1978; Fehlmann et al., 1981). Consistent with the cAMP second-messenger hypothesis for glucagon action, the increased transport that develops over several hours following exposure to hormone is realized even if the "free" glucagon is removed from the medium after the first 15-30 min of incubation (Pariza et al., 1976; Fehlmann et al., 1979). Furthermore, cAMP alone can stimulate transport through protein synthesis-dependent (Auberger et al., 1983) and protein synthesis-independent mechanisms (Moule et al., 1987).

Our current understanding of transport regulation suggests that transcriptional control of a System A-associated gene is required for most of the hormone induction observed. Modest stimulation of System A activity (20-40%) occurs within a few minutes after addition of glucagon, but a much larger increase in activity is seen after a time lag of 60 min. The increase detected during the first 60 min of exposure to hormone is cycloheximide-insensitive (Edmondson and Lumeng, 1980) and probably is the result of an increased membrane potential (Moule et al., 1987), whereas the remainder of the elevated transport activity is blocked by inhibitors of either RNA or protein synthesis (Pariza et al., 1976; Crettaz et al., 1983; Christensen and Kilberg, 1987). The 60 min lag period prior to the cycloheximide-sensitive increase in transport is compatible with the time required for the processing, transit, and insertion of newly synthesized plasma membrane proteins. Supportive evidence for de novo synthesis of a plasma membrane protein comes from the results of Barber et al. (1983) who showed that the inhibitor of asparagine-linked glycoprotein biosynthesis, tunicamycin, blocked the hormone induction of System A. As shown previously for cycloheximide or puromycin (Pariza et al., 1976; Christensen and Kilberg, 1987), if tunicamycin was added to hepatocytes after the hormone-mediated induction of transport activity had begun, inhibition of further stimulation occurred within 1 hr (Barber et al., 1983). These results demonstrate that the System A-associated glycoprotein responsible for the increased uptake activity must be continuously synthesized if the stimulation of transport is to be maintained. Whether this glycoprotein is the carrier itself is not known, but that hypothesis is the least complicated one imaginable. Stimulation by glucagon does not occur in enucleated hepatocytes, despite that fact that glucagon binding to the resulting cytoplasts is slightly increased compared to control cells (Crettaz et al., 1983). Glucagon-induced transport is retained in the cytoplasts if the enucleation is performed after the hormone treatment.

The protein synthesis-dependent hormone induction of transport coincides with the ability to isolate plasma membrane vesicles that retain the stimulated activity (Schenerman and Kilberg, 1986). This observation also holds for the enhancement of System A transport activity following amino acid starvation of cells in culture (Fong et al., 1990). Furthermore, all of the elevated transport activity following glucagon treatment of rats can be recovered following solubilization of plasma membrane proteins and reconstitution into proteoliposomes (Fafournoux et al., 1989). Thus, liver tissue from hormone-treated animals serves as excellent starting material for purification of the carrier protein (Kilberg, 1989).

Cariappa and Kilberg (1990) have shown that glucagon-induced System A activity can be detected in transport-competent Golgi vesicles. The low basal activity and the ability to stimulate the transport activity by 5- to 10-fold permitted the investigators to monitor the new "wave" of carrier biosynthesis following glucagon treatment. A lag of 30-45 min prior to increased activity in cis-Golgi was observed. The acquisition of enhanced transport activity then continued through the cellular machinery to include medial- and trans-Golgi, and finally the plasma membrane. This time-course is consistent with the 1 h lag reported for intact cells or tissue (Christensen and Kilberg, 1987), and is consistent with the proposal that hormone action is via carrier biosynthesis. Further studies demonstrated that the newly synthesized transport activity arrives at the basolateral plasma membrane prior to its transfer, presumably by transcytosis (Bartles and Hubbard, 1988), to the canalicular surface (Cariappa and Kilberg, unpublished results). Tunicamycin treatment did not prevent delivery of the transporter activity to the Golgi, but blocked the subsequent transfer to the plasma membrane consistent with the observations in whole cells (Barber et al., 1983).

After removal of the hormone, elevated System A activity rapidly decays to a basal level with a half-time of about 1.5 h (Handlogten and Kilberg, 1984). This inactivation does not occur if the hepatocytes containing the glucagon-stimulated activity are cultured in medium lacking amino acids (Handlogten et al., 1985). This counter-balance of hormone stimulation and amino acid repression has been shown at substrate levels known to occur in vivo (Fafournoux et al., 1985). For cultured cells, the addition of a single amino acid, even a non-metabolizable amino acid analog such as 2-aminoisobutyric acid (AIB), is sufficient to initiate the decay. Thus, this process is strikingly similar to the amino acid-dependent repression associated with adaptive regulation for System A (Bracy et al., 1986). Interestingly, inhibitors of either RNA or protein synthesis can block the decay of the glucagon-enhanced transport (Kilberg, 1985). In the case of protein synthesis inhibitors, even if they are added several hours after the decay has begun, further loss of transport activity will be prevented (Handlogten and Kilberg, 1984; Handlogten et al., 1985). These data suggest that a newly synthesized mRNA and a corresponding regulatory protein are necessary for the turnover of the hormone-stimulated System A activity. This putative regulatory protein appears to have a half-life of less than 1 h (Pariza et al., 1976; Handlogten and Kilberg, 1984). Moffett and Englesberg (1986) have produced CHO cell mutants which constitutively express high levels of System A, possibly due to a defect in this regulatory process. Interestingly, in contrast to general RNA synthesis inhibitors such as α-amanitin or actinomycin D, inhibitors of poly(A) polymerase such as ara-A (Handlogten et al., 1985) or cordycepin (Kilberg, 1986) do not block the amino acid-dependent repression of the glucagon-induced System A activity.

Insulin

The basic characteristics for insulin-dependent activation of hepatic System A are similar to those for glucagon. Insulin action on transport appears to be mediated via the insulin receptor because anti-receptor antibodies suppress the hormonal activation of transport (Shimoyama et al., 1985). System A represents the only well characterized cellular activity for which insulin and glucagon fail to show antagonistic actions (LeCam and Freychet, 1978a), and indeed, both stimulate at physiological levels. Like glucagon, the stimulation of transport by insulin is preceded by a lag period of 30-60 min (Kletzien et al., 1976b; LeCam and Freychet, 1978a), the enhancement is blocked by either cycloheximide or actinomycin D, (Kletzien et al., 1976b; Fehlmann et al., 1979; LeCam and Freychet, 1978a) and is primarily the result of an increase in the V_{max} (Morin et al., 1982; Kletzien et al., 1976b; Fehlmann et al., 1979). In contrast to glucagon action on System A, the synergistic effect of glucocorticoids does not occur with insulin (Kletzien et al., 1976b; Fehlmann et al., 1979), but may for System N (Gebhardt and Kleeman, 1987). Given that the molecular mechanism by which insulin regulates gene transcription is unclear, considerable research will be required before further insight into transport control by this hormone is completely understood. Varela et al. (1990) have shown that an insulin-induced phospho-oligosaccharide stimulates AIB uptake by isolated hepatocytes. Insulin treatment in vivo can suppress the diabetes-associated increase in hepatic System A activity, presumably by reducing the level of circulating glucagon.

Glucocorticoids

The effect of glucocorticoid on the activity of System A has been investigated using both hepatocyte suspensions and primary monolayer cultures. Significant differences between these two preparations with regard to amino acid transport have been reported (Gurr and Potter, 1980; Kilberg et al., 1981; Morin et al., 1982). Glucocorticoids, including the analog dexamethasone, stimulate Na^+-dependent AIB transport by a RNA- and protein synthesis-dependent process (LeCam and Freychet, 1977). In agreement with data from cell suspensions (LeCam and Freychet, 1977), but in contrast to data from long-term (> 24 h) monolayer cultures (Kletzien et al., 1975), System A activity in hepatocytes cultured in monolayers for less than 24 h is stimulated by dexamethasone alone (Kilberg et al., 1983). Furthermore, in long-term cultures the combined effect of dexamethasone and glucagon is synergistic. This "permissive effect" for glucocorticoids in cultured cells was originally described by Potter and his associates (Kletzien et al., 1976a). In contrast, stimulation of transport by a combination of dexamethasone and glucagon does not produce a synergistic response in freshly isolated hepatocytes (LeCam and Freychet, 1977).

System A activity in hepatocyte cultures older than 24 h does not respond to dexamethasone alone (Kletzien et al., 1975; Kletzien et al., 1976a). If hepatocytes are prepared from adrenalectomized rats and then cultured for less than 24 hr, the basal rate of transport by System A is actually enhanced compared to uptake control animals (Kilberg et al., 1983). However, the stimulation of System A-mediated transport by glucagon, insulin, or dexamethasone treatment of these cultured cells is significantly reduced by prior adrenalectomy of the donor animal (Kilberg et al., 1983). The decreased competence of these cells to respond in vitro to either insulin or glucagon was not reversed by including dexamethasone in the culture medium, even if the cells from adrenalectomized rats were held in culture for 24 hr before the challenge by insulin or glucagon.

Catecholamines

Catecholamine binding to beta-adrenergic receptors increases cellular cAMP, whereas their mechanism of action via alpha-adrenergic receptors is cAMP-independent (Lefkowitz et al., 1983). Pariza et al. (1977) have shown that epinephrine treatment of cultured hepatocytes increases cAMP levels and stimulates Na^+-dependent AIB transport. However, exposure of the cells to the beta-adrenergic agonist isoproterenol also results in increased levels of cAMP, but does not alter amino acid transport. Furthermore, the beta-antagonist propranolol only partially blocks the increase in transport, yet completely abolishes the elevation in cAMP values (Pariza et al., 1977). Alpha-blockers such as phentolamine and dihydroergotamine have no effect on epinephrine-induced cAMP levels as expected, but inhibit the transport enhancement by catecholamines (Canivet et al., 1980; LeCam and Freychet, 1978b). Collectively, these results suggest that catecholamine-stimulated AIB uptake is mediated via cAMP-independent alpha-adrenergic receptors.

ZONAL HETEROGENEITY IN AMINO ACID TRANSPORT

The functional units of the liver are the so-called acini (Rappaport, 1976) which extend along the sinusoids from the terminal portal venule to the terminal hepatic venule. Periportal hepatocytes (near the sinusoidal inflow) can be distinguished biochemically from more downstream located perivenous hepatocytes (near the sinusoidal outflow), although the borderline between the periportal and the perivenous compartment is not defined in general anatomical terms. The definition used here is a comparative-functional one and the size of a periportal or perivenous compartment exhibiting a given transport system or metabolic pathway will depend on the transport systems and metabolic pathways under consideration. Periportal and perivenous hepatocytes differ in their complement of enzymes and metabolic functions. Several reviews have appeared on "functional hepatocyte heterogeneity" or "metabolic zonation" in liver (Traber et al., 1988; Jungermann and Katz, 1982, 1989; Katz, 1989; Häussinger, 1990). Although only a few studies have addressed the question of liver cell heterogeneity in amino acid transport, it is beyond a doubt that transport plays an important role in hepatic nitrogen handling. A small hepatocyte population surrounding the terminal hepatic venule as a layer of 1-3 cells in thickness and amounting to about 7% of all hepatocytes deserves special attention. These cells are a subpopulation of perivenous cells and are immunohistochemically and functionally characterized by the exclusive presence of glutamine synthetase as well as the lack of urea cycle enzymes and probably glutaminase (Gebhardt and Mecke, 1983a; Häussinger, 1983; Gaasbeek-Janzen et al., 1984; Smith and Campbell, 1988). In view of their function, these cells were named perivenous scavenger cells (Häussinger, 1990); this term is also used here.

Zonal Transport of Glutamate, Aspartate and Related Dicarboxylates

Studies with the intact isolated perfused rat liver have demonstrated a marked hepatocyte heterogeneity with respect to glutamate and aspartate transport (Häussinger and Gerok, 1983; Taylor and Rennie, 1987; Häussinger, 1989; Häussinger et al., 1989b; Stoll et al., 1991). Vascular glutamate taken up by the intact liver is almost completely utilized for glutamine synthesis when the latter pathway is stimulated to V_{max} values by NH_4Cl addition (Häussinger et al., 1989b). Because glutamine synthetase is exclusively found in perivenous scavenger cells, these data suggested that vascular glutamate is taken up by this small cell population, whereas other hepatocytes do not measurably contribute to overall hepatic glutamate uptake in the intact organ. Such a strict hepatocyte heterogeneity regarding transport was confirmed in histoautoradiographical studies not only for glutamate (Fig. 1), but also for aspartate uptake (Stoll et al., 1991). In line with a predominant glutamate uptake by perivenous scavenger cells is also the finding that glutamate uptake by perfused liver is largely abolished after destruction of perivenous hepatocytes by CCl_4 (Häussinger and Gerok, 1983; Taylor and Rennie, 1987). Furthermore, glutamate delivered in vivo via the portal vein is largely taken up by perivenous hepatocytes as shown in studies on the short-term fate of injected (^{13}N)glutamate (Cooper et al., 1988).

Whereas metabolic label incorporation studies and histoautoradiography of the intact liver showed little if any significant glutamate uptake by periportal hepatocytes, sizeable rates of Na^+-independent glutamate uptake without zonal differences were found in isolated hepatocyte suspensions being enriched for periportal or perivenous cells (Burger et al., 1989). The Na^+-dependent glutamate uptake was smaller than the Na^+-independent component, but was 6- to 7-fold higher in perivenous compared to periportal cells. After 48h of culture and dexamethasone treatment, only the Na^+-dependent glutamate transport was markedly stimulated in both cell populations, but significantly higher uptake rates for glutamate by perivenous cells were still observed

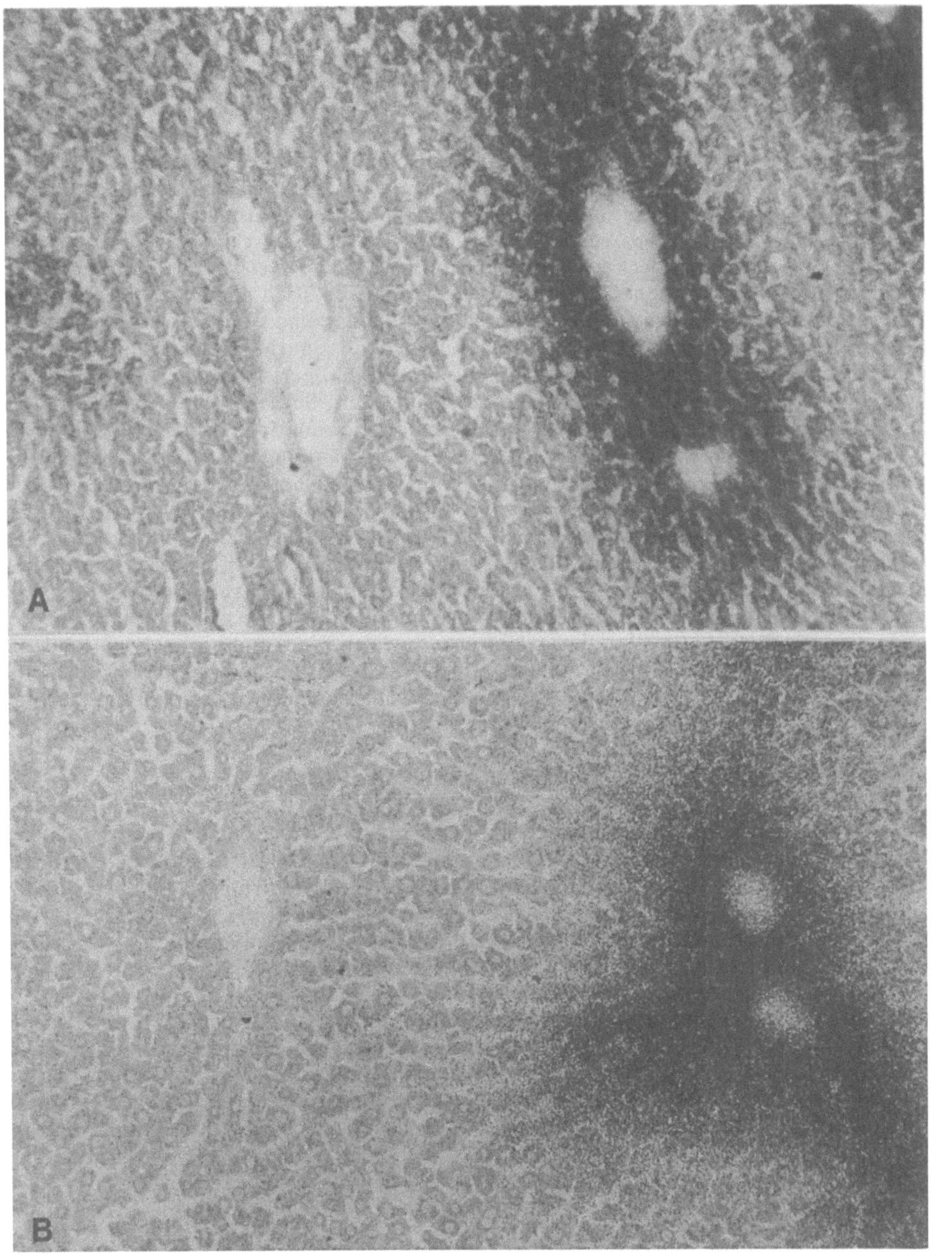

Fig. 1. Histoautoradiograms of perfused rat liver after bolus injection of (^{3}H)glutamate and (^{3}H)aspartate. The histoautoradiographic distribution patterns were not affected by the direction of liver perfusion, which was either from portal to hepatic vein (antegrade) or from hepatic to portal vein (retrograde). For experimental details, see Stoll et al., 1991. (A) (^{3}H)glutamate injected into retrogradely perfused rat liver. (B) (^{3}H)aspartate injected into antegradely perfused rat liver.

(Burger et al., 1989). These data are difficult to reconcile with the findings in the intact liver, but it is likely that the high rates of Na^+-independent glutamate transport described for isolated periportal cells are at least in part due to the experimental system used.

In isolated plasma membrane vesicles from rat liver, Na^+-dependent uptake of aspartate and glutamate was found (Sips et al., 1982) and studies with plasma membrane vesicles isolated from sinusoidal and canalicular domains of the cell membrane revealed that Na^+-dependent glutamate transport is restricted to the canalicular membrane, whereas sinusoidal glutamate transport is Na^+-independent (Ballatori et al., 1986; Cariappa and Kilberg, unpublished results). Because cellular polarity is lost upon isolation of hepatocytes, the question is still open whether dexamethasone induction of Na^+-dependent glutamate transport by intact hepatocytes (Gebhardt and Mecke, 1983a) is also restricted to the canalicular domain of the hepatocyte plasma membrane. In line with a Na^+-independent glutamate transporter in the sinusoidal membrane (Balatori et al., 1986), uptake of vascular glutamate by the intact perfused rat liver was suggested to involve an exchange mechanism (Häussinger et al., 1989b). Addition of labeled glutamate to perfused rat liver stimulates the release of unlabeled glutamate, and conversely, when intracellular glutamate was prelabeled, infusion of unlabeled glutamate resulted in a stimulation of labeled glutamate release (Häussinger et al., 1989b). Evidence has been presented that both glutamate influx and efflux involve perivenous scavenger cells, pointing to a glutamate/glutamate exchange mechanism.

In isolated perfused rat liver and isolated rat hepatocytes, a high intra/extracellular glutamate gradient of about 500 is maintained and glutamate release into the effluent perfusate is normally less than 80 nmol/g liver (Hems et al., 1968; Häussinger and Gerok, 1983, 1984; Häussinger et al., 1989a). Interestingly, glutamate efflux from the liver is stimulated markedly by α- ketoisocaproate, α-keto-4-methyl-valerate, α-keto-4-methylthiobutyrate, benzoate (Fig. 2), phenylpyruvate or α-keto-butyrate (Häussinger and Gerok, 1984; Häussinger et al., 1989a). Maximal rates of stimulated glutamate release are about 1 μmol/min/g liver. The effect is independent of

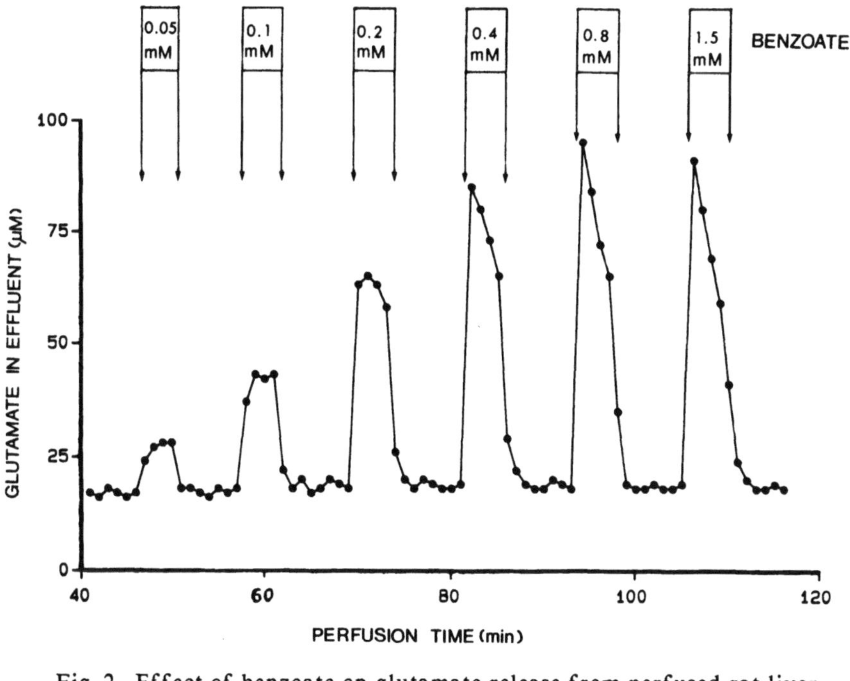

Fig. 2. Effect of benzoate on glutamate release from perfused rat liver.

the intracellular metabolism of the above-mentioned ketoacids and is independent of extracellular Na^+ (Häussinger and Gerok, 1984). Benzoate-stimulated glutamate release from the liver (Fig. 2) is SITS-sensitive and a glutamate/anion exchange system has been proposed (Häussinger et al., 1989a). The subacinar localization of this exchange system is less clear. Benzoate-induced glutamate efflux was not affected by addition of NH_4Cl, but was markedly stimulated after inhibition of glutamine synthetase, indicating the involvement of perivenous scavenger cells (Häussinger et al., 1989a). In line with this observation was the finding that after prelabeling of intracellular glutamate by addition of (1-^{14}C)oxoglutarate (which is predominantly taken up by scavenger cells), the specific radioactivity of the glutamate released under the influence of benzoate was similar to that of simultaneously released and newly synthesized glutamine (Häussinger, unpublished result). However, benzoate-induced glutamate efflux was also stimulated during periportal glutamine breakdown (which leads to glutamate accumulation inside periportal cells), pointing to an involvement of periportal cells at least under these experimental conditions. On the other hand, the histoautoradiographic radiolabel distribution obtained after injection of (^{3}H)glutamate was slightly, if at all, different between control livers and those in which periportal glutamine breakdown occurred at high rates (data not shown). Thus, available experimental evidence suggests that vascular glutamate is largely taken up in the intact liver by perivenous scavenger cells, probably involving a glutamate/anion exchange mechanism. The presence of this transport system in periportal hepatocytes, is not ruled out, although they seem to exhibit little activity under most experimental conditions studied so far. This could for example be due to low intracellular glutamate concentrations in periportal but not in perivenous scavenger cells, favoring glutamate/glutamate exchange in the latter cell population.

Like glutamate and aspartate, in isolated perfused rat liver, dicarboxylates such as α-ketoglutarate, malate, succinate and fumarate are predominantly taken up by perivenous scavenger cells as shown by histoautoradiography and the effective incorporation of these compounds into newly synthesized glutamine (Stoll and Häussinger, 1989, 1991; Stoll et al., 1991). When glutamine synthesis is stimulated to V_{max} values in the intact liver, oxoglutarate taken up from the vascular space is completely incorporated into glutamine. Uptake of α-ketoglutarate is Na^+-dependent in both, intact perfused rat liver and isolated plasma membrane vesicles originating from the sinusoidal domain (Stoll and Häussinger, 1989; Meier et al., 1990). This Na^+-dependent α-ketoglutarate transporter also accepts malate, fumarate and succinate, but not glutamate, as shown in vesicle studies and competition studies in perfused rat liver. In addition to this Na^+-dependent dicarboxylate transporter, which seems to be predominantly localized in perivenous scavenger cells, a dicarboxylate exchange system with minor activity has been suggested, whose subacinar localization is not yet clear (Stoll and Häussinger, 1989; Meier et al., 1990). Because gluconeogenesis predominantly occurs in periportal hepatocytes (Matsumura et al., 1984, for reviews see Jungermann and Katz 1982, 1989), the predominant uptake of vascular glutamate, aspartate and citric acid cycle dicarboxylates by perivenous scavenger cells also explains why these compounds are poor gluconeogenic substrates when added to perfused rat liver or isolated hepatocytes (Ross et al., 1967; Hems et al., 1968; Garrison and Haynes, 1973). On the other hand, in perfused rat liver, gluconeogenesis from methyl esters of malate and succinate is high (Rognstad, 1984), indicating rapid transport of these compounds into periportal hepatocytes.

Hepatocyte Heterogeneity in Glutamine Transport

Whereas glutamine synthetase is restricted to perivenous scavenger cells, glutaminase is found in periportal hepatocytes (Häussinger, 1983; Watford and Smith 1990). This reciprocal distribution of glutamine-metabolizing enzymes in the liver acinus implies opposite net fluxes of glutamine across the cell membrane in both cell types. It appears safe to assume that glutamine uptake into periportal hepatocytes largely involves the Na^+-dependent System N (Kilberg et al., 1980, Fafournoux et al., 1983a; Häussinger et al., 1985a), whereas glutamine efflux from perivenous scavenger cells occurs via facilitated diffusion (Fafournoux et al., 1983a). Both transport systems are inhibited by histidine (Kilberg et al., 1980; Fafournoux et al., 1983a). In line with this, histidine inhibits, in isolated perfused rat liver, not only periportal uptake of added glutamine, but also causes accumulation of newly synthesized glutamine inside the perivenous scavenger cells (Häussinger et al., 1985a). Studies with rat hepatocytes isolated from periportal and perivenous ends of the acinus showed no difference regarding Na^+-dependent transport of histidine, another substrate for System N, whereas V_{max} of Na^+-independent histidine transport was slightly higher in perivenous than in periportal cells (Burger et al., 1989). Given that preparations of perivenous hepatocytes contain only up to 30% perivenous scavenger cells, these findings may indicate that probably all urea synthesizing hepatocytes express this Na^+-dependent transport system.

Zonal Transport of Other Amino Acids

Only a few studies are available on the zonal transport of amino acids other than glutamate, aspartate and glutamine. A slightly faster uptake of cysteine by perivenous than by periportal cells was shown in studies with isolated rat hepatocytes (Penttilä, 1990), whereas the Na^+-dependent and Na^+- independent uptake of 2-aminoisobutyrate (AIB) was not significantly different between freshly prepared periportal and perivenous cells (Burger et al., 1989). Again, it should be noted that these findings do not rule out marked differences in amino acid transport via System A between perivenous scavenger cells and other hepatocytes, because the enrichment of scavenger cells in perivenous cell preparations is usually less than 30%. After 24-30h in primary culture in the presence of dexamethasone plus glucagon, significantly higher rates of AIB uptake were observed in perivenous cells. This indicates that hormonal induction of amino transport systems may also change the subacinar distribution of transport systems, as is already known for the zonation of metabolic pathways. On the other hand, (^{13}N)alanine uptake was found to be higher in the periportal area of the acinus in studies on the intact rat liver in vivo (Cooper et al., 1988). Like glutamine, in perfused rat liver proline is predominantly taken up by periportal hepatocytes, and glutamate derived from both of these precursors enters a glutamate pool which is distinct from that glutamate pool which is entered by vascular glutamate (Häussinger and Gerok, 1983). It is likely that similar to alanine, glutamine and proline, most gluconeogenic amino acids are predominantly taken up by periportal hepatocytes, i.e. into a compartment exhibiting a high capacity for gluconeogenesis and ureogenesis.

INTERACTION BETWEEN AMINO ACID TRANSPORT AND METABOLISM IN LIVER

The velocity of amino acid transport is frequently determined as the initial rate of uptake when an amino acid is present in the extracellular space, but not in the intracellular compartment. Such initial uptake rates are usually much faster than the known maximal rates of amino acid metabolism. This observation has led sometimes to the erroneous conclusion that transport does not control amino acid metabolism. It now is clear that amino acid transporters in the membrane cannot merely be seen as simple translocating systems, but rather as an important site controlling cellular function in general. Amino acid transporters, especially those driven by the electrochemical Na^+ gradient, build up intra/extracellular amino acid concentration gradients. Here, the activity and/or the energization of the transporter is one determinant for the steady state intracellular amino acid concentration, and thus, the intra/extracellular amino acid concentration gradient. This steady state amino acid concentration inside the cell is of crucial importance for flux through those amino acid metabolizing enzymes for which the K_m for the amino acid is far above the extracellular amino acid concentration. An example would be liver glutaminase. Thus, amino acid transporters can control amino acid metabolism by modifying the steady state intracellular amino acid concentration, although the unidirectional rate of initial uptake may far exceed the rate of metabolism.

Another consequence arising from the concentrative amino acid uptake into the hepatocyte is the generation of an osmotic gradient, resulting eventually in cell swelling and activation of volume-regulatory responses (Bakker-Grunwald, 1983; Kristensen and Folke, 1984; Häussinger et al., 1990; Wettstein et al., 1990). Cell volume changes were recently recognized as important modulators of liver cell function (for review see Häussinger and Lang, 1991). Accordingly, the activity of Na^+ dependent amino acid transporters can influence overall cell function by modulating cell volume. Likewise, cell volume changes have been shown to cause an activation of the System N amino acid transporter (Bode and Kilberg, 1991). The degree of amino acid-induced cell swelling seems to be largely related to the steady state intra/extracellular amino acid concentration gradient which is modified by hormones and the nutritional state in a complex way. Current evidence, obtained in experiments with rat liver, suggests that hepatocyte swelling acts like an anabolic signal which inhibits glycogenolysis and proteolysis, and stimulates protein synthesis, glycogen synthesis, amino acid transport and utilization. Accordingly, it was suggested that cell volume alterations may represent a new principle of metabolic control (Häussinger and Lang, 1991) and evidence also has been presented that hormones such as insulin and glucagon exert their metabolic effects in part via modulation of liver cell volume. Several long-known, but mechanistically poorly understood, effects of amino acids which could not be related to their metabolism, such as stimulation of amino acid uptake and glycogen synthesis or inhibition of proteolysis, can be mimicked by equipotent hypoosmotic swelling of the cells.

Control of Amino Acid Metabolism by Transport

Increasing evidence suggests that amino acid transport across the plasma and the mitochondrial membrane exerts significant control on amino acid metabolism. When isolated hepatocytes were incubated with alanine concentrations up to 3 mM, the intracellular alanine concentration was always lower than the extracellular concentration (Sips et al., 1980; Groen et al., 1982). When, however, intracellular alanine metabolism was inhibited, the intracellular alanine concentration increased to values above the extracellular concentration. This observation, as well as work from other laboratories (McGivan et al., 1981; Fafournoux et al., 1983b), led to the conclusion that alanine transport across the plasma membrane is a major site controlling the metabolism of this amino acid under different nutritional conditions. Alanine metabolism is increased in livers from rats fed a high protein diet; this effect was associated with an increase of alanine aminotransferase activity and increased gluconeogenesis from pyruvate in parallel with an enhanced rate of alanine transport across the plasma membrane (Fafournoux et al., 1983b, 1990). Adaptation to a high protein diet not only led to induction of transport System A, but also of most other transport systems except Systems ASC and L. Simultaneously, enzymes involved in the catabolism of the respective amino acids are induced; thus, the adaptation to high protein diet seems to involve a coordinated increase in both transport and metabolizing enzymes. Studies on the repartitioning of amino acids between plasma and liver under these conditions indicated that transport could modulate the hepatic utilization of alanine, serine, threonine, glycine, glutamine, aspartate and aromatic amino acids (Rémésy et al., 1988; Fafournoux et al., 1990).

Pogson et al. (1991) have recently applied the control strength theory (Kacser and Burns, 1973; Heinrich and Rappaport, 1974) to transport and metabolism of tryptophan, phenylalanine and tyrosine in isolated rat hepatocytes. Their data show that control is shared between both transport and metabolism with flux control coefficients for transport being 0.2-0.5. These values rose to 0.6-0.9 after induction of the corresponding amino acid metabolizing enzymes by dexamethasone, indicating a prevalence for control exerted by transport over that exerted by metabolizing enzymes.

Control of Zonal Nitrogen Handling by Glutamine Transport

The role of glutamine transport in regulating hepatic nitrogen handling is discussed here separately, because the physiological implications of transport and metabolism of this amino acid have been studied extensively, and because of the known zonal heterogeneity for glutamine metabolism (for review see Häussinger, 1990).

Fig. 3 depicts the structural and functional organization of glutamine and ammonia metabolism in the liver acinus (Häussinger, 1983). The two major hepatic ammonia-detoxicating systems, urea and glutamine synthesis, are anatomically located in series, one behind the other. In functional terms, this organization represents the sequence of a periportal low-affinity, but high capacity process (ureogenesis) and a perivenous high-affinity process for ammonia detoxication (glutamine synthesis), due to a 10-fold higher $K_m(NH_4^+)$ of carbamoylphosphate synthetase compared to that of glutamine synthetase and to the ammonia concentration found under physiological conditions in portal venous blood (Häussinger, 1983,1990; Sies and Häussinger, 1984). Both in vitro and in vivo, a considerable fraction of the ammonia delivered via the portal vein or being generated during protein breakdown inside the periportal cells escapes periportal urea synthesis and reaches the perivenous area (Häussinger, 1983; Häussinger et al., 1984; Cooper et al., 1987). Here, perivenous glutamine synthetase acts as a high-affinity scavenger for the ammonia which escaped periportal detoxication by urea synthesis. Thus, an intraacinar substrate transfer occurs: ammonia produced during protein breakdown in periportal cells is in part delivered to perivenous cells and is used there for glutamine synthesis (Häussinger, 1983). It is yet unknown whether intraacinar transfer of amino acids also occurs. The capacity of perivenous scavenger cells to synthesize glutamine from NH_4^+ is remarkable; highest rates of glutamine synthesis found in perfused rat liver correspond to 8-10 µmol/min/g perivenous cells. The carbon skeleton required for glutamine synthesis is in part provided by an almost exclusive uptake of vascular α-ketoglutarate, glutamate and malate (Stoll and Häussinger, 1989,1991; Stoll et al., 1991) into these perivenous cells. The maximal rates of glutamate, malate and α-ketoglutarate uptake by perivenous scavenger cells in the intact perfused rat liver are far below the maximal rates of ammonia-stimulated glutamine synthesis by this cell population (Stoll and Häussinger, 1989, 1991; Häussinger et al., 1989b). Although

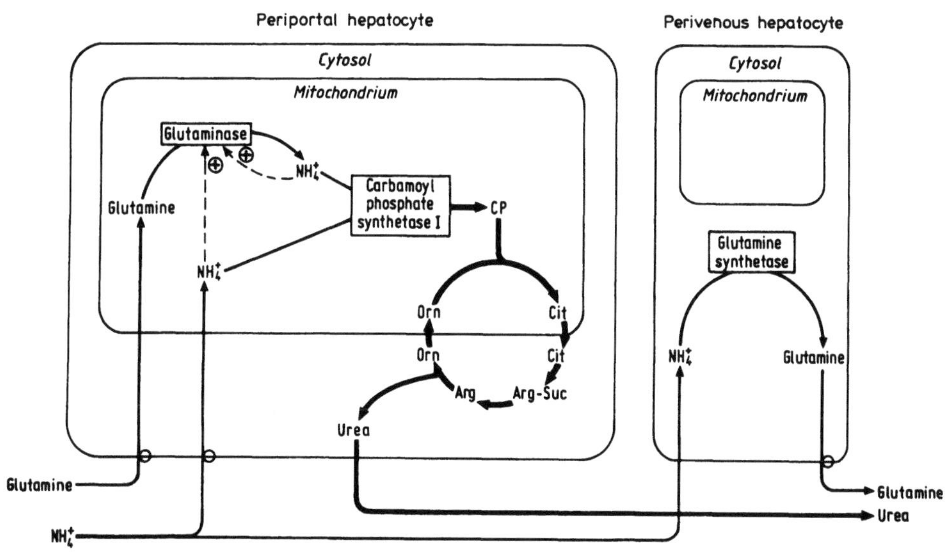

Fig. 3. Intercellular glutamine cycling and ureogenesis. Periportal glutaminase is activated by ammonia and acts as a pH- and hormone-modulated ammonia amplifier inside the mitochondria. The activity of this amplifier determines flux through the urea cycle ("low affinity system for ammonia detoxication"). Glutamine synthetase in perivenous cells acts as scavenger for the ammonia escaping periportal urea synthesis ("high affinity system for ammonia detoxication"). This anatomical sequence of low and high affinity detoxication systems uncouples urea synthesis from the primary need to maintain non-toxic ammonia levels and provides the basis for acid-base control of urea synthesis without threat of hyperammonemia. A complete conversion of a portal ammonia load into urea occurs at a well balanced acid-base situation. Under these conditions periportal glutamine consumption (ammonia amplifying) and perivenous glutamine synthesis (ammonia scavenging) match each other: there is no net glutamine turnover by the liver, but portal ammonia is converted efficiently into urea despite the low ammonia affinity of carbamoylphosphate synthetase. Transport significantly contributes to the regulation of glutamine metabolism in both cell types. From (Häussinger 1990).

most of the carbon skeleton required for glutamine synthesis in perivenous scavenger cells is generated intracellularly from lactate, transport of glutamate and dicarboxylates into these cells may well contribute to the control of glutamine formation. Indeed, evidence has been presented recently that the import of α-ketoglutarate into these cells can become rate-controlling for glutamine synthesis (Stoll et al., 1991). It has been calculated that the intracellular glutamine concentration in perivenous scavenger cells is far above that found in the extracellular compartment (Häussinger, 1985a). Accordingly, glutamine exit from perivenous scavenger cells, which probably occurs via System L, exhibits a K_m(glutamine) of about 10 mM (Fafournoux et al., 1983a) and may be driven by the intra/extracellular glutamine concentration gradient. There also is some evidence that the glutamine efflux pathway is regulated. Glutamine efflux from scavenger cells is inhibited by histidine at physiological concentrations, resulting in an increased intracellular glutamine concentration (Häussinger et al., 1985a). Under these conditions, the glutamine transporter might exert control on glutamine synthesis in perivenous scavenger cells, because glutamine synthetase was reported to be inhibited by glutamine (Deuel et al., 1978). Indeed, histidine inhibited glutamine production from glutamate and NH_4Cl so that the tissue level of glutamine increased simultaneously (Häussinger et al., 1985a).

Whereas glutamine synthetase is perivenous, glutaminase is found in periportal hepatocytes (Häussinger, 1983) and has a joint mitochondrial localization together with carbamoylphosphate synthetase. Liver glutaminase is activated by its product ammonia in physiological concentrations (Häussinger et al., 1975; Joseph and McGivan, 1978; Häussinger and Sies, 1979) and the function of the enzyme is now seen to amplify ammonia inside the mitochondria in parallel to that delivered via the portal vein or arising during hepatic amino acid breakdown (Fig. 3). Because urea synthesis normally is controlled by flux through carbamoylphosphate synthetase (Wanders et al., 1984; Meijer et al., 1985), which largely depends on the actual ammonia concentration inside the mitochondria, amplification of mitochondrial ammonia via glutaminase flux control becomes an important determinant of urea cycle flux. This is especially relevant in presence of physiologically low ammonia concentrations, which are about one order of magnitude below the K_m(ammonia) of carbamoylphosphate synthetase. Modulation of this ammonia amplifying system by extracellular pH and hormones was shown to play an important role in maintaining ammonia and bicarbonate homeostasis in the body (for reviews see Sies and Häussinger, 1984; Häussinger, 1990). Glutamine transport appears to be one major factor governing mitochondrial ammonia amplification, apart from regulatory mechanisms affecting directly enzyme activity of glutaminase (for reviews see McGivan et al., 1984; McGivan 1989). Whereas plasma membrane transport of glutamine occurs via the Na^+-dependent System N (Kilberg et al., 1981) glutamine transport system across the mitochondrial membrane is driven by the H^+ gradient across the mitochondrial membrane (Soboll, 1990). Both transport systems build up glutamine concentration gradients. At a physiological extracellular glutamine concentration of 0.6 mM, in rat liver measured in vivo or in vitro (Häussinger et al., 1985a; Lenzen et al. 1987), the cytosolic and mitochondrial concentrations are about 7 and 20 mM, respectively. Therefore, mitochondrial glutaminase operates under physiological conditions at glutamine concentrations near its K_m(glutamine) of about 28 mM (McGivan et al., 1984). Stimulation of glutaminase by ammonia or glucagon decreases the mitochondrial glutamine concentration from 19 to 9 mM when livers are perfused with physiological glutamine concentrations of 0.6 mM (Häussinger et al., 1985a). In addition, stimulation of glutamine transamination, which largely takes place in the cytosol, markedly decreased glutamine tissue levels and switched glutamine degradation from mitochondrial glutaminase (Häussinger et al., 1985b). Furthermore, histidine, which is known to inhibit competitively glutamine transport via Na^+-dependent System N (Kilberg et al., 1980), decreased both the rate of glutamine breakdown and the intracellular glutamine concentration (Häussinger et al., 1985a). These findings suggested that glutamine transport into periportal cells may become rate-controlling for glutamine breakdown. In line with this observation, application of the control strength theory to hepatic glutamine metabolism revealed significant flux control coefficients for both glutamine entry via System N and glutamine exit via System L (Pogson et al., 1991). However, in that study, hepatocyte heterogeneity in glutamine transport could not be taken into account due to technical limitations. Control of the mitochondrial steady state glutamine concentration by the activity of glutamine transporters also contributes to the pH-sensitivity of flux through glutaminase, and accordingly, to mitochondrial ammonia amplifying. In the perfused rat liver, at a constant extracellular glutamine concentration of 0.6 mM, an increase of the extracellular pH from 7.3 to 7.7 increases the mitochondrial steady state glutamine concentration from 15 to 50 mM (Lenzen et al., 1987). This is paralleled by a 3- to 4-fold increase of flux through glutaminase (Häussinger et al., 1983; Lenzen et al., 1987). Likewise, as shown in isolated hepatocytes, lowering of the extracellular pH diminished the initial rate of glutamine entry and lowered the intra/extracellular distribution ratio of glutamine (Fafournoux et al., 1983a). It is of interest to note that the activity of glutamine transport increases under most conditions known to stimulate urea synthesis, i.e. conditions requiring enhanced mitochondrial ammonia amplifying. These circumstances include dexamethasone and glucagon - treatment (Gebhardt and Kleemann, 1987), feeding of a high-protein diet (Rémésy et al., 1987; Fafournoux et al., 1990), increased delivery of amino acids (Weissbach and Kilberg, 1984) and alkalosis (Fafournoux et al., 1983a; Lenzen et al., 1987). Amino acid transport systems may not only regulate flux through periportal glutaminase by setting the intramitochondrial steady state glutamine concentration, but also by activating glutaminase via alterations of cellular and mitochondrial swelling. Concentrative amino acid uptake into liver cells creates an osmotic gradient, resulting in cell swelling (Bakker-Grunwald, 1983; Kristensen and Folke, 1984; Häussinger et al., 1990; Wettstein et al., 1990). Hypoosmotic cell swelling was shown to stimulate periportal glutamine breakdown (Häussinger et al., 1990), probably due to mitochondrial swelling and subsequent glutaminase activation (McGivan et al., 1980). Concentrative uptake of amino acids into the liver is decreased in acidosis (Fafournoux et al., 1983a,b; Lenzen et al., 1987; Boon and Meijer, 1988; Sestoft et al., 1982), which could contribute to glutaminase inactivation via a decrease in cell volume (Häussinger et al., 1990).

In the intact liver, periportal glutaminase and perivenous glutamine synthetase are simultaneously active resulting in periportal uptake and breakdown, and perivenous resynthesis and export of glutamine (Häussinger and Sies, 1979; Häussinger, 1983; Häussinger et al., 1983). This energy-consuming cycling of glutamine is called the "intercellular glutamine cycle" (Häussinger, 1983) and its importance for overall nitrogen and bicarbonate homeostasis has been stressed (for review see Sies and Häussinger, 1984; Häussinger, 1990). Glutamine cycling is under complex metabolic and hormonal control, involving flux changes through both periportal glutaminase and perivenous glutamine synthetase (Häussinger et al., 1983). Control at the transport step across the respective membranes contributes significantly to this simultaneous regulation of periportal breakdown and perivenous resynthesis of glutamine, and allows the liver to switch from net glutamine uptake to net glutamine output as the metabolic needs of the animal change.

REFERENCES

Auberger, P., Samson, M., LeCam, G., and LeCam, A., 1983, J. Cell. Physiol. 117: 204-210

Bakker-Grunwald, T., 1983, Biochim. Biophys. Acta 731: 239-242

Balatori, N., Moseley, R.H. and Boyer, J.L., 1986, J. Biol. Chem. 261: 6216-6221

Barber, E.F., Handlogten, M.E., and Kilberg, M.S., 1983, J. Biol. Chem. 258:11851-11855

Bartles, J.R. and Hubbard, A.L., 1988, Trends Biochem. Sci. 13: 181-184

Bode, B. and Kilberg, M., 1991, J. Biol. Chem. 266: 7376-7381

Boon, L. and Meijer, A.J., 1988, Eur. J. Biochem. 172: 465-469

Bracy, D.S., Handlogten, M.E., Barber, E.F., Han, H.-P. and Kilberg, M.S., 1986, J. Biol. Chem. 261: 1514-1520

Burger, H.J., Gebhardt, R., Mayer, C. and Mecke, D., 1989, Hepatology 9: 22-28

Canivet, B., Fehlmann, M., and Freychet, P., 1980, Mol. Cell. Endocrinol. 19: 253-261

Cariappa, R. and Kilberg, M.S., 1990, J. Biol. Chem. 265: 1470-1475

Cheeseman, C.I., 1991, Prog. Biophys. Molec. Biol. 55: 71-84.

Christensen, H.N., Biological Transport W.A. Benjamin, Inc., 1975

Christensen, H.N., 1990, Physiol. Rev. 70: 43-77.

Christensen, H.N., and Kilberg, M.S., Amino acid transport across the plasma membrane: role of regulation in interogran flows. in Amino acid transport in animal cells. (Yudilevich, D.L. and Boyd, C.A.R., eds) Manchester University Press, 1987.

Cooper, A.J., Nieves, E., Coleman, A.E., Filc-DeRicco, S. and Gelbard, A.S., 1987, J. Biol. Chem. 262: 1073-1080
Cooper, A.J., Nieves, E., Rosenspire, K.C., Filc-DeRicco, S., Gelbard, A. and Brusilow, S.W., 1988, J. Biol. Chem. 12268-12273
Crettaz, M., Kahn, C.R., and Fehlmann, M., 1983, J. Cell. Physiol. 103: 159-168
Deuel, T.F., Louie, M. and Lerner, A., 1978, J. Biol. Chem. 253: 6111-6118
Edmondson, J.W. and Lumeng, L., 1980, Biochem. Biophys. Res. Commun. 96: 61-68
Englesberg, E. and Moffett, J., 1986, J. Memb. Biol. 91: 199-212
Fafournoux, P., Dudenhausen, E., and Kilberg, M.S., 1989, J. Biol. Chem. 264: 4805-4811
Fafournoux, P., Demigné, C., Rémésy, C. and LeCam, A., 1983a, Biochem.J. 216: 401-408
Fafournoux, P., Rémésy, C. and Demigné, C., 1983b Biochem. J. 210: 645-652
Fafournoux, P., Remesy, C. and Demigne, C., 1985 Biochem. J. 231: 315-320
Fafournoux, P., Rémésy, C. and Demigné, C., 1990, Am. J. Physiol. 259: E614-625
Fehlmann, M., LeCam, A., and Freychet, P., 1979, J. Biol. Chem. 254: 10431-10437
Fehlmann, M., Morin, O., Kitabgi, P., and Freychet, P., 1981, Endocrinology 109: 253-261
Fong, A.D., Handlogten, M.E., and Kilberg, M.S., 1990, Biochim. Biophys. Acta 1022: 325-332
Freychet, P., and LeCam, A., 1978, Proc. Ciba Found. pp. 247-268.
Gaasbeek-Janzen, J.W., Lamers, W.H., Moorman, A.F.M., De Graaf, A., Los, A.J. and Charles, R., 1984, J. Histochem. Cytochem. 32: 557-564
Garrison, J.C. and Haynes, R.C., 1973, J. Biol. Chem. 248: 5333-5343
Gebhardt, R. and Kleemann, E., 1987, Eur. J. Biochem. 16: 339-344
Gebhardt, R. and Mecke, D., 1983a, Embo J. 2: 567-57
Gebhardt, R. and Kleemann, E., 1987, Eur. J. Biochem. 166: 339-344
Groen, A.K., Sips, H.J., Vervoorn, R.C. and Tager, J.M., 1982, Eur. J. Biochem. 122: 87-93
Gurr, J.A., and Potter, V.R., 1980, Ann. N.Y. Acad. Sci. 349: 57-66
Handlogten, M.E., Barber, E.F., Bracy, D.S. and Kilberg, M.S., 1985, Mol.Cell. Endocrin. 43: 61-69
Handlogten, M.E., and Kilberg, M.S., 1984, J. Biol. Chem. 259: 3519-3525
Häussinger, D., Weiss, L. and Sies, H., 1975, Eur. J. Biochem. 52: 421-431
Häussinger, D. and Sies, H., 1979, Eur. J. Biochem. 101: 179-184
Häussinger, D., 1983, Eur. J. Biochem. 133: 269-274
Häussinger, D. and Gerok, W., 1983, Eur. J. Biochem. 136: 421-425
Häussinger, D., Gerok, W. and Sies, H., 1983, Biochim. Biophys. Acta 755: 272-278
Häussinger, D. and Gerok, W., 1984, Eur. J. Biochem. 143: 491-497
Häussinger, D., Sies, H. and Gerok, W., 1984, Chem. Biol. Interact. 48: 191-194
Häussinger, D., Soboll, S., Meijer, A.J., Gerok, W., Tager, J.M. and Sies, H., 1985a, Eur. J. Biochem. 152: 597-603
Häussinger, D., Stehle, T. and Gerok, W., 1985b, Biol. Chem. Hoppe-Seyler 366: 527-536
Häussinger, D., 1989 In: Hepatic Transport of organic solutes (Petzinger, E., Kinne, R.K.H. and Sies, H., eds.), Springer Verlag Heidelberg,pp. 157-166.
Häussinger, D., Stehle, T. and Colombo, J.P., 1989a, Biochem. J. 264: 837-843
Häussinger, D., Stoll, B., Stehle, T. and Gerok, W., 1989b, Eur. J. Biochem. 185: 189-195
Häussinger, D., 1990, Biochem. J. 267: 281-290
Häussinger, D., Lang, F., Bauers, K. and Gerok, W., 1990, Eur. J. Biochem. 193: 891-898
Häussinger, D. and Lang, F., 1991, Biochem. Cell. Biol. 69: 1-4
Heinrich, R. and Rappaport, T., 1974, Eur. J. Biochem. 42: 89-95
Hems, R., Stubbs, M. and Krebs, H.A., 1968, Biochem. J. 107: 807-815
Ishikawa, E., 1975, Adv. Enz. Reg. 14: 117-136
Jefferson, L.S., Schworer, C.M., and Tolman, E.L., 1975, J. Biol. Chem. 250: 197-204
Joseph, S.K. and McGivan, J.D., 1978, Biochim. Biophys. Acta 543: 16-28
Jungermann, K. and Katz, N., 1982, In: Metabolic Compartmentation (Sies, H., ed.) Academic Press London, pp. 411-435.
Jungermann, K. and Katz, N., 1989, Physiol. Rev. 69: 708-764
Kacser, H. and Burns, J.A., 1973, Symp. Soc. Exp. Biol. 32: 65-104
Katz, N., 1989, Histochem. J. 21: 517-529
Kilberg, M.S., 1986, Trends Biochem. Sci. 11: 183-186
Kilberg, M.S., 1989, Regulation of Hepatic Amino Acid Transport and Partial Purification of The System A Carrier. In "Hepatic Transport of Orgainc Solutes, (Petzinger, E., Kinne, R.K.H., and Sies, H., eds.) Springer-Verlag, Berlin, pp. 167-176.

Kilberg, M.S., Barber, E.F., and Handlogten, M.E., 1985, Curr. Topics Cell. Regul. 25: 133-163.
Kilberg, M.S., Handlogten, M.E. and Christensen, H.N., 1980, J. Biol. Chem. 255: 4011-4019
Kilberg, M.S., Handlogten, M.E. and Christensen, H.N., 1981, J. Biol. Chem. 256: 3304-3312
Kilberg, M.S., Vida, T.A., and Barber, E.F., 1983, J. Cell Physiol. 114: 45-52
Kletzien, R.F., Pariza, M.W., Becker, J.E., and Potter, V.R., 1975, Nature (London) 256: 46-47
Kletzien, R.F., Pariza, M.W., Becker, J.E., and Potter, V.R., 1976a, J. Cell. Physiol. 89: 641-646
Kletzien, R.F., Pariza, M.W., Becker, J.E., and Potter, V.R., 1976b, J. Biol. Chem. 251: 3014-3020
Kristensen, L.O. and Folke, M., 1984, Biochem. J. 221: 265-268
LeCam, A., and Freychet, P., 1976, Biochem. Biophys. Res. Commun. 72: 893-901
LeCam, A., and Freychet, P., 1977, Mol. Cell. Endocrinol. 9: 205-214
LeCam, A., and Freychet, P., 1978a, Diabetologia 15: 117-123
LeCam, A., and Freychet, P., 1978b, Endocrinology 102: 379-385
Lefkowitz, R.J., Stadel, J.M., and Caron, J.G., 1983, Ann. Rev. Biochem. 52:159-186
Lenzen, C., Soboll, S., Sies, H. and Häussinger, D., 1987, Eur. J. Biochem. 166: 483-488
Matsumura, T., Kashiwagi, T., Meren, H. and Thurman, R.G., 1984, Eur. J. Biochem. 144: 409-415
McGivan, J.D., Lacey, J.H. and Joseph, S.K., 1980, Biochem. J. 192: 537-542
McGivan, J.D., Ramsell, J.C. and Lacey, J.H., 1981, Biochim. Biophys. Acta 644: 295-304
McGivan, J.D., Bradford, N.M., Verhoeven, A.J. and Meijer, A.J., 1984, in Glutamine Metabolism in Mammalian Tissues (Häussinger, D. and Sies, H., eds.) Springer Verlag Heidelberg, pp. 122-137.
McGivan, J.D., 1989, In: Glutamine and Glutamate in Mammals (Kvamme, E., ed.) Vol. 1, CRC Press, Inc. Boca Raton, pp. 183-202.
Meier, P.J., Zimmerli, B. and O'Neil, B., 1990, Biol. Chem. Hoppe-Seyler 371: 301
Meijer, A.J., Lof, C., Ramos, I. and Verhoeven, A.J., 1985, Eur. J. Biochem. 148: 189-196
Morin, O., Fehlmann, M., and Freychet, P., 1982, Mol. Cell. Endocrinol. 25: 339-352
Moule, S.K., Bradford, N.M., and McGivan, J.D., 1987, Biochem. J. 241: 737-743
Pariza, M.W., Butcher, F.R., Kletzien, F.R., Becker, J.E., and Potter, V.R., 1976, Proc. Natl. Acad. Sci. U.S.A. 73: 4511-4515
Pariza, M.W., Butcher, F.R., Becker, J.E., and Potter, V.R., 1977, Proc. Natl. Acad Sci. U.S.A. 74: 234-237
Penttilä, K.E., 1990, Biochem. J. 269: 659-664
Pogson, C., Low, S.Y., Knowles, R.G., Salter, M. and Rennie, M.J., 1991, In: Regulation of hepatic function (Grunnet, N. and Quistorff, B., eds.)Munksgaard Copenhagen, pp. 262-272.
Rappaport, A.M., 1976, Beitr. Pathol. 157: 215-243
Rémésy, C., Morand, C., Demigné, C. and Fafournoux, P., 1988, J. Nutr. 118: 569-578
Rognstad, R., 1984, Arch. Biochem. Biophys. 230: 605-609
Ross, B.D., Hems, R. and Krebs, H.A., 1967, Biochem. J. 102: 942-951
Schenerman, M.A. and Kilberg, M.S., Biochim. Biophys. Acta 856: 428-436
Schimassek, H. and Gerok, W., 1965, Biochem. Z. 343: 407-413
Sestoft, L., Bartels, P.O. and Folke, M., 1982, Clin. Physiol. 2: 51-58
Shimoyama, R., Savage, C.R., Jr., and Boden, G., 1985, J. Clin. Endocrin. Metab. 60: 928-933
Shotwell, M.A., Kilberg, M.S., and Oxender, D.L., 1983, Biochim. Biophys. Acta 737: 267-284
Sips, H.J., Groen, A.K. and Tager, J.M., 1980, FEBS Lett 119: 271-274
Sips, H.J., DeGraaf, P.A. and van Dam, K., 1982, Eur. J. Biochem. 122: 259-264
Sies, H. and Häussinger, D., 1984 in: Glutamine Metabolism in Mammalian Tissues (Häussinger, D. and Sies, H., eds.) Springer Verlag Heidelberg, pp. 78-97.
Smith, D.D. and Campbell, J.W., 1988, Proc. Natl. Acad. Sci. 85: 160-164
Soboll, S., 1990, Biol. Chem. Hoppe-Seyler 371: 304
Stoll, B. and Häussinger, D., 1989, Eur. J. Biochem. 181: 709-716
Stoll, B. and Häussinger, D., 1991, Eur. J. Biochem. 195: 121-129
Stoll, B., McNelly, S., Buscher, H.P. and Häussinger, D., 1991, Hepatology 13: 247-253
Taylor, P.M. and Rennie, M.J., 1987, FEBS Lett. 221: 370-374
Traber, P.G., Chianale, J. and Gumucio, J.J., 1988, Gastroenterology 95: 30-43

Varela, I., Avila, M., Mato, J., and Hue, L., 1990, Biochem. J. 267: 541-544
Wanders, R.J.A., Van Roermund, C.W.T. and Meijer, A.J., 1984, Eur. J. Biochem. 142: 247-254
Watford, M. and Smith, E.M., 1990, Biochem. J. 267: 265-267
Weissbach, L. and Kilberg, M., 1984, J. Cell. Physiol. 121: 133-138
Wettstein, M., vom Dahl, S., Lang, F., Gerok, W. and Häussinger, D., 1990 Biol. Chem. Hoppe-Seyler 371: 493-501

AMINO ACID TRANSPORT IN INTESTINE

Bruce R. Stevens

Department of Physiology, College of Medicine
University of Florida, Gainesville, FL 32610-0274 USA

During the past 30 years many of the universal principles of membrane transport have been established or confirmed using the readily accessible and testable small intestine. Historically, amino acid transport studies have been conducted in a variety of species at various stages of ontogenetic development. The intact epithelium, isolated enterocytes, and isolated brush border and basolateral membranes have each been used to investigate the physiologic, thermodynamic and kinetic events of absorption. Recently, our laboratory (Pan et al., 1991) and others (Hidalgo & Borchardt, 1990) have used cultured intestinal cell lines to study amino acid transport regulation in monolayers that emulate the *in vivo* epithelium (Zweibaum et al., 1991).

Many enterocyte transport systems operate with unique substrate selectivities and modes of regulation, consistent with the small intestine's distinctive physiological role of extracting amino acids from the environment. Amino acid transporters of the intestinal brush border, in particular, extend and diversify the menu of systems classically encountered in non-epithelial cells.

This chapter examines the most current views regarding the functional role of intestinal amino acid transport, the regulation of enterocyte membrane amino acid transport activity, the multiplicity of intestinal transport systems, and the kinetic and molecular mechanism of a prototype intestinal Na^+/amino acid cotransporter. This chapter covers only free amino acid absorption. The reader is referred elsewhere (Matthews, 1987; Ugolev, 1990; Webb, 1990) regarding the related topic of oligopeptide absorption.

INTESTINAL REGULATION OF AMINO ACID TRANSPORT

Amino acid metabolism in animals begins with the initial step of amino nitrogen entry into the body via intestinal transport. Absorptive epithelial cells lining the intestine are responsible for the initial introduction of amino acids into the pool of the whole animal. Once nutritional amino nitrogen enters the portal blood, then the liver, kidney and the other organs each play a significant role in regulating the individual organ and whole-body inter-organ metabolism of amino acids.

The intestinal lumen encounters daily variations in amino acid concentrations that occur as a function of both time and food composition. In stark contrast to cells of many internal organs, the small intestine adaptively up-regulates its amino acid transport capacity in response to the increase in amino acid substrate

Mammalian Amino Acid Transport, Edited by M.S. Kilberg and
D. Häussinger, Plenum Press, New York, 1992

concentrations (Pan et al., 1991; Stevens, 1991; Salloum et al., 1990; Scharrer et al., 1981; Stein et al., 1987; Ferraris et al., 1988; Diamond & Karasov, 1987; Ferraris & Diamond, 1989). The up-regulation of intestinal amino acid uptake occurs on the time scale of 0.5-1 day, while the reversal to a constitutive baseline uptake level takes several days. As stated by Ferraris and Diamond (1989), "[Intestinal transport] regulation serves to match transporter activity to a running average of diet composition over many days; it does not switch transporters on and off with each meal." The physiologic function of intestinal adaptive regulation is to maintain a modest safety margin between the intestine's uptake capacity and the animal's dietary intake, without wasting energy synthesizing and maintaining unneeded enterocytes and extraneous transporters in the membranes (Diamond, 1991). The intestinal capacity for initial entry of amino acids into the body must not be the rate-limiting step controlling whole-body intermediary metabolism of amino acids. Fig. 1 demonstrates this principle.

After an animal ingests and digests protein, concentrations of individual amino acids in the intestinal lumen rise to high levels in the range of 1-25 mM. At these concentrations, the amino acid transporters are generally running just below their maximal turnover rates (V_{max}), and net uptake via simple passive diffusion may actually be great enough to exceed the rates via mediated pathways (Stevens et al., 1984). At lower luminal concentrations, as in the postabsorptive state, relative uptake via transporter-mediated systems exceeds uptake via simple passive diffusion, as calculated from measured diffusion and transporter kinetic parameters (Stevens et al., 1982, 1984). At very low luminal concentrations (μM range), relative membrane transport is virtually completely dependent on the activities of mediated transporters that are running below their maximal rates. This chapter concerns the intestinal absorption of amino acids only via the cellular route; recent arguments (Diamond, 1991) suggest that paracellular solvent drag across the epithelium tight junctions contributes little to total nutrient uptake under normal physiological conditions.

The absorptive capacity of the intestinal mucosa reversibly increases by invoking three mechanisms: (1) mucosal hyperplasia non-selectively increases uptake, (2) individual enterocytes selectively increase the transport capacity by increasing copies of specific transporter systems in their membranes, and (3) individual transporters themselves are modulated locally as a consequence of manipulating the transporter kinetics by way of transmembrane electrochemical gradients of activator ions, amino acid substrates, and transmembrane electrical potentials. Each of these three modes of regulation is discussed in this chapter.

The mucosa undergoes a reversible hyperplastic response that increases the height of the villus in response to hyperalimentation (Laganiere et al., 1986), corticosteroid treatment (James et al., 1987), or hyperphagia as a consequence of diabetes, hyperthyroidism, neoplasia, pregnancy and lactation, or adolescent rapid growth (Levine, 1991). This trophic response nonspecifically increases amino acid absorption by up to 5-fold in the adult mammal (Diamond, 1991). The mucosa contains an epithelial layer one cell thick comprised primarily of enterocytes. As crypt cells rapidly proliferate (24 hours mean generation time in adult humans) and migrate up the villus (range 1-5 days from crypt to villus tip), the epithelium quickly gains mature absorptive cells. As the crypt to villus tip distance increases during hyperplasia, the resulting larger membrane surface area of the epithelium permits additional simple passive diffusion to occur, and also increases the number of copies of all transporter systems. There has been a recent surge in interest concerning the luminal and extraluminal triggers, and the genetic and cellular mechanisms that underlie general mucosal hypertrophy and increased absorption of amino acid (Levine, 1991). Likely trigger candidates may include circulating hormones and local peptide growth factors (Ugolev et al., 1990; Henning, 1987), and the amino acid substrates themselves (Pan et al., 1991). Our laboratory is currently investigating the interactive role of luminal growth factors, discrete amino acid substrates, and intracellular second messengers in up-regulating specific amino acid transport systems in cultured intestinal epithelial cells.

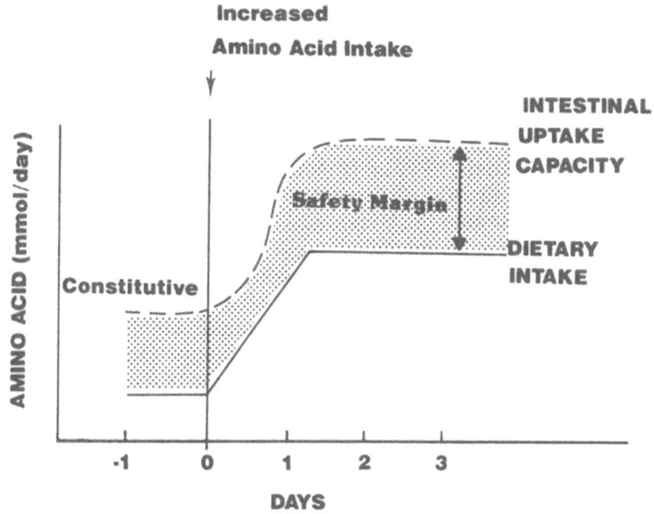

Fig. 1. Functional significance of intestinal up-regulation of amino acid uptake. The intestinal mucosa maintains a constitutive capacity for amino acid uptake, even when dietary intake of protein is low. With augmented dietary intake of protein or amino acids, the uptake capacity of the entire integrated small intestine increases slowly to a new baseline. Specific amino acids invoke the up-regulation of specific transport systems and stimulate a non-specific mucosal hypertrophy. The net physiological effect is to maintain a safety margin of absorptive capacity that is slightly more than the dietary intake. Thus, nutrient extraction from the environment is prevented from being the rate-limiting step in whole-body amino acid intermediary metabolism. (Adapted from Diamond, 1991, with permission.)

The second level of adaptive up-regulation involves reversibly increasing the capacity of specific membrane transport systems by a factor of about 2- to 10-fold. James et al. (1987) showed that dexamethasone treatment selectively increased Na^+-dependent alanine transport, while not affecting Na^+-independent alanine or lysine transport. Monosaccharide and amino acid transporters are independently regulated (Stein et al., 1987). When dietary carbohydrate is isocalorically replaced by protein, essential and nonessential amino acid uptakes are increased, while monosaccharide uptake is concomitantly attenuated (Stein et al., 1987). This phenomenon has been demonstrated in intact intestinal mucosa and in apical membrane vesicles isolated from diet-treated animals (Diamond & Karasov, 1987; Scharrer et al., 1981). Specific amino acids can trigger the increased transport activity of specific transporter systems in enterocyte membranes. The transport system substrates are often the best inducers of their particular transporter in the monolayer epithelium of *in vitro* cultured Caco-2 cells (Pan et al., 1991) and in intact intestine (Stein et al., 1987). However, the specificity pattern is not always clearly delineated in intact intestine experiments. For example, in the mouse intact everted intestine (Stein et al., 1987), system X^-_{AG} is induced best by Asp^- and is unaffected by Lys^+, while Asp^- is the best inducer of system y^+. The mechanism underlying this paradox has not been explored; perhaps a countertransport inter-dependence of the affecter and affected transport systems occurs in the intact intestinal mucosa.

DIFFERENTIAL REGULATION: GENETIC CONTROL AND LOCAL EFFECTORS

Fixed genetic controls may govern the constitutive capacity and potential for adaptive up-regulation in a given portion of the gut. This may be true for individual transport systems or for the mucosa in general. Amino acid transporters are independently controlled ontogenetically, as well as anatomically along the intestinal oral/aboral axis and along the crypt/villus axis. The relative amino acid uptake capacity is collectively greatest at birth and during the early growth period, and then irreversibly declines after adolescence (Buddington & Diamond, 1989), regardless of the dietary level. After the suckling period in most species, there is a greater decline in relative uptake of essential amino acids than nonessential amino acids (Buddington & Diamond, 1989). In fetal and neonatal mammals (and in adult birds), Na^+-dependent and Na^+-independent amino acid transport occurs in the colon (Munck 1989; Buddington & Diamond, 1989), but a few days after birth the mammalian colon diminishes its competency to transport amino acids, and the small intestine becomes the primary absorptive organ.

During the early developmental period, uptake along the oral/aboral axis of the small intestine also changes with age in many species. Upon reaching adulthood, uptake in the jejunum dominates over the extreme proximal and distal regions. There is a need to explore the void in our understanding of the signals, transduction systems, and genetic mechanisms that control developmental and regional changes (oral/aboral axis and crypt/villus axis) in the capacities of intestinal amino acid transport systems. Recent evidence suggests that luminal growth factors such as EGF may play a trigger role (Schwartz & Storozuk, 1988; Morisset & Solomon, 1991; Pan et al., 1991), but the timing and receptor control of these signals is still under investigation.

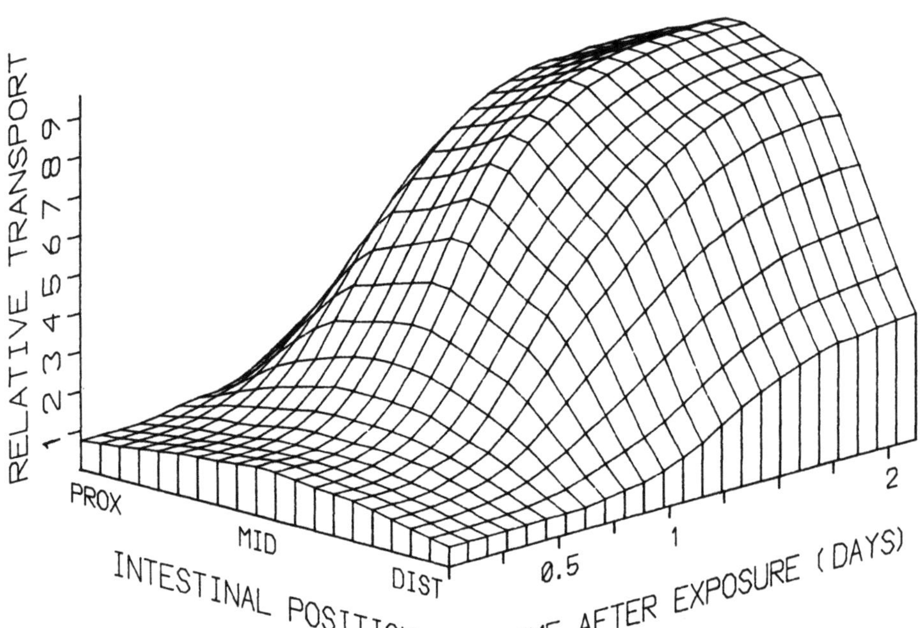

Fig. 2. The concept of regional up-regulation of dipolar amino acid uptake capacity in adult mammalian small intestine. Over a given length of adult intestine, the constitutive uptake capacity is generally greater in the jejunal regions compared to the distal ileum. Upon exposure to particular amino acids, the relative absorptive capacity of the small intestine rises slowly as new copies of specific amino acid transporters are inserted into enterocytes that mature and increase the villus mass. Generally, the jejunum may be the most competent region, responding to dietary exposure with the greatest increase in relative transport capacity.

As an integrated organ, the entire small intestine can increase its capacity to absorb amino acids. In response to dietary amino acid, the enhanced relative transport competence of the jejunum likely results in a differential up-regulation response along the oral/aboral axis. The concept of regional differential regulation of transport is depicted in Fig. 2. As discussed above, the time frame is about a day for inserting new copies of specific transport systems into the membranes of enterocytes.

INTESTINAL DOWN-REGULATION

Intestinal down-regulation of amino acid transport is essentially the return to a constitutive baseline absorptive capacity in the absence of stimulating substrate. When exposure to amino acid ceases, the newly differentiating and developing enterocytes are not stimulated to insert extra copies of transporters into their membranes, and also the numbers of existing absorptive cells decreases as the mucosal crypt : villus tip ratio declines. The down-regulation occurs over several days because transporter-enhanced enterocytes are slowly sloughed off from the villus tip into the lumen, and are replaced by constitutive-transporter enterocytes migrating up the villus from the crypt. In the absence of elevated amino acid, growth factors, or other trophic agents, the net effect is a return to a baseline transport level. Why does the transport capacity not shut off completely? It may be that the continual sloughing of enterocytes, plus plasma protein that constantly enters the lumen across the leaky proximal epithelial junctions, contribute to a moderately consistent source of amino acids in contact with the villus absorptive cells. In addition, the mucosal cells may be genetically programmed to maintain a fixed low level absorptive capacity regardless of the presence of substrate. More studies must address the constitutive maintenance state of intestinal transporters.

LUMEN TO BLOOD TRANSPORT: STUDIES USING TISSUES, CELLS AND MEMBRANE VESICLES

Intestinal transport involves solute vectorial movement through a composite cell layer that separates the outside environment from the animal's internal milieu; it is not simply the trafficking of metabolic reactants and products into and out of a given cell. The intestinal mucosa contains a single layer of histologically polarized epithelial cells that are joined by junctions to form a continuous sheet. Amino acid transport across the mucosa — from the lumen to portal blood — involves movement through a series of membranes and aqueous barriers. To understand intestinal absorption, one must be aware of how each of these barriers contributes to the overall trans-epithelial transport, because the non-membrane pathways affect substrate availability for the membrane pathways. Each barrier serves as a compartment that influences transport in the abutting barriers of the contiguous series. In the case of carrier-mediated transporter systems, the binding and uptake kinetic measurements and countertransport events are thus strongly influenced by the "true" concentrations of substrates and competitive inhibitors that are actually present at the solute/membrane interface.

Tissue and transcellular transport measurements are physiologic and anatomic composite phenomena, while membrane vesicle studies reflect only events within the membrane-proper. Regulation studies of intestinal amino acid transport therefore involve both membrane-level and mucosa-level events. Most amino acid transport occurs via the transcellular route across absorptive enterocytes (Diamond, 1991); other cells types (goblet cells, panath cells, Peyer's patch cells) and paracellular pathways offer a relatively small contribution to total trans-epithelial transport, and are not discussed further here.

Beginning in the lumen, amino acids first encounter a water layer and a mucous/glycocalyx layer that together form an unstirred compartment overlying the enterocytes. The unstirred layer is about 40 μm thick in human and dogs (Strocchi

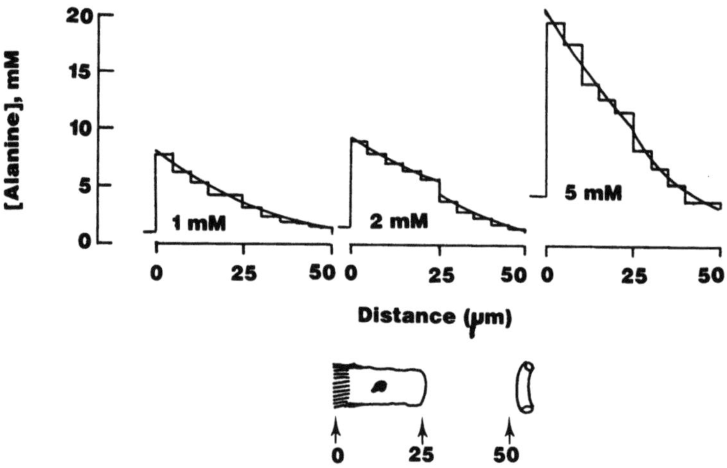

Fig. 3. Amino acid concentration gradients across enterocytes. Rabbit distal ileum was incubated for 45 sec with 1, 2 or 5 mM [^{3}H]-L-alanine at the apical side, then fixed with glutaraldehyde and processed for autoradiography. Transported alanine was determined by measuring optical densities determined in 5 µm increments from the brush border membrane to the core of the villus. A steep gradient was established across the apical membrane from the cytoplasmic side to the luminal side, and also across the cytoplasm from the apical membrane to the basolateral membrane. The distribution was concentration-dependent from 1-5 mM. (Adapted from data of Paterson et al., 1982, with permission.)

& Levitt, 1991), and is maintained *in vivo* by a unique contraction mechanism of the villi; previous erroneous overestimates of the unstirred layer ranged from 150-600 µm. The pH of the portion of the layer nearest to the plasma membrane surface ranges from about pH 6.1 in the proximal and jejunal intestine, to pH 7.3 in the distal ileum; this microclimate surface pH is maintained even when the luminal bulk pH is 7-7.3 (Lucas, 1983). Amino acids must diffuse through this layer, and are presented to the next barrier (the apical microvillus, or brush border, plasma membrane) at a concentration and charged state often different from that in the lumen bulk phase. The reader is referred elsewhere (Preston, 1981; Anderson & Munck, 1987) concerning further discussion of the effect of unstirred layers on solute concentrations at epithelial membrane surfaces. During transcellular movement of amino acid, the brush border membrane of the absorptive enterocyte then transports amino acid into the cytoplasm by passive diffusion and mediated transport systems. The cytoplasm of these columnar epithelial cells is about 25 µm thick and serves as an important unstirred compartment (actually a series of microscopic compartments) that forms a steep gradient between the brush border and basolateral membrane. An example of this gradient for alanine is presented in the profile of Fig. 3, as measured using microdensitomety of villi autoradiographs (Paterson et al., 1982). In this example, the net transport events of the brush border and basolateral membranes permits alanine to form a steep gradient at the apical surface, extending serosally. From the cytoplasmic surface of the apical membrane the gradient extends in two directions: to the lumen, and to the blood. Alanine leaves the enterocyte via the basolateral membrane mainly via Na^+-independent system L plus simple passive diffusion (Stevens et al., 1982, 1984; Wilde & Kilberg, 1991). The high capacity transporters and high permeability of the basolateral membrane (relative to the brush border) (Stevens et al., 1984; Hopfer, 1987) permits rapid exodus of amino acid to the serosal fluid. In the postabsorptive state, enterocyte nutrition may be insured from the basolateral side via Na^+-dependent systems A and ASC (Mircheff et al., 1980; Wilde & Kilberg, 1991) and countertransport involving Na^+-independent systems L and y^+.

MENU OF INTESTINAL AMINO ACID TRANSPORT SYSTEMS

The functionally and biochemically distinct brush border and basolateral membranes act in concert to move nutrients from the lumen to the portal blood. Many amino acid transport systems in the brush border membrane of the small intestine are unique, not found in the membranes of other internal organs. On the other hand, intestinal basolateral membranes possess transporters in common with non-epithelial plasma membranes. Intestinal amino acid transport studies have included radiolabeled substrate uptake kinetics and analogue cross inhibition profiles, activation by inorganic ions, electrogenic ion-dependent uptake using microelectrodes and potential-sensitive fluorescent dyes, *cis* and *trans*-effects using isolated membrane vesicles, and specific inborn diseases of amino acid malabsorption shared by the renal proximal tubule (Stevens et al., 1984; Wright et al., 1986).

Assessment of data and assignment of system names has been confounded by numerous factors, including overlapping specificities of parallel pathways, lack of unique probes which define single systems, variations among species and stages of ontogenetic development, Ca^{+2} masking of some system activities in isolated vesicles, and the interaction of fluxes across both membranes in intact enterocytes (Munck, 1985a,b; Anderson & Munck, 1987; Miyamoto et al., 1990). The enterocyte transport systems reported most often, for a wide range of species, are depicted in Fig. 4. Several distinctive brush border transporters, particularly systems NBB (renamed below as system B) and IMINO, simply defy Procrustean conformity to the conventional non-epithelial categories (Bannai et al., 1984; Christensen 1985,1990; Stevens et al., 1984). Uptake systems in the mouse and guinea pig intestine are also reticent to clear-cut characterization (Karasov et al., 1986; Hayashi et al., 1980). The other systems shown in Fig. 4 are same as those found also in non-intestinal cells.

A strictly Na^+-dependent transporter, broadly selective for dipolar amino acids, was initially described for the brush border (apical) membrane of rabbit small intestine, and was denoted "system NBB" by Stevens et al. (1982, 1984). The letters for this unique system were originally designed to stand for Neutral Brush Border; in retrospect the word "dipolar" or "zwitterionic" would have been chemically more accurate than "neutral" (Christensen & Handlogten, 1977). I now propose that "system NBB" be renamed as "system B" to be consistant with, and to reflect the remarkable analogy with the well characterized system $B^{0,+}$ found in blastocysts (Christensen, personal communication; Van Winkle, 1988, 1990). The substrate selectivities and functions of these systems appear to be related, yet the brush border (system B) and blastocyst (system $B^{0,+}$) forms possess unique properties, and are thus distinguished by the superscripts (Christensen, 1990; Van Winkle, 1988, 1990).

In addition to its existence in the adult rabbit (Stevens et al., 1984) and dog (Bulus et al., 1989), system B (NBB) also occurs in apical membranes of human fetal small intestine (Malo 1991), differentiated and undifferentiated enterocytic Caco-2 adenocarcinoma cells (Pan et al., 1990), and lower vertebrate and invertebrate gut (reviewed by Ahearn et al., 1991; Stevens, 1992). System B (NBB) of intestine is closely related to system $B^{0,+}$ found in blastocysts and certain oocytes (Van Winkle 1990): both systems are Na^+dependent, possess a broad selectivity for most dipolar amino acids, and both interact with the bicyclo amino acids BCH or BCO. However, at neutral pH a discriminating factor is that system $B^{0,+}$ is inhibited by cationic amino acids such as lysine, while system B is not interactive with lysine (Stevens et al., 1982). In the light of studies in invertebrates, conceptuses, eel intestines, fetal intestines, transformed intestinal cells, and adult enterocytes, $B^{0,+}$ may be an ontogenetic and even phylogenetic precursor of the intestinal epithelium apical membrane system B (NBB) (Van Winkle, 1988, 1990; Christensen 1990). These two transporters are functionally related in their ability to extract nutrient amino acids from the environment, in contrast to system A which is an integral participant in regulating intracellular metabolism. To properly examine the relationship between these systems, it will be necessary to determine their molecular structures and/or

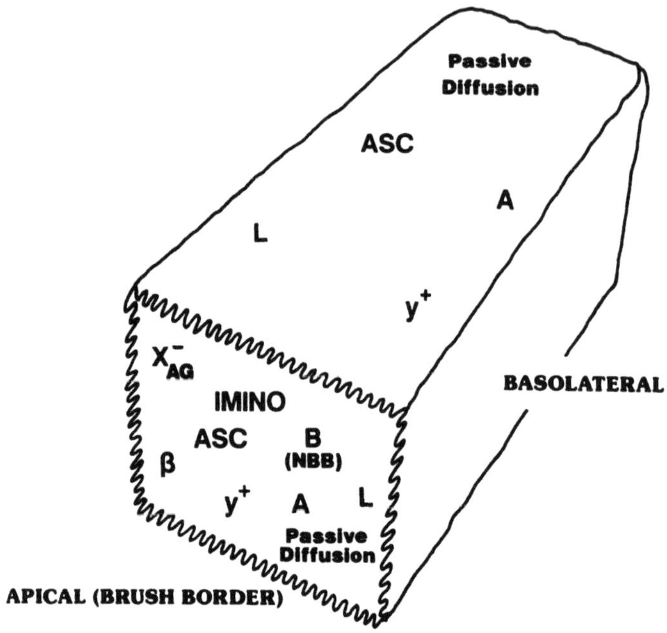

Fig. 4. Amino acid transport pathways reported in mammalian small intestine of various species. This summary represents a compilation of systems observed in various studies using isolated brush border and basolateral membranes, isolated enterocytes, or intact mucosa. All the systems are not necessarily present simultaneously in a given enterocyte, in the same region of the intestine, nor during all stages of development in all species. The original designation for system "NBB" is now renamed system "B." Simple passive diffusion may be a significant pathway at high concentrations of amino acid. The passive permeability in basolateral membranes is generally greater than brush border membranes.

DNA coding sequences. Some progress has been made in this direction using rat intestinal unfractionated mRNA expression in *Xenopus* oocytes. With this expression system (Hediger et al., 1987), general Na^+-dependent and Na^+-independent transport activity has been observed for dipolar and basic amino acids (Aoshima et al., 1988) cystine (McNamara, 1991).

Another Na^+-dependent system found exclusively in intestinal apical membrane is system IMINO, which transports the imino acids proline and pipecolate with a high degree of selectivity (Stevens & Wright, 1985,1987; Wright et al., 1985). Characteristic IMINO system uptake has been reported in invertebrate and mammalian (rabbit, guinea pig, mouse, cows) adult intestines (Ahearn et al., 1991; Stevens et al., 1984; Stevens, 1991; Munck, 1983; Karasov et al., 1987; Moe et al., 1987), and in human fetal small intestine (Malo, 1991). The highly restricted substrate selectivity and Na^+ dependency have permitted the IMINO system to serve as an intestinal prototype for kinetic/mechanism studies, as discussed below in this chapter.

Many brush border systems conform to those found in non-epithelial membranes. These are depicted in Fig. 4. The Na^+-dependent system X_{AG} serves aspartate and glutamate, and may also be present on the basolateral membrane; some reports suggest roles for Cl^- and K^+ interacting with X^-_{AG} in the brush border (reviewed by Hopfer, 1987). Kinetic and electrophysiological studies in eel intestinal BBMV's suggest a system X^-_{AG} activation stoichiometry of 1:2:1 (glutamate:Na^+:Cl^-) (Romano et al., 1989).

Na^+-dependent, Cl^--dependent transport systems separately exist for β-alanine and for taurine in brush border vesicles and intact tissue (Munck & Munck, 1990; Munck & Munck, 1992; Miyamoto et al., 1990a,b). In the presence of Na^+, the β-system is activated by anions ranked in the potency order $Cl^-=SCN^->F^->NO_3^->SO_4^{-2}$ (Miyamoto et al., 1990a). The β-system in BBMV's prepared using Mg^{++}/EGTA is inhibited by Ca^{++} in a dose-dependent manner; this accounts for the absence of β-system activity in BBMV's prepared using Ca^{++} (Miyamoto et al., 1990b; Stevens et al., 1984). In the face of these findings, intestinal membrane preparations should employ a divalent cation substitute for Ca^{++} to avoid the possibility of masking other transport systems.

Na^+-dependent systems A and ASC have been reported in guinea pig ileum brush border membrane vesicles (Hayashi et al., 1980), and in enterocytes isolated from guinea pig whole intestine (Del Castillo & Muniz, 1991). However, cross inhibition profiles, using a wide range of test substrates and inhibitor analogues, have not been suitably conducted yet to establish definitively the existence of these putative systems independent from the system B (NBB), IMINO, or other unique and unclassified systems (Hopfer, 1987; Karasov et al., 1986).

Some evidence suggests that glutamine may be transported in the brush border membrane via a system N-like transporter (Salloum et al., 1990, 1991), but others (Bulus et al., 1989) suggest instead that a general B-like (NBB-like) system is responsible. Brush border glutamine transport is strictly Na^+-dependent (no Li^+ substitution), competitively inhibited by system N substrates, and inhibited at pH < 7. Glutamine uptake across the brush border membrane is apparently up-regulated in the presence of augmented luminal glutamine levels (Salloum et al., 1990, 1991). Fig. 4 does not include system N because definitive identification of an apical or basolateral system N transporter remains unresolved (Salloum et al., 1990, 1991; Ghishan et al., 1990; Said et al., 1989) pending additional cross inhibition studies.

In contrast to the brush border systems, all the basolateral membrane transporters have been also reported in non-epithelial membranes. To ensure postabsorptive nutrition to the enterocytes, some of the basolateral systems can accumulate amino acids by Na^+-dependent secondary active transport from the serosal (blood) side. For example, glutamine is an important metabolic fuel of the enterocyte, and its uptake from blood is ensured by the basolateral Na^+-dependent system A and by Na^+-independent system L (Wilde & Kilberg, 1991; Mircheff et al., 1980; Salloum et al., 1990; Ghishan et al., 1989). The basolateral membrane also retrieves a variety of small dipolar amino acids from the blood via the Na^+-dependent system ASC and a possible Na^+-independent companion system asc (Mircheff, 1980; Lash & Jones, 1984).

Some data (Stevens et al. 1982, 1984; Wilson et al., 1990; Diamond & Karasov, 1987; Karasov et al., 1986; Ferraris & Diamond, 1989; Desjeux et al., 1987) reinforce the concept that apical membranes may possess several discrete Na^+-dependent transport systems that preferentially serve the essential amino acids (e.g., methionine, lysine, phenylalanine); incomplete findings prevent their concrete assignment in Fig. 4. However, strong evidence indicates that the essential cationic amino acids are transported by Na^+-independent system y^+ present on both the brush border and basolateral membranes, as depicted in Fig. 4. Interestingly, it was recently discovered that the mouse cell-surface MuLV retrovirus receptor (ecoR) is one and the same as the essential system y^+ transporter protein (Wang et al., 1991; Kim et al., 1991)! The essential bulky lipophilic amino acids such as phenylalanine, and most dipolar amino acids, are transported across both the brush border and basolateral membranes (Fig. 4) by the ubiquitous and broadly-tolerant system L (Stevens et al., 1984; Davies et al., 1987; Hopfer, 1987).

The phenomonological studies of the past 30 years are in need of a cloning and sequencing breakthrough to definitively characterize all the mammalian transporter proteins, as has been done directly for the intestinal brush border membrane SGLT1 glucose cotransporter (Hediger et al., 1987), or indirectly for the system y^+ protein (Wang et al., 1991; Kin et al., 1991). Only with the DNA and/or protein primary sequences can we clarify the differences among the transporter systems. Cloning genes for the transporter proteins will permit greater understanding of the ontogenetic and phylogenetic relationships between the brush border and basolateral membranes during the life cycle of various species. Furthermore, the recently discovered dual role for the system y^+ protein (as nutrient transporter and as retrovirus receptor) opens the possibility that the other amino acid transporters may also be exploited by viruses to gain entry into cells of the developing intestinal epithelium, and thereby play a role in the etiology of intestinal disease.

THE INTESTINAL Na^+/AMINO ACID COTRANSPORTER MECHANISM

Transporter activities can be controlled locally in the membrane as a consequence of the transporter's inherent kinetic properties of binding and translocation. The kinetic events can be influenced by transmembrane electrochemical gradients of amino acid substrates and activator ions, and by membrane electrical potentials. Enterocyte transport systems that share common substrates can therefore interact indirectly; for example, countertransport (Christensen 1990) influences the net flux rates and distribution of amino acids across the brush border and basolateral membranes.

After Christensen's laboratory (Riggs et al., 1958) initiated the seminal concept that ion gradients (not necessarily Na^+) might energize amino acid accumulation in single cells, Crane proposed and developed a model explicitly describing Na^+-dependent glucose transport in the intestine (reviewed by Crane, 1977). Crane's Na^+-gradient model was extended by Curran, Schultz and others (reviewed by Schultz, 1987) to account for the intestinal transport of free amino acids in an unmetabolized state. It is now established that a variety of ions - Na^+, K^+, Cl^-, H^+ - can drive various amino acid cotransport or countertransport mechanisms in intestines of various species (Ahearn et al., 1992; Giordana et al., 1989; Stevens, 1992; Schultz, 1987; Stein, 1896).

Refinement of the Na^+/solute cotransport kinetic model of mammalian intestine remains a priority in the literature, perhaps as a paradigm for a "universal" cotransport mechanism. Sodium electrochemical gradients provide the driving force for secondary active transport in small intestinal enterocytes, non-epithelial cells, and even prokaryote membranes (Kaback, 1987). The small intestinal and *E. coli* membrane transporter proteins show close homologies at the levels of nucleotide coding and amino acid primary sequences; the prokaryote Na^+/proline cotransporter (PutP) peptide and DNA sequences are about 50% homologous with those of the mammalian intestinal Na^+/glucose cotransporter (Hediger et al., 1989; Pajor et al., 1992; Nakao et al., 1987). Historically, the intestinal brush border membrane has provided a readily accessible and testable system to explore the fundamental principles of ion-activated transport applicable to all cell types.

Fig. 5 describes the current prototype kinetic mechanism for Na^+/amino acid cotransport in the small intestinal brush border membrane. This model is based on recent data compiled from a variety of sources and substrates, and may likely also apply to the mechanism for intestinal SGLT1 Na^+/glucose cotransport (Stevens, 1992).

Most current kinetic data (Stevens, 1992; Stevens & Wright 1985, 1987; Centelles et al., 1991) are consistent with the model of Fig. 5 describing a "preferred random" binding order; either Na or amino acid may bind first, but lumen → enterocyte vectorial movement and accumulation are favored when the activator Na^+ binds preferentially to the carrier *cis* binding sites (C_C) before amino acid at the extracellular surface of the brush border membrane. Initial Na^+ binding increases the affinity of the *cis*-orientated cotransporter binding sites for subsequent amino acid binding in a positive cooperative manner; thus a Na^+/cotransporter/amino acid complex (C_C2NaA) is formed. A conformation shift of the complex to form C_T2NaA results in placing the Na^+ and amino acid at the cytoplasmic (*trans*) side. Physically coupled transport of the activator ion (Na^+) with the substrate amino acid is promoted because the rate coefficient for the Na^+/cotransporter/amino acid complex conformational shift and subsequent translocation of amino acid across the membrane, is greater than the translocation rate coefficients for the cotransporter that is either empty (C) or partially loaded with amino acid only (CA) or Na^+ only (C2Na) (Centelles et al., 1991). Arguments against alternative models of binding order and symmetry are reviewed elsewhere (Centelles et al., 1991; Stevens, 1992).

In the case of ion-activated amino acid transporters, the substrate movement is coupled to the electrochemical potential established across the membrane. For Na^+-activated intestinal transport, the driving force is $\Delta\tilde{\mu}_{Na}$ with a 1:1 or 2:1 (Na^+ : amino acid) coupling stoichiometry. The $\Delta\tilde{\mu}_{Na}$ across the brush border membrane is established by the Na/K-ATPase primary active transporter in the basolateral membrane of the absorptive enterocytes. Thus, according to the Onsager reciprocity relationship for energetic coupling, intestinal amino acid transport systems provide an important pathway for absorbing Na^+ (followed osmotically by water absorption) from the intestinal lumen (Wright et al., 1983; Gerencser & Stevens, 1989). The physiological significance of this is that vectorial salt and water movement occurs from the lumen to the serosal fluid. The World Health Organization and UNICEF have recognized the importance of this physiologic phenomenon in treating cholera-

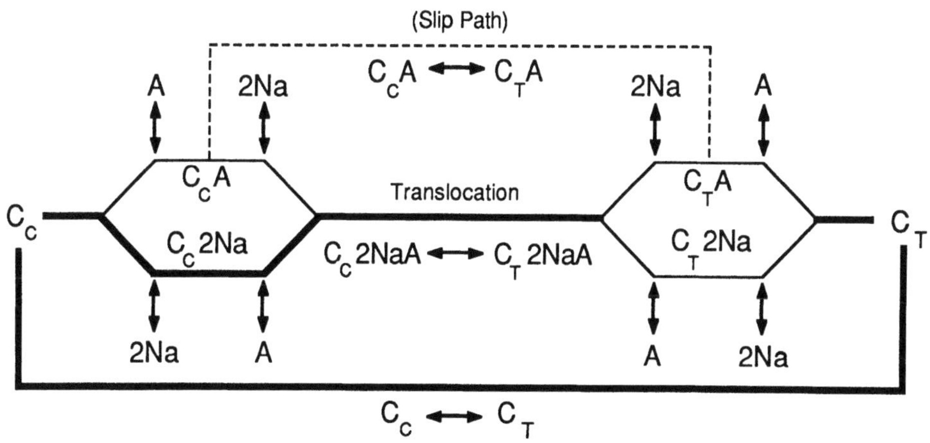

Fig. 5. Cleland diagram of the "preferred random" kinetic paradigm for intestinal brush border membrane Na^+/amino acid cotransport. The diagram shows the preferred pathway (bold lines), the random alternatives (thin lines), and the inner leak (slippage) pathway for amino acid. The model does not account for an inner leakage path for Na^+ alone. Net vectorial transport (*cis* → *trans*) of one amino acid molecule (A) is activated with a coupling stoichiometry of 2:1 (Na^+ : amino acid), assuming $\Delta\tilde{\mu}_{Na}>0$. This simplified model shows binding of both Na^+ ions as a single event. Na^+ and amino acid (A) each combine with the cotransporter's free binding sites at the *cis* (lumen) side (C_C) to form the loaded complex (C_C2NaA). The loaded complex undergoes a conformational shift to C_T2NaA that translocates Na^+ and amino acid to the *trans* (cytoplasm) side of the membrane. The unloaded carrier (C_T) then reverts to a conformation which is ready to accept new substrates at the *cis* side (C_C). Formation of a partially bound alternative complex without Na^+ (CA) discourages energetic coupling to $\Delta\tilde{\mu}_{Na}$ and prevents secondary active transport; amino acid movement via the uncoupled path is termed "slippage." (Adapted from Stevens 1992; Stevens & Wright, 1985, with permission.)

induced secretory diarrhea, and have suggested the use of orally administered NaCl solutions containing amino acids and starch for rapid rehydration of patients (Hirschhorn & Greenough, 1991). In addition to the important role of Na^+/amino acid cotransporters in intestinal absorption of whole-body water, Na^+/amino acid cotransport systems may critically influence volume regulation in individual jejunal enterocytes (MacLeod & Hamilton, 1991).

An interesting property observed for intestinal transport systems is that lanthanide cations with strong third ionization potentials mimic the transporter activating field effects of Na^+, but are probably not themselves cotransported with amino acid. For example, Sm^{3+} stimulates proline transport via system IMINO up to 6-fold compared Na^+ (Stevens & Kneer, 1988).

MEMBRANE-BOUND FUNCTIONAL ARRANGEMENT OF THE IMINO SYSTEM

Thorough understanding of transporter regulation requires a working knowledge of the functional arrangement of the system *in situ* within the membrane. Using the technique of high-energy electron radiation inactivation in conjunction with site-directed antibodies made against the cloned intestinal Na^+/glucose cotransporter, we previously determined (Stevens et al., 1990) that the

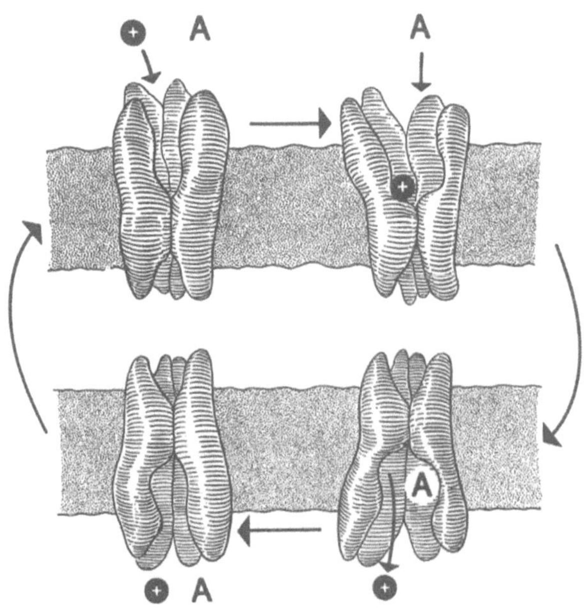

Fig. 6. Tentative brush border membrane-bound *in situ* functional arrangement of intestinal system IMINO. Based on radiation inactivation experiments, the transporting functional unit size is collectively 382 ± 58 kDa and may be comprised of four identical monomer subunits, each about 100 kDa. In the "preferred random" binding model, *cis* Na^+ binds first, which results in a conformational shift that permits proline (or amino acid A) to subsequently bind to the aminoacid-binding site. The fully loaded cotransporter complex then undergoes a new conformational shift that places the substrates near the inner (cytoplasmic) membrane surface. The order of *trans* release to the cytoplasm is arbitrarily shown with Na^+ coming off first. The IMINO cotransporter finally reverts to the conformation that is optimal for binding extracellular Na^+ and proline. The reaction kinetics are described above in Fig. 5. (Adapted from Stevens et al., 1990, with permission.)

SGLT1 membrane-bound functional size was 290,000 daltons, comprised of four intact identical monomer expression products of 73,000 each. In a manner similar to the SGLT1 system, the reaction mechanism and functional asymmetric orientation of the IMINO transporter enables $\Delta\tilde{\mu}_{Na}$ to drive proline from the lumen into the enterocyte, as discussed above. At the DNA sequence and monomer primary sequence levels, there is a high degree of homology between the mammalian SGLT1 and prokaryote Na+/proline cotransporters (Hediger et al., 1989; Pajor et al., 1992). By extending our radiation inactivation studies to include amino acid transport activity, we have estimated the brush border IMINO system *in situ* membrane-bound functional size as 382,000 ± 58,000 daltons (Stevens, Fernandez & Kempner, unpublished). The IMINO monomer size has been estimated at 95,000 - 110,000 daltons on reducing PAGE gels (Wright and Peerce, 1984). Therefore, by analogy with the Na^+/glucose cotransporter, it is likely that system IMINO is arranged as a multimer within the membrane, most likely as a tetramer. The putative tetramer physical arrangement of the functional IMINO system is shown in Fig. 6. As discussed above, two Na^+ ions bind and activate the system. For the sake of clarity, Fig. 6 shows only the net transfer of positive charge (collectively 2 Na^+ ions) with each transported proline molecule.

CONCLUDING REMARKS AND FUTURE DIRECTIONS

The small intestinal mucosa continues to serve as a fruitful tissue for exploring the basic principles of amino acid transporter mechanisms and adaptive regulation. Molecular biology approaches are needed to firmly discern the system selectivities and up-regulation events that are unique to the apical membrane, the ubiquitous nature of the basolateral membrane systems, and the phylogenetic and ontogenetic relationships among transport systems of all cell types. Adaptive regulation of specific transport systems needs to be further explored by expanding studies of enterocyte second and third messengers, and by examining extra-luminal and luminal regulation triggers in cultured epithelial cell models and in intact tissue. In the light of recent discoveries that the system y^+ protein is exploited by retroviruses as a membrane receptor, we are beginning to investigate the relationship between the host intestinal adaptive regulation of amino acid transport and the epithelial susceptibility to retrovirus infection. This new information will build upon existing models of transport to permit greater understanding of the intestine's adaptive control of amino acid transport in health and disease states.

ACKNOWLEDGMENT

The author is supported by National Institutes of Health Grant DK-38715.

REFERENCES

Ahearn, G.A., Gerencser, G.A., and Thamotharan, M., Behnke, R.D., Lemme, T.H., 1992, *Am. J. Physiol.*, 000:R000 (in press)

Anderson, V., and Munck, B.G., 1987, *Biochim. Biophys. Acta*, 902:145.

Aoshima, H., Tomita, K., and Sugio, S., 1988, *Arch. Biochem. Biophys.* 265:73.

Bannai, S., Christensen, H.N., Vadgama, J.V., Ellory, C., Englesberg, E., Guidotti, G.G., Gazzola, G.C., Kilberg, M.S., Lajtha, A., Sacktor, B., Sepulveda, F.V., Young, J.D., Yudilevich, D., and Mann, G., 1984, *Nature*, 311:308.

Buddington, R.K., and Diamond, J.M., 1989, *Annu. Rev. Physiol.*, 51:601.

Bulus, N.M., Abumrad, N.N, and Ghishan, F.K., 1989, *Am. J. Physiol.*, 257:G80.

Centelles, J.J., Kinne, R., and Heinz, E., 1991, *Biochim. Biophys. Acta*, 1065:239

Christensen, H.N., 1990, *Physiol. Rev.*, 70:43.

Christensen, H.N., 1985, *J. Membrane Biol.*, 84:97.

Christensen, H.N., and Handlogten, M.E., 1977, *Biochim. Biophys. Acta*, 469:216.

Crane, R.K., 1977, *Rev. Physiol. Biochem. Pharmacol.*, 78:99.

Del Castillo, J.R., and Muniz, R., 1991, *Am. J. Physiol.*, 261:G1030.

Davies, S., Maenz, D.D., and Cheeseman, C.I., 1987, *Biochim. Biophys. Acta*, 896:247.
Diamond, J., 1991, *News In Physiol. Sci.*, 6:92.
Diamond, J.M., and Karasov, W.H., 1987, *Proc. Natl. Acad. Sci. USA*, 84:2242.
Ferraris, R.P., and Diamond, J.M., 1989, *Annu. Rev. Physiol.*, 51:125.
Ferraris, R.P., Kwan, W.W., and Diamond, J., 1988, *Am. J. Physiol.*, 255:G151.
Gerencser, G.A., and Stevens, B.R., 1989, *Am. J. Physiol.*, 257:R461.
Ghishan, F.K., Arab, N., Bulus, N., Said, H., Pietsch, J., and Abumrad, N., 1990, *Am. J. Clin. Nutr.*, 51:612.
Ghishan, F.K., Sutter, W., Said, H., Leonard, D., Pietsch, J., and Abumrad, N., 1989, *Biochim. Biophys. Acta*, 979:77.
Giordana, B., Sacchi, V.F., Parenti, P., and Hanozet, G.M., 1989, *Am. J. Physiol.*, 257:R494.
Hayashi, K., Yamamoto, S., Ohe, K., Miyoshi, A., and Kawasaki, T., 1980, *Biochim. Biophys. Acta*, 601:654.
Hediger, M.A., Turk, E., and Wright, E.M., 1989, *Proc. Natl. Acad. Sci USA*, 86:5748.
Hediger, M.A., Coady, M.J., Ikeda, T.S., and Wright, E.M., 1987, *Nature (London)*, 330:379.
Henning, S., 1987, *in*: "Physiology of the Gastrointestinal Tract, 2nd ed.," L.R. Johnson, ed. Raven Press, New York.
Hidalgo, I.J., and Borchardt, R.T., 1990, *Biochim. Biophys. Acta*, 1028:25.
Hirschhorn, N., and Greenough, W.B., 1991, *Sci. Am.*, 246:50.
Hopfer, U., 1987, in: "Physiology of the Gastrointestinal Tract, 2nd ed.," L.R. Johnson, ed., Raven Press, New York.
James, P.S., Smith, M.W., Tivey, D.R., and Wilson, T.J., 1987, *J. Physiol. (London)*, 393:569.
Kaback, H.R., 1987, *in*: "Membrane Transport Processes in Organized Systems," T.E. Andreoli, J.F. Hoffman, D.D., Fanestil, and S.G. Schultz, eds, Plenum, New York.
Karasov, W. Solberg, D., Carter, S., Hughes, M., Phan, D., Zollman, F., and Diamond, J., 1986, *Am. J. Physiol.*, 251:G501.
Kim, J.W., Closs, E.I., Albritton, L.M., and Cunningham, J.M., 1991, *Nature*, 352:725
Laganiere, S., Maestracci, D., and Berteloot, A., 1986, *Clin. Invest. Med.*, 9:176.
Lash, L.H., and Jones, D.P., 1984, *Am. J. Physiol.*, 247:G394.
Levine, G.M., 1991, *in*: "Growth of the Gastrointestinal Tract: Gastrointestinal Hormones and Growth Factors," J. Morisset and T.E. Solomon, eds., CRC Press, Boca Raton.
Lucas, M., 1983, *Gut*, 24:734.
MacLeod, R.T., and Hamilton, J.R., 1991, *Am. J. Physiol.*, 260:G26.
Malo, C., 1991, *Gastroenterology*, 100:1644.
Matthews, D.M., 1987, *in*: "Dipeptides as New Substrates in Nutrition Therapy," S.A. Adibi, W. Fekl, P. Furst, and M. Oehmke, eds., Karger, London.
McNamara, P.D., 1991, *J. Biol. Chem.*, 266:986.
Mircheff, A.K., van Os, C.H., and Wright, E.M., 1980, *J. Membrane Biol.*, 52:83.
Miyamoto, Y., Nakamura, H., Hoshi, T., Ganapathy, V., and Leibach, F.H., 1990a, *Am. J. Physiol.*, 259:G372.
Miyamoto, Y., Kulanthaivel, P., Ganapathy, V., Whitford, G.M., and Leibach, F.H., 1990b, *Biochim. Biophys. Acta*, 1030:189.
Moe, A.J., Pocius, P.A., and Polan, C.E., 1987, *J. Dairy Sci.*, 68:504.
Morisset, J., and Solomon, T.E., 1991, *in* "Growth of the Gastrointestinal Tract: Gastrointestinal Hormones and Growth Factors," CRC Press, Inc., Boca Raton.
Munck, B.G., 1985, *J. Membrane Biol.*, 83:15.
Munck, B.G., 1985, *J. Membrane Biol.*, 83:1.
Munck, B.G., 1983, *in*: "Intestinal Transport," M. Gilles-Baillien and R. Gilles, eds., Springer-Verlag, Berlin.
Munck, L.K., and Munck, B.G., 1990, *Biochim. Biophys. Acta*, 1027:17
Munck, B.G., 1989, *Am. J. Physiol.*, 256:G532.
Munck, L.K., and Munck, B.G., 1992, *Am. J. Physiol.* 262:G609.
Nakao, T., Yamato, I., and Anraku, Y., 1987, *Mol. Gen. Genet.*, 208:70.
Pajor, A.M., Hirayama, B.A., and Wright, E.M., 1992, *Am. J. Physiol.*, 000:R000 (in press).
Pan, M., Neu, J., and Stevens, B.R., 1991, *FASEB J.*, 5(4):A760.

Paterson, J.Y.F., Sepulveda, F.V., and Smith, M.W., 1982, *J. Physiol.*, 331:523.
Preston, R.L., 1981, *Biochim. Biophys. Acta*, 688:422.
Riggs, T.R., Walker, L.M., and Christensen, H.N., 1958, *J. Biol. Chem.*, 233:1484.
Romano, P.M., Ahearn, G.A., Storelli, C., 1989, *Am. J. Physiol.*, 257:R180.
Said, H.M., Van Voorhis, K., Ghishan, F.K., Abumrad, N., Hylander, W., Redha, R., 1989, *Am. J. Physiol.*, 256:G240.
Salloum, R.M., Souba, W.W., Fernandez, A., and Stevens, B.R., 1990, *J. Surg. Res.*, 48:635.
Salloum, R.M., Stevens, B.R., and Souba, W.W., 1991, *Am. J. Physiol.*, 261:G22.
Scharrer, E., Wolfram, S., Raab, W., Amann, B., and Agne, N., 1981, *in*: "Mechanisms of Intestinal Adaptation, Falk Symposium 30" J.W.L. Robinson, R.H. Dowling, and R.O. Riecken, eds., MIT Press, Lancaster.
Schultz, S.G., 1987, *in*: "Membrane Physiology," T.E. Andreoli, J.F. Hoffman, D.D., Fanestil, and S.G. Schultz, eds, Plenum, New York.
Schwartz, M.Z., and Storozuk, R.B., 1988, *Am. J. Surg.*, 155:18.
Smith, M.W., Sepulveda, F.V., and Paterson, J.Y.F., 1983, *in*: "Intestinal Transport," M. Gilles-Baillien and R. Gilles, eds., Springer-Verlag, Berlin.
Stein, W.D., 1986, "Transport and diffusion across cell membranes," Academic Press, Orlando.
Stein, E.D., Chang, S.D., and Diamond, J.M., 1987, *Am. J. Physiol.*, 252:G626.
Stevens, B.R., 1984, *Ann. Rev. Physiol.*, 46:417.
Stevens, B.R., 1992, *Am. J. Physiol.*, 263:R000. (in the press)
Stevens, B.R., and Wright, E.M., 1985, *Ann. N.Y. Acad. Sci.*, 456:115.
Stevens, B.R., and Kneer, C., 1988, *Biochim. Biophys. Acta*, 942:205.
Stevens, B.R., and Wright, E.M., 1985, *J. Membrane Biol.*, 87:27.
Stevens, B.R., and Wright, E.M., 1987, *J. Biol. Chem.*, 262:6546.
Stevens, B.R., Fernandez, A., Hirayama, B., Wright, E.M., and Kempner, E.S., 1990, *Proc. Natl. Acad. Sci. USA*, 87:1456.
Strocchi, A., and Levitt, M.D. 1991, *Gastroenterology*, 101:843.
Ugolev, A.M., Timofeeva, N.M., Roshchina, G.M., Smirnova, L.F., Gruzdkov, A.A., and Gusev, S.A., 1990, *Comp. Biochem. Physiol.*, 95A:501.
Van Winkle, L.J., 1990, *Biochim. Biophys. Acta*, 1025:225.
Van Winkle, L.J., 1988, *Biochim. Biophys. Acta*, 947:173.
Wang, H., Kavanaugh, M.P., North, R.A., and Kabat, D., 1991, *Nature*, 325:729.
Webb, K.E., 1990, *J. Anim. Sci.*, 9:3011.
Wilde, S.W., and Kilberg, M.S., 1991, *Biochem. J.*, 277:687.
Wilson, J.W., and Webb, K.E., 1990, *J. Anim. Sci.*, 68:504.
Wright, E.M., and Peerce, B.E., 1984, *J. Biol. Chem.*, 259:14993.
Wright, E.M., Schell, R.E., and Stevens, B.R., 1985, *Biochim. Biophys. Acta*, 818:271.
Wright, E.M., Stevens, B.R., and Peerce, B.E., 1986, *in*: "Amino acid transport in eukaryotic cells and tissues," *Fed. Proc.*, 45:2450.
Wright, E.M., Gunther, R.D., Kaunitz, J.D., Stevens, B.R., Harms, V., Ross, H.J., and Schell, R.E., 1983, *in*: "Intestinal Transport," M. Gilles-Baillien and R. Gilles, eds., Springer-Verlag, Berlin.
Zweibaum, A., Laburthe, M., Grasset, E., and Louvard, D., 1991, *in*: "Handbook of Physiology. The Gastrointestinal System," vol. IV, S.G. Schultz, ed., American Physiological Society, Bethesda.

AMINO ACID TRANSPORT IN BRAIN

Quentin R. Smith and Arthur J.L. Cooper

Laboratory of Neurosciences, National Institute on Aging,
National Institutes of Health, Bethesda, Maryland, and
Departments of Biochemistry and Neurology, Cornell University
Medical College, New York, New York, USA

INTRODUCTION

Amino acids serve special roles in brain contributing to cerebral protein synthesis, intermediary metabolism and interneuronal synaptic communication. At least four amino acids (i.e., GLU, ASP, GLY and GABA) function as neurotransmitters within the central nervous system (CNS), whereas several others (e.g., TRP, TYR, and HIS) contribute as precursors to neurotransmitters, including serotonin, histamine and the catecholamines (Pycock and Taberner, 1981). Amino acids are also required as substrates for cerebral protein synthesis and the formation of neuroactive peptides, including substance P, the enkephalins, vasopressin, neurotensin and somatostatin. Imbalances in amino acids profoundly influence brain function, as evidenced by the severe irreversible mental retardation that occurs in phenylketonuria and maple-syrup urine syndrome. Further, uncontrolled release of excitatory amino acids has been linked to the neuronal degeneration and death that occurs in ischemia, hypoxia, hypoglycemia and seizures (Rothman and Olney, 1987).

Although the brain requires over 20 separate amino acids for normal development and growth, only approximately half can be synthesized within the CNS. Those that can be formed in brain or other tissues of the body at adequate rates include GLY, PRO, ALA, SER, ASN, GLN, GLU, ASP, CYS and TYR. The remaining amino acids - PHE, TRP, LEU, ILE, VAL, MET, HIS, THR, LYS and ARG - cannot be formed at adequate rates and are termed nutritionally "indispensable" or "essential". These latter amino acids must be obtained ultimately from the diet through gastrointestinal absorption, distribution, and transport across the brain capillary, neuronal and glial cell membranes (Harper and Tews, 1988). Transport into brain is regulated and mediated by a limited number of saturable carriers at both the blood-brain barrier (BBB) and brain cell membranes (neurons and glia).

Interest in amino acid uptake into brain derives in part from the fact that a number of key brain neurotransmitter and amino acid metabolic pathways are influenced by the availability of precursor amino acids. For example, the rates of synthesis and brain concentrations of the monoamine neurotransmitters - serotonin, norepinephrine, and histamine - are now known to be directly affected by brain levels of precursor amino acids (i.e., TRP, TYR and HIS, respectively), especially in rapidly firing neurons (Schwartz et al., 1972; Fernstrom, 1983; Acworth et al., 1988). Though less well documented, several studies suggest that brain synthesis of S-adenosyl methionine from MET (Rubin et al., 1974), of phenethylamine and tyramine from PHE and TYR (Young, 1983), and of the kynurenines - quinolinic acid and kynurenic acid - from TRP

Mammalian Amino Acid Transport, Edited by M.S. Kilberg and
D. Häussinger, Plenum Press, New York, 1992

(Moroni et al., 1988; During et al., 1989, Fukui et al., 1991), are also influenced by precursor availability. This dependence arises in part because the key rate-limiting enzymes within the CNS are not saturated with precursor amino acids at normal brain concentrations. As a result, rates of synthesis vary with brain amino acid supply and exchange. Transport within the CNS is also critical for the removal of neurotransmitter amino acids from the synaptic cleft, for the maintenance of cellular compartmentation, for the disposal of amino acid metabolites and for the maintenance of sufficient cellular levels of amino acids to sustain cerebral protein, peptide and neurotransmitter synthesis. Imbalances in brain amino acid concentrations occur in a number of conditions, including liver disease, diabetes, uremia, severe dietary disturbances and the aminoacidurias (McKean et al., 1968; James et al., 1978; Hommes, 1983; Mans et al., 1987).

This chapter will review the factors that regulate amino acid uptake and distribution within brain. Special emphasis will be placed on transport mechanisms at the BBB and at brain cell membranes. Finally, some comments will be made on the relation between transport and regulation of brain amino acid pools, metabolic pathways and neurotransmitter function.

AMINO ACID TRANSPORT INTO BRAIN

Amino acid uptake into brain from the circulation depends critically on the type of amino acid, essential vs nonessential, and on the presence of the BBB. Nonessential amino acids are derived primarily from intracerebral synthesis and show minimal uptake from blood. Synthesis is from glucose and Krebs cycle intermediates, and is influenced, at least in part, by cerebral energy metabolism. Essential amino acids, on the other hand, are derived exclusively from the circulation (except for those released from protein breakdown; Smith et al., 1988) and cross the BBB readily by carrier-mediated transport. Flux rates for essential amino acids are appreciable so that half times for equilibration of free pools are on the order of minutes (Kirikae et al., 1988; Hawkins et al., 1989). Transfer rates for both groups are restricted and regulated by the BBB and by neurons and glia.

Blood-Brain Barrier

The BBB is a regulatory interface that helps protect brain from circulating toxins and helps maintain a constant internal environment for optimal brain function (Fenstermacher and Rapoport, 1984). In most vertebrates, the BBB is formed by the cerebrovascular endothelium, which comprises the largest single surface for solute uptake into the CNS (capillary surface area = ~120 cm^2/g in rat cerebral cortex; Gross et al., 1986). Additional sites include the arachnoid membrane, which surrounds and envelopes the brain, and the choroid plexus epithelium, which secretes cerebrospinal fluid (CSF). At each of these barrier sites there is at least one layer of cells that are joined by multiple bands of tight junctions (*zonulae occludens*)(Brightman and Reese, 1969) which essentially block intercellular diffusion . As there are no alternative paracellular routes, the barrier displays many of the properties of a continuous "tight" epithelium (Crone, 1986), including high electrical resistance (>1500 $Ohm.cm^2$) (Crone and Olesen, 1982; Butt et al., 1990), limited passive ion transport (Smith and Rapoport, 1986), and selective permeability for lipid-soluble as compared to water-soluble compounds (Ohno et al., 1978; Takasato et al., 1984). The BBB is apparently critical for higher CNS function as it is found in all vertebrates and most invertebrates (Abbott et al., 1986), and is formed during the first trimester of human life.

Critical hydrophilic nutrients that are required for cerebral metabolism are transported into brain by saturable carriers at the BBB (Pardridge and Oldendorf, 1977). Oldendorf, in a classical series of studies in the 1970's, identified three carriers for amino acids at the BBB; one for large neutral amino acids (primarily PHE, TRP, LEU, MET, ILE, TYR, HIS, VAL and THR), one for basic amino acids (LYS, ARG and ORN), and one for acidic amino acids (GLU and ASP)(Oldendorf, 1971; Pardridge and Oldendorf, 1975; Oldendorf and Szabo, 1976). Transfer activities were greatest for the large neutral and basic amino acid carriers (Pardridge and Oldendorf, 1977), consistent with the fact that the carriers mediate the uptake of those amino acids that cannot be synthesized within the CNS.

During the past ten years, considerable progress has been made in the technology for studying transport at the BBB. Intravenous administration and quantitative autoradiographic methods are now available for the accurate determination of rates of amino acid influx and metabolism in small brain regions under differing physiological and pathological conditions (Hawkins et al., 1982; Kirikae et al., 1988; Keen et al., 1989). Similarly, *in situ* brain perfusion methods have been developed that allow absolute control of perfusate composition and flow for the characterization of amino acid transport and metabolism over a wider range of conditions than could be tolerated *in vivo* (Takasato et al., 1984; Smith et al., 1984; Greenwood et al., 1989; Hargreaves and Pardridge, 1990; Fukui et al., 1991). In addition, isolated brain microvessel and cultured brain endothelial preparations are available for the *in vitro* study of endothelial transport carriers and for examination of the polarity of the BBB (i.e., transport differences between the luminal and abluminal endothelial cell membranes) (Betz and Goldstein, 1978; Cancilla and DeBault, 1983; Audus and Borchardt, 1986; Hargreaves and Pardridge, 1988; Dehouck et al., 1990). These latter preparations, when combined with the techniques of molecular biology, will allow the future isolation and cloning of the individual amino acid transport proteins. [Such work is being attempted now using Xenopus oocytes as an expression system (e.g., Tate et al., 1989).] Finally, brain amino acid uptake and metabolism can be studied in experimental animals using brain microdialysis techniques (Benveniste, 1989; During et al., 1989) and in humans using positron emission tomography (Hawkins et al., 1989; Koeppe et al., 1990) or magnetic resonance imaging.

Amino Acid Transport Systems at the Brain Capillary Endothelium

Table 1 summarizes the characteristics of the known amino acid transport carriers of the brain capillary endothelium. A brief description of each system is given below.

Table 1. Brain Capillary Amino Acid Transport Systems

Transport System	Activity *In Vivo*	*In Vitro*	Na^+ Dependence	Representative Substrates
Neutral Amino Acid				
L1	+++	++	--	PHE, TRP, LEU, MET, ILE, TYR, VAL, BCH
ASC	+	++	+	ALA, SER, CYS, THR
A	-	++	+	ALA, GLY, PRO, MeAIB
Basic Amino Acid				
y^+	++	++	--	LYS, ARG, ORN, GPA
Acidic Amino Acid				
x^- or X^-	+	++	(?)	GLU, ASP

Refer to text for appropriate references. Scales of transport activity are relative and meant only for comparative purposes. Defining substrates are underlined. Note: differences in *in vivo* and *in vitro* transport activity may reflect asymmetric distribution of transporters between the capillary luminal and abluminal membranes (Betz and Goldstein, 1978).

Large Neutral Amino Acid Carrier - System L1. The brain capillary large neutral amino acid carrier (System L1) is the primary amino acid transporter of the BBB. It has the greatest *in vivo* influx capacity of all cerebrovascular amino acid carriers and mediates the brain uptake of the largest number of amino acids (i.e., ~14 of the 16 most abundant plasma neutral amino acids). In addition, it is the principal transport mechanism for the 8 essential neutral amino acids that cannot be synthesized at adequate rates within the CNS (Oldendorf, 1971; Oldendorf and Szabo, 1976; Smith et al., 1987; Tovar et al., 1988).

Transport by the BBB large neutral amino acid carrier has been demonstrated both *in vivo* and *in vitro*, and it is believed that the mechanism is present at both the capillary luminal and abluminal membranes (Table 1) (Betz and Goldstein, 1978). Uptake is saturable, stereospecific and Na^{+}/energy independent (Oldendorf, 1971; Smith et al., 1987; Momma et al., 1987; Greenwood et al., 1989). The carrier has many of the characteristics of the L System described by Oxender and Christensen (1963) including preference for neutral amino acids with large, bulky side chains (e.g., PHE and LEU), trans-stimulation, and inhibition by 2-aminobicyclo[2.2.1]heptane-2-carboxylic acid (BCH) - the L System-defining substrate (Wade and Katzman, 1975; Cangiano et al., 1983; Smith et al., 1987; Aoyagi et al., 1988; Hargreaves and Pardridge, 1988). The two carriers differ, however, in one important respect - the affinity of the cerebrovascular large neutral amino acid transporter for substrates exceeds that reported for the L System in most other tissues by 100-1000 fold. For example, the K_m for L-PHE uptake into rat brain is ~10 μM (Momma et al., 1987), whereas the value in most other organs is 1-10 mM or greater (Lerner and Larimore, 1986). Recent studies by Hargreaves and Pardridge (1988) suggest that the K_m may be even lower (~0.3 μM for PHE) in isolated human brain capillaries. High affinity L System transporters have also been reported to be present in some blood cells, tumor cells and cultured tissue cells (Shotwell et al., 1982; Vistica, 1983), and Weissbach et al. (1982) have suggested the designations L1 and L2 to distinguish the respective high affinity and low affinity forms. The properties of the BBB large neutral amino acid carrier appear to correspond most closely with the L1 form, and it is that designation (i.e., System L1) which will be used in this review.

Table 2 summarizes plasma concentrations and BBB V_{max} and K_m values for ten neutral amino acids as measured using the *in situ* rat brain perfusion technique (Smith et al., 1987). V_{max} differs among amino acids by ~2-3 fold, but averages ~51 nmol/min/g wet brain. K_m, on the other hand, varies to a far greater extent (~100 fold) with some values as low as ~10 μM and some values close to 1 mM. Affinity varies with the size and hydrophobicity of the amino acid side chain (Fig. 1). However, other factors influence affinity as well. For example, the transporter is stereoselective and prefers L- to D-amino acids (Oldendorf, 1973; Smith et al., 1985; Momma et al., 1987). In addition, substrates must have a free carboxyl group and an unsubstituted amino group on the α carbon (Smith et al., 1989).

Neutral amino acid transport at the BBB is bidirectional with rate coefficients for brain amino acid efflux exceed that those for influx by >5-10 fold (Knudsen et al., 1990). This difference likely arises because of the ~5-10-fold lower concentration of amino acids in brain interstitial fluid as compared to plasma. Based on reported plasma and CSF concentrations (Perry et al., 1975; McGale et al., 1977), the L1 carrier is predicted to be ~96% saturated with amino acids at the capillary luminal membrane and ~75% saturated at the abluminal membrane (Smith et al.,1985, 87). Net flux rates for most neutral amino acids are quite small, except for LEU, ILE, and VAL, which are oxidized within the CNS, and for GLN, which has a net efflux from the CNS (Pell and Bergman, 1983).

The L1 carrier is capable of facilitating the uptake of amino acid drugs into the CNS, as has been shown for L-DOPA, α-methyl-DOPA, azaserine, melphalan, 6-diazo-5-oxo-L-norleucine and acivicin (van Bree et al., 1988; Chastain and Borchardt, 1990; Takada et al., 1991). D,L-NAM, a nitrogen mustard derivative of amino-naphthoic acid, has the highest demonstrated affinity (K_m ~0.2 μM) for the carrier of any known agent (Takada et al,. 1992). The carrier can also mediate the brain uptake of dietary neurotoxins, such as ß-methylamino-L-alanine (Duncan et al., 1991; Smith et al., 1992)

Table 2. Plasma Concentrations and BBB Transport Kinetic Parameters for Influx of some Neutral Amino Acids into Rat Brain.

Amino Acid	Plasma Concentration (μM)	K_m (μM)	$\frac{C}{K_m}$	V_{max} (nmol/min/g)	K_m(app) (μM)	Influx (nmol/min/g)
PHE	81	11	7.36	41	170	13.10
TRP	82	15	1.09[a]	55	330[a]	2.61[a]
LEU	175	29	6.03	59	500	15.45
MET	64	40	1.60	25	860	1.74
ILE	87	56	1.55	60	1210	4.04
TYR	63	64	0.98	97	1420	4.14
HIS	95	100	0.95	61	2220	2.51
VAL	181	210	0.86	49	4690	1.83
THR	237	220	1.08	17	4860	0.79
GLN	485	880	0.55	43	19900	1.03
		$\Sigma C/K_m$	22.05	Average V_{max} 51 ±7		Total Influx 47.24

Values are for the anesthetized rat and are taken from Smith et al. (1987). [a]Estimated assuming only free (non-albumin bound) TRP competes for transport (free fraction ~20%). The flux rte measured in vivo for TRP may exceed the calculated value due to in vivo dissociation (Smith et al., 1990). V_{max} is the maximal saturable transport capacity, K_m is the half-saturation concentration in the absence of competitors, K_m(app) is the "apparent" K_m under normal physiological conditions (i.e., in the presence of normal concentrations of plasma amino acids), and influx is the unidirectional amino acid flux rate from plasma to brain.

which is found in the seeds of certain cycad plants and has been implicated in the pathogenesis of the high incidence amyotrophic lateral sclerosis of Guam and the western Pacific (Kurland, 1988).

Small Neutral Amino Acid Carriers - Systems A and ASC. While for many years the L1 carrier was considered the only neutral amino acid transporter at the BBB, recent studies have forced a re-evaluation of that view. Betz and Goldstein (1978) demonstrated that isolated brain microvessels contain amino acid transport System A, which mediates Na^+-dependent uptake of small neutral amino acids (e.g., ALA, SER and PRO) and is inhibited selectively by methylaminoisobutyric acid (MeAIB). Later, Tayarani et al. (1987) presented evidence for a brain capillary System ASC carrier, which is also Na^+ dependent, transports primarily small neutral amino acids (ALA, SER, CYS and THR) and is not inhibited by either BCH or MeAIB. The estimated K_m for L-ALA uptake by the brain microvessel System ASC transporter was 1.3 mM (Tayarani et al., 1987).

Based on the low reported rates of small neutral amino acid influx into brain (Oldendorf, 1971; Baños et al., 1973, 75; Oldendorf and Szabo, 1976), both Betz and Goldstein (1978) and Tayarani et al. (1987) suggested that Systems A and ASC are located primarily at the brain capillary abluminal membrane and may function to transport amino acids out of brain, not in. However, some ASC System transport activity may be located at the capillary luminal membrane as Sershen and Lajtha (1979) and more recently Tovar et al. (1988) reported small components of brain ALA, SER and THR uptake that display characteristics of System ASC. From the study of Tovar et al. (1988), it would appear that System ASC mediates ~25% of brain THR uptake and System L1 contributes ~40% at tracer THR concentration and in the absence of competitors. Such percentages, however, may not reflect *in vivo* contributions at normal plasma concentrations (Smith et al., 1985). Consistent with the presence of System ASC transport at the BBB, Tews et al. (1987a, 87b) noted that brain THR uptake appears to be influenced by plasma concentrations of both small and large neutral amino acids, whereas brain valine influx depends only on the plasma concentrations of large neutral amino acids.

Wade and Brady in 1981 found no evidence for a significant component of System ASC uptake of L-CYS into brain. However, they may have missed a small contribution of System ASC due to the low sensitivity of the transport method that they employed. Most studies agree that the neutral amino acids with the largest side chains, such as PHE, LEU and TRP, are taken up into brain almost exclusively by the L1 carrier (Oldendorf, 1971; Momma et al., 1987; Tovar et al., 1988; Aoyagi et al., 1988). [^{14}C]MeAIB, the System A specific analogue, shows very little transport across the BBB in adult animals (PA ~3 x 10^{-5} ml/s/g; Blasberg et al., 1983), consistent with the hypothesis that System A contributes minimally to brain amino acid uptake under normal conditions. Other neutral amino acid transporters (e.g., Systems N, T or Gly) have not been demonstrated to be present at the brain capillary endothelium.

Basic Amino Acid Carrier - System y^+. Next to the cerebrovascular L1 carrier, the basic amino acid carrier (System y^+) exhibits the greatest transport capacity for amino acids at the BBB. Preliminary estimates with the brain perfusion technique indicate that the V_{max} and K_m of L-LYS influx are ~25 nmol/min/g wet brain and 65 µM, respectively, for the rat (Q. Smith, unpublished observations). Reported affinities for ARG and ORN are comparable to that of LYS (Pardridge, 1983; Miller et al., 1985). As plasma levels of LYS, and ARG are each on the order of 100-500 µM (Peters and Harper, 1987), the transporter is heavily saturated (>85%) with cationic amino acids at normal concentrations.

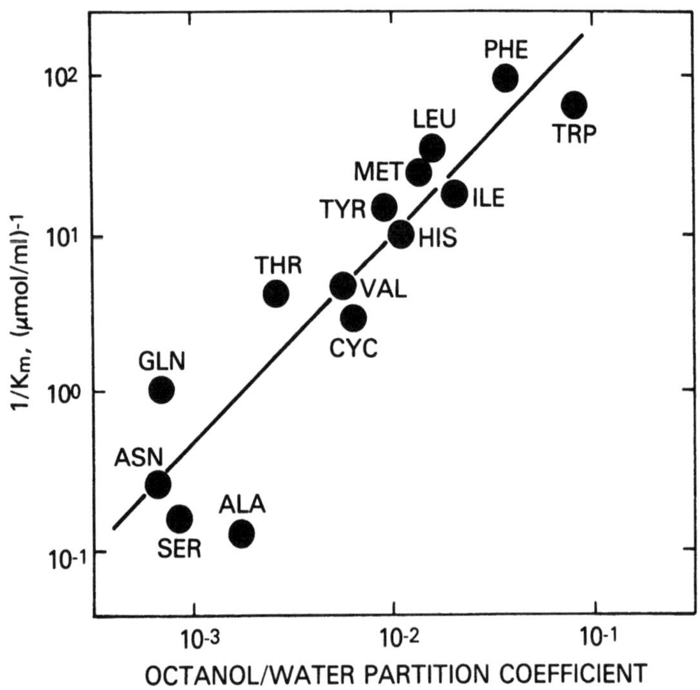

Fig. 1. Relationship of $1/K_m$ to octanol/water partition coefficient for transport of 14 neutral amino acids into brain. $1/K_m$ is a measure of BBB transport affinity, whereas the octanol water partition coefficient is a measure of amino acid side chain hydrophobicity (Yunger and Cramer, 1981). CYC: cycloleucine. The line represents the least-squares fit of the equation, $\log(1/K_m) = a \cdot \log(\text{partition coefficient}) + b$, where a = 1.31 ± 0.17 (S.E.) and b = 3.63 ± 0.40 (S.E.) with a correlation coefficient (r) of 0.910 ($p < 0.001$). From Smith et al. (1987), with permission.

Like the L1 carrier, System y^+ transport is stereospecific and Na^+/energy independent (Oldendorf, 1971; Baños et al., 1974; Cangiano et al., 1988). Affinity is greatest for amino acids with cationic side chains, though short chain neutral amino acids can also interact with moderate affinity in some tissues (White, 1985). 4-Amino-1-guanylpiperidine-4-carboxylic acid (GPA) was developed as a specific model substrate for this system (White, 1985). Cationic HIS apparently binds but is not transported by the BBB y^+ carrier (Oldendorf et al., 1988). Instead, HIS is taken up into brain as the neutral zwitterion by the cerebrovascular L1 carrier.

System y^+ transport has been demonstrated in brain capillaries both *in vivo* and *in vitro* (Baños et al., 1974; Mans et al., 1982; Cangiano et al., 1988), and it is thought that the carrier is located at both the capillary luminal and abluminal membranes. The y^+ carrier is widely distributed throughout the body and is considered relatively unresponsive to hormonal and dietary regulation (White, 1985). Recent reports suggest that System y^+ may be homologous to a previously identified cell-surface receptor for mouse ecotropic retroviruses (Kim et al., 1991; Wang et al., 1991). Injection of mRNA for the receptor into Xenopus oocytes resulted in enhanced sodium-independent transport of basic, but not acid or neutral, amino acids, with transport K_m's for LYS and ARG of ~70 μM. The receptor was detected in various mouse tissues, including brain, and had a molecular weight of ~67,000 with 14 putative membrane-spanning domains (Kim et al., 1991; Wang et al., 1991).

Acidic Amino Acid Carrier (System x^- or X^-). The acidic amino acids, GLU and ASP, are taken up into brain to a limited extent by the acidic amino acid carrier of the BBB (Oldendorf and Szabo, 1976). The transport capacity of this carrier is markedly less than that of either the L1 or y^+ Systems, as might be expected since both GLU and ASP are excitatory neurotransmitters and can be synthesized within the CNS. Preliminary estimates by Davson et al. (1990) with a brain perfusion technique place the V_{max} of saturable GLU influx at 0.01-0.03 nmol/min/g wet brain and the K_m at 3-5 μM. As plasma GLU concentration is ~ 20-150 μM and plasma ASP concentration is on the same order (Peters and Harper, 1987), the transporter is quite saturated with acidic amino acids at normal plasma concentrations. Little is known about the exact characteristics of this carrier (Na^+ dependence, selectivity, etc...) as no detailed kinetic studies have yet been performed.

The acidic amino acid carrier has been suggested to function primarily to transport amino acids out of the CNS (Pardridge, 1979). This hypothesis was based on the reported net efflux of GLU from the isolated, perfused dog brain (Drewes et al., 1977). However, as significant net brain GLU efflux was not observed by Pell et al. (1983) in a separate study, the hypothesis may require further evaluation.

Brain acidic amino acid concentrations have been noted to be rather independent of plasma levels following acute changes in plasma acidic amino acid concentrations (Price et al., 1984; Peters and Harper, 1987). Such homeostasis may reflect saturated influx, regulated efflux or regulated intracerebral synthesis and catabolism. Evaluation of the homeostatic mechanisms for acidic amino acids at the BBB is further complicated by the fact that brain GLU and ASP concentrations exceed plasma concentrations by some 20-100 fold (Price et al., 1984; Peters and Harper, 1987). Such large concentration gradients likely mask small changes in brain interstitial fluid levels produced by altered plasma concentrations and transport across the BBB.

Nonsaturable Uptake. Most studies of the concentration dependence of BBB amino acid transport have reported a "nonsaturable" component of uptake that varies in magnitude depending on the amino acid and the method employed to measure BBB transport (e.g., Daniel et al., 1977; Pardridge, 1983; Smith et al., 1984, 87; Greenwood et al., 1989). In some reports, the nonsaturable component comprises up to 50% of total amino acid uptake from plasma (Miller et al., 1985), whereas in others, the contribution is much less (5-10%)(Daniel et al., 1977; Smith et al., 1985).

The nonsaturable component may represent simple passive diffusion of amino acids across the BBB. Amino acids are known to cross black lipid membranes by passive diffusion (Klein et al., 1971), and cerebrovascular permeability coefficients for nonsaturable uptake at the BBB match well those predicted on the basis of octanol/water partition coefficients (i.e., lipid solubility) (Smith et al., 1985; Momma et al., 1987). Alternatively, the component may reflect a transport measurement error or may arise due to uptake via a presently undefined, low affinity - high capacity mechanism (Christensen, 1989). The component does not result as an artifact of cerebral tracer metabolism, as the nonmetabolizable amino acid - ACHC also has a nonsaturable component of uptake into brain (Aoyagi et al., 1988). Regardless, for many of the large neutral amino acids under physiological conditions, the contribution of the nonsaturable component is minor (<15%) and thus can be ignored (Smith et al., 1985; Momma et al., 1987).

Transport Carriers at Other Barrier Sites

Although the cerebral capillaries are the primary interface for solute exchange between plasma and brain interstitial fluid, other barrier sites also contribute and have important roles in the production of CSF and in the removal of toxins from the CNS. Preston et al. (1989) recently demonstrated the presence of Systems L1 and ASC at the choroid plexus epithelium. Similarly, earlier studies by Wright and colleagues suggested that LEU and other amino acids may be transported out of CSF by saturable processes at the arachnoid membrane (Wright et al., 1971; Wright, 1974).

Facilitated transport systems for essential large neutral amino acids (System L or L1) have been demonstrated at the blood-nerve and blood-retinal barriers (Fernstrom et al., 1986; Wadhwani et al., 1990). Further, Abbott et al. (1988) showed that the L System is present at the "glial" BBB of the dogfish. As at the brain capillary endothelium, facilitated transport at these other sites is likely required to overcome the limitations of passive permeability so as to allow sufficient amino acids into neural tissue to meet the needs of metabolism and protein synthesis. The choroid plexus and arachnoid membrane may also have important roles in regulation of CSF amino acid concentrations.

The enzymes of the γ-glutamyl cycle are well represented in brain, especially in the choroid plexus (Tate et al., 1973; Okonkwo et al., 1974), and it has been proposed that the cycle may be important for the translocation of amino acids into the brain. However, the γ-glutamyl cycle has not been equated with any of the classical amino acid carriers and the role of the cycle at the barrier awaits further study.

Factors that Influence Amino Acid Transport into Brain

Numerous studies have shown that brain concentrations of essential amino acids are critically influenced by plasma concentrations and transport across the BBB. Unidirectional influx rates for amino acid uptake from plasma can be estimated using the Michaelis-Menten equation, with a correction for multiple substrate competition (Pardridge, 1983; Smith et al., 1987), as

$$\text{Influx} = (V_{max}C)/[K_m(1 + \Sigma(C_i/K_i)) + C]$$

where C = plasma concentration of the amino acid of interest, V_{max} and K_m are the transport constants of the amino acid, and C_i and K_i are the plasma concentration and corresponding half-saturation constant of each competing amino acid.

Correction for competition is critical because the BBB large-neutral, basic and acidic amino acid carriers each transport multiple substrates at high affinity and are each essentially saturated at normal plasma concentrations. For example, the brain capillary L1 transporter expresses affinity for at least 14 primary plasma neutral amino acids, with nine having plasma concentrations equal to or greater than K_m (Table 2). As a consequence, $\Sigma(C_i/K_i)$ is > 9, and saturation is > 90% (Smith et al., 1987). Similar calculations can be performed for the BBB basic and acidic amino acid carriers from reported K_m values (Pardridge, 1983; Davson et al., 1990). For both systems, $\Sigma(Ci/Ki)$ > 7 and thus saturation exceeds 85%.

Competition has two important consequences. First, it makes the summed influx for each group of amino acids (i.e., large neutral amino acids, basic amino acids or acidic amino acids) essentially independent of plasma concentrations, due to transport saturation. Second, it makes the influx of each individual amino acid a sensitive function not only of its own concentration but of the concentrations of all competitors. Thus, an increase in the concentration of one amino acid will decrease the flux rates of other amino acids by competitive inhibition. Such competition has been demonstrated extensively both *in vivo* and *in vitro* following dietary or pharmacological manipulation of plasma amino acid concentrations (McKean et al., 1968; Fernstrom and Faller, 1978; Pratt, 1982; Smith et al., 1987). Competition makes the apparent *in vivo* K_m for BBB amino acid transport ($K_m(app) = K_m [1 + \Sigma(C_i/K_i)]$), substantially exceed the true K_m as measured in the absence of other amino acids (Table 2). Plasma amino acid concentrations are influenced by multiple dietary, hormonal and metabolic factors and can become imbalanced in several disease states (Harper and Tews, 1988).

From inspection of Eq. 1 it can be seen that brain influx for an amino acid depends primarily on four factors: (1) the plasma concentration of the amino acid, (2) the concentrations of competitors, (3) the transport V_{max} and (4) the transport K_m's. At normal plasma concentrations, influx for an amino acid is proportional to its plasma concentration and inversely related to the concentration of competitors. This latter relation was first recognized by Fernstrom and Wurtman in 1972 and termed the "plasma ratio", defined as $C/\Sigma Ci$. The plasma ratio has since proved a valuable predictor of changes in brain amino acid concentrations, especially for the essential neutral amino acids (Fernstrom and Faller, 1978). The accuracy of the ratio, however, may be improved under certain conditions by weighting the concentration of each amino acid by its respective K_m to obtain a normalized ratio $[(C/K_m)/(\Sigma C_i/K_i)]$. Such correction is important because transport affinity differs among amino acids by >10 fold.

The transport capacity of the BBB for amino acids is also a variable and can change with anesthesia, capillary perfusion, and disease (James et al., 1978; Miller et al., 1985; LaManna and Harik, 1986). For example, in hepatic encephalopathy the V_{max} of the BBB L1 carrier increases by 1-2 fold, whereas the capacity of the BBB y^+ carrier decreases by a similar magnitude (James et al., 1978; Cardelli-Cangiano et al., 1981; Mans et al., 1982). The change in transport capacity for the L1 carrier in liver disease may be related to trans-stimulation from increased brain GLN concentration (Cangiano et al., 1983; Jonung et al., 1985; Rigotti et al., 1985) or to some other effect of altered blood ammonia (Mans et al., 1983).

Infant rats appear to express increased transport activities for Systems L1, ASC, A and y^+ for uptake into brain (Lefauconnier et al., 1983; Nagashima et al., 1987). Chronic hyperglycemia, which has been reported to reduce BBB glucose transport capacity, apparently does not alter the transport capacity or affinity of the BBB amino acid carriers (McCall et al., 1982; Brosnan et al., 1984). Similarly, chronic hypertyrosinemia (i.e., increased plasma TYR of 10-30 fold) also does not alter the BBB amino acid transport capacity, at least in the adult rat, though it markedly reduces influx rates of competing neutral amino acids due to transport inhibition (Q. Smith, unpublished observation).

Finally, protein binding has been noted to influence the brain uptake of TRP, the one amino acid that binds appreciably and reversibly to plasma proteins (Fernstrom, 1983). Generally, 70-90% of plasma TRP is bound to serum albumin . However, the observed brain extraction of TRP substantially exceeds that predicted by the free fraction, suggesting that some TRP may be "stripped off" albumin as it passes through the cerebral circulation (Pardridge, 1983; Fenerty and Lindup, 1989). This "stripping" effect is flow dependent and is greater at low flow rates than at high flow rates (Smith et al., 1990).

AMINO ACID TRANSPORT INTO NEURONS AND GLIA

Once amino acids traverse the brain capillary endothelia, they are available for transport into neurons and glia. Amino acid uptake mechanisms have been described for brain cells, but are not as well characterized as those at the BBB. Most work has centered on uptake mechanisms for neurotransmitter amino acids, such as GLU, ASP, GLY and GABA. However, before such work is discussed, it may be useful to review the metabolism, distribution and cycling of amino acids (especially neurotransmitter amino acids) within the CNS.

Metabolic Compartmentation of Amino Acid and Ammonia Metabolism

Amino acid metabolism is compartmented within the CNS, as was first shown for GLN/GLU with [^{14}C]GLU (Berl et al., 1961) and [^{15}N]ammonia (Berl et al., 1962), and later with [^{13}N]ammonia (Cooper et al., 1979, 85, 88). Such studies demonstrated that ammonia upon entering the brain is converted to GLN in a small, rapidly turning over GLU-containing compartment that is distinct from a larger more slowly turning over GLU-containing compartment. The small GLU pool is estimated to comprise ~20% of total cerebral GLU (Berl et al., 1961). Tracer studies have also revealed the existence of two distinct tricarboxylic acid cycles within the CNS (Van Den Berg et al., 1966; Berl and Frigyesi, 1968, 69; Berl et al., 1970).

It is now generally agreed that astrocytes and neuronal perikarya largely represent the small and large GLU compartments, respectively (Patel and Balazs, 1970, 71; Van Den Berg and Van Den Velden, 1970; Schousboe et al., 1977b; Cooper et al., 1979; Nicklas et al., 1979; Krespan et al., 1982; Nicklas, 1983; Rothstein and Tabakoff, 1984). The division of brain into two metabolic compartments, however, may be simplistic. Based on the metabolic fate of various labeled compounds in brain, Van Den Berg et al. (1975) have suggested that the large and small compartments may be further subdivided into two and eight metabolic "spaces", respectively. Although this division is based on kinetic considerations the suspicion of multiple compartments has some support. Astrocytes from different regions posses different biochemical profiles (e.g., Schousboe and Divac, 1979; Patel et al., 1985). Moreover, distinct populations of mitochondria can be isolated from rat brain (Neidle et al., 1969; Lai and Clark, 1976; Dienel et al., 1977; Dennis et al., 1977; Lai et al., 1977). In the ensuing discussion, however, we will take the simple view that astrocytes and neurons represent the small and large compartments, respectively.

In brain, many blood-derived amino acids appear to be preferentially metabolized in astrocytes (e.g. Hawkins and Mans, 1983). For example, after administration of labeled PHE, TYR or PRO to rats the specific activity of brain GLN exceeds that of GLU (Van Den Berg, 1970), suggesting that the labelled amino acids are preferentially degraded and incorporated into GLN in the rapidly-turning over, astrocytic GLN/GLU pool. Other workers have shown that, following administration of labeled LEU to rats (Roberts and Morelos, 1965; Berl and Frigyesi, 1968; Patel and Balázs, 1970,71; Cremer et al., 1975) considerable label is present in the brain GLU/GLN pools. In the work of Cremer et al. (1975), the cerebral [^{3}H]GLN-to-[^{3}H]GLU specific activity ratio exceeded 1.0 at early time points following administration of L-[4,5-^{3}H]leucine to rats. This finding suggests that the carbon of blood-derived LEU is incorporated into GLU preferentially in astrocytes. Nevertheless, LEU is also taken up and metabolized in neuronal preparations (Peterson and Raghupathy, 1972; Chaplin et al., 1976; Yudkoff et al., 1983).

Some evidence suggests that utilization of blood-derived amino acids for protein synthesis occurs preferentially in neuronal perikarya (e.g. Balazs et al., 1973). Many studies have been directed toward measuring whole brain protein synthesis. For example, Dwyer et al. (1982) estimated that rat brain protein turns over at 2.8 %/h. Estimates of rates of LEU degradation versus LEU incorporation into protein have been obtained with radiolabelled-LEU. An early study with primate brain placed LEU oxidation at 2-3 times that of protein incorporation (Barrio et al., 1982). More recent estimates for rat brain suggest that the ratio may be closer to 1.0 (Keen et al., 1989). Smith et al. (1988) demonstrated that there is significant recycling of amino acids in cerebral protein synthesis.

In summary, both neurons and glia play important roles in cerebral amino acid metabolism and in the maintenance of carbon and nitrogen homeostasis in brain. The metabolism of astrocytes is intimately connected with that of neurons, and several workers (e.g., Bradbury, 1979, 1985) have emphasized the special anatomical relationship between these cells. Astrocytic end-feet ensheathe cerebral capillaries, and their processes abut onto neurons, especially in their synaptic zones. Astrocytic processes also underlie the ventricular ependymal layer. This anatomical arrangement allows astrocytes to act as a metabolic intermediary between blood (or cerebrospinal fluid) and neurons and to contribute to the extracellular environment of the neurons. Such considerations have led to the proposal that astrocytes participate in neurotransmitter cycling.

Neurotransmitter Cycles

Glutamate. Astrocytes contain active uptake systems for GLU, ASP and GABA (reviewed in Hertz, 1979). It has been proposed that GLU released from nerve endings is taken up by astrocytes and 5-carbon units are returned to neurons as electrophysiologically neutral GLN - "The Glutamine Cycle" (Benjamin and Quastel, 1975). Glutaminase within nerve endings converts this GLN to GLU completing the cycle (Fig. 2). Some evidence supports this idea (e.g., Ramaharobandro et al., 1982; Waniewski and Martin, 1986). Indeed, cultured astrocytes avidly accumulate GLU (McLennan, 1976; Hertz et al., 1978a). However, it appears unlikely that all released neuronal GLU carbon can be replenished by astrocyte GLN. Thus, a) GLN is accumulated at least as actively into cultured astrocytes as into cultured neurons (Schousboe et al., 1988a); b) the carbon of GLU taken up into cultured astrocytes is only partly released as GLN (Yu et al.,1982), and c) pyruvate carboxylase which catalyzes the major anaplerotic reaction of brain is present in astrocytes but not in neurons (Yu et al.,1983; Shank et al., 1985). Shank and colleagues showed that neuronal preparations can take up and metabolize α-ketoglutarate and malate (Shank and Campbell, 1981, 82, 84a; Shank et al., 1989) and proposed that these compounds are involved in an astrocyte to neuron metabolic shuttle. The same authors argued that both α-ketoglutarate and GLN are precursors of the transmitter pool

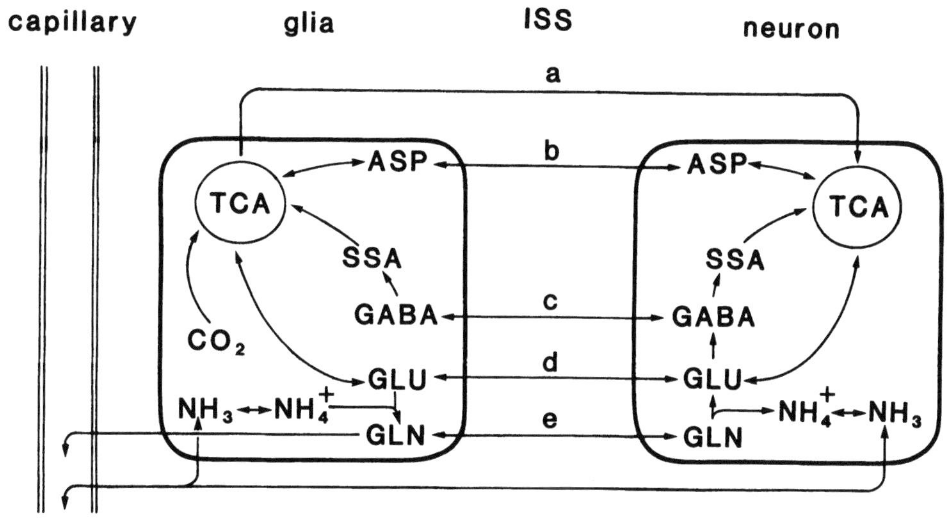

Fig. 2. Metabolic Compartmentation of Amino Acid Neurotransmitter Metabolism; d - e, the "Glutamine Cycle"; c - e, the "Glutamine, Glutamate, GABA Cycle"; d - a, release of GLU from neurons to glia followed by return of carbon to neurons as α-ketoglutarate and malate; b - e, cycling of ASP carbon. Note also that astrocytes are rich in the anaplerotic enzyme, pyruvate carboxylase, so that CO_2 fixation provides a means of replenishing carbon lost from the small TCA cycle (as α-ketoglutarate, oxaloacetate or malate). No attempt is made to distinguish between metabolic and "releasable" (i.e., neurotransmitter) pools of amino acids although such pools are probably distinct (see the text). SSA, succinic semialdehyde; ISS, interstitial fluid. From Cooper (1988), with permission.

of GLU throughout the brain. Other workers have provided strong evidence from *in vitro* studies that GLN is an excellent precursor of the releasable (i.e., neurotransmitter) pool of GLU in nerve endings (Hamberger et al., 1969, 79; Bradford et al., 1978, 89). The presence of distinct, storage (astrocytes) and newly synthesized (nerve endings) pools of GLU has been inferred from tissue slice experiments (de Barry et al., 1986). Lesions of glutamatergic pathways to different regions of the brain are often accompanied by loss of 20-40% of endogenous GLU content, possibly indicative of the approximate size of the neurotransmitter pool of GLU (reviewed in Fonnum, 1984 and Nicholls, 1989). Immunohistochemical studies suggest a particularly high concentration of GLU in hippocampus and other glutamatergic regions of the brain (Storm-Mathisen et al., 1983). Recently, Peng et al. (1991) showed that α-ketoglutarate added to cultures of cerebellar granule cells is a precursor of neurotransmitter GLU. Interestingly, synthesis of this neurotransmitter appears to occur in a pool distinct from the mitochondrial GLU pool (see also McMahon and Nicholls, 1990).

In summary, there is strong, but indirect, evidence that glutamatergic neurons possess a neurotransmitter (releasable) pool of GLU which is distinct from the metabolic pool and which may be repleted, in part, from 5-C and 4-C units derived from astrocytes.

Aspartate. Astrocytes in culture contain a high capacity, high affinity uptake system for L-ASP (Drejer et al., 1983a). The capacity of astrocytes is considerably greater than that of cultured cerebellar granule cells (Drejer et al., 1983a). Thus, it is conceivable that neurotransmitter pools of ASP are also replenished at least, in part, by astrocytic GLN, α-ketoglutarate and malate (Cooper, 1988; Fig.2).

GABA. Van Den Berg and colleagues have proposed that GABA released from nerve terminals is taken up by astrocytes and neurotransmitter carbon is replenished at least in part by astrocytic GLN - "The GABA-glutamine-glutamate cycle." (Fig. 2) (Van Den Berg, 1973; Van Den Berg et al., 1978). Generally, GABA uptake is faster into cultured neurons than into astrocytes, suggesting that of the GABA released from nerve terminals, more is taken up into neurons than into astrocytes (cf. Hertz, 1979). Nevertheless, several specific blockers of glial GABA uptake are also anticonvulsants suggesting a functional role of glia in neurotransmitter GABA metabolism (reviewed in Schousboe et al., 1988b). Indeed, the anticonvulsant drug, valproate was recently shown to increase the K_m of the GABA carrier in cultured astrocytes (Nilsson et al., 1990). Other studies support a role of astrocytic GLN in neurotransmitter GABA metabolism (cf. Shank and Aprison, 1977; Tapia and Gonzalez, 1978; Reubi et al., 1978; Paulsen et al., 1988; Battaglioli and Martin, 1990, 91).

After carotid artery infusion of [^{13}N]ammonia to rats for 20 mins the specific activity of GABA relative to that of GLU is 0.56 (Cooper et al., 1988). This figure is remarkable when it is considered that the rat brain contains about 12 mM GLU and 2 mM GABA and it is unlikely that GABA is readily obtained from total brain GLU. Either a small amount of [^{13}N]ammonia traverses the astrocytic end feet to the GABAergic nerve endings to be incorporated into GLU in a distinct GABAergic compartment, or [^{13}N]GLU (or [amine-^{13}N]GLN) in astrocytes is a precursor of neuronal [^{13}N]GABA. Recently, Rosenberg and colleagues presented evidence from NMR data suggesting that GABA is synthesized from precursors derived in part from the anaplerotic pyruvate carboxylase reaction (Brainard et al., 1989), but this conclusion has been criticized (Shank and Nicklas, 1990). Cerdan et al. (1990) used ^{13}C-NMR to show that GABA is synthesized in the large GLU compartment (= neurons) in brain *in vivo*, and that carbon of this GABA is derived in part from astrocytic GLN. Finally, Drejer et al. (1983b) showed that granule cell-conditioned media selectively enhances GABA uptake into cultured cerebellar astrocytes whereas cerebral cortex neuron-conditioned media selectively enhances GLU uptake into astrocytes. There seems to be a mechanism for neurotransmitter amino acid uptake into astrocytes that is dependent on stage of development, regional origin and surrounding neuronal type (Minchin and Fonnum, 1979; Drejer et al., 1983b).

The pool of GLU which functions as a precursor of GABA can be estimated by selective destruction of GABAergic neurons (Minchin and Fonnum, 1979; Korf and

Venema, 1981; Fonnum, 1988). Losses in GLU are relatively modest. Fonnum (1988) has summarized the relative GLU pool sizes in neostriatum as follows: GLU neurotransmitter pool (20-30%), the precursor pool of GABA (~5%), the glial pool (10-30%), and the metabolic pool (~50%).

Location of Enzymes Involved in Neurotransmitter Carbon Cycling. Further evidence for neurotransmitter cycles and for the pivotal role of GLN in these processes comes from the findings that nerve endings are enriched in glutaminase (Bradford et al., 1978; Reubi et al., 1978; Szerb and O'Regan, 1985; Yudkoff et al., 1989; Erecińska et al., 1990), whereas brain glutamine synthetase is located primarily in astrocytes (Martinez-Hernandez et al., 1977; Riepe et al.,1977; Norenberg and Martinez-Hernandez, 1979). Immunohistochemical (Martinez-Hernandez et al., 1977; Norenberg and Martinez-Hernandez, 1979) and Northern blot analyses (Mearow et al., 1989) suggest that glutamine synthetase and its mRNA are especially abundant in glutamatergic areas of the brain. Several workers have provided immunohistochemical evidence that glutaminase is localized in glutamatergic neurons (e.g., Altschuler et al., 1984, 85; Donoghue et al,. 1985; Kaneko et al., 1987a, 88a; Aoki et al., 1991). However, the findings of immunohistochemical staining in glia, neuronal perikarya and dendrites (Kaneko et al., 1987b, 88b; Aoki et al., 1991) and the direct measurement of this enzyme in glia (reviewed in Hertz and Schousboe, 1988) suggests that glutaminase is involved in general cerebral metabolism in addition to its role in neurotransmission.

The brain contains a relatively large amount of glutamate dehydrogenase. The K_m exhibited by this enzyme for ammonia is high (10-30 mM) (Chee et al., 1979). Thus, at physiological levels of ammonia (0.2 mM), the enzyme is far from saturated, and relatively little label is incorporated into brain GLU when rats are administered [^{13}N]ammonia *in vivo* (Cooper et al., 1979, 1985). Additionally, relatively little label is incorporated into GLU when cultured neurons are incubated with 0.3 mM [^{15}N]ammonia (Yudkoff et al., 1990). It is possible that glutamate dehydrogenase is involved in the catabolism rather than the synthesis of brain GLU (Cooper et al., 1979). Immunohistochemical staining of glutamate dehydrogenase is more intense in astrocytes than in neurons (e.g., Wenthold et al., 1987; Aoki et al., 1987a, 1987b). It was suggested that the enzyme may be important in the metabolism of GLU taken up into glia (Wenthold et al., 1987). However, glutamate dehydrogenase is present in both astrocytes and neurons (reviewed in Hertz and Schousboe, 1988); it seems probable that the enzyme in brain is involved in general metabolic processes as well as in neurotransmitter metabolism.

Altschuler et al. (1981, 1982) have provided immunocytochemical evidence that aspartate aminotransferase is enriched in certain glutamatergic/aspartergic regions of the brain and retina. Moreover, inhibition of aspartate aminotransferase with aminooxyacetate is associated with a large reduction in GLU release from glutamatergic cells upon depolarization (Palaiologos et al., 1988a, 88b; Kihara and Kubo, 1989; Peng et al., 1991). However, aminooxyacetate is an indiscriminate inhibitor of all pyridoxal 5'-phosphate-containing enzymes and reacts with alpha-keto acids (Smith et al., 1977). Moreover, inhibition of aspartate aminotransferase will disrupt the malate-aspartate shuttle and energy metabolism, and thereby may indirectly impede neurotransmitter GLU release. It seems probable that different metabolic strategies may be involved in the production and breakdown of the excitatory amino acid transmitters, and that the pathways may be used to varying degrees in different populations of neurons (Wenthold and Altschuler, 1983).

GABA and glutamate decarboxylase were once thought to occur exclusively in neurons, but are now known to occur in e.g., glia, kidney, pancreas and oviduct. Immunohistochemical techniques have been used to locate the enzyme to GABAergic regions of brain and retina (reviewed in Wu et al., 1988; Voaden, 1988). Immunohistochemical studies have also revealed a predominant localization of GABA transaminase to neurons (e.g., McGeer et al., 1983), but some immunoreactivity can be detected in astrocytes particularly those enclosing GABA-containing neurons (Chan-Palay et al., 1979). This is in accord with the observation that GABA transaminase activity is present in primary cultures of astrocytes (e.g., Bardakdjian et al., 1979) but is more active in synaptosomes and cultured neurons (for review see Hertz, 1982).

Ornithine aminotransferase has been suggested to be a marker for glutamatergic neurons (Wong et al., 1981), and indeed, ornithine aminotransferase is present in synaptosomes (Yoneda et al., 1982). However, later work showed the enzyme to be present in astrocytes, GABAergic neurons and glutamatergic neurons in roughly comparable levels (Drejer and Schousboe, 1984). Other studies seem to rule out the possibility that ORN can contribute significantly to GABA and GLU neurotransmitter pools (Shank and Campbell, 1983).

In conclusion, there is much evidence for considerable cycling of GLU, ASP and GABA carbon (and nitrogen) between neurotransmitter pools (vesicles) and astrocytes. Evidently, the pathways by which released neurotransmitters are removed and neurotransmitter pools are replenished vary among different cell types. Further, several enzymes of intermediary metabolism (e.g., glutamate dehydrogenase, glutaminase, glutamine synthetase, aspartate aminotransferase) are intimately involved in these neurotransmitter cycling processes.

Receptors for Amino Acid Neurotransmitters

The mechanism by which GLU, ASP and GABA are released from nerve terminals is a subject of controversy. There are two main theories: (1) The direct cytoplasmic release hypothesis may be depicted as: vesicular amino acid ⇝ cytosolic amino acid → Ca^{2+}-dependent release; (2) The exocytotic release hypothesis may be depicted as: cytosolic amino acid ⇝ vesicular amino acid → Ca^{2+}-dependent release (see Nicholls, 1989 for a recent review). Exocytotic release requires ATP. It is now thought that normal nerve endings utilize pathway (2). However, glial cells can only release GLU via a Ca^{2+}-independent mechanism, and this pathway may predominate in nerve cells under pathological conditions with low ATP (McMahon and Nicholls, 1990; McMahon et al., 1990; see Erecińska and Silver, 1990 for a discussion). In the case of L-GLU, which is the principal neurotransmitter that mediates fast excitatory synaptic transmission in the vertebrate CNS, the neurotransmitter released diffuses across the synaptic cleft and reaches postsynaptic receptors linked to ion channels. GLU in the extracellular space is rapidly removed (see below). Activation of these receptors leads to (a) stimulation of phosphatidylinositol metabolism (e.g., Sugiyama et al., 1987; Recasens et al.,1988; Sugiyama et al., 1989) or (b) cationic conductance increase (e.g., Jahr and Stevens, 1987; Cull-Candy and Usowicz, 1987), and subsequent neuronal excitation (Ly and Michaelis, 1991). GLU receptors are classified into three main groups according to whether they are activated by (1) quisqualate/alpha-amino-3-hydroxy-5-methyl-4-isoxazolepropionic acid (AMPA), (2) kainic acid, or (3) N-methyl-D-aspartate (NMDA) (e.g., Watkins et al., 1990).

The amino acid sequence of the kainate receptor (MW ~98,000) has been deduced from the cDNA and the receptor-ion channel complex has been expressed in frog oocytes following injection of mRNA formed from transcription of the cDNA (Hollman et al., 1989). As with GABA, GLY and nicotinic acetylcholine receptors, native kainate-quisqualate-sensitive GLU receptors form a family of heteromeric proteins (Boulter et al., 1990). NMDA has no effect on, whereas quisqualate and AMPA activate the receptor-ion channels (Boulter et al., 1990; Keinanen et al., 1990). The non-N-methyl-D-aspartate GLU receptors (GluRs) are encoded by a gene family, known members of which are designated GluR-1,-2,-3,-4,-5; -1 through -4 undergo alternative splicing to yield multiple receptors (Pellegrini-Gampietro et al., 1991) Michaelis and colleagues have identified synaptosomal proteins (63 kDa, 67 kDa (Chen et al., 1988), 58 kDa (Cunningham and Michaelis, 1990)) that bind NMDA. These may be related to the four proteins Ikin et al. (1990) suggest make up the NMDA receptor-ion channel complex of rat brain. The L-GLU-activated cation channel proteins from rat brain synaptosomes have been solubilized, partially purified and incorporated into liposomes (Ly and Michaelis, 1991). Finally, the NMDA receptor (Moriyoshi et al., 1991) and a GLU-binding subunit of the NMDA receptor (Kumar et al., 1991) have recently been cloned and characterized.

The CNS contains at least two forms of GABA receptors. The $GABA_A$ receptors are ligand-regulated ion channels mediating inhibitory neurotransmission. They are thought to be largely postsynaptic and to be concentrated in central and peripheral sympathetic neurons (McGeer and McGeer, 1989). The receptors are associated with Cl^- influx and possess binding sites for GABA, benzodiazepines and barbiturates. There is considerable microheterogeneity among α- and ß-subunits (e.g., Sato and Neale, 1989). A partial structure of the $GABA_A$ receptor, based on the model of the nicotinic acetylcholine receptor, has been proposed (Kosower, 1988). During the resting stage, the channel is depicted as a closed "flower"; on activation by GABA, the "flower" opens allowing influx of Cl^- (Kosower, 1988). Apparently, there are strong homologies among the $GABA_A$, GLY and nicotinic receptors suggesting a common evolutionary origin of this super family of receptors (Kosower, 1988; Betz, 1990). $GABA_B$ receptors are widespread in mammalian nervous tissue where they occur on presynaptic autonomic and central nerve terminals (Bowery et al., 1987; McGeer and McGeer, 1989). In brain, $GABA_B$ receptors activate potassium channels (see Nicholl, 1988 for review). The response is mediated via pertusis toxin-sensitive G protein (Thalmann, 1987). Agonist binding is stimulated by Ca^{2+} (e.g., Al-Dahan and Thalmann, 1989). There appear to be distinct subsets of $GABA_B$ receptors (Scherer et al., 1989). Procedures for solubilization of $GABA_A$ and $GABA_B$ receptors under relatively mild conditions have been published (Shank et al., 1990).

GLY is the major inhibitory amino acid neurotransmitter of mammalian spinal cord. The spinal cord receptor is a ligand-gated chloride channel consisting of two subunits (48 kDa and 58 kDa) and an associated protein (93 kDa). Strychnine binds to the 48 kDa protein and the amino acid sequence of this protein has been deduced (see Becker et al., 1989 for original references). Deficiencies in the receptor have been implicated in the pathogenesis of spasticity and spinal cord degeneration (e.g., Betz, 1987). The distribution of GLY receptors in brain has been mapped (Triller et al., 1985; Altschuler et al., 1986). In addition, GLY binds with high affinity to the NMDA receptor (e.g., Johnson and Ascher, 1987; Romano et al., 1989). The binding is not sensitive to strychnine. GLY may facilitate excitatory transmission in the rat brain through allosteric interaction of the NMDA receptor (Johnson and Ascher, 1987; Wood et al., 1990).

GLU is present at very high levels in brain (10-12 mM), yet a low concentration must be maintained in the extracellular fluid; excessive GLU in the extracellular fluid is excitotoxic. This excitotoxic action of GLU may be important in initiating brain injury following ischemia and other pathological insults. The mechanism of glutamate neurotoxicity may be related to its prolonged depolarizing action at GLU postsynaptic receptor sites, resulting in plasma membrane alterations, activation of Ca^{2+} channels and impaired ion homeostasis (for review see Rothman and Olney, 1987). Interestingly, external GLU levels may actually determine to some extent the architecture of growing neurons (e.g., pyramidal cells). Thus, very low levels have no effect, low levels lead to stabilization, moderate levels lead to dendrite pruning, and high levels lead to cell death (Mattson et al., 1988). For a discussion of neurotransmitters as developmental signals for astrocytes and neurons see Meier et al. (1991).

Amino Acid Transporters in Astrocytes and Neurons

With this background in neurotransmitter cycles and receptors, we are in a better position to discuss the uptake systems per se of amino acids in astrocytes and neurons. A landmark paper in this area is that by Sershen and Lajtha (1979). These authors described the occurrence of at least ten transport systems for amino acids into mouse neural cells. The first five systems, namely Systems A, L, ASC, x^- (acidic amino acids), and y^+ (basic amino acids) are analogous to transport systems described for the BBB and other tissues. Separate transporters fairly specific for GLY, PRO, GABA (mostly neuronal), TAU (mostly glial) and LYS were also described.

Glutamate Transport. An excellent review of CNS GLU metabolism and transport has recently appeared (Erecińska and Silver, 1990). The salient points of this review plus some additional references are given below. Originally it was thought that neural tissues contain both high-affinity and low-affinity GLU transporters (Logan and Snyder, 1971; Balcar and Johnson, 1972a). However, there is now strong evidence that at high concentrations GLU enters cells predominantly by passive diffusion and not by a low affinity mechanism (Wheeler, 1987). High affinity uptake of GLU across plasma membranes is by either a Na^+-dependent or by a Cl^--dependent carrier (Erecińska and Silver, 1990).

i. Sodium-dependent, high-affinity transport. GLU uptake into brain slices, synaptosomes, bulk isolated astrocytes and neurons, and cultured astrocytes and neurons occurs predominantly via a Na^+-dependent, high-affinity transporter. It is not yet clear whether the same transporter is present in astrocytes and neurons, but the K_m values reported (3-80 μM) do not differ significantly between the two cell types. The V_{max} is between 2 and 60 nmol/min/mg protein at 37° C and is generally greater in cultured astrocytes (see Erecińska and Silver, 1990 for original references). There is some controversy as to whether [^{3}H]GLU is preferentially taken up into glia (McLennan, 1976) or into nerve endings (Storm-Mathisen and Iversen, 1979) in brain slices. Nevertheless, it seems probable that a considerable portion of released neurotransmitter GLU is taken up into astrocytes (Fig. 2) by the Na^+-dependent high-affinity transporter.

This transporter translocates L- (but not D-) GLU. D- and L-ASP, cysteate, D- and L-cysteinesulfinate and threo-3-hydroxy-DL-ASP are either translocated or inhibit L-GLU uptake. The transporter is inactive toward kainate, ibotenate, quisqualate, and neutral amino acids including GLN and GABA (Balcar and Johnson, 1972b; Roberts and Watkins, 1975; Stallcup et al., 1979; Griffiths et al., 1989; Erecińska and Silver, 1990). The transporter is also inhibited by ß-methylene-DL-aspartate and L-aspartate-ß-hydroxamate (Bender et al., 1989).

The K_m exhibited by the transporter for Na^+ is 12-18 mM in brain tissue slices (Balcar and Johnson, 1972a) and in cultured neurons and astrocytes (Schousboe et al., 1977b; Drejer et al., 1982). If one Na^+ enters with each GLU, then the downhill movement of Na^+ ion is the driving force for entry of GLU. If, however, two or more Na^+ ions are transported with each GLU, then the entering complex is positively charged and the transmembrane electrical potential may be used as the driving force. As reviewed by Erecińska and Silver (1990), literature values for the ratio of Na^+ to GLU uptake range from 1 to > 2. This may be a consequence of the different experimental protocols employed, or perhaps, glia and neurons have different transporters. Erecińska et al. (1983) suggest the most reliable estimate of stoichiometry, at least for synaptosomal preparations, is 2 Na^+ and 1 H^+ transported for each GLU. In addition, GLU uptake may be associated with adenosine release in nerve endings (Schousboe et al., 1989; Hoehn and White, 1990) but not in astrocytes (Schousboe et al., 1989).

ii. Chloride-dependent, high-affinity transport. This transporter is present in both neurons and glia and mediates the exchange of external and internal GLU. It requires Cl^- but not a Cl^- gradient. The system has a very high affinity for GLU (K_m = <1 μM), but the velocity is <1% that of the Na^+-dependent carrier in normal brain tissue. The transporter is inhibited by L-ASP, L-cysteate, cystine, quisqualate, ibotenate and DL-2-amino-4-phosphonobutyrate, but not by kainate and N-methyl-D-aspartate (see Erecińska and Silver, 1990 for original references). It has been suggested that the Cl^- -dependent GLU-transporter is, in fact, a cystine transporter, and is necessary for the delivery of intracellular CYS for glutathione synthesis (Murphy et al., 1989). This idea is supported by the findings of Cho and Bannai (1990) who investigated GLU uptake into glioma cells and cultured astrocytes. These authors described (1) a Na^+-dependent transporter (X_{AG}-) which is inhibited by ASP, and (2) a Na^+-independent transporter, which is inhibited by cystine and quisqualate, and possesses properties consistent with the x_c- transporter of various peripheral tissues. In gliomas it was estimated that X_{AG}- and x_c- contributed 35% and 55%, respectively, to uptake in the presence of 50 mM L-GLU. In astrocytes, negligible uptake of GLU occurred by the x_c- transporter. Nevertheless, the x_c- system was absolutely necessary for cell integrity; blocking of cystine uptake with GLU resulted in decreased glutathione levels and cell death.

iii. Glutamate transport into synaptic vesicles. GLU is stored in vesicles in the nerve endings for subsequent release during neuronal activity. In vesicles isolated from brain, the transporter has a relatively low affinity for GLU (K_m = 1.1-1.6 mM), requires ATP,

Mg^{2+} and Cl^-, and is inhibited by uncouplers (e.g., FCCP), thiol reagents and high concentrations of Cl^- and thiocyanate. The transporter is relatively specific for L-GLU. The mechanism for concentration of GLU into vesicles may involve the membrane electrical potential (positive inside) and the pH gradient (acid inside) although evidence suggests that the former is more important than the latter (cf. Erecińska and Silver, 1990). It is possible that levels of GLU in vesicles may reach 0.5-1.0 M. The regional distribution of GLU vesicles correlates with the distribution of the Na^+-dependent high affinity system in synaptosomes. Moreover, the number of rat brain GLU vesicles increase greatly between postnatal days 10 and 30 during the period of synaptogenesis. (See Erecińska and Silver, 1990 for original references and discussion.) The ATP-dependent vesicular GLU uptake system has been solubilized and reconstituted into liposomes in an active form (Carlsson et al., 1989). Recently, it has been shown that synaptic vesicular uptake of GLU appears to be inhibited by a specific protein of MW > 100,000 (Lobur et al., 1990).

iv. Glutamate transport in mitochondria. GLU enters mitochondria via a GLU^-/OH^- antiport (formally equivalent to a GLU^-/H^+ symport) carrier (Dennis et al., 1976; Dennis and Clark, 1977; Erecińska and Silver, 1990). The K_m is ~1.5 mM, the rate is temperature dependent and the uptake is inhibited by non-polar thiol reagents. The second carrier whereby GLU enters mitochondria is via exchange for ASP. This antiport is an integral part of the malate-ASP shuttle which transfers reducing equivalents from the cytosol into mitochondria in lieu of NADH (Brand and Chappell, 1974; Dennis et al., 1976, 77; Minn and Gayet, 1977; Fitzpatrick et al., 1983; Lai et al., 1989).

Aspartate Transport. ASP is taken up into neural tissues by the Na^+-dependent, high-affinity GLU transporter (e.g., Drejer et al., 1983a).

Glutamine Transport. Multiple systems have been described for GLN uptake into neural tissues. A high affinity transporter (K_m 5-50 μM) has been detected in brain slices (Balcar and Johnson, 1975), dorsal roots (Roberts and Keen, 1974) and crude nerve-terminal enriched fractions (Shank and Campbell, 1982). Moderate affinity uptake systems (K_m = 0.25-0.7 mM) are present in synaptosomes (Baldessarini and Yorke, 1974; Weiler et al., 1979), and in brain tissue slices (Benjamin et al., 1980), and low affinity uptake systems (K_m ~2.5 mM) in brain slices (Cohen and Lajtha, 1972; Benjamin et al., 1980). GLN uptake into cultured cerebellar granule cells (Hertz et al., 1980), cortical neurons (Hertz et al., 1980; Ramaharobandro et al., 1982), and cortical astrocytes (Schousboe et al., 1979; Ramaharobandro et al., 1982) has been investigated. Reported K_m (mM) and V_{max} (nmol/min/mg protein) values are (0.7, 10.3), (3.0,1.0-28.2), (0.2-3.3, 2.0-50.2), respectively. When uptake rates are calculated for an external GLN concentration of 100 μM (representing the concentration of glutamine in the extracellular fluid in brain) they are remarkably uniform (0.5-1.5 nmol/min/mg) for all cultured neural cell types (Schousboe et al., 1988a). Such rates are comparable to release rates of GLN from cerebral cortical neurons and astrocytes, but greater than rates of release of GLN from cerebellar granule cells (Schousboe et al., 1988a). Some reports suggest that GLN uptake into neural tissues is only weakly (Baldessarini and Yorke, 1974) or partially (Shank and Campbell, 1982) Na^+ dependent, whereas others show a strong Na^+ dependence (e.g., Ramaharobandro et al., 1982; Erecińska et al., 1990). Depolarization decreases GLN uptake but stimulates the glutaminase reaction; concentrations of phosphate and H^+ appear to be major regulators of glutaminase in nerve endings.

GABA Transport. Iversen and coworkers have shown that rat brain slices and homogenates accumulate [^{3}H]GABA, particularly in the nerve terminals (Iversen and Neal, 1968; Iversen and Johnston, 1971; Iversen and Bloom, 1972). Termination of GABA synaptic action is thought to occur by active removal of GABA from the synaptic cleft (e.g., Martin, 1976). However, the synaptic uptake systems for GABA may be too slow (by two orders of magnitude) to fully account for the rapid (30 μsec) termination of GABA transmission (Cupello and Hyden, 1985). Thus, it has been suggested that the diffusion of GABA molecules away from the vicinity of receptors may also play a part in transmitter inactivation (Shank and Aprison, 1988). Similar arguments hold for glutamatergic transmission.

GABA uptake has been investigated in a large number of neural preparations (see Schousboe et al., 1977a for review). Reported K_m values vary from 0.3 to 40 µM. High affinity uptake systems have been described for GABA in cultured neurons and astrocytes (K_m ~ micromolar range) (Schousboe et al., 1977a; Hertz et al.,1978b; Balcar et al., 1979), with influx into neurons generally being faster (Hertz, 1979). Glial and neuronal GABA transport is strongly inhibited by homo-ß-proline, cis-hydroxynipecotic acid and related compounds (Schousboe et al., 1978; Larsson et al., 1981). GABA transport in cortical astrocytes is reported to be accompanied by one equivalent of Na^+, whereas cotransport of GABA and 2 Na^+s may occur in glial cells of bullfrog sympathetic ganglia (Sakai et al.,1990). V_{max} and K_m values of GABA transport change with time in cultured rat cortical neurons, as do some of the properties of the GABA transporter toward inhibitors (Balcar et al., 1989).

Synaptosomes appear to possess both high affinity and low affinity GABA transport systems (Levi and Raiteri, 1973; Hitzemann and Loh, 1975; Raiteri et al., 1975; Sellstrom et al., 1976; Hertz et al., 1978b; Pastuszko et al,. 1981; Shank et al., 1990). One study suggests that even a third carrier may be present (Wood and Sidhu, 1987). Each system is Na^+-dependent, energy-dependent, temperature sensitive and strongly inhibited by nipecotic acid (Wood and Sidhu, 1987).

The ionic coupling of synaptosomal GABA transport is controversial. In an early study Martin (1973) suggested that each GABA molecule is accompanied by 3 Na^+s. Three years later, Sellstrom and colleagues proposed that the uptake of each GABA molecule is accompanied by the influx of 3Na^+s and the efflux of 1 K^+ (Sellstrom et al., 1976). Such a coupling, it was suggested, could maintain through effective use of the electrochemical gradient a nerve-ending/extracellular fluid GABA gradient of $>10^5$, so that even at a nerve-ending GABA concentration of 100 mM, the GABA concentration in the extracellular fluid would be <1 µM. More recently, Pastuszko et al., (1982) reported a GABA to Na^+ coupling ratio of 1: 2. Synaptosomal GABA uptake is also Cl^- dependent, and Kanner (1973) proposed that influx may occur as a $nNa^+:mCl^-$:GABA complex where n > 1 and m = 1. In 1988, Keynan and Kanner reconstituted the GABA transporter in proteoliposomes and found that both the high affinity and low affinity forms require Na^+ and Cl^-. For the high-affinity system n = 2.6 and n= 1.0 (Keynan and Kanner, 1988). Interestingly, the high- and low-affinity transporters may be manifestations of the same carrier. The high affinity and low affinity sites may be the extracellular and cytosolic aspects, respectively, of the same transporter (Mabjeesh and Kanner, 1989), although an additional low affinity transporter may exist. It is also possible that neurons and astrocytes have different transporters for GABA. Indeed, it has recently been reported that phorbol 12-myristate 13-acetate (PAM), an activator of protein kinase C, inhibits GABA uptake into primary cultures of astrocytes but not neurons (Gomeza et al,. 1991). A high affinity Na^+- and Cl^--dependent rat brain GABA transporter has been cloned and expressed in Xenopus oocytes; the inferred protein contains 12 possible membrane spanning regions and is unlike any previously reported protein (Guastella, et al., 1990).

In the past few years, GABA vesicles have been isolated from brain (Fykse and Fonnum, 1988; Fykse et al., 1989; Kish et al., 1989) and the transporter has been incorporated into proteoliposomes (Hell et al., 1988, 90). The K_m exhibited by the vesicular GABA transporter is in the millimolar range (comparable to that exhibited by the vesicular glutamate transporter) and 3 orders magnitude higher than that exhibited by the synaptosomal transporter. The uptake of GABA into vesicles is driven by the electrochemical proton gradient.

Glycine Transport. GLY uptake is reported to be mediated by a high-affinity system (K_m ~40 µM) in tissue slices from rat spinal cord, medullar and pons, whereas a lower affinity system (K_m ~300 µM) is present in slices from cerebral cortex, cerebellum and midbrain (Johnston and Iversen, 1971). Snyder and coworkers identified a high-affinity uptake system for GLY in synaptosomes prepared from spinal cord (Bennett et al., 1972; Logan and Snyder, 1972), and frog-retina and frog retinal synaptosomes are also reported to contain a high affinity GLY uptake system that is saturable, energy-dependent and Na^+-dependent (Salceda et al., 1974).

GLY vesicles are more numerous in the spinal cord than in the brain. As found for GABA and GLU vesicles, the vesicular GLY carrier has a rather low affinity for substrate (K_m ~8 mM) and is readily distinguished from the synaptic transporter (Kish et al., 1989; Christensen et al., 1990). Vesicular GLY uptake is both energy- and proton gradient-dependent, but varies little with Na^+ and Cl^-. GABA has little effect on GLY uptake except at high concentration. Immunohistochemical studies suggest that the Golgi cell terminals of rat cerebellum contain high levels of both GLY and GABA (Ottersen et al., 1988). Thus, vesicular stores of the two inhibitory neurotransmitters may coexist in the same nerve terminals.

Taurine and Beta-Alanine Transport. TAU is present in the CNS and retina in high concentration, and an excellent review of TAU in the CNS has recently been published (Huxtable, 1989). As pointed out by Huxtable (1989), the sulfonate group of TAU is highly acidic (pK_a of 1.5), far lower than the pK_a of the carboxylate groups of ß-ALA and GABA (3.55 and 4.03, respectively). The zwitterionic nature of TAU prevents its passage through lipid layers and allows very steep gradients between cellular and extracellular spaces. There has been considerable debate as to whether TAU is a neurotransmitter within the CNS, with some authors strongly in favor of a neurotransmitter role (Hosli and Hosli, 1978). However, Huxtable suggests that for most neurons "there is little convincing evidence that endogenous TAU is released with sufficient speed or in sufficient quantity from nerve terminals to initiate rapid changes in ion permeabilities and membrane potentials". TAU coreleased with neurotransmitters may have a modulatory role, and in some specialized cases (e.g., retina, pyramidal cell interneurons in the hippocampus) it may function as a neurotransmitter. However, such a role would involve only a small percentage of the TAU within the CNS (Huxtable, 1989). TAU's primary role may be to stabilize nerve endings and modify the time course of the coreleased primary transmitter.

It is generally assumed that high affinity transport systems for neurotransmitter amino acids are Na^+-dependent whereas high-affinity receptor binding is Na^+-independent. Thus, to support a neurotransmitter function for TAU, it would be important to characterize a high-affinity, Na^+-independent binding of TAU to CNS preparations, by analogy to that for GABA and GLY. Although several groups have described such binding to brain synaptosomes, others have not (see Table 4 in Huxtable, 1989). Thus, the presence of high-affinity Na^+-independent binding sites for TAU in the CNS has not yet been unequivocally demonstrated. The interaction of TAU with GABA receptors has also been the subject of much debate. However, Huxtable (1989) suggests that there is no convincing evidence that the electrophysiological actions of TAU are mediated via binding to the GABA receptor. Instead TAU may modulate GABA release and binding, and may perhaps bind to a distinct TAU receptor.

The presence of a high-affinity, Na^+-dependent uptake system for TAU (and ß-ALA) is on a much firmer footing. TAU transport has been studied in many neural preparations including synaptosomes, retina, brain slices, astrocytes and neurons (Huxtable, 1989). Reported K_m values vary greatly, but at least in neuronal and glial cell lines K_m values of the Na^+-dependent transporter are more consistent (5-40 µM). There has been some controversy regarding the status of the transporter(s) for GABA, ß-ALA and TAU. Some authors suggest that ß-ALA is transported solely by the astrocytic GABA transporter and thus its uptake is a marker for astrocyte function. ß-ALA has also been considered to be a neurotransmitter, but the evidence is not yet compelling and no defined role for ß-ALA in the CNS has been identified (Larsson et al., 1986). However, careful studies suggest that cultured neurons and astrocytes possess a high-affinity carrier for GABA with K_m values of ~25 µM and 31 µM, respectively, and a single distinct high affinity carrier for both TAU (K_m ~20 µM and ~25 µM, respectively) and ß-ALA (K_m ~73 µM and 71 µM, respectively) (Larsson et al., 1986). TAU and ß-ALA inhibit GABA uptake weakly in a noncompetitive fashion whereas GABA is a weak noncompetitive inhibitor of TAU. On the other hand, ß-ALA and TAU strongly and competitively inhibit each other's uptake (Larsson et al., 1986). Thus, it is unlikely that ß-ALA is transported on the GABA carrier and its uptake cannot be used as a measure of glial function.

Other transporters. In addition to those amino acids mentioned above, uptake systems for most of the commonly occurring amino acids have been studied in various neural preparations (e.g., Peterson and Raghupathy, 1972; Shank and Campbell, 1984b). In the report of Shank and Campbell (1984b), high-affinity uptake systems were identified for LEU (K_m ~10-54 μM) and for TYR (K_m ~10-13 μM) in synaptosomes, astrocytes and granule cells. Because TRP is a precursor of serotonin, its uptake into various neural tissues has been extensively examined (e.g., Mandell and Knapp, 1972; Denizeau and Sourkes, 1977; Aragon et al., 1987 and refs. cited therein). TRP uptake into nerve-endings is Na^+-dependent and is mediated by a high-affinity mechanism (K_m ~7 μM). Uptake is inhibited by PHE and TYR. Interestingly, the V_{max} of the synaptosomal TRP transporter is increased by prior treatment of the experimental animal with lithium.

The uptake of α-aminoisobutyric acid (AIB, a synthetic substrate for Na^+-dependent System A transport) has been extensively studied using mouse astrocyte cultures (Brookes, 1988). Astrocytic AIB uptake exhibits many of the properties of System A transport, including Na^+-dependence, pH sensitivity, regulatory control and pattern of analogue inhibition. The K_m of the astrocyte A System was 2.5 mM for AIB. In addition, an unusually strong Na^+-independent system for the uptake of AIB was also found. Although the saturability of the Na^+-independent component and its pattern of analogue inhibition (strongly inhibitable by BCH) were consistent with System L transport, the carrier exhibited some unusual properties, including high-affinity for substrate and concentrative ability (Brookes, 1988; Speciale et al., 1989). The K_m of the Na+-independent L carrier toward L-kynurenine was 32 μM. Inhibition studies showed that the L transporter demonstrates a clear preference for the large neutral amino acids, LEU, TRP, and PHE (Speciale et al., 1989). The astrocytic L transporter appears to possess many of the same properties of the L1 carrier of the BBB (Smith et al., 1987; Fukui et al., 1991). Its distribution within the CNS has not been closely documented.

Finally, a high affinity (K_m ~ 1 μM) transport carrier has been described for LYS in synaptosomes (Shank and Campbell, 1984b). In the same study, a lower affinity carrier was also detected (K_m 85-134 μM) in synaptosomes, astrocytes and granule cells.

SUMMARY AND CONCLUSIONS

Unanswered Questions

In summary, it is now well established that amino acid transport has a critical role in the homeostatic regulation of amino acid pools within the CNS. This is especially so for the amino acid neurotransmitters, for which storage, release, uptake and recycling are crucial for synaptic function. However, it is also true for the essential neutral and basic amino acids which cannot be formed in brain and must be transported into brain by saturable carriers at the BBB and brain cell membranes.

Although considerable progress has been made during the past 20 years in the study of brain amino acid transport, very little is known of the molecular properties of the transporters or their selective expression within the CNS. How are they related to transporters in other tissues? Are some of them unique to the CNS? In some cases at least, do they exist in different neuronal, astrocytic or BBB forms? Where are they expressed within the CNS? How are they regulated and how are they involved in responses to disease states? Some work in this area has been carried out on reconstitution *in vitro* of vesicular transporters. However, many of the above questions will not be answered until the individual genes for the transporters are isolated and cloned.

Further, the implications of the polarity of the BBB have not been fully addressed and an effective model of BBB amino acid transport has not been presented. CSF and brain interstitial fluid concentrations of amino acids have long been known to be a small fraction of respective plasma concentrations (except for GLN) (Perry et al., 1975; Hutson et al., 1985). Yet, the mechanism that maintains these gradients has not been elucidated. Davson et al. (1986) in a recent study showed that the gradients are not the result of cerebral metabolism because similar low CSF concentrations are maintained for the nonmetabolizable amino acids, aminoisobutyric acid and cycloleucine. This suggests that other mechanisms must be operative, such as active efflux across the BBB.

Finally, while considerable effort has been extended towards characterizing the kinetics of amino acid uptake, little is known of the degree to which transport is modulated or controlled *in vivo*. Preliminary reports suggest that there may be some hormonal and neural regulation of amino acid transport at the BBB (Brust, 1986; Brust and Zicha, 1988; Eriksson et al., 1988). Such effects would have important implications for the regulation of amino acid pools in brain.

REFERENCES

Abbott, N. J., Lane, N. J., and Bundgaard, M., 1986, Ann. N. Y. Acad. Sci., 481: 20.
Abbott, N. J., Hart, J., Rogac, L., Taylor, M., and Zlokovic, B. V., 1988, J. Physiol. (Lond.), 407: 25P.
Acworth, I. N., During, M. J., and Wurtman, R. J., 1988, in: "Amino Acid Availability and Brain Function in Health and Disease," G. Huether, ed., Springer-Verlag, Berlin, p. 117.
Al-Dahan, M. I., and Thalmann, R. H., 1989, J. Neurochem., 52: 313.
Altschuler, R. A., Nieses, G. R., Harmison, G. G., Wenthold, R. J., and Fex, J., 1981, Proc. Natl. Acad. Sci. USA, 78: 6553.
Altschuler, R. A., Mosinger, J. L., Harmison, C. G., Parakkal, M. H., and Wenthold, R. J., 1982, Nature, 298: 657.
Altschuler, R. A., Wenthold, R. J., Schwartz, A. M., Haser, W. G., Curthoys, N. P., Parakkal, M. H., and Fex, J., 1984, Brain Res., 291: 173.
Altschuler, R. A., Monaghan, D. T., Haser, W. G., Wenthold, R. J., Curthoys, N. P., and Cotman, C. W., 1985, Brain Res., 330: 225.
Altschuler, R. A., Betz, H., Parakkall, M. H., Reeks, K. A., and Wenthold, R. J., 1986, Brain Res., 369: 316.
Aoki, C., Milner, T. A., Sheu, K. -F. R., Blass, J. P., and Pickel, V. M., 1987a, J. Neurosci., 7: 2214.
Aoki, C., Milner, T. A., Sheu, K. -F. R., Blass, J. P., and Pickel, V. M., 1987b, J. Neurosci. Res., 18: 305.
Aoki, C., Kaneko, T., Starr, A., and Pickel, V. M., 1991, J. Neurosci. Res., 28: 531.
Aoyagi, M., Agranoff, B. W., Washburn, L. C., and Smith, Q. R., 1988, J. Neurochem., 50: 1220.
Aragon, M. C., Herrero, E., and Giminez, C., 1987, Neurochem. Res., 12: 439.
Audus, K. L., and Borchardt, R. T., 1986, J. Neurochem., 47: 484.
Baldzs, R., Patel, A. J., and Richter, D., 1973, in: "Metabolic Compartmentation in the Brain," R. Baldzs, and J. E. Cremer, eds., John Wiley and Sons, New York, p. 167.
Balcar, V. J., and Johnston, G. A. R., 1972a, J. Neurobiol., 3: 295.
Balcar, V. J., and Johnston, G. A. R., 1972b, J. Neurochem., 19: 2657.
Balcar, V. J., and Johnston, G. A. R., 1975, J. Neurochem., 24: 875.
Balcar, V. J., Mark, J., Borg, J., and Mandel, P., 1979, Neurochem. Res., 4: 339.
Balcar, V. J., Hauser, K. L., and Demieville, H., 1989, Neurochem. Res., 14: 229.
Baldessarini, R. J., and Yorke, C., 1974, J. Neurochem., 23: 839.
Baños, G., Daniel, P. M., Moorhouse, S. R., and Pratt, O. E., 1973, Proc. R. Soc. Lond. B., 183: 59.
Baños, G., Daniel, P.M. and Pratt, O.E., 1974, J. Physiol. (Lond.), 236: 29.
Baños, G., Daniel, P.M., Moorhouse, S.R. and Pratt, O.E., 1975, J. Physiol. (Lond.), 246:539.
Bardakdjian, J., Tardy, M., Pimoule, C., and Gonnard, P., 1979, Neurochem. Res., 4: 517.
Barrio, J. R., Phelps, M. E., Huang, S. -C., Keen, R. E., and MacDonald, N. S., 1982, J. Labelled Comp. Radiopharm., 19: 1271.
Battaglioli, G., and Martin, D. L., 1990, J. Neurochem., 54: 1179.
Battaglioli, G., and Martin, D. L., 1991, Neurochem. Res., 16: 151.
Becker, C. -M., Moch, W., and Betz, H., 1989, J. Neurochem., 53: 124.
Bender, A. S., Woodbury, D. M., and White, H. S., 1989, Neurochem. Res., 14: 641.
Benjamin, A. M., and Quastel, J. H., 1975, J. Neurochem., 25: 197.
Benjamin, A. M., Verjee, Z. H., and Quastel, J. H., 1980, J. Neurochem., 35: 67.
Bennett, J. P., Logan, W. J., and Snyder, S. H., 1972, Science, 178: 997.

Benveniste, H., 1989, J. Neurochem., 52: 1667.
Berl, S., and Frigyesi, T. L., 1968, J. Neurochem., 15: 965.
Berl, S., and Frigyesi, T. L., 1969, Brain Res., 12: 444.
Berl, S., Lajtha, A., and Waelsch, H., 1961, J. Neurochem., 7: 186.
Berl, S., Takagaki, G., Clarke, D. D., and Waelsch, H., 1962, J. Biol. Chem., 237: 2562.
Berl, S., Clarke, D. D., and Nicklas, W. J., 1970, J. Neurochem., 17: 999.
Betz, A. L., and Goldstein, G. W., 1978, Science, 202: 225.
Betz, H., 1987, Trends Neuro. Sci., 10: 113.
Betz, H., 1990, Neuron, 5: 383.
Blasberg, R. G., Fenstermacher, J. D., and Patlak, C. S., 1983, J. Cereb. Blood Flow Metab., 3: 8.
Boulter, J., Hollman M., O'Shea-Greenfield, A., Hartley, M., Deneris, E., Maron, C., and Heinemann, S., 1990, Science, 249: 1033.
Bowery, N. G., Hudson, A. L., and Price, G. W., 1987, Neuroscience, 20: 365.
Bradbury, M. W., 1979, "The Concept of the Blood-Brain Barrier." John Wiley and Sons, New York.
Bradbury, M. W. B., 1985, Circ. Res., 57: 213.
Bradford, H. F., Ward, H. K., and Thomas, A. J., 1978, J. Neurochem., 30: 1453.
Bradford, M.F., Ward, H. K., and Foley, P., 1989, Brain Res., 476: 29.
Brainard, J. R., Kyner, E., and Rosenberg, G. A., 1989, J. Neurochem., 53: 1285.
Brand, M. D., and Chappell, J. B., 1974, Biochem. J., 140: 205.
Brightman, M. W., and Reese, T. S., 1969, J. Cell Biol., 40: 648.
Brookes, N., 1988, J. Neurochem., 51: 1913.
Brosnan, J. T., Forsey, R. G. P., and Brosnan, M. E., 1984, Am. J. Physiol., 247: C450.
Brust, P., 1986, J. Neurochem., 46: 534.
Brust, P., and Zicha, J., 1988, Biomed. Biochim. Acta, 12: 1013.
Butt, A. M., Jones, H. C., and Abbott, N. J., 1990, J. Physiol. (Lond.), 429: 47.
Cancilla, P. A., and DeBault, L. E., 1983, J. Neuropath. and Exp. Neuro., 42:191.
Cangiano, C., Cardelli-Cangiano, P., James, J. H., Rossi-Fanelli, F., Patrizi, M. A., Brackett, K. A., Strom, R., and Fischer, J. E., 1983, J. Biol. Chem., 258: 8949.
Cangiano, C., Cardelli-Cangiano, P., Cascino, A., Ceci, F., Fiori, A., Mulieri, M., Muscaritoli, M., Barberini, C., Strom, R., and Fanelli, F. R., 1988, J. Neurochem., 51: 1675.
Cardelli-Cangiano, P., Cangiano, C., James, J. H., Jeppsson, B., Brenner, W., and Fischer, J. E., 1981, J. Neurochem., 36: 627.
Carlsson, M.D., Kish, P.E., and Ueda, T., 1989, J. Biol. Chem., 264: 7369.
Cerdan, S., Kunnecke, B., and Seelig, J., 1990, J. Biol. Chem., 265: 12916.
Chan-Palay, V., Wu, J.- Y., and Palay, S. L., 1979, Proc. Natl. Acad. Sci USA., 76: 2067.
Chaplin, E. R., Goldberg, A. L., and Diamond, I., 1976, J. Neurochem., 26: 701.
Chastain, J. E., and Borchardt, R. T., 1990, Neurosci. Res. Commun., 6: 51.
Chee, P. Y., Dahl, J. L., and Fahien, L. A., 1979, J. Neurochem., 33: 53.
Chen, J. -W., Cunningham, M. D., Galton, N., and Michaelis, E. K., 1988, J. Biol. Chem., 263: 417.
Cho, Y., and Bannai, I., 1990, J. Neurochem., 55: 2091.
Christensen, H. N., 1989, Methods of Enzymology, 173: 576.
Christensen, H., Fykse, E. M., and Fonnum, F., 1990, J. Neurochem., 54: 1142.
Cohen, S. R., and Lajtha, A., 1972, in: "Handbook of Neurochemistry, Vol. 7," A. Latjtha, ed., Plenum Press, New York, p. 543.
Cooper, A. J .L., 1988, in: "Glutamine and Glutamate in Mammals. Vol. 1," E. Kvamme, ed., CRC Press, Boca Raton, Fl., p. 7.
Cooper, A. J. L., McDonald, J. M., Gelbard, A. S., Gledhill, R .F., and Duffy, T. E., 1979, J. Biol. Chem., 254: 4982.
Cooper, A. J. L., Mora, S. N., Cruz, N. F., and Gelbard, A. S., 1985, J. Neurochem., 44: 1716.
Cooper, A. J. L., Lai, J. C. K., and Gelbard, A. S., 1988, in: "The Biochemical Pathology of Astrocytes," M. D. Norenberg, L. Hertz, and A. Schousboe, eds., Alan R. Liss, New York, p. 419.
Cremer, J. E., Heath, D. F., Patel, A. J., Balazs, R., and Cavanagh, J.B., 1975, in: "Metabolic Compartmentation and Neurotransmission. Relation to Brain Structure and Function," S. Berl, D. D. Clarke, and D. Schneider, eds., Plenum Press, New York, p. 461.
Crone, C., 1986, Ann. N. Y. Acad. Sci., 481: 174.

Crone, C., and Olesen, S. P., 1982, Brain Res., 241: 49.
Cull-Candy, S. G., and Usowicz, M. M., 1987, Nature, 325: 525.
Cunningham, M. D., and Michaelis, E. K., 1990, J. Biol. Chem., 265: 7768 .
Cupello, A., and Hyden, H., 1985, Brain Res., 342: 176.
Daniel, P. M., Pratt, O. E., and Wilson, P. A., 1977, Proc. Roy. Soc.(Lond.) B., 196: 333.
Davson, H., Begley, D. J., Chain, D. G., Briggs, F. O., and Shepherd, M. T., 1986, Exp. Neurol., 91: 163.
Davson, H., Lipovac, M. N., Mackic, J. B., Preston, J. E., Segal, M. B., Tang, G., and Zlokovic, B. V., 1990, J. Physiol. (Lond.), 423: 36P.
de Barry, J., Gombos, G., and Vizi, E. S., 1986, Neurochem. Res., 14: 1053.
Dehouck, M. -P., Meresse, S., Delorme, P., Fruchart, J. -C., and Cecchelli, R., 1990, J. Neurochem., 54: 1798.
Denizeau , F., and Sourkes, T.L., 1977, J. Neurochem., 28: 951-959.
Dennis, S. C., and Clark, J. B., 1977, Biochem. J., 168: 521.
Dennis, S. C., Land, J. M., and Clark, J. B., 1976, Biochem. J., 156: 323.
Dennis, S. C., Lai, J. C. K., and Clark, J. B., 1977, Biochem. J., 164: 727.
Dienel, G., Ryder, E., and Greengard, O., 1977, Biochim. Biophys. Acta, 496: 484.
Donoghue, J. P., Wenthold, R. J., and Altschuler, R. A., 1985, J. Neurosci., 5: 2597.
Drejer, J., and Schousboe, A., 1984, J. Neurochem., 42: 1194.
Drejer, J., Larsson, O.M., and Schousboe, A., 1982, Exp. Brain Res., 47: 259.
Drejer, J., Larsson, O. M., and Schousboe, A., 1983a, Neurochem. Res., 8: 231.
Drejer, J., Meier, E., and Schousboe, A., 1983b, Neurosci. Lett., 37: 301.
Drewes, L. R., Conway, W. P., and Gilboe, D. D., 1977, Am. J. Physiol., 233: E320.
Duncan, M. W., Villacreses, N. E., Pearson, P. G., Wyatt, L., Rapoport, S. I., Kopin, I. J., Markey, S. P., and Smith, Q. R., 1991, J. Pharmacol. Exp. Ther., 258:27.
During, M. J., Heyes, M. P., Freese, A., Markey, S, P., Martin, J. B., and Roth, R. H., 1989, Brain Res., 476: 384 .
Dwyer, B. E., Donatoni, P., and Wasterlain, C. G., 1982, Neurochem. Res., 7: 1507.
Erecińska, M., and Silver, I. A., 1990, Progr. Neurobiol., 35: 245.
Erecińska, M., Wantorsky, D., and Wilson, D.F., 1983, J. Biol. Chem., 258: 9069.
Erecińska, M., Zaleska, M. M., Nelson D., Nissim, I., and Yudkoff, M., 1990, J. Neurochem., 54: 2057.
Eriksson, T., and Carlsson, A., 1988, Life Sci., 42: 1583.
Fenerty, C. A., and Lindup, W. E., 1989, J. Neurochem., 53: 416.
Fenstermacher, J. D., and Rapoport., S. I., 1984, in: "Handbook of Physiology: The Cardiovascular System IV," S. R. Geiger, ed., American Physiological Society, Bethesda, p. 969.
Fernstrom, J. D., 1983, Physiol. Rev., 63: 484.
Fernstrom, J. D., and Faller, D. V., 1978, J. Neurochem., 30: 1531.
Fernstrom, J. D., and Wurtman, R. J., 1972, Science, 178: 414.
Fernstrom, M. H., Volk, E. A., and Fernstrom, J. D., 1986, Am. J. Physiol. 251: E393.
Fitzpatrick, S. M., Cooper, A. J. L., and Duffy, T. E., 1983, J. Neurochem., 41: 1370.
Fonnum, F., 1984, J. Neurochem., 42: 1-8.
Fonnum, F., 1988, in: "Glutamine and Glutamate in Mammals. Vol. 2, E. Kvamme , ed., CRC Press, Boca Raton, Fl., p. 57.
Fukui, S., Schwarcz, R., Rapoport, S. I., Takada, Y., and Smith, Q. R., 1991, J. Neurochem., 56: 2007.
Fykse, E. M., and Fonnum, F., 1988, J. Neurochem., 50: 1237.
Fykse, E. M., Christensen, H., and Fonnum, F., 1989, J. Neurochem., 52: 946.
Gomeza, J., Casado, M., Giminez, C., and Aragon, C., 1991, Biochem. J., 275: 435.
Greenwood, J., Hazell, A. S., and Pratt, O. E., 1989, J. Cereb. Blood Flow Metab., 9: 226.
Griffiths, R., Grieve, A., Dunlop, J., Damgaard, I., Fosmark, H., and Schousboe, A., 1989, Neurochem. Res., 14: 333.
Gross, P. M., Sposito, N. M., Pettersen, S. E., and Fenstermacher, J. D., 1986, Blood Vessels, 23: 261.
Guastella, J., Nelson, N. Nelson, H., Czyzyk, L., Keynan, S., Miedel, M. C., Davidson, N., Lester, H. A. and Kanner, B. I., 1990, Science 249: 1303.
Hamberger, A., Chiang, G. H., Sandoval, E. and Cotman, C. W., 1969, Brain Res., 168: 531.
Hamberger, A.C., Chiang, G. H., Nylen, E. S., Scheff, S.W., and Cotman, C. W., 1979, Brain Res., 168: 513.

Hargreaves, K. M., and Pardridge, W. M., 1988, J. Biol. Chem., 263: 19392.
Hargreaves, K. M., Buciak, J. L., and Pardridge, W. M., 1990, J. Cereb. Blood Flow Metabol., 10: 162.
Harper, A. E. and Tews, J. K., 1988, in: "Amino Acid Availability and Brain Function in Health and Disease," G. Huether, ed., Springer-Verlag, Berlin, p. 3.
Hawkins, R. A., and Mans, A. M., 1983, in: "Handbook of Neurochemistry, 2nd Edition, Vol. 3," A. Lajtha, ed., Plenum Press, New York, p. 259.
Hawkins, R. A., Mans, A. M., and Biebuyck, J. F., 1982, Am. J. Physiol., 242: E1.
Hawkins, R. A., Huang, S. -C., Barrio, J. R., Keen, R. E., Feng, D., Mazziotta, J. C. and Phelps, M. E., 1989, J. Cereb. Blood Flow. Metab., 9: 446.
Hell, J. W., Maycox, P. R., Stadler, H., and Jahn, R., 1988, EMBO J., 7: 3023.
Hell, J. W., Maycox, P. R., Stadler, H., and Jahn, R., 1990, J. Biol. Chem., 265: 2111.
Hertz, L., 1979, Prog. Neurobiol., 13: 277.
Hertz, L., 1982, in: "Handbook of Neurochemistry. 2nd Edition, Vol. 1," A. Lajtha, ed., Plenum Press, New York, p. 319.
Hertz, L., and Schousboe, A., 1988, in: "Glutamine and Glutamate in Mammals. Vol. 2.," E. Kvamme, ed., CRC Press, Boca Raton, Fl., p. 39.
Hertz, L., Schousboe, A., Boerchler, N., Mukerji, S., and Fedoroff, S., 1978a, Neurochem. Res., 3: 1.
Hertz, L., Wu, P. H., Schousboe, A., 1978b, Neurochem. Res., 3: 313.
Hertz, L., Yu, A., Svenneby, G., Kvamme, E., Fosmark, H., and Schousboe, A., 1980, Neurosci. Lett., 16: 103.
Hitzemann, R. J., and Loh, H. H., 1975, J. Neurochem., 30: 471.
Hoehn, K., and White, T.D., 1990, J. Neurochem., 54: 1716.
Hollman, M., O'Shea-Greenfield, A., Rogers, S.W. and Heinemann, S., 1989, Nature, 342: 643.
Hommes, F. A., 1983, in: "Handbook of Neurochemistry, Vol. 10, Pathological Neurochemistry," A. Lajtha, ed., Second Edition, Plenum, New York, p.15.
Hosli, L., and Hosli, E., 1978, Rev. Physiol. Biochem. Pharmacol., 81: 135.
Hutson, P. H., Sarna, G. S., Kantamaneni, B. D., and Curzon, G., 1985, J. Neurochem., 44: 1266.
Huxtable, R. J., 1989, Progr. Neurobiol., 32: 471.
Ikin, A. F., Kloog, Y., and Sokolovsky, M., 1990, Biochemistry, 29: 2290.
Iversen, L. L., and Bloom, F. E., 1972, Brain Res., 41: 131.
Iversen, L. L., and Johnston, G. A. R., 1971, J. Neurochem., 18: 1939.
Iversen, L. L., and Neal, M. J., 1968, J. Neurochem., 15: 1141.
Jahr, C. E., and Stevens, C. F., 1987, Nature, 325: 522.
James, J. H., Escourrou, J., and Fischer, J. E., 1978, Science, 200: 1395.
Johnson, J. W., and Ascher, P., 1987, Nature, 325: 529.
Johnston, G. A. R., and Iversen, L. L., 1971, J. Neurochem., 18: 1951.
Jonung, T., Rigotti, P., James, J. H., Brackett, K. and Fischer, J. E., 1985, J. Neurochem., 45: 308.
Kaneko, T., Urade, Y., Watanabe, Y., and Mizuno, N., 1987a, J. Neurosci., 7: 302.
Kaneko, T., Akiyama, H., and Mizuno, N., 1987b, Neurosci. Lett., 77: 171.
Kaneko, T., Urabe, Y., and Mizuno, N., 1988a, J. Histochem. Cytochem., 36: 997.
Kaneko, T., Shigemoto, R., and Mizuno, N., 1988b, Brain Res., 457: 160.
Kanner, B. I., 1973, Biochim. Biophys. Acta, 726: 293.
Keen, R. E., Barrio, J. R., Huang, S.-C., Hawkins, R. A., and Phelps, M. E., 1989, J. Cereb. Blood Flow Metab., 9: 429.
Keinanen, K., Wisden, W., Sommer, B., Werner, P., Herb, A., Verdoorn, T. A., Sakmann, B., and Seeburg, P. H., 1990, Science, 249: 556.
Keynan, S., and Kanner, B. I., 1988, Biochemistry, 27: 12.
Kihara, M., and Kubo, T., 1989, J. Neurochem., 52: 1127.
Kim, J. W., Closs, E. I., Albritton, L. M. and Cunningham, J. M., 1991, Nature 352: 725.
Kirikae, M., Diksic, M., and Yamamoto, Y. L., 1988, J. Cereb. Blood Flow Metab., 8: 598.
Kish, P. E., Fischer-Bovenkerk, C. and Ueda, T., 1989, Proc. Natl. Acad. Sci. USA, 86: 3877.
Klein, R. A., Moore, M. J., and Smith, M. W., 1971, Biochim. Biophys. Acta, 233: 420.
Knudsen, G. M., Pettigrew, K. D., Patlak, C. S., Hertz, M. M., and Paulson, O. B., 1990, J. Cereb. Blood Flow Metab., 10: 698.
Koeppe, R. A., Mangner, T., Betz, A. L., Shulkin, B. L., Allen, R., Kollros, P., Kuhl, D. E., and Agranoff, B. W., 1990, J. Cereb. Blood Flow Metabol., 10: 727.

Korf, J., and Venema, L., 1981, J. Neurochem., 40: 1171.
Kosower, E. M., 1988, FEBS Lett., 231: 5.
Krespan, B., Berl, S., and Nicklas, W. J., 1982, J. Neurochem., 38: 509.
Kumar, K. N., Tilakaratne, N., Johnson, P. S., Allen, A. E. and Michaelis, E. K., 1991, Nature 354: 70.
Kurland, L. Y., 1988, Trends Neurosci., 11: 51.
Lai, J. C .K., and Clark, J. B., 1976, Biochem. J., 154: 423.
Lai, J. C. K., Walsh, J. M., Dennis, S. C., and Clark, J. B., 1977, J. Neurochem., 28: 625.
Lai, J. C. K., Murthy, C. R. K., Cooper, A. J. L., Hertz, E. and Hertz, L., 1989, Neurochem. Res., 14: 377.
LaManna, J. C., and Harik, S. I., 1986, J. Cereb. Blood Flow Metab., 6: 717.
Larsson, O. M., Thorbek, P., Krogsgaard-Larsen, P., and Schousboe, A., 1981, J. Neurochem., 37: 1509.
Larsson, O.M., Griffiths, R., Allen, I. C. and Schousboe, A., 1986, J. Neurochem., 47: 426.
Lefauconnier, J.-M., and Trouve, R., 1983, Dev. Brain Res., 6: 175.
Lerner, J., and Larimore, D. L., 1986, Comp. Biochem. Physiol., 84B: 235.
Levi, G., and Raiteri, M., 1973, Life Sci., 12: 81.
Lobur, A. T., Kish, P.E., and Ueda, T., 1990, J. Neurochem., 54: 1614.
Logan, W. J., and Snyder, S. H., 1971, Nature, 234: 297.
Logan, W. J., and Snyder, S. H., 1972, Brain Res., 42: 413.
Ly, A. M., and Michaelis, E. K., 1991, Biochemistry, 30: 4307.
Mabjeesh, N. J., and Kanner, B. I., 1989, Biochemistry, 28, 7694.
Mandell, A. J., and Knapp, S., 1977, Fed. Proc., 36: 2142.
Mans, A. M., Biebuyck, J. F., Shelly, K., and Hawkins, R. A., 1982, J. Neurochem., 38: 705.
Mans, A. M., Biebuyck, J. F., and Hawkins, R. A., 1983, Am. J. Physiol., 245: C74.
Mans, A. M., DeJoseph, M. R., Davis, D. W., and Hawkins, R. A., 1987, Am. J. Physiol., 253: E575 .
Martin, D. L., 1973, J. Neurochem., 21: 345.
Martin, D. L., 1976, in: "GABA in Nervous System Function," E. Roberts, T. N. Chase, and D. B. Tower, eds., Raven Press, New York, p.347.
Martinez-Hernandez, A., Bell, K. P., and Norenberg, M. D., 1977, Science, 195: 1356.
Mattson, M. P., Dou, P., and Kater, S. B., 1988, J. Neurosci., 8: 2087.
McCall, A. L., Millington, W. R., and Wurtman, R. J., 1982, Proc. Natl. Acad. Sci. USA, 79: 5406.
McGale, E. H. F., Pye, I. F., Stonier, C., Hutchinson, E. C., and Aber, G. M., 1977, J. Neurochem., 29: 291.
McGeer, E. G., McGeer, P. L., and Thompson, S., 1983, in: "Glutamine, Glutamate and GABA in the Central Nervous System," L. Hertz L., E. Kvamme, E. G. McGeer, A. Schousboe, eds., Alan R. Liss, New York, p. 3.
McGeer, P. L., and McGeer, E. G., 1989, in: "Basic Neurochemistry, 4th Edition," G. Siegel, B. Agaranoff, R. W. Albers, and P. Molinoff, eds., Raven Press, New York, p. 311.
McKean, C. M., Boggs, D. E., and Peterson, N. A., 1968, J. Neurochem., 15: 235.
McLennan, H., 1976, Brain Res., 115: 139.
McMahon, H. T., and Nicholls, D. G., 1990, J. Neurochem., 54: 373.
McMahon, H. T., Rosenthal, L., Meldolesi, J., and
Nicholls, D. G., 1990, J. Neurochem., 55: 2039.
Mearow, K. M., Mill, J. F., and Vitkovic, L., 1989, Mol. Brain Res., 6: 223.
Meier, E., Hertz, L., and Schousboe, A., 1991, Neurochem. Int., 19: 1.
Miller, L. P., Pardridge, W. M., Braun, L. D., and Oldendorf, W. H., 1985, J. Neurochem., 45: 1427.
Minchin, M. C. W., and Fonnum, F., 1979, J. Neurochem., 32: 203.
Minn, A., and Gayet, J., 1977, J. Neurochem., 29: 873.
Momma, S., Aoyagi, M., Rapoport, S. I., and Smith, Q. R., 1987, J. Neurochem., 48: 1291.
Moriyoshi, K., Masu, M., Ishii, T., Shigemoto, R., Mizuno, N. and Nakanishi, S., 1991, Nature 354: 31.
Moroni, F., Russi, P., Lombardi, G., Beni, M., and Carla, V., 1988, J. Neurochem., 51: 177.
Murphy, T. H., Miyamoto, M., Sastre, A., Schnaar, R. L., and Coyle, J. T., 1989, Neuron, 2: 1547.
Nagashima, T., Lefauconnier, J. M., and Smith, Q. R., 1987, J. Cereb. Blood Flow Metab., 7: S501.

Neidle, A., Van Den Berg, C. J., Grynbaum, A., 1969, J. Neurochem., 16: 225.
Nicholl, R. A., 1988, Science, 241: 545.
Nicholls, D. G., 1989, J. Neurochem., 52: 331.
Nicklas, W. J., 1983, in: "Glutamine, Glutamate and GABA in the Central Nervous System," L. Hertz, E. Kvamme , E. McGeer, and A. Schousboe, eds., Alan R. Liss, New York, p. 219.
Nicklas, W. J., NuÑez, R., and Duvoisin, R., 1979, J. Neurochem., 33: 839.
Nilsson, M., Hansson, E., and Ronnback, L., 1990, Neurochem. Res., 15: 763.
Norenberg, M. D., and Martinez-Hernandez, A., 1979, Brain Res., 161: 303.
Ohno, K., Pettigrew, K. D., and Rapoport, S. I., 1978, Am. J. Physiol., 235: H299.
Okonkwo, P. O., Orlowski, M., and Green, J. P., 1974, J. Neurochem., 22: 1053.
Oldendorf, W. H., 1971, Am. J. Physiol., 221: 1629.
Oldendorf, W. H., 1973, Am. J. Physiol., 224: 967.
Oldendorf, W. H., and Szabo, J., 1976, Am. J. Physiol., 230: 94.
Oldendorf, W. H., Crane, P. D., Braun, L. D., Gosschalk, E. A., and Diamond, J. M., 1988, J. Neurochem., 50: 857.
Ottersen, O.P., Storm-Mathisen, J., and Somogyi, P., 1988, Brain Res., 450: 342.
Oxender, D. L., and Christensen, H. N., 1963, J. Biol. Chem., 238: 3686.
Palaiologos, G., Hertz, L., and Schousboe, A., 1988a, J. Neurochem., 51: 317.
Palaiologos, G., Hertz, L, and Schousboe, A., 1988b, Neurochem Res., 14: 359.
Pardridge, W. M., 1983, Physiol. Rev., 63: 1481.
Pardridge, W. M., 1979, in: "Glutamic Acid: Advances in Biochemistry and Physiology," L. J. Filer, Jr., ed., Raven Press, New York, p. 125.
Pardridge, W. M., and Oldendorf, W. H., 1975, Biochim. Biophys. Acta, 401: 128.
Pardridge, W. M., and Oldendorf, W. H., 1977, J. Neurochem., 28: 5.
Pastuszko, A., Wilson, D. F., and Erecińska, M., 1981, Proc. Natl. Acad. Sci. USA, 78: 1242.
Pastuszko, A., Wilson, D. F., and Erecińska, M., 1982, J. Biol. Chem., 257: 7514.
Patel, A. J., and Balazs, R., 1970, J. Neurochem., 17: 955.
Patel, A. J., and Balazs, R., 1971, Biochem. J., 121: 469.
Patel, A. J., Weir, M. D., Hunt, A., Tahourdin, C. S. M., and Thomas, D. G. T., 1985, Brain Res., 331: 1.
Paulsen, R. E., Odden, E., and Fonnum, F., 1988, J. Neurochem., 51: 1294.
Pell, J. M., and Bergman, E. N., 1983, Am. J. Physiol., 244: E282.
Pellegrini-Gampietro, D.E., Bennett, M. V. L., and Zukin, R. S., 1991, Proc. Natl. Acad. Sci. U.S.A., 88: 4157.
Peng, L., Schousboe, A., and Hertz, L., 1991, Neurochem. Res., 16: 29.
Perry, T. L., Hansen, S., and Kennedy, J., 1975, J. Neurochem., 24: 587.
Peters, J. C., and Harper, A. E., 1987, Am. J. Physiol., 252: R902.
Peterson, N. A., and Raghupathy, E., 1972, J. Neurochem., 19: 1423.
Pratt, O. E., 1982, J. Inherited Metabol. Dis., 5 (Suppl. 2): 75.
Preston, J. E., Segal, M. B., Walley, G. J., and Zlokovic, B. V., 1989, J. Physiol. (Lond.), 408: 31.
Price, M. T., Pusateri, M. E., Crow, S. E., Buchsbaum, S., Olney, J. W., and Lowry, O. H., 1984, J. Neurochem., 42: 740.
Pycock, C. J., and Taberner, P. V., 1981, "Central Neurotransmitter Turnover", University Park Press, Baltimore (1981).
Raiteri, M., Federico, R., Coletti, A., and Levi, G., 1975, J. Neurochem., 24: 1243.
Ramaharobandro, N., Borg, J., Mandel, P., and Mark, J., 1982, Brain Res., 244: 113.
Recasens, M., Guiramand, J., Nourigat, A., Sassetti, I., and Devilliers, G., 1988, Neurochem. Int., 13: 463.
Reubi, J. -C., Van Den Berg, C., and Cuenod, M., 1978, Neurosci. Lett., 10: 171.
Riepe, R. E., and Norenberg, M. D., 1977, Nature, 268: 645.
Rigotti, P., Jonung, T., Peters, J. C., James, J. H., and Fischer, J. E., 1985, J. Neurochem., 44:929.
Roberts, P. J., and Keen, P., 1974, Brain Res., 67: 352.
Roberts, P. J., and Watkins, J. C., 1975, Brain Res., 85: 120.
Roberts, S., and Morelos, B. S., 1965, J. Neurochem., 12: 373.
Romano, C., Williams, W. V., Fischberg, D. J., Cocero, N., Weiner, D. B., Greene, M. I., and Molinoff, P. B., 1989, J. Neurochem. 53: 362.
Rothman, S. M. and Olney, J. W., 1987, Trends Neuro. Sci., 10: 299.

Rothstein, J. D., and Tabakoff, B., 1984, J. Neurochem., 43: 1438.
Rubin, R. A., Ordonez, L. A., and Wurtman, R. J., 1974, J. Neurochem., 23: 227.
Sakai, S., Tasaka, J., and Tosaka, T., 1990, Neurochem. Res., 15: 843.
Salceda, R., 1980, Neurochem Res., 14: 49.
Sato, T. N., and Neale, J. H., 1989, J. Neurochem., 53: 1089.
Scherer, R. W., Ferkany, J. W., Karbon, E. W., and Enna, S. J., 1989, J. Neurochem., 53: 989.
Schousboe, A., and Divac, I., 1979, Brain Res., 177, 407-409.
Schousboe, A., Hertz, L., and Svenneby, G. S., 1977a, Neurochem. Res., 2: 217.
Schousboe, A., Svenneby, G., and Hertz, L., 1977b, J. Neurochem., 29: 999.
Schousboe, A., Krogsgaard-Larsen, P., Svenneby, G., and Hertz, L., 1978, Brain Res., 153: 623.
Schousboe, A., Hertz, L., Svenneby, G., and Kvamme, E., 1979, J. Neurochem., 32: 943.
Schousboe, A., Drejer, J., and Hertz, L., 1988a, in: "Glutamine and Glutamate in Mammals, Vol. 2," E. Kvamme, ed., CRC Press, Boca Raton, Fl. p. 21.

Schousboe, A., Larsson, O. M., Krogsgaard-Larsen, P., Drejer, J., and Hertz, L., 1988b, in: "The Biochemical Pathology of Astrocytes," M. D. Norenberg, L. Hertz, and A. Schouseboe, eds., Alan R. Liss, New York, p. 381.
Schousboe, A., Frandsen, A., and Drejer, J., 1989, Neurochem. Res., 14: 871.
Schwartz, J. C., Lampart, C., and Rose, C., 1972, J. Neurochem., 19: 801.
Sellstrom, A., Venema, R., and Henn, F., 1976, Nature, 264: 652.
Sershen, H., and Lajtha, A., 1979, J. Neurochem., 32: 719.
Sershen, H., and Lajtha, A., 1979, J. Neurochem., 32: 719.
Shank, R. P., and Aprison, M. H., 1977, J. Neurochem., 28: 1189.
Shank, R. P., and Aprison, M. H., 1988, in: "Glutamine and Glutamate in Mammals. Vol. 2," E. Kvamme, ed., CRC Press, Boca Raton, Fl., p.3.
Shank, R. P., and Campbell, G .LeM., 1981, Life Sci., 28: 843.
Shank, R. P., and Campbell, G. LeM., 1982, Neurochem Res., 7: 601.
Shank , R. P., and Campbell, G. LeM., 1983, J. Neurosci. Res., 9: 47.
Shank, R. P., and Campbell, G. LeM., 1984a, J. Neurochem., 42: 1153.
Shank, R. P., and Campbell, G. LeM., 1984b, J. Neurosci., 4: 58.
Shank, R. P., and Nicklas, W. J., 1990, J. Neurochem. 55: 352.
Shank, R. P., Bennett, G. S., Freytag, S. O., and Campbell, G. LeM., 1985, Brain Res., 329: 364.
Shank, R. P., Baldy, W. J., and Ash, C. W., 1989, Neurochem. Res., 14: 371.
Shank, R. P., Baldy, W. J., Mattucci, L. C., and Villani, F. J. Jr, 1990, J. Neurochem., 54: 2007.
Shotwell, M. A., Mattes, P. M., Jayme, D. W., and Oxender, D. L., 1982, J. Biol. Chem., 257: 2974.
Smith, C. B., Deibler, G. E., Eng, N., Schmidt, K., and Sokoloff, L., 1988, Proc. Natl. Acad. Sci. USA, 85: 9341.
Smith, Q. R., and Rapoport, S. I., 1986, J. Neurochem., 46: 1732.
Smith, Q. R., Takasato, Y., and Rapoport, S. I., 1984, Brain Res., 311: 167.
Smith, Q. R., Takasato, Y., Sweeney, D. J., and Rapoport, S. I., 1985, J. Cereb. Blood Flow Metab., 5: 300.
Smith, Q. R., Momma, S., Aoyagi, M., and Rapoport, S. I., 1987, J. Neurochem., 49: 1651.
Smith, Q. R., Aoyagi, M., and Rapoport, S. I., 1989, Soc. Neuroscience Abstr., 15: 1025.
Smith, Q. R., Fukui, S., Robinson, P., and Rapoport, S. I., 1990, in: "Amino Acids: Chemistry, Biology and Medicine," G. Lubec and G. A. Rosenthal, ede., ESCOM, Amsterdam, p. 364.
Smith, Q. R., Nagura, H., Takada, Y., and Duncan, M. W., 1992, J. Neurochem., 58: 1330.
Smith, S. B., Briggs, S., Triebwasser, K. C. and Freedland, R. A., 1977, Biochem. J., 162: 453.
Speciale, C., Hares, K., Schwarcz, R., and Brookes, N., 1989, J. Neurosci., 9: 2066.
Stallcup, W. B., Buloch, K., and Baetge, E.E., 1979, J. Neurochem., 32: 57.
Storm-Mathisen, J., and Iversen, L. L., 1979, Neuroscience, 4: 1237.
Storm-Mathisen, J., Leknes, A. K., Bore, A. T., Vaaland, J. L., Edminson, P., Haug, F. M. -S., and Ottersen, O. P., 1983, Nature, 301: 517.
Sugiyama, H., Ito, I., and Hirono, C., 1987, Nature, 325: 531.
Sugiyama, H., Ito, I., and Watanabe, M., 1989, Neuron, 3: 129.

Szerb, J. C., and O'Regan, P. A., 1985, J. Neurochem., 44: 1724.
Takada, Y., Greig, N. H., Vistica, D. T., Rapoport, S. I., and Smith, Q. R., 1991, Cancer Chemother. Pharmacol., 28: 89.
Takada, Y., Vistica, D., Greig, N., Rapoport, S. I., and Smith, Q. R., 1992, Cancer Res., in press.
Takasato, Y., Rapoport, S. I., and Smith, Q. R., 1984, Am. J. Physiol., 247: H484.
Tapia, R., and Gonzalez, R. M., 1978, Neurosci. Lett., 10: 165.
Tate, S. S., Ross, L. L., and Meister, A., 1973, Proc. Natl. Acad. Sci. USA, 70:1447.
Tate, S. S., Urade, R., Getchell, T. V., and Udenfriend, S., 1989, Arch. Biochem. Biophys., 275: 591.
Tayarani, I., Lefauconnier, J. -M., Roux, F., and Bourre, J. -M., 1987, J. Cereb. Blood Flow Metab., 7: 585 .
Tews, J. K., Greenwood, J., Pratt, O. E., and Harper, A. E., 1987a, Am. J. Physiol., 252: R78.
Tews, J. K., Greenwood, J., Pratt, O. E. and Harper, A. E., 1987b, J. Neurochem., 48: 1879.
Thalmann, R. H., 1987, Neurosci. Lett., 82: 41.
Tovar, A., Tews, J. K., Torres, N., and Harper, A. E., 1988, J. Neurochem., 51: 1285.
Triller, A., Cluzeaud, F., Pfeiffer, F., Betz, H., and Korn, H., 1985, J. Cell. Biol., 101: 683.
van Bree, J. B. M. M., Audus, K. L., and Borchardt, R. T., 1988, Pharm. Res., 5: 369.
Van Den Berg, C. J., 1970, J. Neurochem., 17: 973.
Van Den Berg, C. J., 1973, in: "Metabolic Compartmentation in the Brain," R. Balazs and J. E. Cremer, eds., MacMillan, London, p. 137.
Van Den Berg, C. J., and Van Den Velden, J., 1970, J. Neurochem., 17, 985.
Van Den Berg, C. J., Mela, P., and Waelsch, H., 1966, Biochem. Biophys. Res. Commun., 23: 479.
Van Den Berg, C. J., Matheson, D. F., Ronda, G., Reijnierse, G. L. A., Blokhuis, G. G. D., Kroon, M. C., Clarke, D. D., and Garfinkel, D., 1975, in: "Metabolic Compartmentation and Neurotransmission: Relation to Brain Structure and Function," S. Berl, D. D. Clarke, and D. Schneider, eds., Plenum Press, New York, p. 515.
Van Den Berg, C. J., Matheson, D. F., and Nijemanting, W. C., 1978, in: "Amino Acids and Chemical Transmitters," F. Fonnum , ed., Plenum Press, New York, p. 709.
Vistica, D. T., 1983, Pharmacol. Ther., 22: 379.
Voaden, M. J., 1988, in: "Glutamine and Glutamate in Mammals, Vol. 2," E. Kvamme, ed., CRC Press, Boca Raton, Fl., p.71.
Wade, L. A., and Brady, H. M., 1981, J. Neurochem., 37: 730.
Wade, L. A., and Katzman, R., 1975, J. Neurochem., 25: 837.
Wadhwani, K. C., Smith, Q. R., and Rapoport, S. I., 1990, Am. J. Physiol., 258: R1436.
Wang, H., Kavanaugh, M. P., North, R. A. and Kabat, D., 1991, Nature 352: 729.
Waniewski, R. A., and Martin, D. L., 1986, J. Neurochem., 47: 304.
Watkins, J.C., Krogsgaard-Larsen, P. and Honore, T., 1990, Trends Pharmacol. Sci., 11: 25.
Weiler, C. T., Nystrom, B., and Hamberger, A., 1979, J. Neurochem., 32: 559.
Weissbach, L., Handlogten, M. E., Christensen, H. N., and Kilberg, M. S., 1982, J. Biol. Chem., 257: 12006.
Wenthold, R. J., and Altschuler, R. A., 1983, in: "Glutamine, Glutamate and GABA in the Central Nervous System," L. Hertz, E. Kvamme, E. G. McGeer, E. Schousboe, eds., Alan R. Liss, New York, p.33.
Wenthold, R. J., Altschuler, R. A., Skaggs, K. K., and Reeks, K. A., 1987, J. Neurochem., 48: 636.
Wheeler, D. D., 1987, Neurochem. Res., 12: 667.
White, M. F., 1985, Biochim. Biophys. Acta, 822: 355.
Wong, P. T. -H., McGeer, E. G., and McGeer, P. C., 1981, J. Neurochem., 36: 501.
Wood, J. D., and Sidhu, H. S. , 1987, J. Neurochem. 49: 1202.
Wood, P. L., Rao, T.S., Iyengar, S., Lanthorn, T., Monahan, J., Cordi, A., Sun, E., Vazquez, M., Gray, N., and Contreras, P., 1990, Neurochem. Res., 15: 217.
Wright, E. M., 1974, Brain Res., 76: 354.
Wright, E. M., Nogueira, G. J., and Levin, E., 1971, Exp. Brain Res., 13: 294.
Wu, J. -Y., Denner, L. A., Lin, C. -T., and Hwang, B., 1988, in: "Glutamine and Glutamate in Mammals, Vol.1.," E. Kvamme, ed., CRC Press, Boca Raton, Fl., p.101.

Yoneda, Y., Roberts, E., and Dietz, G. W. Jr., 1982, J. Neurochem., 38, 1686.
Young, S. N., 1983, in: "Handbook of Neurochemistry, Vol. 3, Metabolism in the Nervous System," A. Lajtha, ed., Second Edition, Plenum, New York, p. 559.
Yu, A. C. H., Schousboe, A., and Hertz, L., 1982, J. Neurochem., 39: 954.
Yu, A. C. H., Drejer, J., Hertz, L., and Schousboe, A., 1983, J. Neurochem., 41: 1484.
Yudkoff, M., Nissim, I., Kim, S., Pleasure, D., Hummeler, K., and Segal, S., 1983, Biochem. Biophys. Res. Commun., 115: 174.
Yudkoff, M., Zaleska, M. M., Nissim, I., Nelson, D., and Erecińska, M., 1989, J. Neurochem., 53: 632.
Yudkoff, M., Nissim, I., and Hertz, L., 1990, Neurochem. Res., 15: 1191.
Yunger, L. M., and Cramer, R. D., 1981, Mol. Pharmacol., 20: 602.

MUSCLE AMINO ACID METABOLISM AND TRANSPORT

B. Mackenzie, A. Ahmed and M.J. Rennie

Department of Anatomy and Physiology
University of Dundee
Dundee DD1 4HN, Scotland

INTRODUCTION

In mammals, including man, skeletal muscle is the largest part of the lean body mass and contains most of the protein in the body; it is responsible for 40-60% of whole-body protein turnover and in addition it has important functions in the diurnal flow of fuel and nitrogen between the periphery and the viscera and in the supply of amino acids during starvation, injury and disease (Daniel et al, 1977; Rennie, 1985; Rennie and Harrison, 1984).

It has only been relatively recently accepted (Daniel et al, 1977; Cahill, 1970; Marliss et al, 1971) that the metabolic activities of muscle include a function as a store of protein which can be drawn on by the rest of the body as a source of amino acids for various purposes including the supply of fuel. About one-sixth of the total of muscle protein is probably dispensable without irretrievable loss of function and this is, in energy equivalents, a considerable store. It seems to be important for the short term regulation of fuel supply during the diurnal cycle and fasting, especially in the supply of gluconeogenic substrates (Felig et al, 1970; Marliss et al, 1971; Aoki et al, 1973), and also for the acute response to injury and disease, when the liberation of amino acids from muscle is also necessary to adequately mount the immune response and for the supply of substrates for wound healing (Newsholme and Parry-Billings, 1990; Newsholme et al, 1989). It is somewhat surprising, therefore, that until recently relatively little was known about muscle amino acid transport, especially of the natural amino acids, and how important it was in serving, and possibly regulating, the processes of intermediary metabolism of amino acids, the interorgan transfer of substrates and protein turnover.

This article will review what is presently known about this subject, substantially more for skeletal than for cardiac or smooth muscle.

THE FREE AMINO ACID POOLS IN MUSCLE

Of the total mass of amino acids in the body, most is found chemically combined into protein and the free amino acid pool constitutes only a small proportion of the total (Table 1). Nevertheless, the free amino acid pool itself appears to be structured in the sense that the extent to which amino acids are concentrated inside muscle cells varies with the type of amino acids and the distribution ratio of the different amino acids across the muscle membranes alters during different circumstances (Table 1, Figure 1); the pool itself may consist of more than one compartment (Kipnis et al, 1961; Hider et al, 1969).

Mammalian Amino Acid Transport, Edited by M.S. Kilberg and
D. Häussinger, Plenum Press, New York, 1992

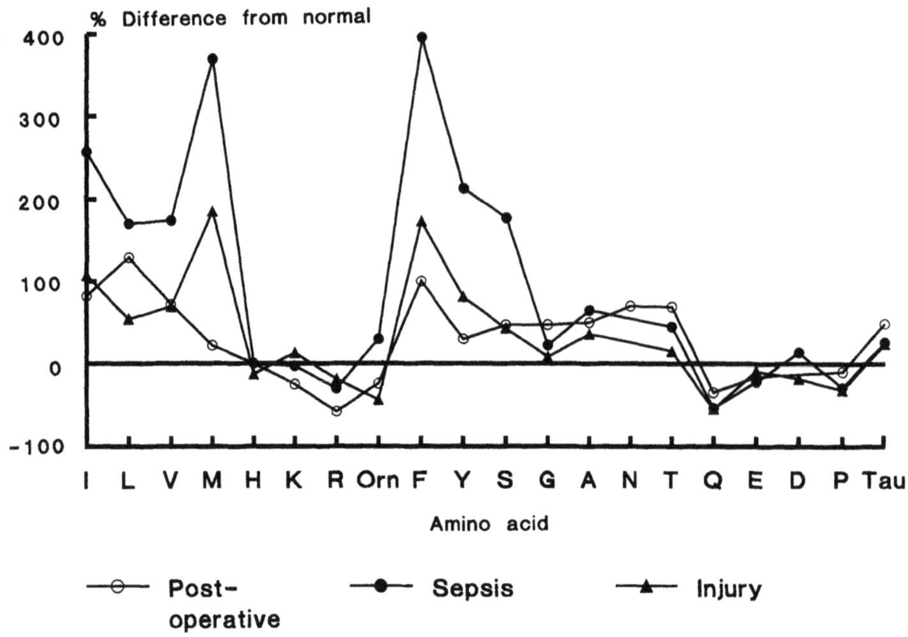

Fig. 1. Amino acid intramuscular concentration profiles from patients in various clinical conditions. Data are expressed as percentage change from normal; values are calculated from original data of Vinnars et al (1975) (post-operative) and Askanazi et al (1980) (injury and sepsis). Amino acids are identified by their IUPAC conventional one-letter symbols.

Van Slyke and Meyer (1913) provided the first evidence for concentrative uptake of amino acids when they showed that amino acids injected into rats rapidly disappeared from the bloodstream and their concentration in various tissues increased. However, it was not until the advent of easily available radioactive analogues of amino acids (and the means of quantifying them) and the development of automated amino acid analysis that much progress was made in understanding the basis of the concentrative effect or the changes in the free amino acid pools which occur under physiological and pathophysiological circumstances.

The distribution ratio for amino acids across the muscle membrane (Table 1) differs between the different physiologically important amino acids by up to 30 fold (eg., compare glutamine with tyrosine in human muscle; Bergström et al, 1974). A major task of present research will be to explain this behaviour.

The free amino acid pool shows changes (Figure 1) which are presumably the net result of alterations in intermediary metabolism transport, or of variation in the rates of withdrawal of amino acids into protein via synthesis and liberation of amino acids from protein by breakdown (see Waterlow and Fern, 1981, and Waterlow et al, 1978, for a full discussion of these points). We are still not in a position to fully explain the changes which occur physiologically in the free amino acid pool (Bergström et al, 1990; Alvestrand et al, 1988), eg., in going from the fed state to the postabsorptive state and on to the fasted state, or the changes which occur with, for example, provision or withdrawal of insulin (but for recent results see Bennet et al, 1990a; Bennet et al, 1991). Even so, we do have a much better understanding than we did 40 years ago.

Evidence for more than one intramuscular free pool. The fate of the amino acids which enter the muscle tissue is not entirely clear. Kipnis and colleagues showed, many years ago, that the rate of incorporation of amino acids into diaphragm muscle was faster than would be expected if the amino acids mixed entirely with the total free intracellular amino acid pool (Kipnis et al, 1961). Later more supportive evidence was provided by

Table 1. Plasma, muscle free and protein bound amino acid concentrations in normal, fasting and septic humans, and protein bound amino acids in rat.*

Amino acid	Plasma (μM)	Intracellular (ic) (μmol.l ic water^{-1})	Protein (μmol.g^{-1}) Rat	Human	Muscle protein (M%)	Normal fasting release (M%)	Septic fasting release (M%)
Alanine	419	3096	111	120	8.7	31.5	22.0
Glutamine	761	21020	148 + amide	189	13.7	9.6	13.6
Glutamate	40	3800					
Valine	271	324	83	84	6.1	4.9	4.8
Leucine	104	236	109	116	8.4	3.4	5.5
Isoleucine	67	106	50	59	4.3	1.5	3.0
Tyrosine	97	115	36	29	2.1	2.6	2.6
Phenylalanine	69	76	45	45	3.3	2.5	2.1
Methionine	29	52	36	25	1.8	1.5	2.1
Glycine	303	1131	117	134	9.7	9.6	8.6
Proline	231	995	–	73	5.3	6.9	4.2
Lysine	212	1086	58	112	8.1	8.7	9.2
Threonine	172	807	60	75	5.4	6.1	6.1
Serine	153	916	74	70	5.0	0.4	5.0
Arginine	99	719	67	60	4.3	2.9	3.9
Histidine	104	459	26	34	2.5	3.7	2.5
Taurine	104	17780	–	–	–	–	–
Asparagine	58	405	110 + amide	125	9.0	3.9	3.9
Aspartate	13	1746					

*Data for humans are drawn from Vesterberg et al (1987), Askanazi et al (1980), Scharff and Wool (1965) and Waterlow and Millward (1989), with values for rat muscle from E.B. Fern, PhD Thesis (cited in Waterlow et al, 1978).

showing that when incubated muscle was exposed to first one, then another, tracer amino acid, that the incorporation of the first tracer into protein rapidly diminished during the exposure to the second amino acid (Hider et al, 1969). Thus, either amino acids are activated for incorporation into protein directly from the extracellular pool or there is a small intracellular compartment that is preferentially rapidly labelled. If protein synthesis is for any reason inhibited then the small precursor pool for protein synthesis should become saturated with external amino acid and the aminoacyl tRNA into which precursor amino acid is normally complexed, dissociates or is degraded so that the flux of amino acids will be redirected towards the internal free pool.

Animal experiments *in vivo*, and with perfused tissues (Stirewalt et al, 1985; Everett et al, 1981; Chikenji et al, 1983; Watt et al, 1989) all suggest that in fact the true protein precursor specific activity lies somewhere between that of the extracellular fluid and the intracellular fluid, and a similar situation has recently been described for human muscle (Watt et al, 1991).

Physiological and pathophysiological modulation of the free pool. To some extent the behaviour of the free amino acid pool can be understood in terms of the relative proportions (Table 1) of amino acids in the free pool to the protein-bound pool (Millward and Waterlow, 1978; Millward et al, 1983). For example, when protein synthesis is stimulated in anabolic circumstances, eg., after a meal or during provision of insulin to young animals, the intramuscular concentration of branched chain amino acids (BCAA) falls as they are taken into protein, whereas in starvation or during the feeding of a low protein diet these amino acids rise in the free pool. The changes for the BCAA are greater than for some other amino acids because the ratio between their concentration in protein and that in the free pool is so much greater than for other amino acids. Since they form a substantial part of muscle protein (together they account for a fifth of muscle protein), when synthesis is inhibited (or possibly breakdown stimulated) they are being released into the free pool in large amounts.

Nevertheless such theories do not explain the very large distribution ratio for glutamine or the fact that alanine has a much smaller distribution ratio than glutamine. These phenomena must ultimately be understood in terms of the relative rates of *de novo* synthesis of these metabolically important amino acids and the rates of their transport across the muscle membrane by various carriers but we are still some way off this happy condition.

OVERVIEW OF MUSCLE AMINO ACID AND PROTEIN METABOLISM

Intermediary Metabolism

Intermediary amino acid metabolism in muscle is normally thought of as being limited to the transamination of alanine, the BCAA, and glutamate (plus the oxidative deamination of the latter) (Goldberg and Chang, 1978), the oxidation of BCAA via the branched chain keto acid dehydrogenase (Paxton et al, 1986), the involvement of aspartate in the purine nucleotide cycle (Lowenstein and Goodman, 1978) and the synthesis of glutamine (Ruderman and Lund, 1972; Goldstein et al, 1983). Glutaminase activity is normally found in mitochondria in those tissues where it exists and might thus be thought to exist in muscle mitochondria since there is measurable activity in "muscle" as a tissue. However the total tissue activity, at least in rat (Hundal et al, 1990a; Hundal et al, 1991; Wu et al, 1991) and probably also in human muscle, is low. It is also likely that most "muscle" glutaminase activity (except possibly in animals like the chicken, the muscle of which seems to have substantial activity; Wu et al, 1991) actually exists in non-muscle cells such as endothelial cells, fibroblasts and cells of the immune system (Kelso et al, 1989). However other glutamine metabolizing enzymes may play some part in muscle amino acid metabolism. Wu and colleagues have recently discovered in rat and chicken muscle an aminotransferase that accepts glutamine as a substrate (Wu et al, 1991); its activity in whole rat muscle is apparently substantially higher than total glutaminase activity and is suggested to be of the same order as muscle glutamine production (ie presumably glutamine synthetase activity). The implication is that glutamine carbon could be channelled via 2-oxoglutaramate via ω-amidase activity to 2-oxoglutarate and thus enter the Krebs cycle. For the pathway to be important there would have to be keto acids available for transamination. The only keto acids of consequence in muscle are the branched-chain keto acids so the pathway would produce BCAA only, and would be limited by BCAA transamination! The cell-specific localization (ie muscle or accessory cells) and the physiological and pathophysiological importance of this route *in vivo* needs urgently to be investigated.

Muscle's capacity for intermediary amino acid metabolism is demonstrated by the fact that during fasting or exercise the export of glutamine and alanine from human limbs is much greater than might be expected if their source were the catabolism of muscle protein, in which they are found in a lower proportion than in the mixture of amino acids exported (Tables 1 and 4) (Ruderman and Lund, 1972; Felig, 1975; Felig and Wahren, 1971). During the feeding of a protein meal, (especially one containing meat) the output of glutamine and alanine also increases (Elia and Livesey, 1983); oddly, fasting, exercise and feeding share the common characteristic that the availability of BCAA rises, compared to the post-absorptive state and glutamine production increases. The BCAA transaminase reaction has a high K_m; i.e., transamination is supply-driven (Krebs, 1972), and it is likely that all of the nitrogen in alanine and at least half of the α-amino nitrogen in glutamine is transferred just because of an increase in the availability of the BCAA.

Human muscle differs from rat muscle in having a larger capacity than the liver for complete oxidation of the BCAA (Elia and Livesey, 1981; Khatra et al, 1977), so that only small amounts of keto acids escape human muscle (Elia and Livesey, 1981; Cheng et al, 1985) for oxidation in the liver, unlike in the rat in which it is the major organ of their disposal. The source of the NH_3 added as amide into glutamine (i.e., whether from the glutamate dehydrogenase reaction or liberated from the purine nucleotide cycle) is not yet certain, and the provenance of the carbon for alanine and glutamine also remains obscure. The work of Wagenmakers and colleagues (Wagenmakers et al, 1990; Wagenmakers et al, 1989) in normal subjects and patients with McArdle's disease suggests that during exercise of moderate to severe intensity the purine nucleotide cycle plays a less important role than glutamate dehydrogenase and that the ammonia produced originally comes from the BCAA.

Some workers have suggested (Newsholme et al, 1989; Snell and Duff, 1981) that apart from leucine, the carbon skeleton of which can only give rise to acetyl-CoA, the carbon from those amino acids which transaminate with 2-oxoglutarate to form glutamate (i.e., isoleucine, valine and aspartate) give rise to oxaloacetate and thence via the phosphoenolpyruvate carboxykinase and pyruvate kinase reactions, to pyruvate; thus alanine carbon may be formed using the carbon of the original amino acids. Similarly, any amino acid carbon which can enter the Kreb's cycle and become incorporated into 2-oxoglutarate will, theoretically, be available for the production of glutamine via the formation of glutamate, as has been directly demonstrated recently for leucine, which however contributes only a small proportion of the total glutamine produced – although it increases in sepsis (Yoshida et al, 1991; Le Marchand-Brustel et al, 1982).

Incomplete oxidation of the branched chain amino acids. There is a possibility that intermediates of metabolism (3-hydroxyisobutyrate, 2-methylbutyrate and 2-methyl-3-hydroxybutyrate) in muscle (Spydevold and Hokland, 1983; Spydevold, 1979) may be important in interorgan carbon transfer (Palmer et al, 1985). Such intermediates, not being esterified with co-enzyme A, could leave mitochondria easily and enter the plasma where they would be available for uptake into gluconeogenic tissues. Definitive evidence seems lacking, but the theory is strengthened by recent findings of Brosnan and Letto (1991). They found that when the total flux entering the branched chain ketoacid pathway in perfused heart was measured as the rate of release of labelled carbon dioxide from carboxyl-labelled valine, the rate of complete oxidation of α-ketoisovalerate was only a fifth, the rest being released into the medium; also, isolated renal cortical tubules and hepatocytes are able to make glucose effectively from 3-hydroxyisobutyrate, thus raising the possibility that valine and possibly isoleucine catabolism may be incomplete in muscle and heart (at least in rats) *in vivo* and that this pathway could be important for gluconeogenesis. If so, then, as Goldberg and Chang found, pyruvate derived from glycolysis would be a more important immediate source of alanine and glutamine than the carbon chains of the BCAA (Goldberg and Chang, 1978), but some of the carbon could have been originally part of muscle protein as valine and isoleucine.

Glucose-alanine cycling between muscle and liver. Since carbon entering the Krebs cycle from glycolysis can give rise to the carbon skeletons of both alanine and glutamate, it was proposed (Felig et al, 1970) that a glucose-alanine cycle operates in such a way that alanine released by muscle is a precursor for hepatic gluconeogenesis; this glucose could be taken up by muscle to provide pyruvate for further alanine synthesis with any of the amino acids oxidized in muscle providing amino nitrogen. The physiological importance for this glucose-alanine cycle in its full form has not been proven, and its biological value can be challenged on the grounds that no new fuel is provided to the organism from protein stores (Snell and Duff, 1981). Nevertheless there is no doubt that under fasting conditions, for example (and probably after injury and burns) gluconeogenesis from alanine is elevated (Felig et al, 1970). The question still to be determined is exactly how much of this is as a result of increased glycolysis in muscle or increased net breakdown of muscle amino acids, and until the involvement of the branched chain intermediates seemed realistic these were incompatible alternatives; now it seems that they could be reconciled; i.e., the glucose-alanine cycle could operate and muscle protein could be a source of new carbon.

Alanine production during glycolysis in cardiac muscle. In cardiac muscle, alanine rather than lactate appears to be the end product of glycolysis under normal circumstances (Schwartz et al, 1985; Mudge et al, 1976). This has the advantage that no protons are produced as is the case for lactate. Under conditions of ischaemia or hypoxia alanine production increases until it is limited by the availability of glutamate when pyruvate is channelled to lactate. Thus, lactate production by the heart is a sign of a considerable metabolic stress upon the system. The extent to which glutamine or glutamate from outside the heart can provide nitrogen for the production of alanine is not well worked out. It does appear however that glutamate at high concentrations may be able to cross into the heart (Choong et al, 1988) to enable alanine production to continue; this would allow the continuation of glycolysis without the need to produce lactate and protons (which inhibit glycolysis) thus allowing a high rate of energy production via glycolysis during hypoxia.

Although methionine can be oxidized in muscle (Elia and Livesey, 1981; Livesey and Lund, 1980; Scislowski et al, 1987) the low free muscle methionine normally available and the high K_m of the transamination reaction mean that this is unlikely to be a physiologically important pathway. Metabolism of methionine to cysteine and the oxidation of methionine to sulphoxide (Scislowski et al, 1987) is also possible but little information exists about the physiological importance of these pathways in muscle.

Amino Acids as Precursors for Nucleotides in Muscle

All cells of the body have the capacity to make purine and pyrimidine ribonucleotides from 5-phosphoribose 1-pyrophosphate (PRPP). The amino acids which are involved in the production of inosine monophosphate for PRPP are glutamine, glycine and aspartate; aspartate is also involved in the production of orotic acid, the other starting material for which is carbamoyl phosphate. The physiological chemistry of nucleotide synthesis in muscle has not been well investigated but it is tantalizing to suggest that glutamine availability in particular could have an important regulatory function. Carbamoyl phosphate synthetase II activity is regulated by the availability of ammonia and therefore the extent of glutaminase or glutamate dehydrogenase reactions might have a regulatory role. Many of the reactions of synthesis of purines and pyrimidines involve the amide nitrogen of glutamine and the hydrolysis of ATP, which suggests the possibility of them being regulated by the availability of these substrates.

Obviously these reactions would normally only be of major importance under circumstances in which there was hyperplasia of muscle (requiring DNA and RNA synthesis), i.e., during development. However it may be that the response to injury when there is mitosis of satellite cells or when there is adaptive hypertrophy (eg., to increased physical work) both requiring substantial nucleotide synthesis. In severe injury and sepsis, when the "sick-cell syndrome" develops (Newsholme et al, 1989) the total pool of adenine nucleotides falls (Liaw et al, 1982) and this correlates well with the fall in muscle glutamine (Askanazi et al, 1980), so it is tempting to speculate that the lack of glutamine (which is certainly partly due to increased outward transport) may limit the synthesis of adenine nucleotides in muscle – possibly a clue to how much is normally made from this source. Unfortunately, the extent to which everyday "housekeeping" functions in muscle require *de novo* synthesis of nucleotides, rather than use of the salvage pathway is unknown.

Protein Synthesis and Breakdown in Muscle

This subject has been reviewed recently (Newsholme et al, 1989; Askanazi et al, 1980; Barrett and Gelfand, 1989; Kettlehut et al, 1988; Smith and Rennie, 1990). For the present purposes it seems sufficient to note that protein synthesis and breakdown proceed so much more slowly than the rates of exchange of amino acids across the muscle membrane that transport is normally unlikely to limit them by virtue of the availability or removal of amino acids (although methionine and tyrosine could be limiting in some circumstances; Bennet et al, 1990c). Nevertheless, there is some evidence that when rates of synthesis and breakdown are markedly altered, amino acid pool sizes can change in ways suggesting that the transporters do not maintain intramuscular amino acid concentrations at a constant level (for example, during starvation or protein deprivation, concentrations of the BCAA may rise considerably; Millward and Waterlow, 1978). Furthermore, it is possible that alteration of intracellular amino acid concentrations can influence processes of protein metabolism in muscle in ways which are not yet understood. Good examples of this are the possible effects of the BCAA and their keto acids (Tischler et al, 1982) and glutamine (MacLennan et al, 1987; MacLennan et al, 1988) on protein synthesis and breakdown and the possibilities that amino acids stimulate insulin sensitivity of the processes (Garlick and Grant, 1988).

The Theoretical Role of Transport Process for the Regulation of Metabolism

Transport steps are, in many ways, similar to other steps in metabolic pathways except that they are vectorial. Some transporters, eg., the glucose transporter, are obvious and proven candidates for participation in regulation of metabolism. Nevertheless, in the case of the amino acids, transport has usually been overlooked as a potential site for regulation because it occurs more rapidly than other steps in their metabolism (Waterlow and Fern, 1981) and because of an insistence in identifying so-called regulatory steps in metabolism as only those where there is a large (compared to the equilibrium constant) mass action ratio between reactants and products (Newsholme and Crabtree, 1981). There is no doubt that particular enzymes do play a substantial part in contributing to the overall control exercised within a metabolic pathway; however, many investigators now accept the theory that within intact metabolizing systems regulation of metabolism is shared between many steps and it is not easy to practically assign their control strengths (Kacser and Porteous, 1987). Furthermore, the balance of control may shift from step to step, under a variety of metabolic circumstances. When control of metabolism is investigated in this light it is possible to obtain evidence of regulation of metabolic pathways via alterations in transport characteristics: this has already been done for hepatic aromatic amino acid (Pogson et al, 1990) and glutamine metabolism (Salter et al, 1986).

The view that only particular kinds of irreversible reactions can be important for the control of metabolism, has in the case of muscle glutamine transport led to the prediction of the existence of novel transporters (Newsholme and Parry-Billings, 1990). Unfortunately the logic used in coming to this conclusion appears to be faulty and does not take account of facts which make the proposed scheme of glutamine efflux unlikely (see discussion of muscle glutamine efflux below).

METHODS FOR THE STUDY OF AMINO ACID TRANSPORT IN MUSCLE

Skeletal muscle is a difficult tissue to investigate because of the intricate network of connective tissue and contractile protein. Isolation of subcellular fractions, especially membrane fractions, from muscle is particularly difficult. Such procedures are easier in cultured muscle but culturing adult muscle is not easy and sometimes what is claimed to be a primary culture from adult muscle in fact turns out to be the result of de-differentiation of muscle and regrowth from satellite cells.

Influence of intramuscular metabolism on results from amino acid transport studies. In all muscle preparations except those of isolated membrane vesicles, or reconstructed systems, there is a danger that intramuscular metabolism will influence the results. This inescapable problem can be minimized by the use of short time periods of uptake or using inhibitors of amino acid metabolism (such as aminooxyacetate, a transaminase inhibitor, or methionine sulphoxamine, an inhibitor of glutamine synthetase, or 6-diazo-5-oxo-norleucine, an inhibitor of many glutamine binding enzymes, or inhibitors of various processes of protein synthesis and breakdown in muscle).

The use of synthetic amino acid analogues to avoid the problem of metabolism. Another approach is to use non-metabolizable model amino acids such as 2-aminoisobutyric acid (AIB) or its *N*-methyl derivative, MeAIB, 1-aminocyclopentane-1-carboxylic acid (cycloleucine) or 2-aminobicyclo[2.2.1]heptane-2-carboxylic acid (BCH) (Christensen, 1988; Christensen, 1987). A major advantage of using these is that their transport may be limited to one or two carrier paths but this can actually be a disadvantage for the understanding of physiological and pathophysiological processes if a single analogue does not adequately probe alternate routes of transport or the transport of other physiologically important amino acids. Also, widespread textbook errors concerning insulin sensitivity of amino acid transport (eg see Norman and Litwack, 1987; Schauf et al, 1990) have resulted from over-interpretation and extrapolation of results principally obtained with AIB and MeAIB, without regard to relative rates of transport of these and natural amino acids. The ease of use of analogues has often (it seems to us) seduced workers into using them without a critical appraisal of how closely relevant or irrelevant they are to the metabolic problem at hand, eg., the use of MeAIB to investigate System A transport in uremia or sepsis. (See below for further discussion of these points).

Studies in Whole Animals

In a number of early studies the intravenous injection of radiolabelled amino acids (and often extracellular markers) was used to measure their muscle distribution ratio *in vivo* under different circumstances (Riggs and Walker, 1960; Riggs et al, 1963; Nallathambi et al, 1972; Christensen, 1982). This might still be useful to detect effects of major transport changes during various pathophysiological challenges. The disadvantage is that the measurements are the net result of transport into and out of muscle and also, apart from identifying concentrative processes, they say little about regulation.

The results of the measurement of arterio-venous (A-V) differences across organs and muscle beds (Marliss et al, 1971; Aoki et al, 1973; Pozefsky et al, 1969; Felig, 1975) have been important in the development of interest in amino acid metabolism in muscle. When tracer amino acids have been used in limb balance and exchange studies (Bennet et al, 1990b; Barrett and Gelfand, 1989; Thompson et al, 1989) they have usually been utilized in a manner which does not give information about unidirectional rates of transport but rather of net uptake or exchange of tracer.

Theoretically, the paired-tracer dilution method, which has proved so useful in studies of perfused organs (Yudilevich and Mann, 1982; Hundal et al, 1987), ought to be applicable to studies of amino acid transport *in vivo*; indeed the transport of glucose and MeAIB (Saccomani et al, 1990) in the human forearm have been examined in this way. The paired-tracer method provides information about the relative appearance of the transportable tracer (i.e., the amino acid) and the extracellular marker in venous blood but another approach is to measure the accumulation of positron-emitting tracer amino acids within muscle. The feasibility of this for amino acid transport has been demonstrated by the use of ^{13}N and ^{11}C compounds in rabbits (Bading et al, 1987; Conlon et al, 1989) and such techniques should be directly applicable to man; glucose transport in muscle has already been examined in this way (Selberg et al, 1991).

Pratt and Daniel and co-workers used a novel approach to study amino acid transport *in vivo* in rats by using a high pressure pump to deliver tracer into the bloodstream at an exponentially decreasing rate until a desired asymptotic concentration of tracer was reached; thus blood tracer availability remained steady over a short period after which tissue accumulation was measured (Banos et al, 1973). In this way valuable information was obtained about the first order rate constants of entry of amino acids into muscle. Unfortunately the technique seems not yet to have been used to investigate transport in any pathophysiological conditions. It would be interesting to see this kind of application with ^{13}C and ^{15}N tracers in man.

Perfused Muscle

Perfused skeletal muscle. Our group adapted the paired tracer isotope dilution technique to studies of unidirectional uptake in skinned hindlimbs of the rat (Rennie et al, 1983; Hundal et al, 1987). This technique is able to provide substantial physiologically relevant information about the characteristics of unidirectional uptake into hindlimb tissue, a supposition supported by the very similar behaviour of ^{13}N-labelled amino acids *in vivo* and in perfused limbs of the rabbit (Conlon et al, 1989). It is assumed that cells of the vasculature itself did not contribute to the uptake because the recovery of transportable tracer was related to a marker (mannitol) which rapidly gained access to the extracellular space. The kinetic constants for unidirectional inward transport obtained by the use of this method are likely to reflect those operating *in vivo* and therefore should be valuable in understanding interorgan fluxes involving muscle. The values of the transport K_m in pure membrane preparations are likely to be lower (Ahmed et al, 1990b) than in perfused preparations probably because, in the perfused preparation, there is a greater influence of unstirred layers and there may be additional effects of tracer passage through gaps in capillary endothelium.

The paired tracer technique can be used to study efflux either by measuring total tracer recovery after a bolus injection or by loading the tissue with tracer and following its washout (Rennie et al, 1991); the first of these is only semi-quantitative but the second provides a good measure of unidirectional efflux, although it is expensive in terms of tracer; of course, the results are difficult to interpret because the intracellular concentrations of tracer and tracee (or those of other possibly interactive substances) are not known.

The perfused hindquarter preparation which has been used to study the accumulation of non-metabolizable amino acids (Zorzano et al, 1985) provides indices of net transport only.

The perfused heart. The perfused heart ought to be a good preparation for the study of amino acid transport but surprisingly little work has been done on cardiac amino acid transport *per se*. Use of this preparation is discussed further in a later section.

Isolated Incubated Muscle

A methods review by Goldberg et al (1975) is useful as an introduction to the use of incubated muscle preparations.

The rat diaphragm. The rat diaphragm (or hemidiaphragm) was a popular preparation in early studies of amino acid transport (Kipnis and Parrish, 1965; Akedo and Christensen, 1962; Wool et al, 1965). Care needs to be exercised in its use, for example in the retention of ribs to avoid leakage of tracers through the cut ends of muscle fibres; also if connective tissue barriers are not removed, their presence could make interpretation sometimes difficult. The diaphragm may be inadequate as a model of all skeletal muscles; it certainly contains fibre types which may be specific to it.

However, some of the same criticism can be levelled at studies of most limb muscles, which often turn out to be unrepresentative of the total musculature on close histochemical and metabolic investigation (Ariano et al, 1973).

Incubated whole muscles. Many workers have used isolated incubated whole muscles of frogs, rabbits, rats and mice for studies of amino acid metabolism and transport and there is a certain amount of concern regarding the viability of such preparations. Often "smaller" is taken to equal "better" because in this way diffusion limitations, eg., of oxygen, will be minimized. For rat muscle the upper limit of size before anoxic effects (eg., central fibre glycogen depletion) become apparent is not large (Maltin and Harris, 1985). The group working in Fischer's laboratory (Hall-Angeras et al, 1991) reproduced some of the results of Maltin and Harris and have shown, in addition, that protein synthesis is depressed in the core of small muscles in the flaccid state but not when stretched to resting length (Hummel et al, 1988). Also AIB transport appears to be somewhat lower in flaccid muscles than in stretched muscles (Warner et al, 1989), a result also reported recently for MeAIB (Tovar et al, 1991). This effect is not seen with the L-system paradigm substrate BCH (Tovar et al, 1991) so that the modulation of AIB and MeAIB transport cannot simply be due to an increase in the surface area:volume ratio. Nevertheless, metabolic challenges such as sepsis or starvation appear to have a similar relative effect in flaccid or stretched muscles (Hummel et al, 1988; Warner et al, 1989).

The use of muscle strips (Leighton et al, 1989) is an alternative approach which has become popular for work on the metabolism of muscles from adult rats. Their viability has been examined and seems to be good (Newsholme et al, 1986). The epitrochlearis muscle is sufficiently small that even in adult animals diffusion limitations may not be important, and it has proved useful for studies of muscle metabolism (eg., see Harris et al, 1974; Maroni et al, 1986b; Maillet and Garber, 1980). The use of muscles from adult animals gets over the objection that results obtained with small muscles from young animals (eg., 70g rats) are specific to particular developmental stages.

Reeds and co-workers have pointed out (Smith et al, 1983) that the use of muscles from very small (usually young) animals may not be logical, since these animals have high metabolic rates and so do their muscles. Small muscles and muscle bundles taken from larger animals may be perfectly viable if their metabolic rates are sufficiently low. This approach has been used for the measurement of glucose transport in human muscle using small fibre bundles (Dohm et al, 1988) and there is no reason why the approach could not be extended to the study of amino acid transport.

Work on isolated incubated amphibian muscles (Narahara and Holloszy, 1974) may obviate many of the problems of working with mammalian muscles. Frog muscles are easy to manipulate, relatively resistant to anoxic damage, and can be kept for a substantial time at low temperatures.

Membrane Vesicles

The preparation of sarcolemmal vesicles from skeletal muscle is difficult. Two recent successful approaches are those of Grimditch et al (1985) and Klip and Walker (1983). In our experience (Ahmed et al, 1990b), the procedure of Grimditch is the more difficult to use but other workers (Hirshman et al, 1988) have not found this to be a problem. There are numerous examples in the literature now of studies of sugar and organic acid transport in isolated sarcolemmal vesicles from the rat (Hirshman et al, 1988; Roth and Brooks, 1990a; Roth and Brooks, 1990b) and both skeletal and cardiac muscle vesicles have been used to investigate amino acid transport (Ahmed et al, 1990b; Young et al, 1987). The same techniques can also be adapted successfully to the study of amino acid transport in isolated sarcolemma from human muscle (Ahmed et al, 1990a).

The kinds of vesicles which are obtained by these conventional techniques are relatively small (intravesicular volume ~5 μm^3) compared to so-called "giant" vesicles (prepared by exposure of muscle fragments or of whole muscles to collagenase, KCl and urea) (Juel, 1991). This technique has been successfully used for the study of lactate transport in rat muscle (Juel, 1991) and offers the attractive possibility of manipulating the intravesicular contents in such a way that the effects of various regulators, inhibitors and stimulants could be investigated.

Tissue Culture of Muscle

There is a substantial body of literature concerning the culture of muscle cells of various kinds (Konigsberg, 1979). Workers wishing to enter this field should be cautious about fully defining their aims; if they are interested in cellular processes *per se* then the use of modified muscle cell lines such as L6 (Smith et al, 1984), L8 (Gulve and Dice, 1989) rat omega myotubes (Emig et al, 1988), or BC3H-1 cells (Farese et al, 1985; Schubert et al, 1974) may be adequate for their needs. If, however, they are interested in physiological regulation then such preparations may be unsuitable as the cells may not represent the physiological behaviour of skeletal muscle *in vivo* (Konigsberg, 1979); for example, they may lack sensitivity to particular hormones or particular transport systems may be missing or over-expressed. Valuable information can, of course, be obtained for such cell types so long as it is interpreted cautiously.

Primary muscle cultures are useful, although they are more difficult to set up and maintain. Nevertheless, we and others, have found it possible to set up primary cultures of neonatal rat muscle, with good results (Airhart et al, 1982; Tadros and Rennie, 1991; Sarabia et al, 1990).

Extracellular Space Measurements

It is important always to have good measurements of extracellular space, ideally with inulin (Kipnis and Parrish, 1965), although mannitol and sucrose have deliberately been chosen where it was important to match diffusional characteristics (Hundal et al, 1987). The ideal extracellular marker for amino acid transport may well be 6-aminohexanoic acid which is not transported and yet behaves physically like an amino acid. Unfortunately it is not available labelled.

Some workers appear to make such measurements in separate experiments from those in which they make penetrant tracer uptake measurements, and this is really unacceptable.

AMINO ACID TRANSPORT SYSTEMS IN MUSCLE: IDENTIFICATION AND KINETIC CHARACTERIZATION

Non-metabolizable Amino Acid Analogues

An overwhelming proportion of the published data deals with artificial amino acids such as AIB and MeAIB, and a smaller amount with cycloleucine. This is understandable, since on first attempting to investigate the transport of the physiologically common amino acids in isolated rat muscles it was discovered that using the long incubation periods required (given the low specific activity of early tracers) a substantial fraction of the radioactivity appeared in protein and that this was markedly stimulated by insulin. The extent of the stimulation was sufficiently large to interfere with measurements of amino acid transport. The use of naturally-occurring radiolabelled amino acids became much less common when the radiolabelled non-metabolizable model amino acid substrates were introduced (Noall et al, 1957; Riggs et al, 1963).

AIB – an insufficiently specific model substrate for System A. The use of AIB to identify the presence of a Na^+-dependent system which transports only small neutral amino acids (ie System A) became flawed when it was discovered that AIB could be transported by another system which was not completely blocked by AIB, ie System ASC (Christensen et al, 1967). The indiscriminate use of AIB (as opposed to its *N*-methyl derivative) as a paradigm substrate for System A has not really been justified for many years, especially now with the availability of labelled MeAIB; unfortunately this does not seem to have been recognized by a large number of investigators.

Defining the routes of uptake for alanine. The relationship between the findings using non-metabolizable model amino acid and naturally-occurring amino acid transport in muscle *in vivo* is a difficult one to sort out. The first attempt to do so, and one which was largely successful, was that of Riggs and McKirahan (1973). They investigated the extent to which alanine accumulation in rat diaphragm could be inhibited by amino acids which were then thought to be paradigm substrates of System A (AIB), System ASC (serine or cysteine) and System L (leucine). It is easy with hindsight to fault the way in which the study was carried out, eg., by measurement of the net accumulation rather than unidirectional transport, by the lack of any inhibitors of alanine transamination, by the lack of concurrent measurement of extracellular space with tracer uptake and by the use of AIB (rather than MeAIB) as a paradigm competitive substrate for System A and of leucine (which can be transported by System A) rather than BCH (which is specific for system L).

Nevertheless, the results obtained are actually very similar to those we have recently obtained (by up-to-date techniques) in the stripped soleus muscle of adult rats (G.W.A. Thompson and M.J. Rennie, unpublished results, Figure 2A). Thus, we find

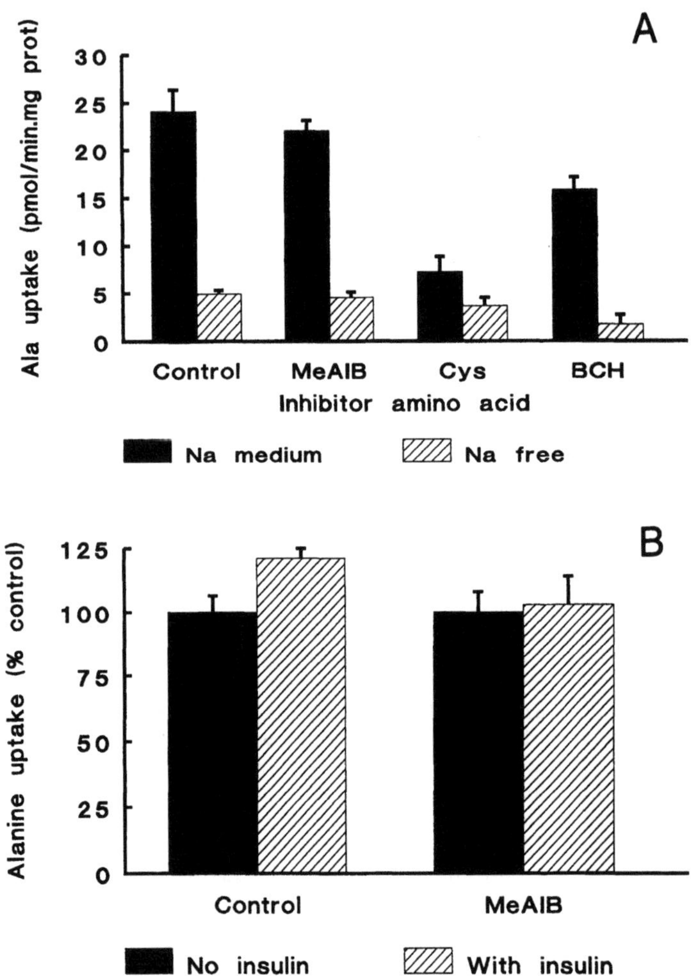

Fig. 2. Alanine uptake in stripped soleus incubated *in vitro*. (A) Inhibition by model substrates of Systems A, ASC and L; 0.3 mM [^{3}H]alanine uptake (mean ± SEM) was measured over 30 min without inhibitor, with 10 mM MeAIB, 10 mM cysteine or 25 mM BCH. **(B)** Effect of insulin (at 1 mU.ml^{-1}) on alanine uptake with or without MeAIB (G.W.A. Thompson and M.J. Rennie, unpublished).

alanine transport appears to occur to a minor extent by System A (about 10%) with the remainder being divided by System ASC and System L (about 60% and 30% respectively). Also, in our hands, alanine transport shows only partial Na^{+}-dependence and only a minor component is insulin-sensitive (Figure 2B). Even though 0.1 $mU.ml^{-1}$ insulin results in a 50% increase in the uptake of the System A specific analogue MeAIB (at 50 μM) in soleus of 90-120 g rats, a 10 fold greater concentration of insulin is required to stimulate 50 μM or 0.3 mM alanine transport by 15-20%; this slight stimulation of alanine transport is not observed in the presence of 10 mM MeAIB (Figure 2B) indicating that System A is the route which is insulin-sensitive. Furthermore, the sensitivity of alanine uptake to insulin generally is much less than the sensitivity of muscle to stimulation of, for example, 2-deoxyglucose uptake for which the same amount of insulin will double the initial rate of transport.

Physiologically most of the System A activity (which has a K_m in rat or mouse limb muscle of about 1.5 mM, see Table 2) will be largely saturated, given the presence of other System A substrates (eg., leucine, methionine, glycine and serine), and the non-insulin-sensitive amino acid transporters (i.e., System ASC and System L) will not be affected by the insulin. The extent of any possible insulin stimulation of System A will be greater in young animals since it appears that they have more System A than older animals (at least if the sheep is representative of other mammals; Kimball and Jefferson, 1988; Forsberg et al, 1988).

It may be that System A is only important during muscle differentiation and development and, possibly, during remodelling of muscle after injury or, as a result of hypertrophy or changes in composition secondary to alterations in the muscle activity.

Altogether, the results suggest that under normal circumstances in adult muscle, System A is not a physiologically important system; certainly insulin-stimulated amino acid transport via this route has little significance. This view is strengthened by the results of studies carried out using measurements of A-V difference of amino acids in human subjects (Eriksson et al, 1983; Lundholm et al, 1987; Arfvidsson et al, 1991), and the first results using the paired tracer technique in the human arm (Saccomani et al, 1990). Thus, the tremendous effort which has been expended on studies of System A in various pathophysiological circumstances may have resulted in data which at best is illustrative of phenomena whose significance will be revealed in the future; at worst, it may be irrelevant to the observed pathophysiology of protein and amino acid metabolism.

The other non-metabolizable amino acid analogues which have been used in various studies are cycloleucine and BCH. Unfortunately, like AIB, cycloleucine (cLeu) appears to be capable of transport by more than one system, in this case via Systems A and L and, therefore, results obtained with it are difficult to interpret. BCH, however, is a model amino acid which appears to be almost entirely transported by System L (Christensen et al, 1969) and has proved useful as a competitive inhibitor in identifying the extent to which System L transport occurs.

So far the only use of labelled BCH in detailed studies of muscle has been an extensive investigation by Harper's group of transport of MeAIB and BCH in baby rat soleus muscle, carried out with a view to obtaining information on kinetic parameters of transport, and on the effects of Na^{+}, glucose and oxygen (Tovar et al, 1991). These workers also examined the extent to which various natural and model amino acids competed with MeAIB or BCH and investigated the effects of insulin and adaptive regulation. The study was unusual in using ^{14}C-labelled BCH (which was presumably synthesized by the authors since it is unavailable commercially). The approach described to measure the kinetic constants of MeAIB and BCH uptake was not ideal since the concentrations (10 mM) used hardly exceeded the K_m values obtained (1.9 mM for MeAIB and 7.9 mM for BCH) (Table 2).

However, new information was presented concerning the competition of MeAIB uptake by naturally-occurring amino acids present at 10 mM: as expected basic and acidic amino acids had no effect, and inhibition by large neutral amino acids (i.e., BCAA and the aromatic amino acids) were poor. Hydroxyacids and the small neutrals had some

Table 2. Transport kinetics for the model amino acids under control and altered conditions. Incubations were each carried out at 37 °C unless otherwise stated; data are given as mean (± SEM where available).

Tracer	Muscle preparation	Conditions	V_{max} (nmol.min^{-1}.g^{-1})	K_m (mM)	Ref*
AIB	Perfused rat hindlimb	Control	15.0 ± 4.0	1.26 ± 0.35	(1)
AIB	Mouse soleus	Lean mice, control	30.1 ± 2.8	1.35 ± 0.16	(2)
		Lean + 33 nM insulin	60.1 ± 6.7	1.28 ± 0.04	
		Obese, control	14.4 ± 2.9	0.73 ± 0.14	
		Obese + insulin	41.1 ± 3.2	1.16 ± 0.08	
AIB	Rat soleus	Control	27.8 ± 1.7	1.68 ± 0.12	(3)
		Endotoxin-treated	9.2 ± 1.0	1.64 ± 0.19	
AIB	Wether lamb biopsied external intercostal	Control	82.6 ± 11.7	11.4 ± 4.2	(4)
AIB	Frog sartorius (at 19 °C)	Control	1.33	1.5	(5)
		With 2 mU.ml^{-1} insulin	2.83	1.6	
MeAIB	Rat epitrochlearis	Sham-operated (SO), control	50.6 ± 20.6	2.63 ± 1.80	(6)
		SO + 10 mU.ml^{-1} insulin	70.6 ± 11.0	0.97 ± 0.36	
		Acute uremia (ARF), control	32.6 ± 12.7	1.60 ± 1.28	
		ARF + insulin	37.6 ± 5.7	0.33 ± 0.16	
MeAIB	Rat soleus	Control	35.0	1.9	(7)
MeAIB	Rat extensor digitorum longus	Control	3.92	0.70	(8)
		With 200 nM insulin	8.10	0.62	
cLeu	Rat omega myotubes	Control	1280 ± 340	0.12 ± 0.04	(9)
BCH	Rat soleus	Control	440	7.9	(7)

*Data are taken from (1) Hundal et al (1989), (2) Le Marchand-Brustel et al (1982), (3) Karlstad and Sayeed (1986), (4) Forsberg et al (1988), (5) Narahara and Holloszy (1974), (6) Maroni et al (1990), (7) Tovar et al (1991), (8) Guma et al (1988) and (9) J.Y. Yeh, B.R. Ou, and N.E. Forsberg (personal communication).

inhibitory effects (threonine 17%, histidine 27%, glutamine 40%, cysteine 42%, serine 43%, proline 44%, alanine 55%; sadly, given its metabolic importance, methionine was not investigated). Unfortunately, the authors were not able to determine which of the inhibitions were competitive and which non-competitive, a question of some importance (Christensen, 1985).

As might have been expected, basic and acidic amino acids had no effect on BCH transport, which was also unaffected by low concentrations of the small neutral amino acid such as proline or alanine. The most effective competitors were typical large neutral amino acids and the aromatic amino acids. Isoleucine was the strongest inhibitor reducing BCH transport by nearly 40%, but the extent of the inhibition cannot be compared with that for MeAIB since it would be unreasonable to expect to achieve any significant inhibition at a concentration (10 mM) of potential inhibitor at little more than the K_m for transport of BCH itself (about 8 mM). All of the inhibition studies were carried out during incubations for periods exceeding those in which the uptake was linear and this also makes their quantitative significance difficult to assign.

Tovar and colleagues also examined the uptake of the model amino acids in preparations to which had been added plasma from rats fed low or high protein diets. MeAIB uptake was reduced by 40% by plasma from rats fed either diet but there were no effects in BCH uptake by plasma from rats fed the low protein diet, in which, presumably, concentrations of BCAA and the other large neutral amino acids was depressed. The uptake was reduced by 23% when plasma from rats fed the high protein diet (and presumably with high large neutral amino acid plasma concentrations) was added to the incubation medium.

Long term incubation of muscles in the absence of glucose or oxygen inhibited MeAIB uptake, as might be expected from the likely effects on membrane potential and the Na^+ gradient, but had no effect on BCH accumulation.

As shown by Gulve and colleagues (1991) preincubation of muscles in amino acid deprived mixtures caused stimulation of MeAIB accumulation by about 4 fold and the effect was also seen by Tovar and colleagues after periods as short as one hour. As found by Gulve et al, adaptive regulation was prevented by the presence of amino acids, especially those which are inhibitory of A system transport in this preparation. Amino acid deprivation caused only a 25% stimulation of BCH accumulation after two hours but no further increase could be seen when this period was extended. Inhibitors of protein and RNA synthesis strongly inhibited the adaptive response in MeAIB transport but tunicamycin (an inhibitor of protein glycosylation) had only about half as much effect; BCH was unaffected.

The stimulatory effects of insulin and phorbol esters on MeAIB uptake were similar to those found by other workers (Guma et al, 1988), but neither phorbol esters nor insulin affected BCH uptake.

Efflux of model amino acids from muscle. The efflux of AIB from muscle has been studied in soleus, hemidiaphragm and extensor digitorum longus muscles by Lewis and Goldspink (1984). These workers identified a fast phase ($k = 0.014 \pm 0.001$ min^{-1}) and a slow phase ($k = 0.007 \pm 0.0005$ min^{-1}) of AIB efflux both distinct from the washout of tracer from the extracellular space. Since AIB appears to be a substrate for Systems A and ASC (both Na^+-dependent) and Kipnis and Parrish (1965) had evidence of a glycine-inhibitable Na^+-independent component of muscle amino acid transport, it is interesting that Lewis and Goldspink found that replacement of extracellular Na^+ with choline caused a marked increase in the fast phase with no change in the slow phase, suggesting that fast phase (likely to be A or possibly ASC) was external Na^+-dependent and the slow phase not. If the transport process is symmetrical (as suggested by homoexchange) and, as for System N^m (Hundal et al, 1987), the effect of Na^+ is to stimulate V_{max} rather than alter the affinity of the carrier for the transported amino acid, then the reversal of the Na^+ gradient (by removing extracellular Na^+) ought to stimulate efflux, as observed by Lewis and Goldspink.

Some interesting studies of AIB efflux have also been carried out by Cooper and Kohn (1980). A number of compounds such as local anesthetics and barbiturates have membrane-stabilizing effects in muscle and exhibit common effects on hexose and ion transport suggesting a common mechanism for these effects; for example, tetracaine inhibits increases in transport at low concentrations (less than 1 mM) but stimulates transport, particularly efflux, at higher concentrations. Unfortunately for the neatness of this picture chlorpromazine, for example, which conforms to the general pattern in its effects on hexose and ion transport, shows a different pattern of action on AIB efflux, with no effect on efflux at all at low concentrations and stimulations at high concentrations. Furthermore, concentrations of tetracaine which stimulate AIB efflux also reduce AIB uptake and the effects of tetracaine on AIB efflux are quite different to those on cycloleucine efflux which it inhibits. There must be some clues here as to the mechanisms of regulation of System A or ASC transport in muscle but obviously more work needs to be done to sort them out.

Cooper and Kohn also showed that the efflux of AIB in the soleus muscle shows homo- and heteroexchange diffusion, i.e., AIB efflux is stimulated by extracellular AIB

Table 3. Summary of kinetic characteristics for amino acid transport in rat skeletal muscle. Kinetic parameters (mean ± SD, $n = 3$-4), and pH and insulin sensitivity data are from Hundal et al (1987, 1989) and Hundal and Watt (1991), and ion dependence from Ahmed et al (1990b) and Hundal et al (1987, 1989).

Substrate	V_{max} (nmol.min^{-1}.g muscle^{-1})	K_m (mM)	Ion dependence	System	Comments
Gln	1156 ± 193	9.25 ± 1.15	Na^+-dependent and Li^+-tolerant	N^m	Insulin-sensitive and pH-insensitive
Asn	1054 ± 274	8 ± 2	Na^+-dependent	N^m	
His	229 ± 19	1.25 ± 0.4	Na^+-dependent	N^m	
Ala	332 ± 53	3.9 ± 0.9	Na^+-dependent and independent	ASC (also A and L)	Insulin- and pH-insensitive
Ser	410 ± 61	3.4 ± 0.5	Na^+-dependent, Li^+-intolerant	ASC	Insulin- and pH-insensitive
Leu	2800 ± 420	20 ± 2	Na^+-independent	L	Insulin- and pH-insensitive
Val	2575 ± 50	20.1 ± 0.1	Na^+-independent	L	
Ile	2550 ± 50	18.3 ± 1.7	Na^+-independent	L	
Phe	3019 ± 384	19.32 ± 1.5	Na^+-independent	L	Insulin- and pH-insensitive
Lys	136 ± 46	2.1 ± 1.3	Na^+-dependent	y^+	Insulin- and pH-insensitive
Pro	190 ± 48	4.1 ± 0.6	Na^+-dependent	Pro	
Glu	86 ± 6	1.05 ± 0.05	Na^+-independent	x^-_{AG}	Insulin-insensitive and pH-sensitive
Tau	0.9 ± 0.35	7.0 ± 2.8	Apparent Na^+-dependence	ß	Insulin-insensitive
3-MeHis	Most likely saturable			N^m (and/or L and ASC)	pH-insensitive

or glycine. Since AIB transport is mediated by Systems A and ASC and that we know that System A and System L can probably handle glycine, these results are not simple to interpret.

Transport of Natural Amino Acids

The problem of amino acid metabolism in the absence of specific metabolic inhibitors (eg., of transamination) was partly overcome by Banos et al (1973) who rapidly measured the rate of entry of amino acids into muscles of living rats using a technique described earlier. These studies, valuable as they were in showing the relative rates of entry of the natural amino acids, were unable to provide a description of the kinetic parameters of the amino acid transporters present in the sarcolemma.

Studies our group performed using the perfused rat hindlimb were designed to fill this gap as far as practicable. They provided, for rat skeletal muscle, kinetic parameters for the transport of a variety of natural neutral, acidic and basic amino acids chosen to be as near as possible paradigm substrates for particular systems (Hundal et al, 1987; Hundal et al, 1989). Detailed kinetic studies were performed for a variety of other neutral, cationic and anionic amino acids (eg., alanine, serine, phenylalanine, lysine, glutamate etc). As a bonus, the work also resulted in the identification in muscle of System N^m (apparently a variant of System N, first described in the liver; Joseph et al, 1978; Kilberg et al, 1980) which prefers glutamine, asparagine and histidine as substrates. It was distinguishable from the liver system by its insulin sensitivity (Hundal et al, 1987; Hundal et al, 1991) and pH independence (Hundal et al, 1987). (Information obtained regarding the kinetics, ion and membrane potential dependence, substrate specificity, etc. for the transport of various amino acids is given in Table 3).

Our group later provided a detailed account of glutamine transport at the membrane level using a purified sarcolemmal vesicle preparation (Ahmed et al, 1990b; Figure 3) devoid of intracellular contaminants with the aim of eliminating many of the problems associated with the perfused hindlimb, such as heterogeneity of transporting tissues. The results, in terms of specificity, were broadly in agreement with those obtained by us in the perfused hindlimb. The presence of Systems N^m, ASC and L in rat sarcolemma was confirmed and, recently, we have identified the presence of System A (using MeAIB uptake as a marker) as only a minor contributor to transport in rat sarcolemmal vesicles.

Direct evidence was also obtained of the dependence of System N^m on membrane potential and the studies provided the first quantitative data concerning the unidirectional efflux of glutamine from rat skeletal muscle. This confirmed, as predicted (Rennie et al, 1989), that the efflux of glutamine was ion linked, certainly involving Na^+ and possibly K^+ in the process. Recently, studies of isolated rat skeletal muscle strips have confirmed that the efflux process has characteristics of System N^m and is sensitive to both membrane potential and intramuscular Na^+ (Figure 4).

To date the above work provides most of our information about native amino acid transport in skeletal muscle. The major conclusions can be summarized as follows:

Glutamine, asparagine and histidine are transported by a Na^+-dependent transporter (System N^m) with 1 Na^+ : 1 AA stoichiometry and the transport of glutamine is membrane potential sensitive and can be stimulated by insulin at high concentrations. The transport of each of the three amino acids shows strong competitive mutual inhibition with near equivalence of K_i and K_m values; System ASC and L substrates appear to have a non-competitive inhibitory effect. The transport of glutamine, asparagine and histidine will tolerate Li^+-for-Na^+ substitution; Li^+-tolerance appears to be a characteristic of System N^m in skeletal muscle, since transport of serine (a substrate for System ASC which is itself Na^+-dependent in muscle) does not tolerate Li^+-for-Na^+ substitution. Alanine, serine and cysteine, all of which are System ASC substrates show mutual inhibition of transport activity. Alanine seems also to be transported via System A, only to a small extent. System L exists as a very low affinity ($K_m \sim 20$ mM) and high capacity transporter which transports aromatic amino acids like phenylalanine and tyrosine as well as BCAA, but seems also to have the ability to transport alanine to some extent. Classical System A seems to be of very low activity in rat skeletal muscle: the results from the studies in sarcolemmal vesicles show that System A activity is less than one third that of other amino acid transporters such as System N^m and in incubated muscles the activity is even less (Figure 2A). Amino acids with acidic side chains (glutamate and aspartate) are transported exclusively via a pH-dependent dicarboxylate amino acid carrier with a high affinity but a very low capacity; this should effectively compartmentalize those amino acids within muscle, a prediction in accord with the tracer studies in people (Darmaun et al, 1986) and in rabbit muscle *in vivo* and in perfused preparations (Conlon et al, 1989). The basic amino acid lysine is transported by a Na^+-dependent carrier (y^+) whose stoichiometry is 2 Na^+ : 1 lysine. The presence of a transporter resembling System Gly which transports glycine in other cells and tissues was detected. This system was not saturable but occurred at a rate faster than could be explained by diffusion.

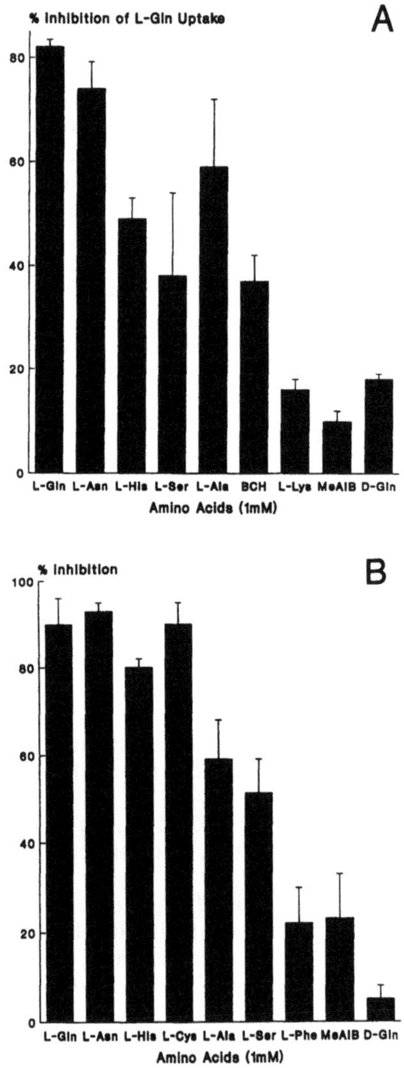

Fig. 3. Inhibition by 1mM amino acids of initial (45s) uptake of glutamine into sarcolemmal vesicles. (A) Uptake of 5 μM L-[^{14}C]glutamine in rat skeletal muscle sarcolemmal vesicles (from Ahmed et al, 1990b). (B) Uptake of 0.1 μM L-[^{3}H]glutamine in sarcolemmal vesicles isolated from human skeletal muscle (from Ahmed, 1991). Values represent the means ± SEM for 3 membrane preparations measured in triplicate.

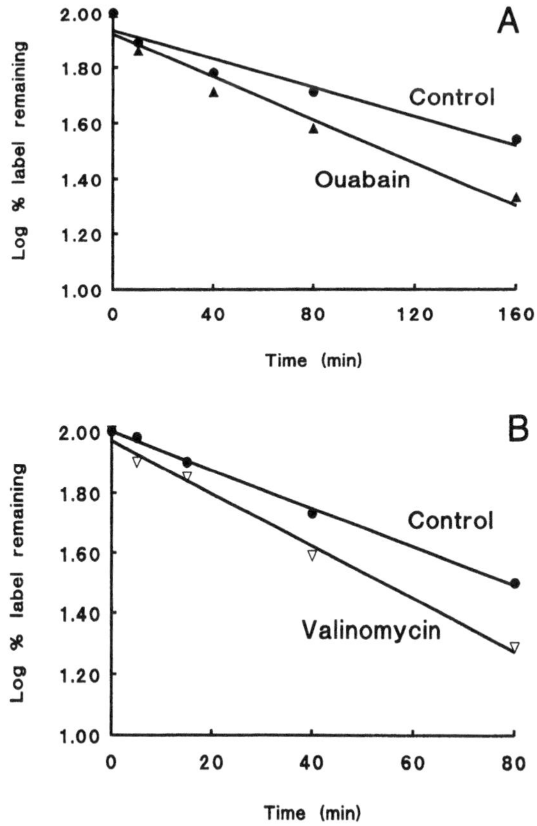

Fig. 4. Increase in glutamine efflux rate in soleus muscle incubated with (A) ouabain or (B) valinomycin. Muscle strips were preloaded with 10 mM [^{3}H]glutamine for 60 min and glutamine efflux measured over 160 min; *n* = 10 (N.M. Willhoft and M.J. Rennie, unpublished).

In the perfused rat hindlimb the ß-amino sulphonic acid taurine is transported by a low capacity, high affinity transporter (Hundal and Watt, 1991), presumably resembling System ß (Christensen, 1985), with only ß-alanine competing for transport. The Na^+-dependence of taurine transport could not be established for certain but it appears to follow a 2 Na^+ : 1 taurine stoichiometry which would explain the high taurine distribution ratio between muscle and plasma. In perfused muscle System ß seems to be insensitive to insulin, and its mechanism appears to require the presence of sulphydryl groups, since the uptake of 10 µM taurine was abolished by 0.4 mM mersalyl acid.

Does the above information help us understand the distribution ratios and behaviour of muscle amino acids? It seems likely that those amino acids which are highly concentrated are either complexed (as taurine may be) or are accumulated in muscle by the actions of a Na^+-dependent transporter of high specificity, eg., glutamine (Hundal et al, 1987; Hundal et al, 1989). Even so, if another Na^+-independent transporter can carry a particular amino acid which is accumulated by an Na^+-dependent transporter (eg., alanine or histidine, which could both be accumulated in this way, eg., by Systems ASC and N^m respectively), then the amino acid can leave down its concentration gradient on the Na^+-independent system (eg., System L); the distribution ratio for such an amino acid will be less than it would be for an amino acid which is not a substrate for the Na^+-independent transporter (eg., glutamine). On the other hand the stimulation of System L carrier movement by substrates like alanine or histidine leaving muscle will cause the slight accumulation of amino acids not *a priori* expected to be concentrated, eg., BCAA and the aromatic amino acids (Table 1). Thus, the energy currency of ATP is used to drive the Na^+ pump, the secondary active co-transport of Na^+ and amino acids (eg., via

Systems A, ASC and N^m) and the tertiary active transport of L system substrates (Christensen, 1987; Christensen, 1990).

PHYSIOLOGICAL MODULATION OF MUSCLE AMINO ACID TRANSPORT

If regulatable membrane transport in muscle participates in the modulation of interorgan amino acid metabolism (Christensen, 1982; Christensen, 1990) then mechanisms of regulation should be capable of being demonstrated. However, we have not yet been able to adequately identify more than a few possible candidates for regulation and the quantitative aspects of their control remain for the most part elusive. Regulation of individual amino acid fluxes could, theoretically, occur via acute interactions between different transport systems communicating via the relative concentrations of their substrates or magnitude of ion gradients or by changes in transporter expression and, in principle, hormones, growth factors, cytokine mediators, etc. could influence any or all of these.

Modulation by Cations

Kipnis and Parrish (1965) found that in lymph node cells (i.e., *not* muscle cells) the effect of Na^+ was to alter the affinity of the carrier for AIB transport. That lymph cells do not behave like muscle is suggested by the fact that the K_m for AIB transport under normal circumstances (i.e., 140 mM Na^+) was only 0.1 mM which is substantially less than that (1-2 mM) found in muscle by other workers (eg., see Table 2). Kipnis and Parrish found that the relationship between Na^+ and AIB entry in lymph cells was linear and concluded that the stoichiometry was 1:1. Unfortunately this seems not to have been tested for skeletal muscle AIB or MeAIB transport, although we found such a 1:1 stoichiometry does exist for glutamine transport in skeletal muscle (see later). Kipnis and Parrish also reported the effects of increasing K^+ on AIB transport in intact rat diaphragm and found that basal and insulin-stimulated transport were inhibited markedly (at 70 mM K^+). Furthermore, efflux of AIB was markedly elevated in the presence of high external K^+; also, exchange diffusion, which was normally negligible for AIB, could be readily demonstrated in the presence of K^+. Despite assertions to the contrary these could have been due to alterations in membrane potential due to high extracellular K^+.

In the perfused rat hindlimb preparation, our group investigated the Na^+-dependence of amino acid transport by progressively replacing perfusate Na^+ (initially at 145 mM) with choline (Hundal et al, 1987; Hundal et al, 1989). The normal rates of uptake of both leucine and glutamate were preserved, even at 44 mM perfusate Na^+, indicating that these amino acids were transported independently of Na^+. Serine and glutamine uptakes fell in line with the reduction in perfusate Na^+, suggesting Na^+-amino acid co-transport. The linear relationship between the reciprocals of glutamine influx (via System N^m) and perfusate Na^+ concentration indicate that Na^+-glutamine co-transport has a stoichiometry of 1:1. A 50% reduction in perfusate Na^+ resulted in a 50% decrease in V_{max} without any effect upon the K_m for glutamine. These results indicate that Na^+ stimulated the translation of the carrier plus glutamine across the membrane rather than the ability of the carrier to bind glutamine. We speculated, on the basis of this observation, that the transfer across the sarcolemma of the carrier+ion+amino acid complex by Na^+ should depend on local Na^+ availability, the Na^+ electrochemical gradient and the membrane potential; direct evidence in favour of this prediction was later provided by our group in showing that the initial rate of glutamine transport in sarcolemmal vesicles was decreased as initial membrane potential was abolished in the presence of the K^+ ionophore valinomycin (Ahmed et al, 1990b).

In another study our group investigated the effects of denervation on glutamine transport in rat skeletal muscle (Hundal et al, 1990a). Denervation was chosen as a model for those kinds of injury (such as crush injury of muscle, and possibly other tissues) in which the membrane depolarizes, Na^+_i rises and K^+_i falls, thus possibly revealing the modulation by changes in these factors on amino acid transport. Hindlimb denervation increased the intramuscular Na^+ 17% and decreased K^+ by 12%. The resting membrane potential in isolated muscles fell from -75 mV in control muscle to -59 mV after denervation, but this could be prevented by adding 3 μM tetrodotoxin (which blocks Na^+ channels and restores membrane potential) to the perfusate; under these conditions the transport of glutamine into the muscle was depressed initially but recovered after the addition of tetrodotoxin.

Intramuscular K^+ concentration was reduced in denervated muscle; these results together with those obtained in sarcolemmal vesicles indicate the likely involvement of K^+ in the glutamine efflux process.

The observation that the effect of Na^+ on glutamine transport is a V_{max} effect suggests that changes in the Na^+ concentration on the *cis* side of the membrane could modulate glutamine transport. Exposure over periods of 30 min or more to ouabain, whose inhibitory effect on the Na^+/K^+ pump increases intracellular Na^+ concentration, has been shown to increase glutamine efflux and decrease influx by about the same extent, 12-15% (Hundal et al, 1987). Recent studies using incubated stripped soleus muscle from 250 g rats have shown that exposure (1 h) to ouabain (which doubles intramuscular Na^+) or valinomycin in the presence of high extracellular K^+ (which depolarizes muscle) both stimulate glutamine efflux, as predicted if System N^m transport were involved in the efflux step (Figure 4).

The increased glutamine export from muscle observed in diabetes (Hundal et al, 1990b), and possibly also in fasting, may include an effect of hypoinsulinaemia in reducing the tonic stimulation of the sarcolemmal Na^+/K^+ ATPase (Clausen and Kohn, 1977) which, by permitting intracellular Na^+ accumulation and depleting intracellular K^+, may enhance glutamine efflux.

Together the above results may help us understand the changes in intramuscular amino acid content and membrane distribution ratios in pathophysiological conditions where there is a fall in membrane potential and K_i and a concurrent increase in the Na^+_i concentration. However this view, in the form stated, has been challenged (Newsholme and Parry-Billings, 1990). For example, it has been been suggested that glutamine transport across the membrane is not symmetrical in the sense that the inward and outward transport steps are not near equilibrium and occur by separate processes, and therefore the work on the N^m transporter is irrelevant to a consideration of efflux. Unfortunately there appear to be some errors in the calculations used to justify this position: for example, the free energy change for glutamine entry is not as Newsholme and Parry-Billings suggest -12.3 $kJ.mol^{-1}$, but actually only -3.6 $kJ.mol^{-1}$ (due to a failure to correctly attribute the sign in that part of the calculation dealing with the product:reactants ratio). The corrected value of -3.6 is less than the value of -4.2 $kJ.mol^{-1}$ which Newsholme and Parry-Billings (for reasons not made clear) use as a marker threshold for the identification of non-equilibrium reactions.

Also, when these authors calculate the energy change for the process of glutamine release by skeletal muscle, they effectively assume that the equilibrium constant for the reaction is unity, without evidence, and that it only involves glutamine inside and outside the membrane with no involvement of ions or E_m, again without evidence. Thus, both calculations which are claimed to show that the processes of entry and exit of glutamine from skeletal muscle are non-equilibrium steps are in fact flawed.

There are further difficulties with the approach. Newsholme and Parry-Billings claim that the concentration at which glutamine exit is half maximal is less than ~0.3 mM but in fact there is no experimental evidence provided in the source quoted (the Oxford PhD thesis of Parry-Billings) that this is in fact the case.

The idea that two transporters are involved in the regulation of the glutamine concentration in muscle is not *a priori* impossible and indeed in liver this does appear to be the case (Pogson et al, 1990). However, in liver the glutamine transporters are System N in the inward direction and System L in the outward direction. System L commonly shows such characteristics as *trans*-stimulation and the lack of sensitivity to transmembrane Na^+ distribution or membrane potential; however, it can be shown that

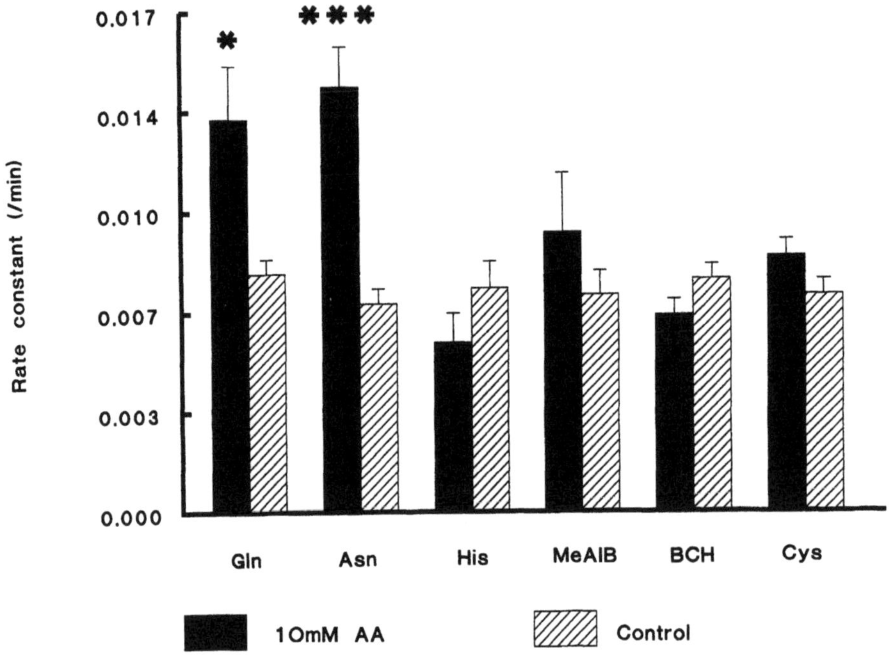

Fig. 5. Effect of amino acid incubation on the unidirectional [^{3}H]glutamine efflux from rat soleus muscle. [^{3}H]Glutamine efflux was measured as for Figure 4, in the presence or absence of various amino acids each at 10 mM. Data (Willhoft and Rennie, 1991) are mean ± SEM, $n = 10$ (*$P < 0.05$; ***$P < 0.001$).

in rat muscle the efflux process *does* show *trans*-stimulation (Rennie et al, 1989; and Figure 5) but only by glutamine (and also asparagine, a System N^m substrate) and not by System L substrates. Furthermore, efflux behaves as if the process were being mediated by System N^m, i.e., abolishing membrane potential stimulates glutamine efflux and so does increasing intracellular Na^+ (Figure 4). All of these considerations cause severe doubts to be raised about the interpretation of Newsholme and Parry-Billings.

Effects of Insulin

Insulin has been known for many years to cause a fall in plasma concentrations of amino acids (Lotspeich, 1949) and to stimulate muscle protein synthesis (Lotspeich, 1949; Jefferson, 1980). At first glance, it may therefore seem that insulin directly stimulates amino acid transport into muscle perhaps in a manner similar to the stimulation of glucose transport by recruitment of transporters to the sarcolemma, but in fact the extent of the effect of insulin on muscle amino acid uptake in human muscle *in vivo* is doubtful and even well accepted effects of insulin in animal muscle may actually be seen only in young animals or certain species eg., rats and mice, but not sheep, rabbits or man (Bading et al, 1987; Eriksson et al, 1983; Lundholm et al, 1987; Arfvidsson et al, 1991).

Nevertheless there have been many reports of insulin effects on rodent muscle. Kipnis and Noall discovered in 1958 that the accumulation of AIB in diaphragm muscle *in vitro* could be stimulated by insulin and that the effect was stronger in fasted animals (Kipnis and Noall, 1958). An effect of insulin (at physiological concentrations; Akedo and Christensen, 1962) could also be seen for some natural amino acids including glycine, proline and methionine (Manchester and Young, 1963; Manchester and Young, 1958). That the effect was a transport effect was shown by its persistence when protein synthesis was inhibited by puromycin (Fritz and Knobil, 1963). The use of puromycin was justified on the basis that although it inhibited protein synthesis there seemed to be little effect on amino acid uptake, although long exposure to it did diminish amino acid accumulation. However, puromycin and possibly other inhibitors do not have totally clear cut effects. For example, in its presence the accumulation of amino acids (such as alanine, phenylalanine, tyrosine, histidine and leucine) which in its absence were not stimulated by insulin became stimulated (Wool et al, 1965).

The results of attempts to discover the kinetic mechanism of the stimulation by insulin of amino acid transport in incubated muscle have been confusing. Akedo and Christensen (1962) claimed that insulin increased the affinity, i.e., decreased the K_m with little change of the maximum capacity (V_{max}) of diaphragm transport of AIB, and recently work from Mitch's group claims the same thing for MeAIB in the epitrochlearis muscle; this work was also unusual in claiming an effect of insulin on the non-saturable component of MeAIB uptake (Maroni et al, 1990).

Because it was accepted that insulin increased glucose transport by a V_{max} stimulating effect (which now turns out to be substantially due to recruitment of glucose transporters into the membrane; Wardzala and Jeanrenaud, 1981; Klip et al, 1987) and the fact that glucose transport was Na^+-independent, whereas AIB transport was Na^+-dependent caused Narahara and Holloszy to carefully examine the nature of the insulin effect (Narahara and Holloszy, 1974). However, they clearly demonstrated a V_{max} effect, and it now seems that there is a greater weight of evidence which suggests that in muscle generally (or possibly limb muscle generally) the effect of insulin on amino acid transport is indeed to stimulate the carrier-amino acid complex transfer across the membrane rather stimulating the binding of the amino acid to the transporter (Table 2).

Recent studies (Guma et al, 1988) on insulin stimulation of MeAIB uptake (reflecting System A activity) in incubated extensor digitorum longus (EDL) muscle of the rat consolidate most of the early and recent findings. Thus, ouabain (which inhibits the sodium pump) and gramicidin D (which depolarizes muscle) both reduced basal MeAIB uptake but did not affect the percentage increase in response to insulin, indicating that the stimulatory action of insulin was independent of the Na^+ electrochemical gradient, although a preserved Na^+ electrochemical gradient seemed to be a requirement for a maximal effect. Cycloheximide did not reduce the fractional insulin stimulation of MeAIB uptake indicating an acute effect of insulin independent of protein synthesis. The authors proposed that the stimulation of MeAIB uptake by insulin (by a V_{max} effect) could be a direct effect of insulin upon System A, perhaps involving translocation of transporter molecules from intracellular pools. The same group has also examined in the detail the effects of phorbol ester, a stimulator of protein kinase C (Guma et al, 1990): they found that no stimulation of basal MeAIB uptake, but phorbol ester added 30 min before 100 nM insulin caused a diminution of the expected effect of insulin. This was odd because similar treatment accelerated glucose uptake, which seems often to be regulated in parallel with MeAIB transport. Closer examination suggested that the inhibitory effect of phorbol ester was not due to effects on insulin action at the receptor or associated tyrosine kinase activity but was instead some other selective post-receptor inhibition of a protein kinase C-stimulated step.

The effects of insulin on the kinetics of AIB transport have also been examined in soleus muscles of lean and gold thioglucose-obese mice (Le Marchand-Brustel et al, 1982). These studies showed that the K_m of saturable AIB influx was 1.35 mM in the absence of insulin and unchanged in the presence of 33 nM insulin; however, insulin doubled the V_{max} (from 4.5 to 9.0 nmol AIB.min^{-1}.mg protein^{-1}). In obese mice, the basal rate of AIB influx was much lower than in control mice, a decrease that was mainly accounted for by a 30% reduction in the V_{max} of the transport. However, there was also a concomitant decrease in K_m in the obese mice, although it was insufficient to counteract the change in V_{max}. The dose responsiveness of insulin on AIB transport was less in the obese mice with the half maximal concentration of insulin being more than doubled.

That System ASC in rat epitrochlearis muscle is not sensitive to insulin was convincingly demonstrated (Maroni et al, 1986a) by the failure to stimulate AIB or cycloleucine transport in the presence of System L and A inhibitors. Using the hindlimb preparation, our group reported no effect of 1 mU.ml^{-1} insulin upon uptake of substrates of System L, ie the branched-chain amino acid leucine and the aromatic amino acid phenylalanine; also, in the perfused hindlimb, insulin failed to stimulate transport of serine (by System ASC) or glutamate (by System x^-_{AG}) or lysine (by System y^+). Under physiological conditions, in the perfused hindlimb at least, only the transport of System N^m substrates appears to be subject to any significant stimulation by insulin: 1 $mU.ml^{-1}$ insulin resulted in a stimulation of glutamine transport, with a 30% rise in V_{max} and no change in the K_m. Use of ^{13}N tracers in studies of the effect of insulin on uptake of glutamate and glutamine in rabbit muscle *in vivo* and in perfused hindlimbs *in vitro* has confirmed that glutamine transport is insulin sensitive but glutamate is not (Conlon et al, 1989). The same group could find no effect of starvation (which depresses insulin) on [^{11}C]AIB uptake in the same system (Bading et al, 1987).

The ability of insulin to stimulate amino acid transport in muscle appears to be markedly less than its ability to stimulate sugar transport. When we examined the transport of alanine and 2-deoxyglucose (a glucose analogue which is metabolizable only to glucose-6-phosphate and not further) in the presence of insulin (0.1 $mU.ml^{-1}$) at physiological alanine and glucose concentrations, a marked stimulation of glucose transport was seen by physiological concentrations of insulin but alanine uptake was only marginally stimulated (G.W.A. Thompson and M.J. Rennie, unpublished results). This accords with the results of Hasselgren and colleagues who showed that in baby rat (60-70 g) soleus muscle even insulin at 10 $mU.ml^{-1}$ only stimulated tracer AIB accumulation by 50% with a stimulation at more physiological concentrations being hardly significant (Hasselgren et al, 1987).

Apart from glutamine transport then, contrary to the generally-held belief that insulin stimulates amino acid transport into muscle, it seems in general to be insulin-insensitive. Net transport *in vivo* of a wide range of amino acids, mediated by various systems, shows no significant sensitivity to insulin under normal physiological concentrations of amino acid or insulin. Reductions in plasma amino acid concentrations following exogenous insulin administration or stimulation of endogenous insulin release as a consequence of carbohydrate loading, may therefore not be a result of a direct insulin effect upon amino acid transport *per se*. Rather, since the rate of amino acid uptake is, for most amino acids, 2-3 fold greater than the rate of incorporation into protein and under normal conditions amino acid transporters are generally unsaturated, increased fluxes via these transporters may be as a result of depletion of the free amino acid pool in muscle due to insulin-stimulated increases in protein synthesis (Bennet et al, 1990a; Bennet et al, 1991) or diminutions of protein breakdown (Bennet et al, 1990a; Flakoll et al, 1989).

Effects of Other Hormones on Amino Acid Transport

The effects of growth hormone on the accumulation of AIB and cycloleucine into many tissues (especially muscle, but surprisingly not heart muscle) were shown by Riggs and co-workers (Riggs and Walker, 1960; Hider et al, 1969). Growth hormone also increased the uptake of AIB and cycloleucine as well as alanine and histidine into the diaphragms of previously hypophysectomized rats (Kostyo, 1968). Now that recombinant human growth hormone of various molecular weights is available it has been possible to investigate the structure-function relationships of growth hormone and the stimulation of amino acid transport and protein synthesis (Cameron et al, 1988). The full length growth hormone molecule (22 kDa) has both growth-promoting and insulin-like activity but a 20 kDa variant lacks the insulin-like stimulation of glucose transport. When AIB uptake into rat diaphragms was examined in the presence of both the 22 kDa and 20 kDa growth hormones, the 20 kDa variant was only a fifth as effective as the bigger molecule in stimulating AIB transport (Cameron et al, 1988). The authors concluded that the rapid stimulatory effect of growth hormone on amino acid transport (and protein synthesis) are expressions of the insulin-like activity of growth hormone and are not components of the responses of target cells to its growth promoting actions. The transport effects of growth hormone have been rather poorly investigated and, given the important clinical effects of growth hormone, they probably require reinvestigation with recombinant human growth hormone. If the transport effects of growth hormone do mimic those of insulin it is difficult to reconcile this with the suggestions by Kostyo that the effects of growth hormone are dependent upon protein synthesis, according to their propensity for inhibition by puromycin and cycloheximide (Kostyo, 1968), since insulin effects are not.

Little is known about the effects of other hormones on skeletal muscle amino acid transport. When the distribution ratio of AIB and cycloleucine were studied *in vivo*, growth hormone increased the distribution ratio of AIB and, to a lesser extent, cycloleucine (Riggs et al, 1963) but other hormones had smaller effects. Cortisol had no effect on the distribution ratio of cycloleucine but decreased that of AIB – an effect probably due to a stimulation of efflux (Lewis and Goldspink, 1982), possibly secondary to an increased availability of intracellular Na^+ (Hundal et al, 1991). However, there are no effects on muscle AIB or cycloleucine accumulation as a result of testosterone or ß-estradiol administration (Riggs et al, 1963).

Amongst the catabolic hormones, glucagon, adrenalin and cortisol are said to have no effect on amino acid uptake as assessed by AIB distribution ratio in incubated rat soleus (Warner et al, 1990a). However, AIB distribution ratio alone is a poor index of natural amino acid uptake *per se*. Lewis and Goldspink showed clearly that in incubated rat EDL muscle, the accumulation of AIB was depressed by 23% in muscle exposed to 1 μM cortisol for 4 h; this was not due to an effect upon AIB influx, but rather it was due to a 25% enhancement of efflux over the first 15 minutes following AIB preloading (Lewis and Goldspink, 1982). A comprehensive study of the effects of a 7-day glucocorticoid treatment on muscle amino acid flux (Muhlbacher et al, 1984) (by an A-V difference technique) showed a 4 fold increase in the glutamine and alanine release with a decline in the plasma concentration of glutamine, indicating increased glutamine consumption by other organs (eg., kidney, intestine). Our group recently presented evidence regarding the acute alterations in glutamine transport characteristics in perfused hindlimb after glucocorticoid treatment (Hundal et al, 1991). It was observed that there was a 25% rise in intracellular Na^+ concentration after chronic (8-day) glucocorticoid treatment; the efflux of glutamine from muscle was increased with alterations in the apparent kinetic characteristics of glutamine influx after both chronic and acute dexamethsone treatment. The V_{max} of glutamine influx fell by 60% in dexamethasone-treated rats; the K_m also fell (i.e., there is an increase in transporter affinity). The acute changes on glutamine transport were measured after addition of the 200 nM dexamethasone to the perfusate. The unidirectional glutamine uptake was reduced by 65% in dexamethasone experiments compared to the controls. This acute effect led us to postulate that there is a direct action of the glucocorticoid on the muscle glutamine transporter System N^m.

The Effects of Fasting on Amino Acid Transport in Muscle

This is an obviously potentially valuable subject for study (see Table 4). When the problem was investigated using the injection of cycloleucine into whole animals, no effects on the distribution ratio of the tracer amino acid could be observed; i.e., the distribution ratio was identical in muscle fed and starved animals (Adibi et al, 1976; Nallathambi et al, 1972). This may well have been because cycloleucine is principally transported by Systems ASC and L and neither of these systems appear to be hormone sensitive. Nevertheless, when a similar experiment was carried out using AIB (Adibi et al, 1976), 12 h of fasting caused a 30% decrease in the muscle distribution ratio, which decreased further (by 60%) after 1 day but was then unchanged over the succeeding 4 days; protein deprivation had no effect over 5 days (Nallathambi et al, 1972). These changes did not appear to be explicable on the basis of alterations of insulin or glucagon and suggest that regulation of amino acid transport in muscle is much more complex than might at first be thought. In particular it may be that alterations in intracellular amino acid concentrations and ions, as well as alterations of substances with biological roles of which we are hardly yet aware, influence transport. For example, it is said that when added to the incubation medium, plasma from starved rats reduces AIB accumulation in muscles from control rats (Warner et al, 1989); the authors concluded that an unidentified circulating factor contributed to a reduction in amino acid accumulation (as a result of reduced uptake or enhanced efflux) in fasting.

Table 4. Plasma amino acid flux across muscle (nmol.100g $leg^{-1}.min^{-1}$): effects of feeding, starvation, infusion of nutrients, exercise and sepsis. All studies involved the acute effects, except *10 day intravenous feeding regimens and †daily exercise (see Albert et al, 1986).

Amino acid	Postabsorptive (mean from a variety of sources)	Fed (Day 10) (1)‡	Starvation (2)	Refeeding* (AA/dextrose/ lipid) (2)	AA alone (3)	AA and insulin (4)	Sepsis (5)	Exercise (40 min at 30% Vo_2max) (6)	Daily exercise† (7)
Ile	-16	84	-12	68	31	102	-48	82	61
Leu	-20	130	-25	98	16	143	-78	132	88
Val	-18	94	7	84	40	146	-61	316	113
Met	-9	—	-13	14	-3	—	-35	-28	12
Phe	-15	11	-14	6	11	33	-48	76	9
Thr	-33	—	-22	3	—	—	-97	22	21
Lys	-35	74	-31	22	-35	—	-139	—	39
Tyr	-13	-4	-13	6	-6	12	-41	90	12
His	-21	11	-12	12	-6	—	-39	—	6
Arg	-20	42	-14	27	-8	—	-70	—	33
Ala	-195	-14	-119	-131	-216	-22	-444	-446	-192
Asp	-1	-35	1	5	-31	—	—	-18	4
Asn	-25	—	-17	-4	-28	—	—	—	1
Glu	79	102	42	212	238	146	—	—	107
Gln	-203	172	-153	-245	-99	-343	—	—	-138
Gly	-52	-11	-75	-2	—	—	-182	76	-16
Orn	0	14	3	28	8	—	5	—	17
Pro	-104	—	-5	111	—	—	—	158	41
Ser	9	—	-6	56	64	—	-61	122	44
Tau	-17	-7	4	-12	-3	—	-5	-98	0
Balance	-709	330	-474	267	35	217	-1344	—	262

‡Data are drawn from (1) Elia and Livesey (1983), (2) Albert et al (1986), (3) Lundholm et al (1987), (4) Bennet et al (1990a), (5) Wahren et al (1988), (6) Ahlborg et al (1974) and (7) Albert et al (1977).

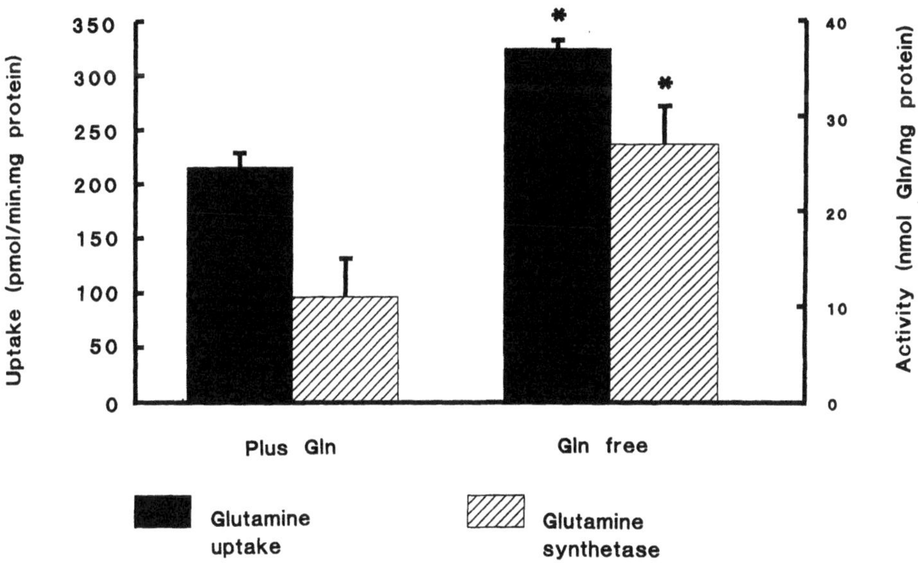

Fig. 6. Glutamine uptake and synthesis in cultured myoblasts. Myoblasts (from neonatal rat skeletal muscle) were cultured in glutamine-free medium or medium containing 2 mM glutamine for 4 days before measuring uptake of [^{3}H]glutamine at 50 μM over 1 min, and glutamine synthetase activity (measuring glutamine production over 60 min); data are mean ± SEM, $n = 3$, $P < 0.05$ (L.B.M. Tadros, P.M. Taylor and M.J. Rennie, unpublished).

It is known that, in a variety of types of animal cells grown in culture, amino acid starvation results in the increased transport activity and principally the activity of System A. These effects are easily demonstrated for chick embryo heart cells (Guidotti et al, 1991), for Chinese hamster ovary (CHO-K1) cells (Englesberg et al, 1986) and for liver cells, in which the effects can also be demonstrated after a period of starvation *in vivo* (Kilberg et al, 1986). A similar effect can be seen in cultured L6 muscle cells (Logan et al, 1982). Recently work from Holloszy's group (Gulve et al, 1991) has shown that when soleus muscle is incubated for periods of 5 h (during which viability can be preserved), there is protein synthesis-dependent induction of System A transport. This induction can be blocked by cycloheximide and also appears to be sensitive to the inhibitory effects of glutamine in the medium. Similar results have been observed for MeAIB, but not BCH, by others (Tovar et al, 1991). The phenomenon of adaptive regulation can also be shown for glutamine transport in the liver (Low et al, 1991) and recently workers in our group have demonstrated that when primary rat muscle cultures are grown in the absence of glutamine, both the muscle glutamine synthetase and the membrane transport of glutamine are enhanced (Figure 6). Obviously, expression of transporters according to environmental conditions could be a generally important homeostatic mechanism but little is yet known about its basis.

Effects of Catabolic States on Glutamine Metabolism and Transport

We now have a substantial amount of information about the effects of catabolic states (injury, burns, diabetes, corticosteroid treatment, sepsis, etc) on amino acid metabolism (Rennie et al, 1986; Rennie, 1985; Smith and Rennie, 1990). Plasma amino acid concentrations may be increased, decreased or unchanged but the total pool of skeletal muscle amino acids is always decreased with glutamine, the most abundant metabolically active amino acid in the body, being decreased the most and the BCAA, tyrosine and phenylalanine and methionine being increased markedly (Figure 1). Alanine concentration is often elevated probably due to increases in glycolysis. Making sense of the changes is not easy and, of course, alterations of protein turnover and amino acid oxidation are likely to be important.

Sepsis causes rapid muscle wasting and a massive rise in amino acid export (Clowes et al, 1980) and transport of amino acids has been studied extensively. Bacterial infected perfusate increased the efflux of glutamine and of serine, and decreased the uptake of glutamine, in the rat hindlimb (Lynn et al, 1986) but this effect was abolished following 0.2 μm filtration of the perfusate; however, 20 $\mu g.ml^{-1}$ of *E. coli* endotoxin inhibited the glutamine uptake by 19%.

Despite the rise (not a fall) in intramuscular alanine concentration seen in septic muscle, inward System A activity seems to be depressed during sepsis induced by endotoxin or by cecal ligation and puncture (Karlstad and Sayeed, 1986; Karlstad and Sayeed, 1985a; James et al, 1990). However, AIB uptake in the presence of specific inhibitors of Systems A and L (MeAIB and BCH respectively) and Na^+-independent uptake of cycloleucine were unaffected, indicating that sepsis does not induce any changes in the activity of Systems ASC and L. The inhibition of AIB uptake in muscles obtained from rats previously injected with endotoxin was not due to any associated endotoxin-induced hypotension, since bleeding normal rats did not reproduce it (Warner et al, 1990b). Also, the effects are not due to some free radical generation at the muscle membrane since they are not abolished by the free radical scavenger superoxide dismutase (Warner et al, 1987). The addition of plasma from septic rats to the incubation medium of isolated muscles also reduced MeAIB uptake, by 31%, suggesting the involvement of a circulating factor and incubation of muscles with whole plasma and a small molecular weight (<10 kDa) plasma fraction from endotoxin-treated rats resulted in a 22% reduction in transport; nevertheless there seems to be no direct effect on amino acid transport in incubated muscle of the obvious cytokine candidates, i.e., tumour necrosis factor or interleukin-1 (see Hasselgren et al, 1990).

During endotoxemia, the responsiveness of insulin-stimulation of AIB uptake is attenuated (Karlstad and Sayeed, 1985b) at a time when plasma insulin concentration is elevated (Karlstad and Sayeed, 1987), although James et al (1990) found that in sepsis, whilst the maximal MeAIB uptake after insulin-treatment was reduced, MeAIB transport could be stimulated by a lower concentration of insulin than that required in control muscles; however, as the ED_{50} values were not determined little can be concluded about the effects of sepsis on insulin-sensitivity.

A reduction in AIB uptake in rat soleus *in vivo* following endotoxic shock (20 $mg.kg^{-1}$ endotoxin injection) was associated with intracellular Na^+ and Cl^- accumulation in the endotoxic soleus muscles, with a drop in the intracellular K^+ concentration and membrane potential from -89 to -53 mV (Karlstad and Sayeed, 1987). These changes support the predictions made concerning the importance of ion and membrane potential changes on glutamine efflux from muscle during injury and sepsis (Rennie et al, 1986).

The efflux of glutamine from skeletal muscle increases with starvation, acidosis, trauma, sepsis and burns and the enhanced effluxes are always associated with a decrease in the tissue/plasma gradient of glutamine with a diminished intramuscular concentration. Hypothetical mechanisms for the observed changes need to account for the increased flux and the decreased distribution ratio at the same time. None of the possible changes in the processes of protein synthesis (usually down) and breakdown (often down and occasionally up) and glutamine synthesis (usually up but sometimes down) or glutaminase activity (unchanged or up), or in the availability of intermediates, eg., aspartate, ammonia, etc. (Jaspers et al, 1989) adequately explain the observed facts. Given incontrovertible observations of an increased net flux of glutamine with a diminished distribution ratio, it is inescapable that the process transporting glutamine from muscle has, in some way, been increased under these circumstances. (See also the work of Ardawi and Majzoub, 1991, and Ardawi, 1988, for elegant demonstrations of the changes which occur with injury and sepsis in the rat).

The possible effects on transport of glutamine under different circumstances have been presented in a number of papers from our group (Rennie et al, 1986; Hundal et al,

1991; Hundal et al, 1990b; Hundal et al, 1990a). We believe that the characteristics of the glutamine transporter are such that alteration in the availability of intracellular ions (principally increases in Na^+, and decreases in K^+) and also in membrane potential (low in many circumstances) will depress the inward transport of glutamine and accelerate its outward efflux. There is also the possibility that the inward transport of glutamine may be inhibited by increased concentrations of BCAA (Hundal et al, 1987a) but these are just as likely to inhibit the efflux, especially since the concentrations inside muscle under most conditions of injury rise more than the concentrations in plasma.

The importance of the intramuscular glutamine concentration is illustrated by the fact that glutamine stimulated protein synthesis and inhibited protein degradation (MacLennan et al, 1987; MacLennan et al, 1988) in perfused rat skeletal muscle. Septic rats given glutamine intravenously maintain more normal rates of muscle protein turnover examined *in vitro* than septic control rats (Ardawi, 1991). That this is likely to be an important mechanism clinically is shown by the fact that parenteral administration of the dipeptide alanyl-glutamine (rapidly hydrolyzed to alanine and glutamine) in post-operative patients reversed the inhibition of protein synthesis consequent to surgery (Barua, 1991).

Effects of Muscle Contractile Activity on Amino Acid Transport

When the effects of prior exercise on the ability of insulin to stimulate AIB accumulation were assessed in the perfused rat hindlimb (Zorzano et al, 1985), the sensitivity and responsiveness to insulin were enhanced. The insulin-responsiveness of AIB accumulation in soleus and red and white gastrocnemius muscles differed substantially with that of soleus being less; also prior exercise at a moderate rate stimulated the responsiveness of soleus and red gastrocnemius but not of white gastrocnemius. These effects did not appear to be due to alterations in the capacity of various muscles to bind insulin. The results are interesting in that in many ways they are the opposite of the decreases in insulin sensitivity seen both for glucose and AIB transport in denervated muscle (Cleland et al, 1990).

Indirect support for the involvement of membrane potential in the transport of glutamine in skeletal muscle also comes from investigation of the effects of electrical stimulation of muscle in perfused hindlimb (Hundal and Rennie, 1987), which showed a small decrease in the unidirectional inward transport of glutamine and a small rise in the fractional tracer efflux.

The effects of exercise in increasing the uptake and oxidation of the BCAA (Rennie et al, 1981; Millward et al, 1982) are unlikely to be related to a modulation of transport but more likely to be related to their increased delivery as blood flow increases.

Experiments carried out recently in our laboratory (G.W.A. Thompson and M.J. Rennie, unpublished results) confirm by more specific methods the suggestion made by Goldberg and Goodman (Goldberg and Goodman, 1969a; Goldberg and Goodman, 1969b) on the basis of AIB distribution after tracer injection *in vivo* that hypertrophy of muscle can cause a marked stimulation of System A transport. Furthermore, the effects are specific to System A since we have found this occurs without alteration of transport of System ASC or System L; the stimulation is not affected by concurrent denervation of the soleus muscle although Goodman and Goldberg found that denervation *per se* decreased AIB accumulation after seven days. It appears that denervation has a biphasic effect on AIB accumulation since Diehl found that AIB uptake was accelerated in the first few days after denervation (Diehl, 1964; Diehl and Jones, 1966).

AMINO ACID TRANSPORT IN CARDIAC MUSCLE

Our present understanding of amino acid transport in cardiac muscle is poor and the subject appears not to have been a major focus of interest. Early work demonstrated that amino acids could be accumulated against a concentration gradient in perfused heart and that despite an inhibition of cardiac muscle protein synthesis by puromycin, insulin is capable of increasing the intracellular concentrations of a number of neutral amino acids (Scharff and Wool, 1965). In the Langendorff perfused rabbit heart, phenylalanine taurine and alanine show high tracer extraction which is inhibitable by non-labelled amino acids, suggesting rapid, saturable carrier mediated transport (Mann and Yudilevich, 1982; Mann and Yudilevich, 1985).

In isolated papillary muscle of the rabbit the Na^+-dependence, and ouabain sensitivity of AIB transport were demonstrated (Lesch et al, 1970). Investigations of ion dependence in a sarcolemmal vesicle preparation from avian heart demonstrated that both alanine and AIB are transported in a Na^+-dependent manner (Paris and Ailhaud, 1980). The initial characterization (without kinetic parameters) of alanine and leucine uptake has been reported in membrane vesicles prepared from dog heart (Young et al, 1987).

Gazzola and co-workers (1972) working with chick embryo heart cells described, for the first time, the phenomenon of adaptive regulation for the known System A substrates. This post-starvation enhancement of the transport activity of System A substrates does not require Na^+ during the incubation period (unlike that observed by Tovar and co-workers in skeletal muscle). The ion and pH-dependence of the effect and its inhibition profile suggested that only System A showed adaptive regulation in chick embryo heart cells. The authors concluded that the rate of carrier activity was modulated by extracellular concentration of the relevant amino acid (i.e., up or down regulation of the carrier protein) and that Na^+ would only acutely affect the binding of the substrate to carrier (a K_m effect). The activity of System A in chick embryo heart cells was increased in the presence of insulin (Guidotti et al, 1974), maximumal after 2 h exposure to the hormone (Elsas et al, 1975), a V_{max} effect believed to involve the synthesis of new carrier protein since the stimulatory effect of insulin was abolished by treatment with cycloheximide.

These studies in chick embryo heart cells only elucidated in detail the transport characteristics of System A substrates (mainly AIB); other metabolically important native amino acids (glutamine, glutamate, BCAA, etc.) were not investigated.

Given the importance of the role of glutamate as a donor to pyruvate in the production of alanine in heart and the metabolic role of glutamine transport in whole body nitrogen balance it is unfortunate that almost no information is available on glutamine or glutamate transport in heart. A role for glutamine in the heart in physiological and pathophysiological states could be hypothesized. For example, under conditions of ischaemia and hypoxia the normal reliance of the heart upon free fatty acids as a major fuel cannot be sustained due to lack of oxygen and glycolysis is accelerated. The extent to which glycolysis can be accelerated in the heart may partially determine its ability to withstand decreased oxygen delivery and thus the possible modulatory effects of glutamine on glycogen metabolism (i.e., increasing glycogen stores; Scislowski et al, 1989) could be important in protecting heart from later damage. During partial ischaemia or increased heart work the build-up of lactate associated with increased glycolytic rate is minimised in heart muscle by production of alanine from pyruvate (Mudge et al, 1976). This requires availability of glutamate as a nitrogen donor. During anoxia alanine production is observed to increase but its production becomes limited by the availability of glutamate, whose intracellular concentration falls (Jefferson et al, 1971). The administration of glutamate intravenously to patients with post-operative cardiac failure resulted in an improvement in cardiac circulation, a cessation of ammonia release and an increase in alanine release from the myocardium (Jefferson et al, 1971). Work in rats (Choong et al, 1988) also showed that reperfusion of ischaemic hearts with 20 mM glutamate led to a 30% improvement in cardiac function; the underlying mechanisms involved in this post-ischaemic recovery is not understood but may include the generation of ATP via anaerobic succinate production. The ability to increase intracellular glutamate may however be limited by the observation that the dicarboxylate amino acids (glutamate and aspartate) appear to have the lowest rate of outward transfer, relative to the distribution gradient, of any amino acids studied, as measured by their accumulation in the recirculating perfusate of isolated perfused rat heart (Morgan et al, 1971).

Our group has recently initiated the investigations of amino acid transport in isolated working heart and cardiac sarcolemmal vesicle preparations. Preliminary results in rat cardiac sarcolemmal vesicles (A. Ahmed and M.J. Rennie, unpublished results) indicate that there is a concentrative accumulation of glutamine across the vesicles which is dependent on the prevailing Na^+-gradient. Studies in isolated working heart show the uptake of glutamine was saturable and during the reversible decrease in aortic output and coronary flow in hypoxic hearts there was a 22% decrease in glutamine uptake (S.E.O. Khogali, A.A. Harper and M.J. Rennie, unpublished results). This observation supports the hypothesis that glutamine transport across cardiac sarcolemma is modulated under pathophysiological conditions and may have an important role in cardiac amino acid metabolism.

Taurine shows a high distribution ratio across the cardiac sarcolemma and comprises up to 50% of the total free amino acid pool (Jacobsen and Smith, 1968). Taurine has been implicated in the osmoregulation and the maintenance of normal electrophysiological function of the heart (Earm et al, 1990) and it is therefore interesting that the intracellular concentrations of taurine are also increased in blood after hypoxia and drug-induced myocardial necrosis and following acute cardiac pain (Welty et al, 1982; Bhatnagar et al, 1990). Investigations of the release of taurine from the oxygen-deficient perfused rat showed that the release of taurine under conditions of oxygen-deficiency increased with time, but no quantitative data on the mechanisms of taurine release was reported (Crass and Lombardini, 1978).

Chapman and co-workers have recently shown that the introduction of glutamate and taurine into isolated cardiac myocytes protects them against the effects of the oxygen and Ca^{2+} paradoxes (Chapman and Rodrigo, 1990; Suleiman and Chapman, 1990). The mechanism by which this protection takes place is not known but it is antagonized by the presence of azaserine, a glutamine analogue with antimetabolite activity (Chapman and Rodrigo, 1990); it is suggested that the protective effect may involve regulation of intracellular sodium, either through changes in cardiac sarcolemmal conductances (Earm et al, 1990) or through activity of Na^+/amino acid co-transporters, both of which may improve heart function in avoiding damage after cardiac ischaemia and hypoxia.

These reports, although indicating an important role for taurine in cardiac homeostasis, have failed to address the question of the importance of taurine transport and to elucidate the mechanisms of its transfer across the cardiac sarcolemma. Unfortunately, as for skeletal muscle, very little information is available regarding taurine transport in heart.

AMINO ACID TRANSPORT IN SMOOTH MUSCLE

There is a notable lack of available information about amino acid transport in smooth muscle. In an investigation of hormonal effects, uptake of AIB into aortic smooth muscle cells grown to confluency was stimulated by physiological concentration of insulin (25 $\mu U.ml^{-1}$); the maximum effect was a two-fold increase with an ED_{50} of 250 $\mu U.ml^{-1}$ (Kaiser et al, 1985). Insulin-like growth factor I (IGF I) was shown to compete for the insulin binding sites, and insulin and IGF I both increased DNA synthesis, measured by [^{3}H]thymidine incorporation into DNA, although growth hormone was inactive in these respects. Thus, insulin-stimulated amino acid uptake, in parallel with a mitogenic effect of insulin, may ensure an adequate supply of amino acids for protein synthesis during growth of smooth muscle.

ACKNOWLEDGEMENTS

Work from our laboratory described in this review was supported by Action Research for the Crippled Child, Ajinomoto GmbH, the British Diabetic Association, the British Heart Foundation, the European Society of Parenteral and Enteral Nutrition, the Medical Research Council, the Science and Engineering Research Council, the Chief Scientist's Biomedical Research Committee of the Scottish Home and Health Department, the Scottish Hospital Endowments Research Trust, the University of Dundee, the Committee of Vice-Chancellors and Principals, The Wellcome Trust, and Wellcome Research Laboratories. We thank our colleagues for access to unpublished data and for helpful criticism. The skills and patience of Mrs Dorothea Watt in preparing the manuscript are particularly worthy of thanks.

REFERENCES

Adibi, S.A., Morse, E.L. and Mirsky, I.A., 1976, *Am. J. Physiol.*, 230:245-250.
Ahlborg, G., Felig, P., Hagenfeldt, L., Hendler, R. and Wahren, J., 1974, *J. Clin. Invest.*, 53:1080-1090.
Ahmed, A., Gibson, J.N.A., Taylor, P.M. and Rennie, M.J., 1990a, *Biochem. Soc. Trans.*, 18:1238-1239.
Ahmed, A., Taylor, P.M and Rennie, M.J., 1990b, *Am. J. Physiol.*, 259:284-291.
Ahmed, A., 1991, *Clin. Nutr.*, in press.
Airhart, J., Arnold, J.A., Stirewalt, W.S. and Low, R.B., 1982, *Am. J. Physiol.*, 243:C81-C86.
Akedo, H. and Christensen, H.N., 1962, *J. Biol. Chem.*, 237:118-127.
Albert, J.D., Matthews, D.E., Legaspi, A., Tracey, K.J., Jeevanandam, M., Brennan, M.F. and Lowry, S.F., 1977, *Clin. Sci.*, 77:113-120.
Albert, J.D., Legaspi, A., Horowitz, G.D., Tracey, K.J., Brennan, M.F. and Lowry, S.F., 1986, *Am. J. Physiol.*, 251:E604-E610.
Alvestrand, A., DeFronzo, R.A., Smith, D. and Wahren, J., 1988, *Clin. Sci.*, 74:155-163.
Aoki, T.T., Muller, W.A., Brennan, M.F. and Cahill, G.F., 1973, *Diabetes*, 22:768-775.
Ardawi, M.S.M., 1988, *Clin. Sci.*, 74:165-172.
Ardawi, M.S.M., 1991, *Clin. Sci.*, 81:215-222.
Ardawi, M.S.M. and Majzoub, M.F., 1991, *Metabolism*, 40:155-164.
Arfvidsson, B., Sachrisson, H., Moller-Loswick, A-C., Hyltander, A., Sandström, R. and Lundholm, K., 1991, *Am. J. Physiol.*, 260:E46-E52.
Ariano, M.A., Armstrong, R.B. and Edgerton, V.R., 1973, *J. Histochem. Cytochem.*, 21:51-55.
Askanazi, J., Carpentier, Y.A., Michelsen, C.B., Elwyn, D.E., Fürst, P., Kantrowitz, L.R., Gump, F.E. and Kinney, J.M., 1980, *Ann. Surg.*, 192:78-85.
Bading, J.R., Corbally, M.T., Fissekis, J.D., DiResta, G.R. and Brennan, M.F., 1987, *J. Nucl. Med.*, 28 No. 398:650.
Banos, G., Daniel, P.M., Moorhouse, S.R. and Pratt, O.E., 1973, *J. Physiol.*, 235:459-475.
Barrett, E.J. and Gelfand, R.A., 1989, *Diabetes Metab. Rev.*, 5:133-148.
Barua, J.M., 1991, *Proc. Nutr. Soc.*, in press.
Bennet, W.M., Connacher, A.A., Scrimgeour, C.M., Jung, R.T. and Rennie, M.J., 1990a, *Am. J. Physiol.*, 259:E185-E194.
Bennet, W.M., Connacher, A.A., Scrimgeour, C.M. and Rennie, M.J., 1990b, *Eur. J. Clin. Invest.*, 20:37-46.
Bennet, W.M., Connacher, A.A., Smith, K., Jung, R.T. and Rennie, M.J., 1990c, *Diabetologia*, 33:43-51.
Bennet, W.M., Connacher, A.A., Jung, R.T., Stehle, P. and Rennie, M.J., 1991, *Diabetes*, 40:499-508.
Bergström, J., Fürst, P., Norée, L.-O. and Vinnars, E., 1974, *J. Appl. Physiol.*, 36:693-697.
Bergström, J., Fürst, P. and Vinnars, E., 1990, *Clin. Sci.*, 79:331-337.
Bhatnagar, S.K., Welty, J.D. and Al Yusuf, A.R., 1990, *Int. J. Cardiol.*, 27:361-366.
Brosnan, M.E. and Letto, J., 1991, *Amino Acids*, 1:29-35.
Cahill, G.F., 1970, *New Eng. J. Med.*, 282:668-675.
Cameron, C.M., Kostyo, J.L., Adamafio, N.A., Brostedt, P., Roos, P., Skottner, A., Forsman, A., Fryklund, L. and Skoog, B., 1988, *Endocrinology*, 122:471-474.
Chapman, R.A. and Rodrigo, G.C., 1990, *J. Mol. Cell. Cardiol.*, 22 (Suppl III):S.133.
Cheng, K.N., Dworzak, F., Ford, G.C., Rennie, M.J. and Halliday, D., 1985, *Eur. J. Clin. Invest.*, 15:349-354.
Chikenji, T., Elwyn, D.E. and Kinney, J., 1983, *J. Surg. Res.*, 34:68-82.
Choong, Y.S., Gavin, J.B. and Armiger, L.C., 1988, *J. Mol. Cell. Cardiol.*, 20:1043-1052.
Christensen, H.N., Liang, M. and Archer, E.G., 1967, *J. Biol. Chem.*, 242:5237-5246.
Christensen, H.N., Handlogten, M.E., Lam, I., Tager, S. and Zand, R.A., 1969, *J. Biol. Chem.*, 244:1510-1520.
Christensen, H.N., 1982, *Physiol. Rev.*, 62:1193-1233.

Christensen, H.N., 1985, *J. Memb. Biol.*, 84:97-103.
Christensen, H.N.and Kilberg,M.S., 1987, in: *Amino Acid Transport in Animal Cells*, D.L. Yudilevich and C.A.R. Boyd, eds., Manchester University Press., Manchester., pp. 10-46.
Christensen, H.N., 1988, *TIBS*, 13:40-42.
Christensen, H.N., 1990, *Physiol. Rev.*, 70:43-77.
Clausen, T. and Kohn, P.G., 1977, *J. Physiol.*, 265:19-42.
Cleland, P.J.F., Abel, K.C., Rattigan, S. and Clark, M.G., 1990, *Biochem. J.*, 267:659-663.
Clowes, G.H.A., Randall, H.T. and Cha, C-J., 1980, *J. Parent. Ent. Nutr.*, 4:195-205.
Conlon, K.C., Bading, J.R., DiResta, G.R., Corbally, M.T., Gelbard, A.S. and Brennan, M.F., 1989, *Life Sci.*, 44:847-859.
Cooper, G.J. and Kohn, P.G., 1980, *J. Physiol.*, 302:89-105.
Crass III, M.F. and Lombardini, J.B., 1978, *Proc. Soc. Exp. Biol. Med.*, 157:486-488.
Daniel, P.M., Pratt, O.E. and Spargo, E., 1977, *Lancet*, ii:446-448.
Darmaun, D., Matthews, D.E. and Bier, D.M., 1986, *Am. J. Physiol.*, 251:E117-E126.
Diehl, J.F., 1964, *Biochim. Biophys. Acta*, 82:611-612.
Diehl, J.F. and Jones, R.R., 1966, *Am. J. Physiol.*, 210:1080-1085.
Dohm, G.L., Tapscott, E.B., Pories, W.J., Dabbs, D.J., Flickenger, E.G., Meelheim, D., Fushiki, T., Atkinson, S.M., Elton, C.W. and Caro, J.F., 1988, *J. Clin. Invest.*, 82:485-494.
Earm, Y.E., Spindler, A.J., Noble, S.J. and Noble, D., 1990, *J. Physiol.*, 426:57.(Abstract)
Elia, M. and Livesey, G., 1981, in: *Metabolism and Clinical Implications of Branched Chain Amino and Ketoacids*, M. Walser and J.R. Williamson, eds., Elsevier/North Holland, New York Amsterdam Oxford, pp. 257-262.
Elia, M. and Livesey, G., 1983, *Clin. Sci.*, 64:517-526.
Elsas, L.J., Wheeler, F.B., Danner, D.J. and DeHaan, R.L., 1975, *J. Biol. Chem.*, 250:9381-9390.
Emig, A.C., Merrill, G.F. and Forsberg, N.E., 1988, *FASEB J.*, 2:A642.
Englesberg, E., Moffett, J. and Perier, F., 1986, *Fed. Proc.*, 45:2441-2443.
Eriksson, L.S., Hagenfeldt, L., Felig, P. and Wahren, J., 1983, *Clin. Sci.*, 65:491-498.
Everett, A.W., Prior, G. and Zak, R., 1981, *Biochem. J.*, 194:365-368.
Farese, R.V., Standaert, M.L., Barnes, D.E., Davis, J.S. and Follet, R.J., 1985, *Endocrinology*, 116:2650-2655.
Felig, P., Pozefsky, T., Marliss, C. and Cahill, G.F., 1970, *Science*, 167:1003-1004.
Felig, P., 1975, *Ann. Rev. Biochem.*, 44:933-955.
Felig, P. and Wahren, J., 1971, *J. Clin. Invest.*, 50:2703-2709.
Flakoll, P.J., Kulaylat, M., Frexes-Steed, M., Hourani, H., Brown, L.L., Hill, J.O. and Abumrad, N.N., 1989, *Am. J. Physiol.*, 257:E839-E847.
Forsberg, N.E., Kaneps, A.J. and Riebold, T.W., 1988, *Can. J. Anim. Sci.*, 68:409-417.
Fritz, G.R. and Knobil, E., 1963, *Nature*, 200:682-683.
Garlick, P.J. and Grant, I., 1988, *Biochem. J.*, 254:579-584.
Gazzola, G.C., Franchi, R., Saibene, V., Ronchi, P. and Guidotti, G.G., 1972, *Biochim. Biophys. Acta*, 266:407-421.
Goldberg, A.L., Martel, S.B., and Kushmerick, M.J., 1975, in: *Meth. Enzymol., Vol 39*, J.G. Hardman and B.W. O'Malley, eds., Academic Press, New York, pp. 82-94.
Goldberg, A.L. and Chang, T.W., 1978, *Fed. Proc.*, 37:2301-2307.
Goldberg, A.L. and Goodman, H.M., 1969a, *Am. J. Physiol.*, 216:1116-1119.
Goldberg, A.L. and Goodman, H.M., 1969b, *Am. J. Physiol.*, 216:1111-1115.
Goldstein, L., Perlman, D.F., McLaughlin, P.M., King, P.A. and Cha, C.-J., 1983, *Biochem. J.*, 214:757-767.
Grimditch, G.K., Barnard, R.J., Kaplan, S.A. and Sternlicht, E., 1985, *Am. J. Physiol.*, 249:E398-E408.
Guidotti, G.G., Franchi-Gazzola, R., Gazzola, G.C. and Ronchi, P., 1974, *Biochim. Biophys. Acta*, 356:219-230.
Guidotti, G.G., Gazzola, G.C., Borghetti, A.M. and Franchi-Gazzola, R., 1991, *Biochim. Biophys. Acta*, 406:264-279.
Gulve, E.A., Cartee, G.D., Youn, J.H. and Holloszy, J.O., 1991, *Am. J. Physiol.*, 260:C88-C95.
Gulve, E.A. and Dice, J.F., 1989, *Biochem. J.*, 260:377-387.
Guma, A., Testar, X., Palacin, M. and Zorzano, A., 1988, *Biochem. J.*, 253:625-629.
Guma, A., Camps, M., Palacin, M., Testar, X. and Zorzano, A., 1990, *Biochem. J.*, 268:633-639.

Hall-Angeras, M., Angeras, U., Von Allmen, D., Higashiguchi, T., Zamir, O., Hasselgren, P.-O. and Fischer, J.E., 1991, *Metabolism*, 40:247-251.
Harris, R.C., Hultman, E. and Nordesjö, L.-O., 1974, *Scand. J. Clin. Lab. Invest.*, 33:109-120.
Hasselgren, P.-O., Warner, B.W., James, J.H., Takehara, H. and Fischer, J.E., 1987, *Arch. Surg.*, 122:228-233.
Hasselgren, P.-O., James, J.H., Benson, D.W., Li, S. and Fischer, J.E., 1990, *Arch. Surg.*, 125:510-514.
Hider, R.C., Fern, E.B. and London, D.R., 1969, *Biochem. J.*, 114:171-178.
Hirshman, M.F., Henriksson, H.W., Wardzala, L.J., Horton, E. and Horton, E.S., 1988, *FEBS Lett.*, 238(2):235-239.
Hummel, R.P.,III, Hasselgren, P.-O., James, J.H., Warner, B.W. and Fischer, J.E., 1988, *Metabolism*, 37:1120-1127.
Hundal, H.S., Rennie, M.J. and Watt, P.W., 1987, *J. Physiol.*, 393:283-305.
Hundal, H.S., Rennie, M.J. and Watt, P.W., 1989, *J. Physiol.*, 408:93-114.
Hundal, H.S., Babij, P., Watt, P.W., Ward, M.R. and Rennie, M.J., 1990a, *Am. J. Physiol.*, 259:E148-E154.
Hundal, H.S., Mackenzie, B., Taylor, P.M. and Rennie, M.J., 1990b, *Biol. Chem. Hoppe-Seyler*, 371:297.
Hundal, H.S., Babij, P., Taylor, P.M., Watt, P.W. and Rennie, M.J., 1991, *Biochim. Biophys. Acta*, 1092:376-383.
Hundal, H.S. and Rennie, M.J., 1987, *J. Physiol.*, 392:28P.
Hundal, H.S. and Watt, P.W., 1991, *J. Physiol.*, in press.
Jacobsen, J.G. and Smith, L.H., 1968, *Physiol. Rev.*, 48:424-448.
James, J.H., Hasselgren, P.-O., Hummel, R.P., Warner, B.W. and Fischer, J.E., 1990, *Metabolism*, 39:335-340.
Jaspers, S.R., Henriksen, E.J, Satarug, S. and Tischler, M.E., 1989, *Life Sci.*, 38:303-310.
Jefferson, L.S., Wolpert, E.B., Giger, K.E. and Morgan, H.E., 1971, *J. Biol. Chem.*, 246:2171-2178.
Jefferson, L.S., 1980, *Diabetes*, 29:487-496.
Joseph, S.K., Bradford, N.M. and McGivan, J.D., 1978, *Biochem. J.*, 176:827-836.
Juel, C., 1991, *Biochim. Biophys. Acta*, 1065:15-20.
Kacser, H. and Porteous, J.W., 1987, *TIBS*, 12:6-14.
Kaiser, N., Tur-Sinai, A., Hasin, M. and Cerasi, E., 1985, *Am. J. Physiol.*, 249:E292-E298.
Karlstad, M.D. and Sayeed, M.M., 1985a, *Fed. Proc.*, 44:1576.
Karlstad, M.D. and Sayeed, M.M., 1985b, *Am. J. Physiol.*, 248:R142-R146.
Karlstad, M.D. and Sayeed, M.M., 1986, *Am. J. Physiol.*, 251:R150-R156.
Karlstad, M.D. and Sayeed, M.M., 1987, *Am. J. Physiol.*, 252:R674-R680.
Kelso, T.B., Shear, C.R. and Max, S.R., 1989, *Am. J. Physiol.*, 257:E885-E894.
Kettlehut, I.C., Wing, S.S. and Goldberg, A.L., 1988, *Diabetes Metab. Rev.*, 4:751-772.
Khatra, B.S., Chawla, R.K., Sewell, C.W. and Rudman, D., 1977, *J. Clin. Invest.*, 59:558-564.
Kilberg, M.S., Handlogten, M.E. and Christensen, H.N., 1980, *J. Biol. Chem.*, 255:4011-4019.
Kilberg, M.S., Bracy, D.S. and Handlogten, M.E., 1986, *Fed. Proc.*, 45:2438-2440.
Kimball, S.R. and Jefferson, L.S., 1988, *Diabetes Metab. Rev.*, 4:773-787.
Kipnis, D.M., Reiss, E. and Helmreich, E., 1961, *Biochim. Biophys. Acta*, 51:519-524.
Kipnis, D.M. and Noall, M.W., 1958, *Biochim. Biophys. Acta*, 228:226-227.
Kipnis, D.M. and Parrish, J.E., 1965, *Fed. Proc.*, 24:1051-1059.
Klip, A., Ramlal, T., Young, D.A. and Holloszy, J.O., 1987, *FEBS Lett.*, 224:224-230.
Klip, A. and Walker, D., 1983, *Archiv. Biochem. Biophys.*, 221(1):175-187.
Konigsberg, I.R. in: *Meth. Enzymol., Vol. 58*, New York:Academic Press, 1979. pp. 511-527.
Kostyo, J.L., 1968, *Ann. N. Y. Acad. Sci.*, 148:389-407.
Krebs, H.A., 1972, *Adv. Enzyme Regulation*, 10:406-413.
Le Marchand-Brustel, Y., Jeanrenaud, B. and Freychet, P., 1982, *Am. J. Physiol.*, 243:E74-E79.
Leighton, B., Dimitriadis, G.D., Parry-Billings, M., Bond, J., de Vasconcelos, P.R.L. and Newsholme, E.A., 1989, *Clin. Sci.*, 77:61-67.
Lesch, M., Gorlin, R. and Sonnenblick, E.H., 1970, *Circ. Res.*, 27:445-460.
Lewis, S.E.M. and Goldspink, D.F., 1982, *Biochem. Soc. Trans.*, 10:171-172.
Lewis, S.E.M. and Goldspink, D.F., 1984, *Biochem. J.*, 221:71-79.
Liaw, K.Y., Askanazi, J., Michelsen, C.B., Fürst, P., Elwyn, D.H. and Kinney, J.M., 1982, *Ann. Surg.*, 195:12-18.
Livesey, G. and Lund, P., 1980, *Biochem. J.*, 188:705-713.

Logan, W.J., Klip, A. and Gagalang, E., 1982, *J. Cell. Physiol.*, 112:229-236.
Lotspeich, W., 1949, *J. Biol. Chem.*, 179:175-180.
Low, S.Y., Taylor, P.M., Hundal, H.S., Pogson, C.I. and Rennie, M.J., 1991, Submitted to *Biochem. J.*.
Lowenstein, J.M. and Goodman, M.N., 1978, *Fed. Proc.*, 37:2308-2312.
Lundholm, K., Bennegård, K., Zachrisson, H., Lundgren, F., Edén, E. and Möller-Loswick, A-C., 1987, *J. Clin. Invest.*, 80:763-771.
Lynn, J.C., Hundal, H.S., Watt, P.W. and Rennie, M.J., 1986, *Biochem. Soc. Trans.*, 14:1068-1069.
MacLennan, P.A., Brown, R.A. and Rennie, M.J., 1987, *FEBS Lett.*, 215:187-191.
MacLennan, P.A., Smith, K., Weryk, B., Watt, P.W. and Rennie, M.J., 1988, *FEBS Lett.*, 237:133-136.
Maillet, C. and Garber, A.J., 1980, *Am. J. Clin. Nutr.*, 33:1343-1353.
Maltin, C.A. and Harris, C.I., 1985, *Biochem. J.*, 232:927-930.
Manchester, K.L. and Young, F.G., 1958, *Biochem. J.*, 70:353.
Manchester, K.L. and Young, F.G., 1963, *Biochem. J.*, 89:202.
Mann, G.E. and Yudilevich, D.L., 1982, *J. Physiol.*, 324:19-20P.
Mann, G.E. and Yudilevich, D.L., 1985, in: *Carrier Mediated Transport of Solutes from Blood to Tissue*, D.L. Yudilevich and G.E. Mann, eds., Longman, London, pp. 323-331.
Marliss, E.B., Aoki, T.T., Pozefsky, T., Most, A.S. and Cahill, G.F., 1971, *J. Clin. Invest.*, 50:814-817.
Maroni, B.J., Karapanos, G. and Mitch, W.E., 1986a, *Am. J. Physiol.*, 251:F81-F86.
Maroni, B.J., Karapanos, G. and Mitch, W.E., 1986b, *Am. J. Physiol.*, 251:F74-F80.
Maroni, B.J., Haesemeyer, R.W., Kutner, M.H. and Mitch, W.E., 1990, *Am. J. Physiol.*, 258:F1304-1310.
Millward, D.J., Davies, C.T.M., Halliday, D., Wolman, S.L., Matthews, D.E. and Rennie, M.J., 1982, *Fed. Proc.*, 41:2686-2691.
Millward, D.J., Bates, P.C., Brown, J.G., Halliday, D., Odedra, B., Emery, P.W., and Rennie, M.J., 1983, in: *Molecular Pathology of Nerve and Muscle: Noxious Agents and Genetic Lesions*, A.D. Kidman, J.K. Tomkins, C.A. Morris and N.A. Cooper, eds., Humana Press, New Jersey, pp. 315-342.
Millward, D.J. and Waterlow, J.C., 1978, *Fed. Proc.*, 37:2283-2290.
Morgan, H.E., Earl, D.C.N., Brodus, A., Wolpert, E.B., Giger, K.E. and Jefferson, L.S., 1971, *J. Biol. Chem.*, 246:2152-2162.
Mudge, G.H., Mills, R.M., Taegtmeyer, H., Gorlin, R. and Lesch, M., 1976, *J. Clin. Invest.*, 58:1185-1192.
Muhlbacher, F., Kapadia, C.R., Colpoys, M.F., Smith, R.J. and Wilmore, D.W., 1984, *Am. J. Physiol.*, 247:E75-E83.
Nallathambi, S.A., Goorin, A.M. and Adibi, S.A., 1972, *Am. J. Physiol.*, 223:13-19.
Narahara, H.T. and Holloszy, J.O., 1974, *J. Biol. Chem.*, 249:5435-5443.
Newsholme, E.A., Leighton, B., Challis, R.A.J. and Lozeman, F.J., 1986, *Biochem. J.*, 238:621.
Newsholme, E.A., Newsholme, P., Curi, R., Crabtree, B., and Ardawi, M.S.M., 1989, in: *Perspectives in Clinical Nutrition*, J.M.Kinney and P.R.Borum, eds., Urban & Schwarzenberg, Baltimore-Munich, pp. 71-98.
Newsholme, E.A. and Crabtree, B., 1981, *TIBS*, Feb:53-56.
Newsholme, E.A. and Parry-Billings, M., 1990, *J. Parent. Ent. Nutr.*, 14:63S-67S.
Noall, M.W., Riggs, T.R., Walker, L.M. and Christensen, H.N., 1957, *Science*, 126: 1002-1005.
Norman, A.W. and Litwack, G. *Hormones*, San Diego:Academic Press Inc, 1987.
Palmer, T.N., Caldecourt, M.A., Snell, K. and Sugden, M.C., 1985, *Biosci. Rep.*, 5:1015-1033.
Paris, S. and Ailhaud, G., 1980, *Biochim. Biophys. Acta*, 601:630-639.
Paxton, R., Scislowski, P.W.D., Davis, E.J. and Harris, R.A., 1986, *Biochem. J.*, 234:295-303.
Pogson, C.I., Low, S.Y., Knowles, R.G., Salter, M., and Rennie, M.J., 1990, in: *Regulation of Hepatic Function: J. Alfred Benzon Symposium 30*, B. Quistorff and N. Grunnett, eds., Munksgaard, Copenhagen, pp. 262-272.
Pozefsky, T., Felig, P., Tobin, J.D., Soeldner, J.S. and Cahill, G.F., 1969, *J. Clin. Invest.*, 48:2273-2282.
Rennie, M.J., Edwards, R.H.T., Krywawych, S., Davies, C.T.M., Halliday, D., Waterlow, J.C. and Millward, D.J., 1981, *Clin. Sci.*, 61:627-639.

Rennie, M.J., Idström, J.P., Mann, G.E., Schersten, T. and Bylund-Fellenius, A-C., 1983, *Biochem. J.*, 214:737-743.
Rennie, M.J., 1985, *Br. Med. Bull.*, 41:257-264.
Rennie, M.J., Hundal, H.S., Babij, P., MacLennan, P.A., Watt, P.W., Jepson, M.M. and Millward, D.J., 1986, *Lancet*, ii:1008-1012.
Rennie, M.J., MacLennan, P.A., Hundal, H.S., Weryk, B., Smith, K., Taylor, P.M. and Egan, C., 1989, *Metabolism*, 38:47-51.
Rennie, M.J., Ahmed, A., Low, S.Y., Hundal, H.S., Watt, P.W., MacLennan, P.A. and Egan, C.J., 1991, *Biochem. Soc. Trans.*, 18:1140-1141.
Rennie, M.J. and Harrison, R., 1984, *Lancet*, ii:323-325.
Riggs, T.R., Sanders, R.B. and Weindling, H.K., 1963, *Enocrinology*, 73:789-792.
Riggs, T.R. and McKirahan, K.J., 1973, *J. Biol. Chem.*, 248:6450-6455.
Riggs, T.R. and Walker, L.M., 1960, *J. Biol. Chem.*, 235:3603-3607.
Roth, D.A. and Brooks, G.A., 1990a, *Archiv. Biochem. Biophys.*, 279:377-385.
Roth, D.A. and Brooks, G.A., 1990b, *Archiv. Biochem. Biophys.*, 279:386-394.
Ruderman, N.B. and Lund, P., 1972, *Israel. J. Med. Sci.*, 8:295-302.
Saccomani, M.P., Cobelli, C., De Fronzo, R.A. and Bonadonna, R.C., 1990, *San Raffaele Proc. Milan*, 1:32P.
Salter, M., Knowles, R.G. and Pogson, C.I., 1986, *Biochem. J.*, 234:635-647.
Sarabia, V., Ramlal, T. and Klip, A., 1990, *Biochem. Cell Biol.*, 68:536-542.
Scharff, R. and Wool, I.G., 1965, *Biochem. J.*, 97:257-271.
Schauf, C.L., Moffett, D.F. and Moffett, S., 1990, *Human Physiology: Foundations and Frontiers*, C.V.Mosby, St. Louis.
Schubert, D., Harris, A.J., Devine, C.E. and Heinemann, S., 1974, *J. Cell Biol.*, 61:309-413.
Schwartz, R.G., Barrett, E.J., Francis, C.K., Jacob, R. and Zaret, B.L., 1985, *J. Clin. Invest.*, 75:1204-1211.
Scislowski, P.W.D., Hokland, B.M., Davis-van Thienen, W.I.A., Bremer, J. and Davis, E.J., 1987, *Biochem. J.*, 247:35-40.
Scislowski, P.W.D., Niblock, A., Lindsay, Y., Weryk, B., Watt, P.W. and Rennie, M.J., 1989, *Clin. Nutr.*, 8:P80.
Selberg, O., Burchert, W., Müller, M.J., Meyer, G.J. and Hundeshagen, H., 1991, *Clin. Nutr.*, 10(Suppl):P52.
Smith, K. and Rennie, M.J., 1990, in: *Muscle Metabolism, Balliere's Clinical Endocrinology and Metabolism, Vol 3, No 3*, J.B. Harris and D.M. Turnbull, eds., Balliere Tindall, London, pp. 461-499.
Smith, R.H., Palmer, R.M. and Reeds, P.J., 1983, *Biochem. J.*, 214:153-161.
Smith, R.J., Larson, S., Stred, S.E. and Durschlag, R.P, 1984, *J. Cell. Physiol.*, 120:197-203.
Snell, K. and Duff, D.A., 1981, in: *Metabolism and Clinical Implications of Branched Chain Amino and Ketoacids*, M. Walser and J.R. Williamson, eds., Elsevier/North Holland, New York, pp. 251-256.
Spydevold, O., 1979, *Eur. J. Biochem.*, 97:389-394.
Spydevold, O. and Hokland, B., 1983, *Int. J. Biochem.*, 15:985-990.
Stirewalt, W.S., Low, R.B. and Slaiby, J.M., 1985, *Biochem. J.*, 227:355-362.
Suleiman, M.S. and Chapman, R.A., 1990, *J. Physiol.*, 430:72P.
Tadros, L. and Rennie, M.J., 1991, *J. Physiol.*, in press.
Thompson, G.N., Pacy, P.J., Merritt, H., Ford, G.C., Read, M.A. and Halliday, D., 1989, *Am. J. Physiol.*, 256:631-639.
Tischler, M.E., Desautels, M. and Goldberg, A.L., 1982, *J. Biol. Chem.*, 257:1613-1621.
Tovar, A.R., Tews, J.K., Torres, N. and Harper, A.E., 1991, *Metabolism*, 40:410-419.
Van Slyke, D.D. and Meyer, G.M., 1913, *J. Biol. Chem.*, 16:197-212.
Vesterberg, K., Vinnars, E., Leander, U. and Fürst, P., 1987, *Clin. Nutr.*, 6:213-219.
Vinnars, E., Bergström, J. and Fürst, P., 1975, *Ann. Surg.*, 182:665-671.
Wagenmakers, A.J.M., Brookes, J.H., Coakley, J.H., Reilly, T. and Edwards, R.H.T., 1989, *Eur. J. Appl. Phys.*, 59:159-167.
Wagenmakers, A.J.M., Coakley, J.H. and Edwards, R.H.T., 1990, *Int. J. Sports Med*, 11:S101-S113.
Wahren, J., von Euler, C. and Eriksson, S., 1988, *Clin. Nutr.*, 7:203-206.
Wardzala, L.J. and Jeanrenaud, B., 1981, *J. Biol. Chem.*, 256:7090-7093.
Warner, B.W., Hasselgren, P., James, J.H., Bialkowska, H., Rigel, D.F., Ogle, C. and Fischer, J.E., 1987, *Arch. Surg.*, 122:1142-1146.

Warner, B.W., Hummel, R.P., Hasselgren, P.-O., James, J.H. and Fischer, J.E., 1989, *J. Par. Ent. Nutr.*, 13:344-348.
Warner, B.W., Hasselgren, P.-O., Hummel, R.P., James, J.H., Pedersen, P. and Fischer, J.E., 1990a, *Am. J. Surg.*, 159:295-300.
Warner, B.W., Hasselgren, P.-O., James, J.H., Hummel, R.P., Rigel, D.F. and Fischer, J.E., 1990b, *Ann. Surg.*, 211:323-328.
Waterlow, J.C., Garlick, P.J. and Millward, D.J. *Protein Turnover in Mammalian Tissues and in the Whole Body*, Elsevier/North Holland, Amsterdam, 1978.
Waterlow, J.C. and Fern, E.B., 1981, in: *Nitrogen Metabolism in Man*, Applied Science Publishers, England, pp. 1-16.
Waterlow, J.C. and Millward, D.J., 1989, in: *Energy Transformations in Cells and Organisms*, W. Wieser and E. Gnaiger, eds., Georg Thieme Verlag, Stuttgart, pp. 277-282.
Watt, P.W., Stenhouse, M.G., Corbett, M.E. and Rennie, M.J., 1989, *Clin. Nutr.*, 8(Suppl):47.
Watt, P.W., Lindsay, Y., Scrimgeour, C.M., Chien, P.A.F., Gibson, J.N.A., Taylor, D.J. and Rennie, M.J., 1991, *Proc. Nat. Acad. Sci. USA*, 88:5892-5896.
Welty, M.C., Welty, J.D. and McBroom, M.J., 1982, *J. Mol. Cell. Cardiol.*, 14:353-357.
Willhoft, N.M. and Rennie, M.J., 1991, *J. Physiol.*, in press.
Wool, I.G., Castles, J.J. and Moyer, A.N., 1965, *Biochim. Biophys. Acta*, 107:333-345.
Wu, G., Thompson, J.R. and Baracos, V.E., 1991, *Biochem. J.*, 274:769-774.
Yoshida, S., Lanza-Jacoby, S. and Stein, T.P., 1991, *Biochem. J.*, 276:405-409.
Young, L.H., Zaret, B.L. and Barrett, E.J., 1987, *Am. J. Physiol.*, 252:H1070-H1076.
Yudilevich, D.L. and Mann, G.E., 1982, *Fed. Proc.*, 41:3045-3053.
Zorzano, A., Balon, T.W., Garetto, L.P., Goodman, M.N. and Ruderman, N.B., 1985, *Am. J. Physiol.*, 248:E546-E552.

AMINO ACID TRANSPORT IN THE KIDNEY

J.S. Schwegler[1], S. Silbernagl[1], B.K. Tamarappoo[2] and T.C. Welbourne[2]

[1]Department of Physiology
University of Würzburg
D-8700 Würzburg, Germany

[2]Department of Physiology and Biophysics
School of Medicine in Shreveport
Louisiana State University
Shreveport, LA 71130-3932, USA

FILTRATION, REABSORPTION, AND EXCRETION

Amino acids are valuable substrates for the organism that can be utilized in a variety of ways. They are used for protein synthesis, gluconeogenesis, lipogenesis, ATP production, as neurotransmitters and hormones, for detoxification reactions, etc. Thus, it seems reasonable that renal reabsorption processes prevent the urinary loss of these compounds. As hypothesized by Ludwig in the mid-19th century (Ludwig, 1844), urine is produced by ultrafiltration from the blood plasma. As amino acids are freely filtered by the glomerulus (*vide infra*), amino acid conservation is achieved by tubular reabsorption processes. This chapter focuses on the nature of these reabsorptive mechanisms in the kidney.

1.1 Glomerular Filtration

The relative molecular mass of naturally occurring amino acids ranges between 75 (glycine) and 241 (cystine). Therefore, the glomerular basal membrane does not constitute any significant filtration barrier for these compounds, regardless of their electrical charge. Thus, the amino acid concentration in the luminal fluid at the very beginning of the proximal tubule is identical to that in the arrterial plasma. As the tubular load of any substance that is filtered by the glomerulus is the product of its concentration in the ultrafiltrate and the glomerular filtration rate (GFR), the tubular amino acid load strongly depends on plasma concentration and GFR. The plasma concentration of amino acids ranges from 2 (aspartate) up to more than 330 µmol/l (alanine) in man (Cusworth and Dent, 1960), yielding a total amino acid concentration of roughly 2.5 mmol/l and a tubular load of about 435 mmol/day (Silbernagl, 1988). Similar concentrations have been measured in dog (Wright and Nicholson, 1966), rat (Lingard et al., 1974) and rabbit (Block and Hubbard, 1962), the traditional experimental animals. Plasma amino acid concentrations are, however, subject to large variations depending on dietary amino acid intake. For example, after an ordinary meat meal amino acid levels rise more than two-fold (DeSanto et al., 1991), this increase being in itself a stimulator for GFR increase (see section 1.2). Hence, after a protein meal the tubular amino acid load increases to a greater extent than plasma levels. Compared with amino acids, the filtered load of oligopeptides (i.e., peptides that are small enough to be freely filtered by the glomerular

Mammalian Amino Acid Transport, Edited by M.S. Kilberg and D. Häussinger, Plenum Press, New York, 1992

basal membrane) is fairly low. Of this group, the quantitatively most important compound, glutathione, occurs at plasma concentrations in the low micromolar range in man (Wendel and Cikryt, 1980) and rat (Häberle et al., 1979). However, the erythrocytes contain intracellular glutathione well in the millimolar range (Wendel and Cikryt, 1980). Therefore, any (even subclinical) hemolysis leads to a massive increase of glutathione in the tubular fluid.

1.2 Influences of Amino Acids and Proteins on GFR

As first demonstrated by Pitts (1943), large systemic infusion of a single amino acid, glycine, or combinations of amino acids, i.e., glycine, proline, alanine, and serine, results in an elevated GFR. The hyperfiltration of amino acids, i.e., glycine, might enhance solute reabsorption in the thick ascending limb thus reducing distal solute loads to the macula densa (Woods et al., 1988). While confirming an increase in proximal reabsorption by direct methods, Brown and Navar (1990) were unable to implicate the tubuloglomerular feedback response. Viewed from the perspective of extrarenal signals, systemic amino acid infusions elicit endocrine responses capable of profoundly effecting renal function. Glucagon levels increase under these conditions and systemic infusion of glucagon results in a 40 percent rise in GFR (Alvcstrand and Bergstrom, 1984). However, direct renal artery infusions do not increase GFR (Premen, 1985), while lower doses infused via the portal vein do (Premen, 1985) suggesting an important role for the liver. Indeed, evidence exists for a complex hepatic role in modulating renal perfusion following a high protein meal. Lang et al. (1991) demonstrated that intraportal glutamine infusions reduce GFR nearly 50 percent, ostensibly reflecting hepatic cell swelling attributed to glutamine accumulation; glucagon reverses swelling apparently by accelerating glutamine metabolism (Staddon et al., 1984), suggesting that modulation of this reflex could affect GFR. The liver also produces somatomedin C, SM-C, a potent "glomerulopressin" (Uranga, 1979) capable of acutely increasing GFR (Hirschberg and Kopple, 1989). In fact, a high protein diet elevated plasma and glomeruli SM-C levels associated with a 40 percent increase in GFR (Hirschberg and Kopple, 1991); in this context, the signal for hepatic SM-C release is of considerable interest (Kachra et al., 1991). Nevertheless, factored into the extrarenal regulatory view must be the observation that the GFR response to large systemic amino acid loads is observed even after excluding the liver from the circulation (Woods et al., 1988). Finally, portal vein peptide loads, i.e., glycylglycine, rise after a high protein meal (Christensen, 1949), constituting a prepackage amino acid load targeted for distal segments.

1.3 Quantitative Net Reabsorption at Normal and Elevated Amino Acid Loads

Normally, about 99% of the filtered amino acid load is reabsorbed during the tubular passage (Cusworth and Dent, 1960). The excreted 1% mainly consists of the non-essential amino acids taurine, 1-methylhistidine, histidine, and glycine (Cusworth and Dent, 1960; Tizianello et al., 1980; Bomsel et al., 1989). The fractional excretion of these four amino acids are 6-9%, 14%, 6%, and 3.5%, respectively (Cusworth and Dent, 1960; Tizianello et al., 1980). In contrast, essential amino acids (especially the branched-chain amino acids valine, leucine, and isoleucine) are reabsorbed to well over 99.5%. The body homeostasis of taurine and the β-amino acids is primarily maintained by renal excretion and not, like that of all α-amino acids, by metabolism (Friedman et al., 1981; Chesney et al., 1982; Chesney et al., 1987). Consequently, renal handling regulates β-amino acids and conserves the α forms. This extremely efficient reabsorption process is a differentiated feature of the mature nephron. In infancy, α-amino acid excretion is markedly higher than in adults (Sereni et al., 1955; Brodehl and Gelissen, 1968a), suggesting a gradual development of the responsible tubular transport systems during childhood. Under physiological conditions, any elevation of the tubular load of amino acids by alterations of GFR and/or amino acid plasma concentrations is fully compensated by the kidney, i.e., the fractional excretion of most amino acids is constant (Silbernagl, 1988). This is possible because more than 90% of the filtered load of most amino acids is reabsorbed in the first 2 millimeters of the nephron. An elevation of the intratubular flow rate due to GFR increase carries amino acids to the later parts of the proximal tubule that are normally not involved with reabsorption. These tubule segments have, nevertheless, retained the capacity for active amino acid transport and provide a reserve capacity preventing excretory loss (Silbernagl, 1980). The same situation develops when the amino acid concentration in the glomerular filtrate is high enough to saturate the transport systems at the proximal beginning of the nephron. Exceptions are taurine, histidine, and glycine. Their excretion rates rise when the tubular load is increased (Kopple and Swendseid, 1981; Silbernagl and Deetjen, 1971). On the single nephron level, these amino acids can be shown to have a relatively "flat" reabsorption curve, i.e., reabsorption occurs normally in all portions of the proximal nephron, so that there is no reserve capacity left for compensating higher amino acid loads (Eisenbach et al., 1975).

1.4 The Clearance Concept and Its Problems

The determination of clearances for exo- and endogenous substances has been the predominant tool in classic renal research, and amino acid clearances (C_{AA}) have been determined already in the 1930s (Kirk, 1936):

$$C_{AA} = \frac{U_{AA} \cdot (dV_U/dt)}{P_{AA}} \qquad (1)$$

with U_{AA} = amino acid concentration in the final urine; dV_U/dt = urine volume per time t; P_{AA} = plasma amino acid concentration.

As amino acids are freely filtered at the glomerulus, it is necessary to compare the amino acid clearance with the glomerular filtration rate in order to assess tubular functions. GFR is accurately measured as inulin clearance (C_{In}). The fractional urinary excretion of amino acids (FE_{AA}) is a parameter for the primary way amino acids are handled by the tubule.

$$FE_{AA} = C_{AA}/C_{In} \qquad (2)$$

If the FE_{AA} is smaller than unity (for most amino acids < 0.01), it indicates net reabsorption. This assumption, however, holds true only if net fluxes are considered. The concept is certainly questionable if the respective substance is metabolized in the tubular lumen. For example, cystine and cysteine are readily interconverted, depending on the ambient redox potential (Crawhall and Segal, 1967; Helbig et al., 1984; Völkl et al., 1984). Moreover, renal brush border membrane exo-enzymes like γ-glutamyltransferase are responsible for generating amino acid interconversion in the tubular lumen. In the presence of the endogenous substrate hippurate, glutamine is readily deamidated to glutamate (Silbernagl, 1986). The same enzyme is capable of cleaving the γ-glutamyl moiety of intraluminal peptides like glutathione (Silbernagl, 1977). Amino acids are not only reabsorbed, but also secreted into the tubular lumen. In the loop of Henle a rather intriguing recycling mechanism for neutral and anionic amino acids has been demonstrated (Dantzler and Silbernagl, 1988; Dantzler and Silbernagl, 1990). These amino acids are both reabsorbed and secreted during their passage from the proximal to the distal convolute. Similarly, the tripeptide glutathione is secreted into the proximal tubule lumen before its degradation (Griffith and Meister, 1979; Scott and Curthoys, 1987). Thus, data from amino acid clearance experiments reflect net fluxes and not necessarily unidirectional reabsorptive fluxes.

Another classical method of studying excretion and reabsorption processes by clearance experiments is the titration technique, first applied to glucose (Shannon et al., 1941). Plasma concentrations are elevated step by step by intravenous infusion and the respective urinary excretion is measured. Typically, excretion stays low at moderately elevated plasma concentration unless the tubular transport is exhausted, then it rises linearly with the plasma load. The difference of solute delivery by the plasma (plasma concentration times renal plasma flow) and the respective urinary load is equal to the net tubular reabsorption. Maximum net reabsorption (T_m) is substrate specific. In the 1940s a variety of amino acids were tested for their T_m in dogs (Pitts, 1943; Pitts, 1944; Wright et al., 1947). Saturation could be demonstrated for glycine, lysine and arginine, but not for other amino acids, probably because plasma concentrations could not be elevated enough to saturate transport (Silbernagl, 1988). From the shape of the titration curve, kinetic parameters like apparent K_m and J_{max} can be calculated (Sugita et al., 1967). Unfortunately, the kinetic parameters derived from whole kidney experiments do not always correspond to data from tubule microperfusion. In the rat, K_m for aspartate reabsorption has been found to be 50 times higher by clearance experiments than by continuous microperfusion of the proximal tubule (Linder and Foulkes, 1985; Silbernagl, 1983).

Despite the unquestioned value of clearance tests in the diagnosis of kidney malfunction, results from whole kidney experiments must be carefully evaluated when drawing conclusions about the underlying reabsorptive processes. This is especially true for hereditary tubular transport defects (see section 6.3).

EXTRACTION OF AMINO ACIDS

2.1 Physiological Plasma Loads

The human kidney receives some 1300 µmoles of amino acids per minute delivered in plasma, filters some 300 $\mu mol \cdot min^{-1}$, yet on balance exhibits no net removal (Fig. 1). Nevertheless, the kidneys are a major site of amino acid metabolism serving both their own nutritive requirements as well as interorgan demands. The profile of arteriovenous plasma concentration differences reflects this dual role (Fig. 1A); prominent are glutamine, glycine and citrulline uptake and corresponding alanine, serine, and arginine release (Shaloub et al., 1963; Squires et al., 1976; Heitmann and Bergmann, 1980; Tizianello et al., 1980). Glutamine extraction meets both local energy demands (Halperin et al., 1982) or may function in an interorgan mode by supporting alanine and serine release (Pitts et al., 1970) bound for hepatic gluconeogenesis. A similar flexibility appears for extracted glycine, on the one hand, supporting ammoniagenesis or, on the other, interorgan serine flux (Brosnan, 1986). In contrast, citrulline extraction is unidimensional, coupled one for one to arginine release, with the kidneys being a major source of this important amino acid (Featherstone et al., 1973). Essential amino acids, in contrast to non-essential, are neither extracted in net amounts nor undergo interconversions under post-absorptive conditions, handling being a mere reflection of conservation.

Amino acid fluxes in the form of peptides may converge at or even emerge from the kidneys and these, of course, are not normally monitored. Indeed, the kidneys play a major role in interorgan glutathione flow, both through plasma uptake (Meister and Anderson, 1983) and red cell release (Dass et al., 1991), yet little is known of their quantitative, much less functional, importance. It is known that, on balance, the sheep kidney (Heitmann and Bergmann, 1980) and kidney of humans with chronic renal disease (Tizianello et al., 1980) exhibit net amino acid release, unaccounted for by amino acid uptake. In ruminants a significant interorgan peptide flux has been demonstrated (McCormick and Webb, 1982) and in man peptide as well as protein catabolism have been suspected (Tizianello et al., 1980). It remains to be seen at what point such cryptic flows impact upon amino acid handling and their quantitative and functional significance.

2.2 Physiological Red Cell Loads

Amino acids delivered to the kidneys within red cells are not normally removed in net amounts (Shaloub et al., 1963; Squires et al., 1976; Heitmann and Bergmann, 1980). Extraction of glutamine, glycine, and citrulline are generally, but not exclusively, limited to removal from the plasma (Fig. 1A vs. 1B), most probably reflecting exposure to filtration and subsequent handling. In contrast, release of amino acids occurs preferentially into red cells (Fig. 1A and 1B). As pointed out, such fluxes are anomalous given the slow amino acid penetration *in vitro* (Winter and Christensen, 1964), and the absence of Na^+-gradient driven transport systems in many species. One proposal for dealing with this paradox is interaction between capillary surface and the red cell in facilitating amino acid penetration (Elwyn, 1966), possibly as envisioned in amino acid movement across cell membranes as γ-glutamylpeptides (Bridges and Meister, 1985; Kalra

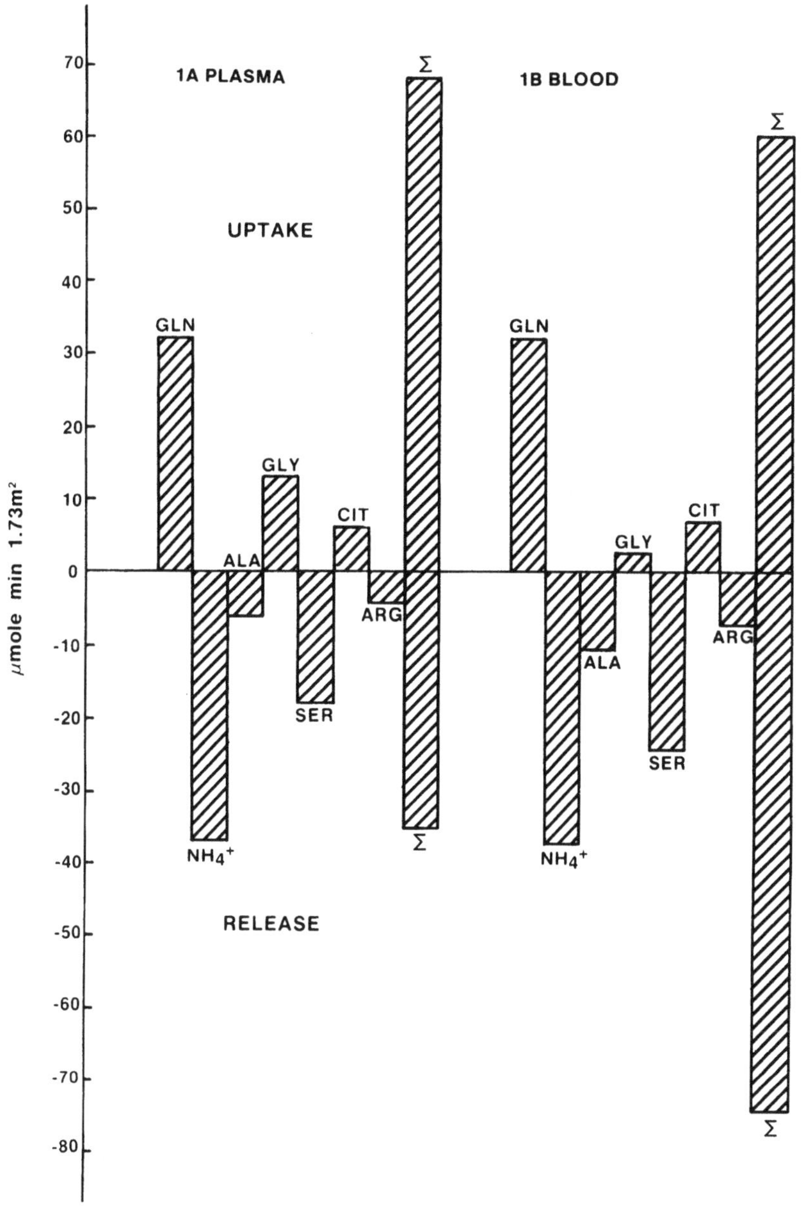

Fig. 1. Net uptake or release of plasma (1A) and whole blood (1B) amino acids by human kidney. Sum (Σ) includes contribution from all 21 amino acids (adapted from Tizianello et al., 1978).

et al., 1981). If such peptides were synthesized within the renal microvasculature by γ-glutamyltranspeptidase, GGTP (Dass and Welbourne, 1980; Dass et al., 1982), and uncoupled by red cell cyclotransferase, the result indeed would be accumulation of the free amino acid and 5-oxoproline (Bridges and Meister, 1985). Note that GGTP is a pluripotent enzyme concentrated in brush borders (Mamelok et al., 1980) but with significant concentrations in loop segments (Albert et al., 1961) and the microvasculature providing multiple sites for peptide synthesis as well as hydrolysis (Dass and Welbourne, 1980). Furthermore, proximal secretion of glutathione (Scott and Curthoys, 1987), and/or back diffusion of glutamine (Lingard et al., 1973), plus NaCl activation of GGTP (Orlowski et al., 1973), make the loop segment an attractive site for peptide formation;

urinary γ-glutamylpeptides of valine, leucine, and isoleucine (Buchanan et al., 1962) and 5-oxoproline (Nakanishi et al., 1991) may reflect this activity. Intravascular red cell peptide accumulation would, of course, be limited by capillary transit time and by the rate of conversion to 5-oxoproline; dissipation of accumulated amino acids entails metabolic conversions, i.e., glutamate, cysteine, and glycine into glutathione, or slow systemic efflux. Interestingly, balance measurements across the kidney have rarely included glutathione and never to our knowledge 5-oxoproline, a little recognized amino acid whose quantitative role is unlikely to reflect its possible functional significance (Vina et al., 1989).

2.3 Elevated Loads

Dietary amino acid loads are normally processed by the liver prior to reaching the kidneys. Consequently, their size and composition are greatly modified before reaching the kidneys as seen in the carefully designed experiments in man (Tizianello et al., 1985), dog (Fukuda and Kopple, 1980) and rat (Yamamoto et al., 1974). In these studies an amino acid load simulating a high protein meal was deployed resulting in a two-fold elevation in the arterial load, due predominantly to increased essential amino acids; note that the glutamine load was little changed despite being a prominent component of the administered load. Curiously, neither GFR nor renal blood flow increased, perhaps reflecting difference in loads and counterbalancing mechanisms (section 1.2). In man, the pre-existing amino acid balance across the kidney swings to a large net uptake constituting about 10 percent of the arterial load and is made up mainly, but not entirely, of those essential amino acids whose arterial levels increased; amino acid excretion was unchanged with the exception of methionine (section 1.3). Nor did glutamine extraction increase despite a two-fold rise in ammonia production, suggesting either more efficient utilization of glutamine or contribution from other amino acids; in favor of the former, alanine release fell, consistent with a shift to glutamine oxidation as a fuel perhaps occasioned by a greater proximal Na^+ reabsorption (Na^+ amino acid cotransport, section 4.2). On the other hand, glycine uptake increases without a corresponding release of serine, consistent with its ammoniagenic precursor role. Finally, branched chain amino acids (BCAA) might also contribute to the ammonia produced; indeed, in the perfused rat kidney high leucine and isoleucine loads enhance ammonia excretion and glutamine production consistent with oxidation in the thick ascending limb of Henle's loop and both ammonia excretion and transfer to the apposed proximal straight segment (Burch et al., 1985). In man BCAA might also increase NH_4^+ excretion were they to reach distal segments and be oxidized. Overall, the renal responses to elevated amino acid loads are increased extraction of essential amino acids, as well as glycine and proline, and enhanced concentrating and acidifying capabilities.

2.4 Interaction of Amino Acids

Viewed from the perspective presented in section 4, disrupting the flow of energy into the Na^+ gradient should expose the downstream sites to unaccustomed large loads of filtered amino acids; this would become even more prominent if proton gradients were also abolished. Indeed, maleate appears to effect just this in mimicking Fanconi's syndrome (section 6.2), a condition resulting in aminoaciduria and an alkaline urine (Berliner et al., 1950; Günther et al., 1979). As depicted in Fig. 2, a large 10-fold increase in amino acid excretion reflects an enormous glutamate excretion, 30-fold increase, contrasting to glutamine excretion, five-fold increase, and most interestingly a nearly two-fold rise in peptide excretion; note that amino acids excreted in the peptide form, predominately essential amino acids plus glutamate and glycine, normally exceed excretion in the free form by five-fold (Asatoor et al., 1978). These peptides are likely the results of interaction of filtered glutamine and/or secreted GSH with luminal GGTP, activated by maleate (Tate and Meister, 1974), forming glutamate and γ-glutamylpeptides; moreover, these would probably be γ-glutamylpeptides of glutamine, leucine, isoleucine, and cystine. Interestingly, patients with Wilson's disease (Fanconi's syndrome, section 6.2) excrete a large amount of amino acids in the form of glutamate-containing peptides similar to this experimental model (Asatoor et al., 1976). Other means of eliciting amino acid interactions involves infusing loads of a competing amino acid; for example, histidine competes with glutamine for proximal reabsorption but has little

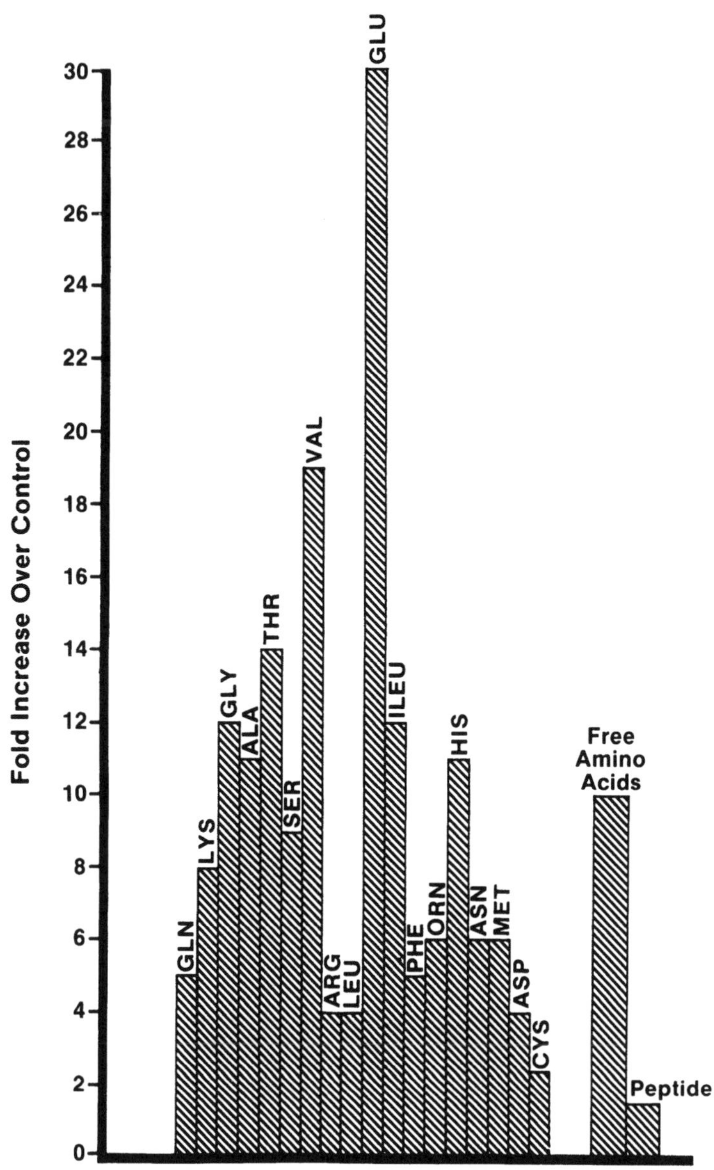

Fig. 2. Increase in 24 hour urinary free amino acids and peptide bound amino acids relative to control levels following maleate administration, 3.5 mmol/kg body weight, to starved rats (adapted from Asatoor et al., 1978).

effect on neutral amino acid excretion, rather elicits aspartate and an especially massive excretion of glutamate (Kamin and Handler, 1951). Since neither histidine nor glutamine interact with the acidic amino acid transporter, the most likely interpretation would be intralumimal conversion to glutamate with the subsequent saturation of the high affinity but low capacity acidic amino acid transporter (Silbernagl, 1983). Noteworthy, in subjects unable to express the dicarboxylic amino acid carrier, glutamate's clearance rate exceeds those for both aspartate and creatinine (Melancon et al., 1977; Teijima et al., 1974). Because glutamine is the largest component of the amino acid load handled by the proximal transport system, competition for, or impairment of the Na^+ gradient, or displacement by other neutral amino acids, or even the transporter's defect (Hartnup disease, section 4.3) results in intraluminal glutamine hydrolysis and glutamate generation (Silbernagl, 1986), offering valuable insights into tubular interactions and perhaps distal function as well. Noteworthy, antiluminal interactions encompass a very heterogeneous system including basolateral membranes of both proximal and distal tubules (site of highest glutaminase activity, Curthoys and Lowry, 1973) as well as the metabolically active renal microvasculature. The presence of GGTP in the antiluminal region and its role in glutamine (Welbourne and Dass, 1988) and glutathione metabolism (Abbott et al., 1984), and γ-glutamyl moiety interconversion (Dass et al., 1988), as well as modulation by PAH, etc. may have relevance for glutamine utilization and ammonium and glutamate release as well as γ-glutamylpeptide uptake (Lash and Jones, 1984). As pointed out (Silbernagl, 1986), the routine deployment of PAH in determining blood flow while monitoring amino acid handling (i.e., glutamine) may influence that extraction rate as well as interactive ones. Basolateral glutamine uptake could also be coupled to alanine efflux, analogous to the Na^+ independent system L, since some 30 percent of BLM vesicle glutamine uptake is Na^+ independent (Windus et al., 1987). Since alanine formation is a late proximal tubule event (Chan et al., 1979), both its production and efflux might be glutamine load dependent.

LOCALIZATION OF AMINO ACID REABSORPTION ALONG THE NEPHRON

3.1 The Proximal Tubule

In the early 1960s, stop-flow experiments revealed that the principal location of amino acid reabsorption is the proximal nephron (Brown et al., 1961; Gayer and Gerok, 1961). Conclusive evidence for proximal tubular amino acid reabsorption came from free flow micropuncture experiments of superficial nephrons *in vivo*. The intraluminal concentration of many amino acids falls below 10% of the plasma level (i.e., about 5% of the filtered load) after passing the first 2 mm of the proximal tubule (see Fig. 3).

Morphologically the proximal tubule consists of a convoluted part (PCT:S_1 and S_2-segments) and a straight part (PST:S_3-segment). There is general agreement that the tubular transport capacity (i.e., the maximum transepithelial flux per millimeter, J_{max}) decreases from the convoluted to the straight part. Based on a leak diffusion model, (section 4.1), a decrease of J_{max} from PCT to PST has been shown, e.g., for glycine (Barfuss and Schafer, 1979), cycloleucine (Lingard et al., 1975), histidine (Lingard et al., 1973), and proline (Roigaard-Petersen and Sheikh, 1984). The capacity of other transport systems like the reabsorption of filtered bicarbonate by luminal Na^+/H^+ exchange (Kragh-Hansen et al., 1985) also decrease from S_1 to S_3. Interestingly, the affinity of amino acids for their transporters increases from PCT to PST (Barfuss and Schafer, 1979; Roigaard-Petersen and Sheikh, 1984). This axial heterogeneity in the proximal tubule makes good sense from a teleological point of view: the early convoluted part of the proximal tubule manages bulk reabsorption while the straight segment mops up.

Amino acid reabsorption can also be considered by focussing on the concomitant sodium fluxes. Solute coupled sodium cotransport is generally believed to be the primary event in water reabsorption in the proximal renal tubule. An interesting aspect to the function of the proximal straight tubule originates from the finding that the apical membrane of PST cells (not PCT cells) possesses sodium conductive channels (Gögelein and Greger, 1986; Jacobsen et al., 1988), but relatively few sodium dependent transporters (Kragh-Hansen et al., 1985; Roigaard-Petersen and Sheikh, 1984). Thus, transepithelial sodium and water fluxes could be driven in the PCT primarily by sodium coupled solute reabsorption, in the PST primarily by solute independent channel-mediated sodium flux. This hypothesis, however, is still rather speculative and requires further thorough investigation.

3.2 Distal Tubule Segments

Micropuncture experiments of superficial rat nephrons have shown that net absorption of an amino acid load does not occur between the last accessible part of the proximal convoluted tubule ("late proximal") and the first accessible part of the distal convolute ("early distal") (Eisenbach et al., 1975; Günther et al., 1979). This is somewhat surprising because the perfused tubule section includes the S_3 segment of the proximal

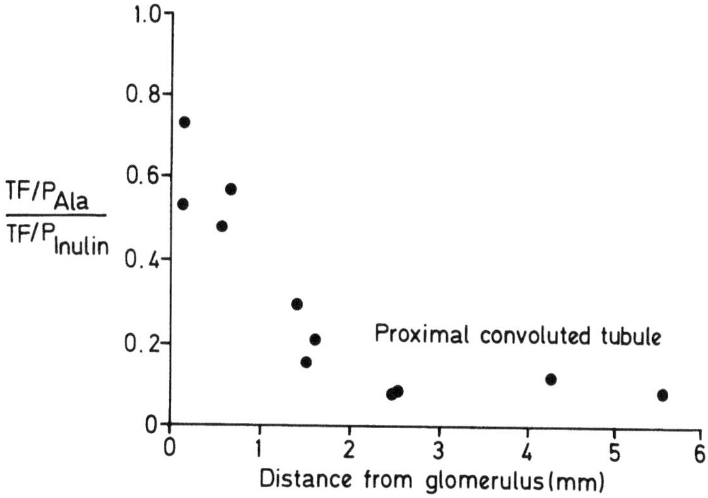

Fig. 3. Fractional recovery of alanine along the proximal convoluted tubule of rat kidney under endogenous free flow conditions (Alanine plasma concentration 400-600 μmol/l) (S. Silbernagl, unpublished results).

tubule that is capable of high-affinity amino acid reabsorption. Therefore, it must be concluded that there is a back leak in the loop of Henle that allows amino acids to move from the peritubular capillary back into the lumen. This notion is further supported by the fact that microinjected radioactively labelled amino acids disappear from the tubular fluid between the late proximal and the early distal puncture site (Bergeron and Morel, 1969). Unidirectional reabsorption is highly stereospecific at least for L-glutamate (Silbernagl, 1983). The amino acids that are found in the lumen after the tubular fluid has passed the loop of Henle are hence not identical to those that have entered the loop.

When radioactively labelled amino acids are microinfused into ascending vasa recta, they appear in the final urine only of the ipsilateral kidney, showing the presence of a direct secretory pathway in the ascending loop of Henle or the collecting duct (Dantzler and Silbernagl, 1990). This is confirmed by the finding that amino acid movement from blood to lumen is highly stereospecific, favoring the L-isomer (Dantzler and Silbernagl, 1991). Earlier microinjection (Bergeron and Morel, 1969) and microinfusion experiments (Günther et al., 1979) show that there is no measurable amino acid flux across the wall of the collecting duct, suggesting amino acid secretion in the ascending part of the loop of Henle. Whether and to what extent this amino acid recycling contributes to the maintenance of the osmotic gradient between the renal cortex and the medulla still awaits to be evaluated.

MECHANISMS AND SPECIFICITY OF TUBULAR AMINO ACID REABSORPTION

4.1 The Pump-Leak Concept: Paracellular Leak *vs.* Transcellular Transport

The intraluminal amino acid concentration in the proximal tubule steadily declines and finally reaches values way below the respective plasma concentration (Silbernagl, 1988). However, this process does not continue until the luminal concentration reaches zero, but halts at a certain minimal luminal concentration (Barfuss and Schafer, 1979; Lingard et al., 1973; Schafer and Barfuss, 1980). Bearing in mind that in Henle's loop there is a significant bidirectional amino acid flux into and out of the tubule lumen (see section 3.2), there is no doubt that the minimal luminal concentration is caused by some kind of "back leak". Under steady state conditions this can be expressed mathematically as an identity of reabsorptive flux, J_{out}, and back leak, J_{in} (Schafer and Barfuss, 1980):

$$J_{out} = J_{in} \qquad (3)$$

As J_{out} can be described by Michaelis-Menten kinetics for a carrier mediated transcellular flux through the epithelium, it is defined by the maximum flux J_{max} and the half-saturation concentration K_m:

$$J_{out} = \frac{J_{max} \cdot C_{lu}}{K_m + C_{lu}} \tag{4}$$

with C_{lu} = luminal concentration of a given amino acid. J_{in} is usually looked upon as a simple first order diffusion process and thus depends solely on the concentration difference between the peritubular (C_{pt}) and the luminal (C_{lu}) amino acid concentration and on the permeability coefficient (P):

$$J_{in} = P \cdot (C_{pt} - C_{lu}) \tag{5}$$

Equations (4) and (5) hold true for equilibrium conditions and for any given flux at particular tubule segments, yielding the net reabsorptive flux (J_{net}) as the difference of J_{out} and J_{in}:

$$J_{net} = J_{out} - J_{in} \tag{6}$$

The intraluminal concentration is the initial concentration of the glomerular filtrate, i.e., the plasma or peritubular concentration (C_{pt}) minus the integrated net flux:

$$C_{lu} = C_{pt} - \int_0^{t^*} J_{net} dt \tag{7}$$

with t* = the contact time of the tubular fluid with the reabsorptive brush border membranes.

Based on this model, attempts have been made to quantitatively describe the reabsorptive process for amino acids (for reviews see Schafer and Barfuss, 1980; Silbernagl, 1988). Comparison of the calculated concentration profile for histidine with experimental micropuncture data (Günther and Silbernagl, 1981) show that in the early proximal convoluted tubule, back leak (J_{in}) does not reduce net reabsorption to any significant extent. The lower the intratubular concentration (in S_2 and S_3), the greater is the influence of back leak permeability P.

However convincing this so-called "pump-leak concept" may be, it does not reveal the mechanisms responsible for this back leak permeability, P. As most amino acids are highly polar molecules, physicochemical diffusion across the lipid bilayer membrane is too slow to account for the observed back leak (Klein et al., 1971). Thus, either there is a high paracellular shunt permeability or back leak occurs carrier mediated via intact cells. The low electrical resistance of the proximal tubular epithelium (Frömter, 1982) has generally been considered a sign for relatively free diffusibility of all sorts of (small) molecules. In cell culture, however, proximal tubular cells are able to maintain an osmotic and hydrostatic gradient despite virtually absent transepithelial resistance (Schwegler et al., 1989b). Also, blood-to-lumen flux of a neutral amino acid like citrulline occurs exclusively via stereospecific transcellular transport in the frog (Gekle and Silbernagl, 1991). Thus, the biochemical and morphological background for leak permeability in the proximal tubule is still not clear.

4.2 Amino Acid Entry Mechanisms across the Luminal Brush Border

Amino acid reabsorption involves transepithelial transport. Thus, there are two major transport barriers: the luminal (apical) brush border and the basolateral cell membrane. In the 1970s, it became feasible to study the transport of radioactively labelled amino acids across the membrane of isolated brush border membrane vesicles (BBMV) of proximal tubular origin (Evers et al., 1976; Fass et al., 1977). This technique has given great impetus to the investigation of amino acid membrane transport on a cellular and molecular level.

4.2.1 Luminal transport of neutral amino acids

Neutral amino acids (i.e., such compounds bearing no electrical net charge at neutral pH: AA^0) have been in the focus of interest because (1) they constitute the greatest amount of filtered amino acids, and (2) their uptake processes work with higher capacity than those for anionic or cationic amino acids (Tanaka et al., 1989). Luminal neutral amino acid uptake follows Michaelis-Menten kinetics showing no apparent cooperativity (Hill-coefficient near unity). The amino acid carriers are capable of uphill transport leading to intracellular amino acid accumulation (section 4.3). Intracellular accumulation of neutral amino acids strictly depends on an inward sodium gradient (Evers et al., 1976; Fass et al., 1977; Hammerman and Sacktor, 1982; Lynch and McGivan, 1987) (see Fig. 4). The coupling ratio is approximated to 1 Na^+ : 1 AA^0 (Paterson et al., 1980). Amino acid transport is electrogenic, i.e., it carries positive net charge into the cell

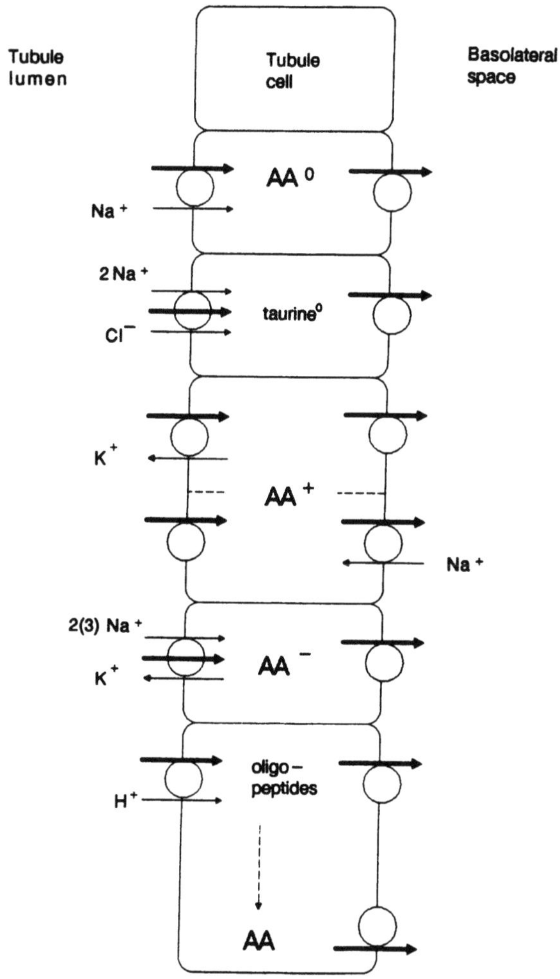

Fig. 4. Proximal tubular transport of neutral amino acids (AA^0), taurine^0, cationic amino acids (AA^+), anionic amino acids (AA^-) and certain di- and tripeptides. Transport of AA^+ is controversial (electroneutral as well as electrogenic transport; sodium dependence as well as sodium independence has been found; see text). Oligopeptide transport comes into play only if the peptide resists intraluminal hydrolysis.

and depolarizes the membrane potential. This fact can be exploited to measure amino acid transport indirectly by recording the cell membrane potential during luminal microperfusion of proximal tubules *in vivo* (Samarzija and Frömter, 1982a). In frog proximal tubule the amino acid induced cell membrane depolarization can exceed 30 mV and has a marked influence on the intracellular acid-base balance (Messner et al., 1985; Lang et al., 1986).

Most nonpolarized cells like fibroblasts and Erlich ascites tumour (EAT) cells possess three distinct transport systems for neutral amino acids, usually termed A, ASC and L (Bannai et al., 1984). In the renal brush border, the situation is not so clear. Few attempts have been made to address the selectivity of luminal transport systems for the "normal" α-amino acids. There are reports of different transport systems for small α-amino acids with overlapping specificity (Mircheff et al., 1982) as well as of a universal system transporting virtually all neutral amino acids (Lynch and McGivan, 1987). Such an unspecific system has also been observed in cultured renal cells where it gradually develops in the course of epithelial differentiation (Boerner et al., 1986). In parallel, the "typical" systems A, ASC, and L are down-regulated by decreasing protein kinase C activity (Boerner et al., 1986; Dawson and Cook, 1987).

β-amino acids, most importantly β-alanine and taurine, however, are reabsorbed by a special luminal 2 Na^+ : 1 Cl^- : 1 AA^0 cotransport (Chesney et al., 1990) (see Fig. 4). This transport is regulated by dietary influences (Friedman et al., 1981) stressing the importance of taurine excretion for the maintenance of taurine homeostasis.

Cystine uptake has been subject to considerable controversy. As early as 1951, it was hypothesized that cystine shares a common transport system with cationic amino acids (Dent and Rose, 1951). This hypothesis was confirmed by the detection of competitive inhibition between cystine and cationic amino acid uptake in renal vesicles (Segal et al., 1977) and in a renal cell line (States and Segal, 1990). Cultured cells from human proximal tubule, however, lack such an interaction (States et al., 1987). Moreover, micropuncture experiments show interaction of cystine uptake not only with cationic amino acids (Silbernagl and Deetjen, 1972b) but also with neutral amino acids (Völkl and Silbernagl, 1982). In order to reconcile these findings, it was suggested that the luminal membrane possesses two parallel transport systems for cystine, one of these being shared with cationic amino acids (Milne et al., 1961). Despite many efforts to prove this hypothesis *in vivo*, final conclusions on the mechanisms for luminal cystine uptake and basolateral cystine exit are not yet possible.

Glycine reabsorption in the proximal convoluted tubule is slower than that of other amino acids. Thus, either the affinity of glycine for the neutral amino acid carrier is lower or there is a special proximal tubular uptake system for these small amino acids. Uptake experiments in dog kidney BBMV show a single specific sodium dependent system (Hammerman and Sacktor, 1982). Others report that glycine uptake in rabbit renal BBMV is driven by a H^+ gradient and not by sodium (Rajendran et al., 1987). Furthermore, the apparent affinity obtained from these vesicle studies is considerably higher than *in vivo*. N-substituted amino acids (proline and hydroxyproline) are reabsorbed both by the general neutral amino acid carrier (Völkl and Silbernagl, 1980) and by a specific high-affinity transporter present in the luminal tubule membrane (Hammerman and Sacktor, 1977).

4.2.2 Luminal transport of cationic amino acids

Roughly 13% of the filtered amino acid load consists of the "dibasic", i.e., at neutral pH positively charged amino acids arginine, lysine, and ornithine (AA^+) (Cusworth and Dent, 1960). The reabsorption of these amino acids utilizes a common carrier system (Silbernagl and Deetjen, 1973). The uptake of cationic amino acids into mammalian tubule cells was found to be electrogenic as observed by intracellular potential measurement (Samarzija and Frömter, 1982b; Hoshi et al., 1976) and vesicle studies (Hammerman, 1982; Hilden and Sacktor, 1981; McNamara et al., 1986). A sodium-independent transport system for cationic amino acids has also been demonstrated in luminal BBMV (Busse, 1978; Hammerman, 1982; Stieger et al., 1983) and cultured proximal tubule cells (Schwegler et al., 1989a). This system seems to be able to accumulate positively charged amino acids simply by the inside negative membrane potential (see Fig. 4). Thus, it resembles system y^+ that is found in nonpolarized cells (White, 1985), where cationic amino acid uptake is directly coupled to the prevailing membrane potential (Bussolati et al., 1989).

Certain inconsistencies *in vivo*, however, blur this overly sharp mechanistic focus. *In vivo* cationic amino acids have a relatively low affinity for their uptake system (K_m = 1-6 mmol/l) (Samarzija and Frömter, 1982b; Silbernagl and Deetjen, 1972a), whereas in BBMV the apparent affinity is 10-500 times higher (Hammerman, 1982; Stieger et al., 1983; McNamara et al., 1986). Furthermore, the reabsorption of cationic amino acids in the rat kidney *in vivo* is completely abolished when luminal sodium is omitted (Ulrich et al., 1974). A possible key for explaining these differences might be the observation that arginine transport in cultured proximal tubule cells shifts from high to low affinity during cellular differentiation. In parallel, J_{max} rises and electrogenicity is reduced (Schwegler et al., 1990).

4.2.3 Luminal transport of anionic amino acids

The "acidic" or anionic amino acids, aspartate and glutamate, are negatively charged at neutral pH (AA^-). They account for 10% of the total tubular amino acid load, the glutamate concentration exceeding that of aspartate 30-fold (Cusworth and Dent, 1960). Micropuncture experiments clearly demonstrate that these compounds are reabsorbed by a specific carrier system that has a very low affinity for neutral or cationic amino acids (Silbernagl, 1983; Silbernagl and Völkl, 1983). In BBMV, uptake of anionic amino acids is driven by cotransport with sodium (Schneider and Sacktor, 1980) together with a countertransport of potassium (Burckhardt et al., 1980; Sacktor and Schneider, 1980) (see Fig. 4). The K^+ efflux can be replaced by NH_4^+ (Fukuhara and Turner, 1985) so that this amino acid reabsorption system might serve as an ammonium extrusion pathway. Stoichiometry was found to be either electrogenic, 3 Na^+ : 1 K^+ : 1 AA^- (Murer et al., 1980) or electroneutral, 2 Na^+ : 1 K^+ : 1 AA^- (Fukuhara and Turner, 1985). In rat proximal tubule the luminal cell membrane is depolarized by anionic amino acid transport (Samarzija and Frömter, 1982c), whereas this was not observed in *Triturus* (Hoshi et al., 1976). According to a mathematical model of rabbit BBMV the electrogenicity depends on the amount of intracellular potassium, favoring the electrogenic configuration under physiological conditions (Heinz et al., 1988).

As for aspartate, the D- and L-isomers are reabsorbed by the same rates *in vivo* (Samarzija and Frömter, 1982c; Silbernagl, 1983) and in BBMV (Schneider and Sacktor, 1980). This is the only known exception from the general rule that renal amino acid carriers are highly stereospecific for the L-isomer. Glutamate, although using the same carrier as aspartate, is handled in a stereospecific way (Silbernagl and Völkl, 1983). It is unknown whether aspartate's uniqueness has some physiological relevance.

4.2.4 Luminal transport of oligopeptides

As discussed in a previous review (Silbernagl, 1988), oligopeptides can cross the luminal brush border membrane directly or after intraluminal cleavage in the form of their constituent amino acids. The latter mechanism seems to be quantitatively more important. Proximal renal tubule cells possess several forms of peptidases in their brush border membrane with the catalytic site oriented towards the lumen (Tate and Meister, 1974; Kenny and Maroux, 1982). For example, the tripeptide glutathione undergoes complete luminal hydrolysis by the membrane-bound enzymes γ-glutamyltransferase, aminopeptidase M and cysteinyl-glycinase (Silbernagl, 1988). Not all of the filtered oligopeptides are, however, hydrolyzed before transport. Certain dipeptides (e.g., glycyl-L-proline) are resistant to membrane hydrolases and transported in their intact form (Ganapathy et al., 1981a; Ganapathy et al., 1981b). This carrier system is driven by an inwardly directed H^+ gradient (Ganapathy and Leibach, 1983; Silbernagl et al., 1987) and thus requires previous acidification of the tubular lumen for active peptide uptake (see Fig. 4). In the absence of luminal dipeptidase activity, tripeptides were also shown to be reabsorbed by a H^+-amino acid symport (Tiruppathi et al., 1990).

Considering the low tubular load of filtered dipeptides (Silbernagl, 1988), it seems unlikely that there is a certain specific transport system just for preventing the loss of these few molecules. Most probably, dipeptide transport helps to reabsorb the cleavage products of intraluminally disintegrated peptides and proteins. This would constitute a salvage pathway avoiding endocytosis and intralysosomal breakdown of filtered low molecular weight proteins. Interestingly, however, dipeptide transport exhibits a much greater activity than amino acid transport at the vesicle membrane of lysosomes (for review see Forster and Lloyd, 1988). Hence, it is reasonable to suggest that the real physiological importance of dipeptide transport might be the transport of cleavage-resistant intralysosomal protein breakdown fragments into the cytoplasm.

4.3 Intracellular Accumulation of Amino Acids

The true intracellular amino acid concentrations are hard to measure because cortical kidney sections always contain a considerable amount of extracellular fluid. This problem is addressed by measuring amino acid content in the homogenate and correcting for ^{3}H-inulin distribution space (Christophe et al., 1971; Toback et al., 1973). This procedure is inexact because several presuppositions have to be taken about the luminal and peritubular amino acid concentrations and the distribution between apical and basolateral volumes. However, some semi-quantitative results follow unambiguously from these studies: The anionic amino acids reach extremely high intracellular levels compared to the tubular fluid. A histochemical study has shown aspartate and glutamate concentrations in rat proximal convoluted tubule to be 8 and 25 mmol/kg cell water, respectively (Chan et al., 1975). The proximal straight tubule accumulates these amino acids to one-third the level of the convoluted part, whereas distal tubule segments contain very high concentrations. Other studies with cortical kidney slices (Christophe et al., 1971; Toback et al., 1973; Blazer-Yost et al., 1979; Duran et al., 1990) report aspartate and glutamate levels of up to 9 mmol/kg. Bearing in mind that the mean intraluminal amino acid concentrations are far less than plasma concentrations, the distribution coefficient (R) is likely to exceed 10^3:

$$R = AA_{in} / AA_{ex} \tag{8}$$

with AA_{in} = the intracellular and AA_{ex} = the extracellular amino acid concentration.

Theoretically, the maximum distribution coefficient can be calculated from the stoichiometry of the carrier and the various driving forces. Assuming the electrogenic $3Na^+ : 1K^+ : 1AA^-$ system (see section 4.2.3), the electrochemical driving force consists of 3 times the chemical gradient for sodium influx (3 x 54 mV (Yoshitomi and Frömter, 1985)) plus the chemical gradient for potasssium efflux (88 mV) plus the electrical driving force for the translocation of one positive charge (70 mV), i.e., 320 mV. This driving force allows (according to the Nernst equation) the buildup of a maximum R of $> 10^5$, way above the observed value. However, an electroneutral $2Na^+ : 1K^+ : 1\ AA^-$ stoichiometry yields a maximum R of 10^3, having the same order of magnitude as the experimental data.

Neutral amino acids are less avidly accumulated. The concentrations in cortical kidney slices reported range from 0.1 mmol/kg (phenylalanine (Blazer-Yost et al., 1979)) up to 1.0 mmol/kg (serine (Christophe et al., 1971)). A more recent study (Duran et al., 1990) yields slightly higher values, probably due to a different correction method for extracellular volume. The "excreted" amino acid taurine (7 mmol/l) and glycine (3-6 mmol/kg (Toback et al., 1973)) are exceptions again. The larger accumulation of taurine can be explained by a higher driving force. In contrast to the usual $1Na^+ : 1AA^0$ stoichiometry for most neutral amino acids, it is transported by a $2Na^+ : 1Cl^- : 1AA^0$ system (see section 4.2.1) that yields an electrochemical driving force of ~220 mV (maximal $R > 10^3$) in contrast to only 124 mV for a $1Na^+ : 1AA^0$ system (maximal $R = 10^2$). The high intracellular glycine levels could be due to the higher luminal concentrations of glycine along the proximal tubule (see section 1.3). A high intracellular glycine concentration might also result from intracellular metabolism rather than transport and this could limit net reabsorption.

The intracellular concentration of the cationic amino acids arginine and lysine has been found to be 20 to 50% of the neutral ones (Christophe et al., 1971). Considering the that membrane potential may be the only driving force for intracellular accumulation (electrochemical driving force of 70 mV, i.e., max R = 14), this finding is consistent with the facilitated diffusion model for cationic amino acids (section 4.2.2). However, similar studies (Blazer-Yost et al., 1979; Duran et al., 1990) show arginine and lysine concentrations that exceed this model.

4.4 Basolateral Amino Acid Exit

As only a small portion of reabsorbed amino acids is used for metabolism in the tubule cell (section 2.1), reabsorbed amino acids must leave the cell at the basolateral (contraluminal) membrane. When an intact kidney is perfused with a labelled amino acid, the label is quickly removed from the tubular fluid, but it takes about 40 seconds until it can be detected in the venous blood (Foulkes, 1985). This "reabsorption delay" is a unique feature of amino acid transport and is not observed, e.g., with glucose (Foulkes, 1987), indicating that the basolateral exit or subsequent processing of amino acids is the rate-limiting step in the release in venous blood (for review see Foulkes, 1990). As carrier-mediated transport is generally reversible, basolateral exit can also be studied by the uptake of a solute from the peritubular blood. This is possible *in vivo* by the double indicator dilution technique (Silverman et al., 1970): A radioactively labelled amino acid is injected into the renal artery together with an extracellular marker. Blood samples from the renal vein are taken before the amino acid has had time to be filtered and reabsorbed. The deficit of amino acid recovery in comparison with the extracellular marker reflects basolateral uptake by the tubule cells. By this and related techniques, four basolateral transport systems have been detected: one for anionic amino acids that is sensitive to heavy metal intoxication (Foulkes, 1971), one for cationic and two for neutral amino acids (Foulkes and Gieske, 1973). Interestingly, at least one of these systems is sensitive to PAH excess and probenecid, raising the question whether basolateral amino acid exit uses the general organic anion transporter (Foulkes, 1990; Foulkes and Blanck, 1990; Foulkes, 1985) or basolateral interconversions occur (section 2.4).

These hypothetical pathways are certainly not a general rule for all amino acids. Electrical measurements *in vivo* have shown that the basolateral uptake of acidic (Samarzija and Frömter, 1982c) but not that of basic amino acids (Samarzija and Frömter, 1982b) affects the membrane potential. In *Triturus*, a basolateral exit mechanism for the cationic amino acid lysine has been found that works as an $Na^+ : AA^+$ antiporter (Nunokawa and Hoshi, 1990). This is until now the only known example of an active, energy-consuming amino acid exit mechanism.

4.5 Basolateral Uptake of Amino Acids

Amino acid transporters in the basolateral membrane are, however, not only responsible for net exit but also for net uptake of certain amino acids. As for glutamine, the entire filtered load of this amino acid is insufficient to support NH_4^+ production during acidosis (Silbernagl, 1988). Thus, glutamine has to be extracted from the peritubular blood. In basolateral membrane vesicles from cortical tubules, a glutamine transporter was found that is subject to regulation by the acid-base balance of the organism (Windus et al., 1984a,b). On the other hand, tubule cells that are located distally to the S_1 and S_2 segments face extremely low luminal amino acid concentrations of < 10% of their plasma levels (Silbernagl, 1988). In order to maintain their domestic amino acid balance these cells depend on basolateral uptake. The skyrocketing intracellular glutamate and aspartate concentrations in distal tubule cells (Chan et al., 1975) are striking evidence for a basolateral amino acid uptake. In this view it is not surprising that in the basolateral membrane, many transport systems have been detected that are also present in nonpolarized cells, e.g., for glycine (Schwab and Hammerman, 1985), glutamate (Sacktor et al., 1981), and serine (Kragh-Hansen and Sheikh, 1984). The latter, however, is somewhat specific for tubule cells because it functions as a non-accumulative facilitated diffusion carrier.

When cultured tubule cells are in their logarithmic growth state, they exhibit primarily basolateral amino acid transport (Boerner and Saier, 1982; Boerner et al., 1986; Rabito and Karish, 1982). Only when the cells have reached confluence, more differentiated properties like Na^+-dependent sugar transport (Rabito, 1986) or low-affinity amino acid transport (Schwegler et al., 1990; Boerner et al., 1986) are observed. Thus, basolateral amino acid uptake seems to be constitutively expressed in an epithelial cell. The specific renal apical transporters obviously originate from "luxury genes" that are only transcribed in the state of metabolic affluence and differentiation. Otherwise, the massive sodium influx associated with active amino acid uptake would endanger the ion homeostasis of the cell.

4.6 Molecular Biology of Renal Amino Acid Transporters

The structural properties of amino acid carriers were thoroughly discussed by D. Oxender in an earlier chapter. Therefore, we give only a brief review on the literature concerning the kidney. The molecular identification of renal amino acid carriers has for a long time been hampered by the fact that there are no inhibitors with an affinity that is high enough for photolabelling. However, since the mid-1980s reconstruction of amino acid transporters in artificial systems is feasible (Koepsell et al., 1984; Wright and Peerce, 1984; Peerce and Wright, 1986; McGivan and Lynch, 1988). More recently, various groups were successful in expressing and identifying renal amino acid carriers in *Xenopus* oocytes by mRNA microinjection (Aoshima et al., 1988; Tate et al., 1989; Coady et al., 1990). Most reports agree that amino acid carriers are very similar to the sodium/glucose cotransporter with respect to their molecular sizes.

A recent study with BBMV subjected to increasing doses of high energy gamma radiation (radiation inactivation technique) reports of a molecular size for the functionally active sodium independent system y^+ of 90 kD. Sodium-coupled glucose, phosphate, and amino acid carriers have roughly twice this MW (205-293 kD), suggesting that they may exist as a dimer, one subunit binding the amino acid, the other one a sodium ion (Beliveau et al., 1990). The same authors found that the rate of destruction of a 120 kD protein relates to the loss of transport activity for neutral amino acid uptake at increasing radiation dose (McCormick et al., 1991).

The concept of several biochemically distinct, secondary active transport systems for the luminal uptake of amino acids, however, had not been generally accepted in the past. It is of considerable historical importance that during the 1970s several active research groups believed the luminal membrane γ-glutamyltransferase plays a critical role in amino acid reabsorption (Orlowski and Meister, 1970; Meister, 1976). Their hypothesis postulated a covalent binding of intracellular glutathione to the amino acid at the luminal membrane and the subsequent hydrolysis by the cytoplasmic enzyme γ-glutamyl-cyclotransferase. The cleavage products, 5-oxoproline and cysteinylglycine, were supposed to be recycled to glutathione again, requiring three ATP molecules for the uptake of one amino acid. This concept has proven to be inadequate for energetic, stoichiometric, and molecular biological reasons.

REGULATION OF THE RENAL HANDLING OF AMINO ACIDS

Regulatory mechanisms have been studied at virtually every level ranging from reconstituted membranes to the whole organ. The approach taken here is to apply a stringent view of adaptive changes gained from the physiological stimuli that elicited them. Towards this end the responses to starvation, metabolic acidosis, chronic renal failure and diabetes offer a range of physiological stimuli for displaying relevant regulatory mechanisms. The potential roles of hormones in regulating transporter expression is subsequently presented in section 5.4.

5.1 Starvation

Removal of dietary amino acids does not reduce the amino acid load delivered to the kidneys because of interorgan regulation (see chapter by H.N. Christensen). In fact, the branched chain amino acids actually rise *in vivo* offsetting a 30 to 40% reduction in glutamine concentration resulting in an unchanged load but one of considerably altered composition (Felig et al., 1969; Heitmann and Bergmann, 1980). Functionally, the kidneys respond to a developing metabolic acidosis and limited fuel availability by increasing glutamine extraction (Heitmann and Bergmann, 1980; Cersosimo et al., 1986); this is coupled to early proximal tubule utilization and ammoniagenesis at the expense of alanine release (Cersosimo et al., 1986). An increased affinity of the dominant brush border glutamine transport system (Daniel et al., 1991) may be an adaptation to the reduced filtered glutamine load presumably aligning reabsorption to the earliest proximal segment (section 3.1) where base generation occurs. At this nephron site more efficient conversion of glutamine nitrogen to ammonia is consistent with enhanced mitochondrial oxidation (Schoolwerth and La Noue, 1978). The fall in net alanine release (Cersosimo et al., 1986) and reversal to net uptake (Felig et al., 1969) may reflect system A induction expressed at the basolateral border (Sepulveda and Pearson, 1984). Increased glycine extraction, on the other hand, is unlikely to reflect transport adaptations since it accompanies an increased load and elevated serine release consistent with subservience to interorgan demands. Likewise, citrulline extraction falls in response to decreased arterial load, in turn, releasing far less arginine in accordance with the slowed urea cycle. The handling of the increased BCAA load presents an enigma, especially during the early phase of starvation when arterial levels increase; specifically, it is unclear whether these loads reach the TAL to support ammoniagenesis. Overall functional responses are those expected given the physiological stimuli and the adaptations in proximal tubule glutamine, and alanine transport may underlie these expressions, as well as possible intranephron regulation involving distal BCAA metabolism.

5.2 Metabolic Acidosis

Arterial plasma glutamine and BCAA fall 30-40% in chronic metabolic acidosis (Heitmann and Bergmann, 1980; Tizianello et al., 1978). Despite the fall in delivered glutamine, uptake is 2 to 3 times the filtered glutamine load consistent with accelerated basolateral transport (Shaloub, 1963; Welbourne, 1990). Interestingly, a paradoxical increase in BBM vesicle glutamine transport was reported in the rat (Forman et al., 1982; McFarlane-Anderson and Alleyne, 1979), while BLM vesicle transport capacity was unchanged (Forman et al., 1983). Subsequent studies, however, demonstrated an adaptive increase in vesicular glutamine hydrolysis resulting from an acidosis-induced increase in BBM GGTP activity and simultaneous glutamate transport (Harrison et al., 1987) accounting for the apparent glutamine transporter adaptation. In the dog the BLM glutamine transporter expresses a 50% increase in activity (Windus et al., 1984a). BLM glutamine uptake may also be regulated by the filtered HCO_3^- load coupled to the electrogenic Na^+-HCO_3^-basolateral efflux and the steady state PD (Eddy, 1985); in this regard, acute metabolic acidosis (lowest HCO_3^- efflux) elevated kidney cortex glutamine levels (Relman and Narin, 1975) consistent with accelerated transport and laggard utilization, the latter requiring time for induction of metabolic processes. The acute restoration of normal HCO_3^- levels reduces uptake (Duong et al., 1984; Preuss et al., 1987) to one that is consistent with luminal transport despite the maintained high mitochondrial utilization rate (Parry and Brosnan, 1978). Thus, possible synthesis of glutamine and glutamate transporters, as well as regulation by accompanying ion fluxes, could explain the transport response in chronic metabolic acidosis. However, in the acute phase, glutamine oxidation rather than transport appears rate limiting; both glycine and ornithine uptake increase but without an increase in serine release, supporting the contention of Tizianello et al. (1980) that glycine and ornithine are ammoniagenic under these conditions (see also section 2.3). The acute response is characterized by the sharpest fall in alkaline reserves and steeper proton gradients with the kidney apparently turning to other alternative base generating fuels due to metabolic rather than transport limitations.

5.3 Chronic Renal Failure

The filtered amino acid load falls in proportion to the reduction in filtration rate in chronic renal failure. Nonetheless, reabsorption rates are depressed (Silbernagl, 1988), fractional excretion rates generally being 10-fold higher at any given load versus control kidneys (Fukuda and Kopple, 1980; Nadvornikova et al., 1978). This is consistent with failure of one or more transport systems (section 6). Accompanying this reduced transport capacity per nephron is a developing metabolic acidosis reflecting the defective glutamine transport. In fact, less than 0.5 percent of the delivered glutamine load was removed (Tizianello et al., 1980) and superimposing an NH_4Cl load on this pre-existing acidosis does not enhance uptake. Healthy subjects in contrast remove 10% of the glutamine, rising to 40% with NH_4Cl loading. Impaired membrane transport is consistent with the disproportionately high fractional excretion rate relative to other neutral amino acids (Tizianello et al., 1980) and the apparent inability to remove glutamine from the antiluminal region (Dass and Martin, 1990); noteworthy is the fact that neither BB nor BLM vesicle preparations from an animal model of CRF displayed defective transport (D. Windus, personal communication) nor was glutaminase activity limiting (Pollack et al., 1965), pointing to altered ion gradients. Noteworthy, ammonium excretion rates factored by GFR are actually severalfold higher in CRF (Welbourne et al., 1972) despite the absence of significant glutamine uptake (Tizianello et al., 1980) nor did exogenous glutamine loads further elevate ammonium excretion (Welbourne et al., 1972). In addition, glycine uptake did not occur; yet both alanine and serine release were increased and sustained respectively, raising the possibility that protein or peptide breakdown might be contributors (Tizianello et al., 1980). Regarding the latter, it has been shown in a rat model that amino acids derived from glutathione breakdown appear to act as ammoniagenic precursors (Dass and Martin, 1990); moreover, release of both amino acids and GSH into red cells diminishes (Dass et al., 1990; Tizianello et al., 1980), suggesting disruption of the antiluminal transfer process. Overall, the CRF kidney appears to depend less upon the Na^+-dependent proximal transport, especially glutamine, and more upon backup systems, presumably the loop segment and/or proton gradient driven uptake; the latter, however, is also likely to be compromised by a proximal acidification defect (Schwartz et al., 1959).

5.4 Hormonal Regulation

Surprisingly, little attention has been focused on hormonal modulation of renal amino acid transporter activity despite their likely complicity in the above responses. A few of these relationships will be explored from this perspective; for indepth coverage the reader is referred to the general review on amino acid transport regulation.

5.4.1 Glucocorticoids

Glucocorticoid levels increase in starvation (Britton et al., 1975), acidosis (Welbourne, 1976), and diabetes (El-Seifi et al., 1987) associated with enhanced glutamine extraction and oxidation as opposed to supporting alanine release (section 5.1). The effect appears to reflect a dual mechanism: enhancing mitochondrial glutamate oxidation, presumably via a Ca^{2+}-dependent mechanism and, secondly, to accelerate glutamine transport (Welbourne, 1990). Regarding transport, the relevant question is whether these effects reflect glucocorticoid dependent synthesis of carriers, as occurs in cultured hepatocytes (Gebhart and Kleeman, 1987) or whether the effect is indirect, i.e., secondary to metabolism. *In vivo* extraction studies suggest that uptake involves the basolateral membrane site where acute acidification accelerates glutamine uptake related to pH (Kuwuhara et al., 1986) or filtered HCO_3^- loads (Welbourne and Salomon, 1989); in contrast, lowering pH *in vitro* reduces system N glutamine uptake (Tamarappoo and Kilberg, 1991). Ideally, glucocorticoid effects on glutamine transport should be discerned in cultured kidney cells, i.e., LLCPK cells, a porcine kidney proximal tubule cell line (Handler et al., 1980), that displays basolateral glutamine transport but appears to lack functional glucocorticoid receptors.

The System A transporter is clearly induced by glucocorticoid in liver (Kilberg et al., 1983). Were similar inductions expressed in the functioning kidney, the most obvious effect would be on alanine handling. Indeed, under the appropriate *in vivo* conditions effects on alanine are discernible; for example, starvation reverses net release to a strong net uptake (Felig et al., 1969) and both acidosis and diabetes are associated with a reduced net uptake. Since the uptake in starvation occurs without an increased load, transport enhancement is suspected and, if so, would probably be expressed on the proximal tubule basolateral membrane (site of gluconeogenesis). Note that changes in net uptake could also reflect a balance between uptake, system A, and release, system L, at different nephron levels; limited glutamine might be expected to slow both alanine production and efflux. Glucocorticoids are also associated with an adaptive increase in glutamate uptake. In cultured hepatocytes Na^+-dependent glutamate transport was enhanced eightfold by dexamethasone (Gebhart and Mecke, 1983); LLCPK cells express glutamate transporter activity predominantly on the apical border (Rabito and Karieh, 1983) and glucocorticoid effects would be of considerable interest given the presence of glutamine synthetase expressed by this cell line. *In vivo* uptake could also be proton gradient (also glucocorticoid dependent) driven, although not necessarily coupled to either ammoniagenesis or to glutamine synthesis.

5.4.2 Glucagon

Effects of glucagon on *in vitro* amino acid transport are biphasic, an acute uptake followed by a sustained delayed response (Edmondson and Lumeng, 1980). The acute phase reflects hyperpolarization of the cell membrane driving neutral amino acid transport, similar to that proposed for basolateral glutamine uptake (section 5.1.1), and the latter dependent upon protein synthesis, presumably representing new carriers. Although elucidated in cultured hepatocytes, this perspective may also be relevant in the renal responses to amino acid loads (section 1.2), possibly acutely affecting basolateral neutral amino acid uptake, metabolism and ammoniagenesis (Aoki et al., 1972), and to events in the thick ascending limb of the loop of Henle involving branched chain amino acid uptake and oxidation (sections 1.2; 2.3; 5.1).

5.4.3 Parathyroid hormone

PTH administration produces a prompt aminoaciduria although the mechanism of action is unclear. There is no doubt, however, that reabsorption in reduced. Decreased reabsorption may result from increased backflux, based on accelerated BLM vesicle uptake and increased tubular permeability following PTH (Silbernagl et al., 1988). In addition, an effect on urinary acidification and proton gradient driven reabsorption is consistent with the increased glutamate, and glycine excretion and, most interestingly, marked reduction in proline's transport maximum (Short et al., 1974) as well. If urinary alkalinization does contribute to reduced reabsorptive capacities by this mechanism, consideration might be given to the elevated PTH levels and reduced amino acid reabsorptive capacity in patients with chronic renal failure.

PATHOPHYSIOLOGY OF HYPERAMINOACIDURIA

Pathological hyperexcretion of amino acids is, except for very extreme forms, not endangering the maintenance of amino acid homeostasis in the organism, because amino acid intake ranges quantitatively way above the possible renal losses. So is hyperaminoaciduria a pathophysiological problem at all, and why? Unfortunately, the intestinal and the renal amino acid transport systems are equally affected by most of the inherited reabsorption deficiencies (e.g., Hartnup disease), leading to reduced absorption of an amino acid by the gut together with the renal hyperexcretion of the same amino acid. If an essential amino acid is concerned, this results in a condition that is equivalent to avitaminosis. Another clinically relevant complication of hyperaminoaciduria (e.g., in cystinuria) is the deposition of amino acid concrements in the urinary tract with secondary obstructive uropathy. Finally, the intratubular accumulation of a specific amino acid due to a defective transport system can affect different reabsorption processes. For example, bicarbonate reabsorption is blocked by high luminal lysine and arginine loads (Gougoux et al., 1978; Batlle and Chan, 1989).

6.1 Hyperaminoaciduria Due to Overflow (Prerenal Aminoaciduria)

Amino acid reabsorption can be exhausted by an excess tubular amino acid load (see the T_m concept, section 1.4). This excess load can be due to (1) extraordinarily high plasma amino acid concentrations or (2) a too fast glomerular filtration.

(1) Extraordinarily high plasma amino acid levels are characteristic for a variety of inborn errors of amino acid metabolism (for review see Brodehl and Bickel, 1968). The most common disease in this group, phenylketonuria, is due to defective phenylalanine hydroxylase. Consequently, phenylalanine itself and atypic metabolites are hyperexcreted. The hyperaminoaciduria in tyrosinemia (Berger, 1985) is even more pronounced. This inborn defect of the catabolic enzyme fumarylacetoacetase is characterized by elevated blood and urine levels for tyrosine and methionine. However, this overflow aminoaciduria must not be confused with the generalized hyperaminoaciduria that is seen in later stages of tyrosinemia due to the development of a Fanconi syndrome (Berger, 1985). Similar forms of overflow hyperaminoaciduria are described for valine, taurine, citrulline, arginine, lysine, glutaric acid (Wellner and Meister, 1981), β-alanine (Scriver et al., 1966), proline (Greene et al., 1973), homocystine (Gerritsen et al., 1962), and many more.

(2) Hyperaminoaciduria due to an increase in GFR is not "prerenal" in the strictest sense. Physiologically, GFR varies only by the factor of 1.5. This variation is not sufficient to exhaust tubular transport processes (Silbernagl, 1988). The situation changes if the number of functioning nephrons is grossly diminished; e.g., in focal segmental glomerulosclerosis or chronic pyelonephritis (Brenner et al., 1982; Bank and Aynedjian, 1966). Now, a largely reduced number of glomeruli (e.g., 10%) has to cope with the total glomerular filtration rate (e.g., 30% of normal). Thus, the GFR per single nephron (SNGFR) increases in the non-affected glomeruli despite an overall reduction of total GFR. Furthermore, because sodium and water excretion have to be maintained, in the few yet working nephrons the fractional reabsorption of salts and water is reduced. Thus, the tubular fluid is more diluted than is normal (Silbernagl, 1988). Obviously, the active uphill reabsorption of solutes from a diluted, fast flowing tubular stream is not as efficient as under normal conditions.

6.2 Generalized Hyperaminoaciduria and the Fanconi Syndrome

In generalized aminoaciduria the urinary excretion of all amino acids is elevated by roughly the same factor. This form accounts for more than two-thirds of all detected aminoacidurias in a pediatric and internal medical collective (Mattingley, 1986) and is regularly associated with other tubular reabsorption deficits. Fanconi was the first to describe this symptomatic entity (Fanconi, 1936) that is characterized by hyperphosphaturia, glucosuria, hpyperaminoaciduria, and hypokaliemic renal tubular acidosis. Many different noxious agents can cause a Fanconi syndrome at later stages, e.g., intoxication with cadmium (Kido et al., 1988), lead (Vacca et al., 1986), copper (Wilson's disease (Barbosa et al., 1985)), glycogen storage diseases, cystinosis, and paraproteinemia (Schneider et al., 1978).

The symptomatology of the Fanconi syndrome is highly suggestive of a generalized impairment of all proximal tubular reabsorption processes. Indeed, morphological changes are localized to the proximal tubule (Clay et al., 1953). However, the development of the aminoaciduria precedes the morphological changes, at least in an experimental animal model (Günther et al., 1979). Therefore, it is most likely that the impairment of solute reabsorption in Fanconi syndrome is due to a reduced electrochemical driving force for intracellular solute accumulation (Silbernagl, 1985) and/or a general change in membrane permeability (Bergeron et al., 1976). As long as appropriate cellular models for the Fanconi syndrome are lacking, final evidence of the molecular pathogenesis cannot be expected.

6.3 Specific Hyperaminoaciduria Due to Inherited Renal Transport Defects

Keeping in mind the complicated carrier-mediated mechanisms for luminal amino acid entry and basolateral exit (see section 4.2. and 4.4), it is not surprising that a large number of genetic defects for renal amino acid transport have been detected. These inherited diseases are all recessively autosomal except for cystinuria (for review see Wellner and Meister, 1981). An encyclopedic list of inherited renal amino acid defects would be far beyond the scope of this chapter. The following examples have been chosen because (1) they are most common in their field; (2) they demonstrate that different biochemical defects can result in more or less similar phenomenological symptoms.

6.3.1 Cystinuria

Besides the Fanconi syndrome, hereditary cystinuria is the most common cause of renal amino acid hyperexcretion (Mattingley, 1986; Feld, 1988). The presence of elevated urinary concentration of cystine, arginine, lysine, ornithine (Dent and Rose, 1951; Crawhall et al., 1967), citrulline (Milne et al., 1962), homoarginine (Cox and Cameron, 1974), and 5-hydroxy-L-Lysine (Roesel et al., 1985) is pathognomonic. The fractional excretion of cystine can exceed unity, i.e., net tubular cystine secretion is possible (Crawhall and Thompson, 1965). Similar results are seen in the gut, where intestinal absorption of cystine and cationic amino acids is reduced (Milne et al., 1961). Therefore, it is reasonable to conclude that a common renal and intestinal cystine/cationic amino acid carrier is impaired (Thier and Segal, 1978). Interestingly, there are reports of an isolated cystinuria without hyperexcretion of cationic amino acids (Brodehl et al., 1968b) as well as of an isolated cationic aminoaciduria without cystinuria (Whelan and Scriver, 1968). From a clinical point of view this illustrates that cystine reabsorption is more complicated than originally considered (see section 4.2.1.).

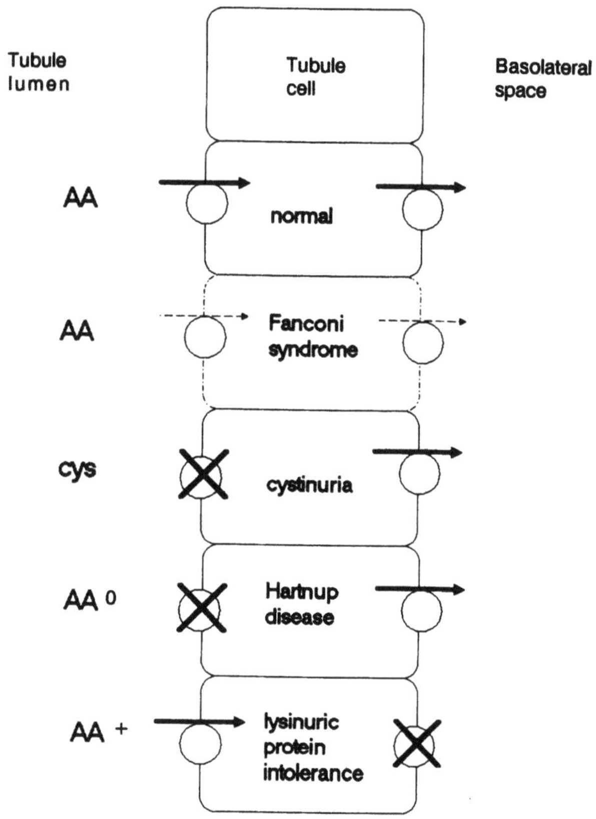

Fig. 5. Principal types of hyperaminoaciduria. The Fanconi syndrome is characterized by a decreased reabsorption not only of amino acids but also of other sodium-dependent transport processes. That's why a diminished sodium gradient or an ATP deficiency is hypothesized to underlay this type of generalized hyperaminoaciduria. In Hartnup disease and in classical cystinuria the defect is specific for a group of amino acids. It is the respective brushborder carrier, which is most probably disturbed in these cases. In cystinuria the energetization of the carrier rather than the (passive) translocation of the amino acid might be defective (Gekle and Silbernagl, 1991). Lysinuric protein intolerance has been shown to be caused by a defect of the basolateral carrier for cationic amino acids.

6.3.2 Hartnup disease

This hereditary disease is characterized by renal hyperexcretion and intestinal malabsorption of a variety of neutral amino acids. The handling of anionic and cationic amino acids is, however, not affected. The symptoms are those of a B avitaminosis (Levy, 1989) and are successfully treated with nicotinamide (Halvorsen and Halvorsen, 1963). The symptomatology is caused by intestinal malabsorption of the essential amino acid tryptophan. The kidney, however, is not damaged by this disease. As with cystinuria, it would be too simple to conclude from Hartnup disease upon a single transport system for all neutral amino acids. In fact, there are reports of renal hyperexcretion and intestinal malabsorption for methionine (Hooft et al., 1964), tryptophan (Drummond et al., 1964), and histidine (Sabater et al., 1976) without effect on the transport of other uncharged amino acids. Thus, it may be speculated that in Hartnup disease the binding site for sodium in various neutral amino acid carriers is defective so that uphill transport into the cells of gut and kidney is no longer possible. Experimental data are, however, lacking. To be complete, it should be stated that there is also an extremely rare hereditary hyperexcretion of anionic amino acids in combination with hyperprolinuria (Teijema et al., 1974).

6.3.3 Lysinuric protein intolerance

A different pathogenesis has been confirmed for the hyperexcretion of cationic amino acids in lysinuric protein intolerance (LPI). Here, the basolateral exit carrier for lysine is defective in kidney tubule cells (Rajantie et al., 1981) and intestinal mucosa (Rajantie et al., 1980). Thus, cationic amino acids regularly enter the cells from the luminal side and accumulate in the cytoplasm, but cannot conclude their transepithelial passage by basolateral exit (see Fig. 5).

As previously stated, basolateral amino acid carriers in cultured proximal tubule cells are more or less identical with the "classical" systems A, ASC, L, and y^+ (Saier et al., 1988; White, 1985). Therefore, it is reasonable to suggest that the basolateral system y^+ is defective in LPI. As all cells possess y^+ carriers for their own metabolism, LPI should be accompanied by impaired cationic amino acid uptake in non-kidney- and non-intestinal cells. Indeed, primary cultured skin fibroblasts from a patient suffering from LPI show reduced cationic amino acid accumulation compared with controls from a normal subject (Smith et al., 1987). This strongly substantiates the suggestion that cationic amino acids leave the proximal tubule cells primarily by the common sodium independent system y^+.

These examples show the great importance that the description of pathophysiological conditions, even of extremely rare ones, has had for the understandingof basic transport mechanisms. On the other hand, clinical diagnosis and treatment equally depend on the investigation of cellular and molecular events. We hope that this mutual stimulation will continue in the future.

ACKNOWLEDGEMENTS

We thank Ms. Sabine Kopp and Ms. Angelika Reuss for their valuable help with the manuscript.

REFERENCES

Abbott, W.A., Bridges, R.J. and Meister, A., 1984, J.Biol. Chem., 259: 15393-15400.
Albert, Z., Orlowski, M. and Folk, J.E., 1961, Nature, 191: 767-770.
Alvestrand, A. and Bergstrom, J., 1984, Lancet, 1: 195-197.
Aoki, T.T., Myller, W.A. and Cahill, G. Jr., 1972, Adv. Enzyme Reg., 10: 145-152.
Aoshima, H., Tomita, K. and Sugio, S., 1988, Archives of Biochemistry and Biophysics, 265: 73-81
Asatoor, A.M., Milne, M.D. and Walshe, J.M., 1975, Clin. Sci. Mol. Med., 51: 369-378.

Asatoor, A,M., Bending, M.R. and Milne, M.D., 1978, Clin. Sci. Mol. Med., 55: 205-212.
Bank, N. and Aynedjian, H., 1966, Journal of Laboratory and Clinical Medicine, 68: 713-727.
Bannai, S., Christensen, H.N., Vadgama, J.V., et al., 1984, Nature Lond., 311: 308.
Barbosa, E.R., Scaff, M., Comerlatti, L.R. and Canelas, H.M., 1985, Arq. Neuropsiquiatr., 43: 234-242.
Barfuss, D.W. and Schafer, J.A., 1979, Am. J. Physiol.,236: F149-F162.
Batlle, D.C. and Chan, Y.L., 1989, Miner. Electrolyte Metab., 15: 187-194.
Beliveau, R., Demeule, M., Jette, M. and Potier, M., 1990, Biochemical Journal, 268: 195-200.
Berger, R., 1985, *Inherited Diseases of Amino Acid Metabolism*, Bickel, H. and Wachtel, U. (eds), Springer, Stuttgart, New York, p. 192-202.
Bergeron, M. and Morel, F., 1969, Am. J. Physiol., 216: 1139-1149.
Bergeron, M., Dubord, L. and Hausser, C., 1976, J. Clin. Invest., 57: 1181-1189.
Berliner, R.W., Kennedy, T.J. and Hilton, J.G., 1950, Proc. Soc. Exp. Bio. Med., 75: 791-794.
Blazer-Yost, B., Reynolds, R. and Segal, S., 1979, Am. J. Physiol., 236: F398-F404.
Block, W.D. and Hubbard, R.W., 1962, Archives of Biochemistry and Biophysics, 96: 557-561.
Boerner, P. and Saier, M.H., 1982, J. Cell Physiol., 113: 240-246.
Boerner, P., Evans-Layng, M., Ü, H.S. and Saier, M.H., 1986, J. Biol. Chem., 261: 13957-13962.
Bomsel, M., Prydz, K., Parton, R.G., Gruenberg, J. and Simons, K., 1989, J. Cell Biology, 109: 3243-3258.
Brenner, B.M., Meyer, T.W. and Hostetter, T.H., 1982, N. Engl. J. Med., 307: 652-659.
Bridges, R.J. and Meister, A., 1985, J. Biol. Chem., 260: 7304-7308.
Britton, G.W., Rotenberg, S. and Adelman, R.C., 1975, Biochem. Biophys. Res. Commun., 64: 184-188.
Brodehl, J. and Bickel, H., 1968, Clinical Nephrology, 1: 148-168.
Brodehl, J. and Gelissen, K., 1968a, Pediatrics, 42: 395-404.
Brodehl, J., Gelissen, K. and Kowalewski, S., 1968b, Pediatric Research, 2: 525-534.
Brosnan, J.T., 1988, Can. J. Physiol. Pharmacol., 65: 2355-2362.
Brown, J.L., Samiy, A.H. and Pitts, R.F., 1961, Am. J. Physiol., 200: 370-379.
Brown, S.A. and Navar, G.L., 1990, Am. J. Physiol., 259: F739-746.
Buch, H.B., Cambon, N. and Lowry, O.H., 1985, Kidney Int., 28: 114-117.
Buchanan, D.L., Haley, E.E. and Markiw, R.T., 1962, Biochem., 1: 612-620.
Burckhardt, G., Kinne, R., Stange, G. and Murer, H., 1980, Biochim. Biophys. Acta, 599: 191-201.
Busse, D., 1978, Archives of Biochemistry and Biophysics, 191: 551-560.
Bussolati, O., Laris, P.C., Nucci, F.A., et al., 1989, Am. J. Physiol., 256: C930-C935.
Cersosimo, E., Williams, P.E., Radoseirtch, R.M., Hoxworth, B.T., Lacy, W.W. and Abumrad, N.N., 1986, Am. J. Physiol., 250: E622-628.
Chan, A.W.K., Burch, H.B., Alvey, T.R. and Lowry, O.H., 1975, Am. J. Physiol., 229: 1034-1044.
Chan, A.W.K., Perry, S.G., Burch, H.B., Fagioli, S., Alvey, T.R. and Lowry, O.H., 1979, J. Histochem. Cytochem., 27: 751-755.
Chesney, R.W., Friedman, A.L. and Albright, P.W., 1982, Proceedings of the Society of Experimental Biology and Medicine, 170: 493-501.
Chesney, R.W., Gusowski, N. and Dabbagh, S., 1987, Pediatr. Nephrol., 1: 9-15.
Chesney, R.W., Zelikovic, I., Jones, D.P., Budreau, A. and Jolly, K., 1990, Pediatr. Nephrol., 4: 399-407.
Christensen, H.N., 1949, Biochem. J., 44: 333-335.
Christophe, J., Winard, J., Kutzner, R. and Hebbelinck, M., 1971, Am. J. Physiol., 221: 453-457.
Clay, R.D., Darmady, E.M. and Hawkins, M., 1953, J. Pathol. Bacteriol., 65: 551-558.
Coady, M.J., Pajor, A.M., Toloza, E.M. and Wright, E.M., 1990, Archives of Biochemistry and Biophysics, 283: 130-134.
Cox, B.D. and Cameron, J.S., 1974, Clinical Science and Molecular Medicine, 46: 173-182.
Crawhall, J.C. and Segal, S., 1967, Biochemical Journal, 105: 891-896.
Crawhall, J.C. and Thompson, C.J., 1965, J. Clin. Invest., 44: 1038-1039.
Crawhall, J.C., Scowen, E.F., Thompson, C.J. and Watts, R.W.E., 1967, J. Clin. Invest., 46: 1162-1171.

Curthoys, N.P. and Lowry, O.H., 1973, J. Biol. Chem., 248: 162-168.
Cusworth, D. and Dent, C.E., 1960, Biochemical Journal, 74: 550-561.
Daniel, H., Minami, H., Morse, E.L. and Adibi, S.A., 1991, Fed. Proc., A745.
Dantzler, W.H. and Silbernagl, S., 1988, Am. J. Physiol., 255: F397-F407.
Dantzler, W.H. and Silbernagl, S., 1990, Am. J. Physiol., 258: F504-F513.
Dantzler, W.H. and Silbernagl, S., 1991, Am. J. Physiol., 261: F495-504.
Dass, P.D. and Bourke, E., 1988, Contr. Nephrol., 63: 77-85.
Dass, P.D. and Martin, D., 1990, Renal Physiol. Biochem., 13: 259-263.
Dass, P.D. and Welbourne, T.C., 1980, Life Sci., 28: 355-360.
Dass, P.D., Holmes, E. and Bermes, E., 1991, FASEB, 5: 923A.
Dawson, W.D. and Cook, J.S., 1987, J. Cell Physiol., 132: 104-110.
Dent, C.E. and Rose, G.A., 1951, Quart J. Medicine, 20: 205-219.
DeSanto, N.G., Coppola, S., Coscarella G., et al., 1991, Renal Physiol. Biochem., 14: 63-70.
Drummond, K.N., Michael, A.F., Ulstrom, R.A. and Good, R.A., 1964, Amer. J. Medicine, 37: 928-948.
Duong, N., Vinay, P., Khoury, N., Guern, C. and Gougoux, A., 1984, Renal Physiol., 7: 32-45.
Duran, M.A., Spencer, D., Weise, M., Kronfol, N.O., Spencer, R.F. and Oken, D.E., 1990, Toxicol. Appl. Pharmacol., 105: 183-194.
Eddy, A.A., 1985, Ann. NY Acad. Sci., 956: 51-62.
Edmondson, J.W. and Lumeng, L., 1980, Biochem. Biophys. Res. Commun., 96: 61-68.
Eisenbach, G.M., Weise, M. and Stolte, H., 1975, Pflügers Arch., 357: 63-76.
El-Seifi, S., Freibert, J.M., Kinsella, J., Cheng, L. and Sacktor, B., 1987, Am. J. Physiol., 252: R40-47.
Elwyn, D.H., 1986, Fed. Proc., 25: 854-861.
Evers, J., Murer, H. and Kinne, R., 1976, Biochim. Biophys. Acta, 426: 598-615.
Fanconi, G., 1936, Dtsch. Med. Wochenschr., 62: 1169-1171.
Fass, S.J., Hammerman, M.R. and Sacktor, B., 1977, J. Biol. Chem., 252: 583-590.
Featherston, W.R., Rogers, Q.R. and Freedland, R.A., 1973, Am. J. Physiol., 224: 127-129.
Feld, R.D., 1988, Crit. Rev. Clin Lab. Sci., 26: 243-261.
Felig, P., Owen, O.E., Wahreu, J. and Cahill, G.F. Jr., 1969, J. Clin. Invest., 48: 584-594.
Foreman, J.W., Reynolds, R.A., Ginkinger, K. and Segal, S., 1983, Biochem. J., 212: 713-720.
Forster, S. and Lloyd, J.B., 1988, Biochim. Biophys. Acta, 947: 465-491.
Foulkes, E.C., 1971, Biochim. Biophys. Acta, 241: 815-822.
Foulkes, E.C., 1985, Am. J. Physiol., 249: F878-F883.
Foulkes, E.C., 1987, Am. J. Physiol., 252: F1042-F1047.
Foulkes, E.C., 1990, Proc. Soc. Exper. Biol. Med., 195: 1-6.
Foulkes, E.C., and Blanck, S., 1990, Proc. Soc. Exper. Biol. Medicine, 193: 56-59.
Foulkes, E.C., and Gieske, T., 1973, Biochim. Biophys. Acta, 318: 439-445.
Friedman, A.L., Albright, P.W. and Chesney, R.W., 1981, Life Sciences, 29: 2415-2419.
Frömter, E., 1982, Pflügers Arch., 393: 179-189.
Fukuda, S. and Kopple, J.D., 1980a, Mineral Electrolyte Metab., 3: 237-247.
Fukuda, S. and Kopple, J.D., 1980b, Mineral Electrolyte Metab., 3: 248-260.
Fukuhara, Y. and Turner, R.J., 1985, Am. J. Physiol., 248: F869-F875.
Ganapathy, V. and Leibach, F.H., 1983, J. Biol. Chem., 258: 14189-14192.
Ganapathy, V., Mendicino, J. and Leibach, F.H., 1981a, Biochim. Biophys. Acta, 642: 381-391.
Ganapathy, V., Mendicino, J. and Leibach, F.H., 1981b, J. Biol. Chem., 256: 118-124.
Gayer, J. and Gerok, W., 1961, Klin. Wochenschr., 39: 1054-1055.
Gekle, M. and Silbernagl, S., 1991, Pflügers Arch., 419:492-498.
Gerritsen, T., Vaughn, J.G. and Waisman, H.A., 1962, Biochem. Biophys. Res. Commun., 9: 493.

Gögelein, H. and Greger, R., 1986, Pflügers Arch., 406: 198-203.
Gougoux, A., Lemieux, G. and Vinay, P., 1978, Kidney Int., 14: 215-227.
Greene, M.L., Lietman, P.S., Rosenberg, L.E. and Seegmiller, J.E., 1973, American Journal of Medicine, 545: 265-271.
Griffith, O.W. and Meister, A., 1979, Proc. Natl. Acad. Sci. USA, 76: 5606-5610.
Günther, R. and Silbernagl, S., 1981, Pflügers Arch., 389: 137-142.
Günther, R., Silbernagl, S. and Deetjen, P., 1979, Pflügers Arch., 382: 109-114.
Häberle, D., Wahlländer, A. and Sies, H., 1979, FEBS Letters, 108: 335-340.

Halperin, M.L., Jungas, R.L., Pichette, C. and Goldstein, M.B., 1982, Can. J. Physiol. Pharmacol., 60: 1431-1435.
Halvorsen, K. and Halvorsen, S., 1963, Pedriatrics, 31: 29-38.
Hammerman, M.R., 1982, Biochim. Biophys. Acta, 685: 71-77.
Hammerman, M.R. and Sacktor, B., 1977, J. Biol. Chem., 252: 591-595.
Hammerman, M.R. and Sacktor, B., 1982, Biochim. Biophys. Acta, 686: 189-196.
Handler, J.S., Perkins, F.M. and Johnson, J.P., Am. J. Physiol., 238: F1-9.
Harrison, D., Joshi, S., Carter, P. and Welbourne, T.C., 1987, Biochim. Biophys. Acta, 902: 301-306.
Heinz, E., Sommerfeld, D.L. and Kinne, K.H., 1988, Biochim. Biophys. Acta, 937: 300-308.
Heitmann, R.N. and Bergman, E.N., 1980, Am. J. Physiol., 239: E248-254.
Helbig, I., Pohl, B. and Busse, D., 1984, Pflügers Arch., 400: R24 (Abstract).
Hilden, S.A. and Sacktor, B., 1981, Arch. Biochem. Biophys., 210: 289-297.
Hirschberg, R. and Kopple, J.D., 1989, J. Clin. Invest., 83: 326-330.
Hirschberg, R. and Kopple, J.D., 1991, J. Am. Soc. Nephrol., 1: 1034-1040.
Hooft, C., Timmenons, J., Snoeck, J., Anterer, I., Ouaert, W. and Vanden-Hends, C.H., 1964, Lancet, 2: 20.
Hoshi, T., Sudo, K. and Suzuki, Y., 1976, Biochim. Biophys. Acta, 448: 492-504.
Hughey, R.R., Rankin, B.B. and Curthoys, N.P., 1980, Am. J. Physiol., 238: F199-204.
Jacobsen, C., Roigaard-Petersen, H. and Sheikh, M.I., 1988, FEBS Letters, 236: 95-99.
Kachra, Z., Barash, I., Yannopoulos, C., Khan, M.N., Guyda, H.J., and Posner, B.I., 1991, Endocrinology, 128: 1723-1730.
Kalra, V.K., Sikka, S.C. and Sethi, G.S., 1981, J. Biol. Chem., 256: 5567-5571.
Kamin, H. and Handler, P., 1951, Am. J. Physiol., 164: 654-659.
Kenny, A.J. and Maroux, S., 1982, Physiol Rev., 62: 91-128.
Kido, T., Honda, R., Tsuritani, I., et al., 1988, Arch. Environ. Health, 43: 213-217.
Kilberg, M.S., Vida, T.A. and Barber, E.F., 1983, J. Cell. Physiol., 114: 45-52.
Kirk, E., 1936, Acta Med. Scand., 89: 450-453.
Klein, R.A., Moore, M.J. and Smith, M.W., 1971, Biochim. Biophys. Acta, 233: 420-433.
Koepsell, H., Korn, K., Ferguson, D., Menuhr, H., Ollig, D. and Haase, W., 1984, J. Biol. Chem., 259: 6548-6558.
Kopple, J.D. and Swendseid, M.E., 1981, J. Nutrition, 111: 931-941.
Kragh-Hansen, U. and Sheikh, M.I., 1984, J. Physiol. (London), 354: 55-67.
Kragh-Hansen, U., Roigaard-Petersen, H. and Sheikh, M.I., 1985, Am. J. Physiol., 249: F704-F712.
Kuwahara, M., Sasaki, S., Shugai, T. and Takekuchi J., 1986. Kidney Int., 30: 340-347.
Lang, F., Messner, G. and Rehwald, W., 1986, Am. J. Physiol., 250: F953-F962.
Lang, F., Tschernko, E., Schulze, E., Öttl, I., Ritter, M., Völkl, H., Hallbrucker, C.H. and Häussinger, D., 1991, Pflügers Arch., 419: 111-113.
Lash, L.H. and Jones, D.P., 1984, J. Biol. Chem., 23: 14508-14514.
Levy, H.L., 1989, *The Metabolic Basis of Inherited Disease*, Scriver, C.R., Beaudet, A.L., Sly, W.S. and Valle, D. (eds), McGraw-Hill, New York, p. 2515-2528.
Linder, G. and Foulkes, E.C., 1985, Environ. Res., 36: 241-247.
Lingard, J.M., Györy, A.Z. and Young, J.A., 1975, Pflügers Arch., 357: 51-61.
Lingard, J.M., Rumrich, G. and Young, J.A., 1973a, Pflügers Arch., 342: 1-12.
Lingard, J.M., Rumrich, G. and Young, J.A., 1973b, Pflügers Arch., 342: 13-28.
Lingard, J.M., Turner, B., Williams, D.B. and Young, J.A., 1974, Australian Journal of Experimental Biology and Medical Science, 52: 687-695.
Ludwig, C., 1844, Wagners Handwörterbuch der Physiologie 2, pp. 629.
Lynch, A.M. and McGivan, J.D., 1987, Biochim. Biophys. Acta, 899: 176-184.
Mamelok, R.D., Groth, D.F. and Prusiner, S.B., 1980, Biochemistry, 19: 2367-2376.
Mattingley, J.M., 1986, Biomed. Chromatogr., 1: 95-100.
McCormick, M.E. and Webb, K.E., 1982, J. Nutr., 112: 276-282.
McCormick, J.I., Jétte, M., Potier, M., Beliveau, R. and Johnstone, R.M., 1991, Biochemistry, 30: 3704-3709.
McFarlane-Anderson, N. and Alleyne, G.A.O., 1979, Biochem J., 182: 295-300.
McGivan, J.D. and Lynch, A., 1988, Contrib. Nephrol., 63: 37-42.
McNamara, P.D., Rea, C.T. and Segal S., 1986, Am. J. Physiol., 251: F734-F742.
Meister, A., 1976, In: *Amino Acid Transport and Uric Acid Transport*, Silbernagl, S., Lang, F. and Greger, R. (eds), Thieme, Stuttgart, p. 40-45.
Meister, A. and Anderson, M.E., 1983, Annu. Rev. Biochem., 52: 711-760.

Melancon, S.B., Dalliare, L., Lemieux, B., Robitaille, P. and Potier, M., 1977, Pediatr., 91: 422-427.
Messner, G., Oberleithner, H. and Lang, F., 1985, Pflügers Arch., 404: 138-144.
Milne, M.D., Asatoor, A.M., Edwards, K.D.G. and Loughridge, L.W., 1961, Gut 2: 323-337.
Milne, M.D., London, D.R. and Asatoor, A.M., 1962, Lancet 1: 49-50.
Mircheff, A.K., Kippen, I., Hiramaya, B. and Wright, E.M., 1982, J. Membr. Biol., 64: 113-122.
Murer, H., Leopolder, A., Kinne, R., and Burckardt, G., 1980, Int. J. Biochem., 12: 223-228.
Nadvornikova, H., Schuck, O., Mally, J., Pechar, J., Dobersky, P. and Tomkowa, D., 1978, Nephron, 20: 83-89.
Nakanishi, T., Shimizer, A., Saiki, K., Fujiuara, R., Funahashi, S. and Hayashi, A., 1991, Clin. Chimica Acta, 197: 249-256.
Nunokawa, T. and Hoshi, T., 1990, Renal Physiol. Biochem., 13: 295-305.
Orlowski, M. and Meister, A., 1970, Proc. Natl. Acad. Sci. USA, 67: 1248-1255.
Orlowski, M., Okonkwo, P.O. and Green J.P., 1973, FEBS Lett.. 31: 237-240.
Oxender, D.L. and Christensen, H.N., 1962, J. Biol. Chem., 238: 3686-3699.
Parry, D.M. and Brosnan, J.T., 1978, Biochem. J., 174: 387-396.
Paterson, J.Y.F., Sepulveda, F.V. and Smith, M.W, 1980, Biochim. Biophys. Acta, 603: 288-297.
Peerce, B.E. and Wright, E.M., 1986, Annals New York Academy of Sciences
Pitts, R.F., 1943, Am. J. Physiol., 140: 156-167.
Pitts, R.F., 1944. Am. J. Physiol., 140: 535-547.
Pollack, V.E., Mattenherimer, H., DeBruin, H. and Weinman, K.J., 1965, J. Clin. Invest., 44: 169-181.
Premen, A.J., 1985, Am. J. Physiol., 249: F319-322.
Preuss, H.G., Slemmer, D., Areas, J. and Conrado, J., 1987, J. Clin. Chem. Clin. Biochem., 25: 4107.
Rabito, C.A., 1986, Miner. Electrolyte Metab., 12: 32-41.
Rabito, C.A. and Karish, M.V., 1982, J. Biol. Chem., 257: 6802-6808.
Rabito, C.A. and Karish, M.V., 1983, J. Biol. Chem., 258: 2543-2547.
Rajantie, J., Simell, O. and Perheentupa, J., 1980, Gut, 21: 519-524.
Rajantie, J., Simell, O. and Perheentupa, J., 1981, J. Clin Invest., 67: 1078-1082.
Rajendran, V.M., Barry, J.A., Kleinman, J.G. amd Ramaswamy, K., 1987, J. Biol. Chem., 262: 14974-14977.
Relman, A.A. and Narin, R.G., 1975, Med. Clin. North Am., 59: 583-589.
Reynolds, R.A., Wald, H. and Segal, S., 1982, Biosci. Rep., 2: 883-890.
Roesel, R.A., Blankenship, P.R., Ganapathy, V. and Leibach, F.H., 1985, J. Inh. Met. D., 8: 123-126.
Roigaard-Petersen, H. and Sheikh, M.I., 1984, Biochem. J., 220: 25-33.
Sabater, J., Ferre, C., Puliol, M. and Maya, A., 1976, Clin. Genet., 9: 117-124.
Sacktor, B. and Schneider, E.G., 1980, Int. J. Biochem., 12: 229-234.
Sacktor, B., Rosenbloom, I.L., Liang, C.T. and Cheng, L., 1981, J. Membr. Biol., 60: 63-71.
Saier, M.H., Daniels, G.A., Boerner, P. and Lin, J., 1988, J. Membr. Biol., 104: 1-20.
Samarzija, I. and Frömter, E., 1982a, Pflügers Arch., 393: 199-209.
Samarzija, I. and Frömter, E., 1982b, Pflügers Arch., 393: 210-214.
Samarzija, I. and Frömter, E., 1982c, Pflügers Arch., 393: 215-221.
Schafer, J.A. and Barfuss, D.W., 1980, Am. J. Physiol., 238: F335-F346.
Schneider, E.G. and Sacktor, B., 1980, J. Biol. Chem., 225: 7650-7656.
Schneider, J.A., Schulman, J.D. and Seegmiller, J.E., 1978, *The Metabolic Basis of Inherited Disease*, Stanbury, J.B., Wyngaarden, J.B. and Fredrickson, D.S. (eds), McGraw-Hill, New York, p. 1660-1682.
Schoolwerth, A.C. and La Noue, K., 1978, J. Biol. Chem.
Schwab, S.J. and Hammerman, M.R., 1985, Am. J. Physiol., 249: F338-F345.
Schwartz, W.B., Hall, P.W., Hays, R.M. and Relman, A.S., 1959, J. Clin. Invest. 39-
Schwegler, J.S., Heuner, A. and Silbernagl, S., 1989a, Pflügers Arch., 414: 543-550.
Schwegler, J.S., Heuner, A. and Silbernagl, S., 1989b, Pflügers Arch., 415: 183-190.
Schwegler, J.S., Schömig, E., Heuner, A. and Silbernagl, S., 1990, *Amino Acids: Chemistry, Biology and Medicine*, Lubec, G. and Rosenthal, G.A. (eds), ESCOM, Leiden, p. 1017-1028.
Scott, R.D. and Curthoys, N.P., 1987, Am. J. Physiol., 252: F877-F882.
Scriver, C.R., Pueschel, S. and Davis, E., 1966, N. Engl. J. Med., 274: 635-645.

Segal, S., McNamara, P.D. and Pepe, L.M., 1977, Science, 197: 169-171.
Sepulveda, F.V. and Pearson, J.D., 1984, J. Cell Physiol., 118: 211-217.
Sereni, F., McNamara, H., Shibuya, M., Kretchmer, N. and Barnett, H.L., 1955, Pediatrics, 15: 575-585.
Shaloub, R. et al., 1963, Am. J. Physiol., 204: 181-186.
Shannon, J.A., Farber, S. and Troast, L., 1941, Am. J. Physiol., 133: 752-761.
Silbernagl, S., 1977, Pflügers Arch., 368: R13 (Abstract).
Silbernagl, S., 1980, Int. J. Biochem., 12: 9-16.
Silbernagl, S., 1983, Pflügers Arch., 396: 218-224.
Silbernagl, S., 1985, *The Kidney: Physiology and Pathophysiology*, Seldin, D.W. and Giebisch, G. (eds) Raven Press, New York, p. 1677-1701.
Silbernagl, S., 1986, Pflügers Arch., 407: S72-S79.
Silbernagl, S., 1988, Physiol. Rev., 68: 911-1007.
Silbernagl, S. and Deetjen, P., 1971, Pflügers Arch., 323: 342-350.
Silbernagl, S. and Deetjen, P., 1972a, Pflügers Arch., 336: 79-86.
Silbernagl, S. and Deetjen, P., 1972b, Pflügers Arch., 337: 277-284.
Silbernagl, S. and Deetjen, P., 1973, Pflügers Arch., 340: 325-334.
Silbernagl, S. and Völkl, H., 1983, Pflügers Arch., 396: 225-230.
Silbernagl, S., Ganapathy, V. and Leibach, F.H., 1987, Am. J. Physiol., 253: F448-F457.
Silverman, M., Aganon, M.A. and Chinard, F.P., 1970, Am. J. Physiol., 218: 743-750.
Smith, D.W., Scriver, C.R., Tenenhouse, H.S. and Simell, O., 1987, Proc. Nat. Acad. Sci. USA, 84: 7711-7715.
Squires, E.J., Hall, D.E. and Brosnan, J.T., 1976, Biochem. J., 160: 125-128.
Staddon, J.M., Bradford, N.M. and McGivan, J.D., 1984, Biochem. J., 217: 855-857.
States, B. and Segal, S., 1990, J. Cell Physiol., 143: 555-562.
States, B., Foreman, J., Lee, J., Harris, D. and Segal S., 1987, Metabolism, 36: 356-362.
Stieger, B., Stange, G., Biber, J. and Murer, H., 1983, Pflügers Arch., 397: 106-113.
Sugita, M., Sugita, K., Furukawa, T. and Abe, H., 1967, Jpn. Circ. J., 31: 405-414.
Tamarappoo, B.K. and Kilberg, M.S., 1991, Biochem. J., 274: 97-101.
Tanaka, H., Akutsu, T., Kobutani, T. and Nishi, H., 1989, Agric. Biol. Chem., 53: 1509-1513.
Tate, S. and Meister, A., 1974, Proc. Nat. Acad. Sci. USA, 71: 3329-3333.
Tate. S.S. and Meister. A., 1974, J. Biol. Chem., 249: 7593-7602.
Tate, S.S., Urade, R., Getchell, T.V. and Udenfriend, S., 1989, Archives of Biochemistry and Biophysics, 275: 591-596.
Teijema, H.L., Gelderen, H.H. van, Giesberts, M.A.H., Serena, M.L. and Angelo, L. De, 1974, Metabolism, 23: 115-123.
Terao, N., Suzuki, M., Asano, Y. and Hosoda, S., 1985, Kidney International, 27: 951-953.
Thier, S.O. and Segal, S., 1978, In: *The Metabolic Basis of Inherited Disease*, Stanbury, J.B., Wyngaarden, J.B. and Fredrickson, D.S. (eds), McGraw-Hill, New York, p. 1578-1592.
Tiruppathi, C., Ganapathy, V. and Leibach, F.H., 1990, J. Biol. Chem., 265: 2048-2053.
Tizianello, A., DeFarrari, G., Garibotto, G. and Guveris, G., 1978, Clin. Sci. Mol. Med., 44: 391-397.
Tizianello, A., DeFarrari, G., Garibotto, G., Guerri, G., Robaudo, C., Acquarone, N. and Ghiggeri, G.M., 1980, J. Clin. Invest., 65: 1162-1173.
Tizianello, A., DeFarrari, G., Garibotto, G., Robaudo, G., Salvidio, G. and Saffioti, S., 1985, Contr. Nephrol., 47: 44-57.
Toback, F.G., Mayers, A.M. and Lowenstein, L.M., 1973, Am. J. Physiol., 225: 1247-1251.
Ullrich K.J., Rumrich, G., and Klöss, S., 1974, Pflügers Arch., 351: 35-48.
Uranga, J., et al., 1979, Horm. Metab. Res., 11: 275-279.
Vacca, C.V., Hines, J.D. and Hall, P.W., 1986, Environ. Res., 41: 440-446.
Vina, J.R., Palacin, M., Puertes, I.R., Hernandez, R. and Vina, J., 1989. Am. J. Physiol., 257: E916-922.
Völkl, H. and Silbernagl, S., 1980, Pflügers Arch., 387: 253-259.
Völkl, H. and Silbernagl, S., 1982, Pflügers Arch., 395: 190-195.
Völkl, H. Geibel, J., Silbernagl, S. and Lang, F., 1984, Pflügers Arch., 402: R5 (Abstract).
Welbourne, T. C. and Dass, P. D., 1988, Pflügers Arch., 411: 573-578.
Welbourne, T.C. and Solomon, S., 1989, Life Sci., 45: 2169-2174.
Welbourne, T.C., 1976, Endocrinology, 99: 1071-1079.
Welbourne, T.C., 1990, Semin. Nephrol., 10: 339-349.

Welbourne, T.C., 1991, Contrib. Nephrol., 50:
Welbourne, T.C., Weber, M. and Bank, N., 1972, J. Clin. Invest., 51: 1952-1960 .
Wellner, D. and Meister, A., 1981, Annu. Rev. Biochem., 50: 911-968.
Wendel, A. and Cikryt, P., 1980, FEBS Letters, 120: 209-211.
Whelan, D.T. and Scriver, C.R., 1968, Pediatric Research, 2: 525-534.
White, M.F., 1985, Biochim. Biophys. Acta, 822: 355-374.
Windus, D.W., Cohn, D.E., Klahr, S. and Hammerman, M.R., 1984, Am. J. Physiol., 246: F78-F86.
Windus, D.W., Klahr, S. and Hammerman, M.R., 1984, Am. J. Physiol, 247: F:403-F407.
Winter, C.G. and Christensen, H.N., 1965, J. Biol. Chem., 239: 872-878.
Woods, L.L., Mizelle, H.L., Montani, J. and Hall, Jr., 1988, Am. J. Physiol., 252: F303312.
Wright, E.M. and Peerce, B.E., 1984, J. Biol. Chem., 259: 14993-14996.
Wright, L.A. and Nicholson, T.F., 1966, Canadian Journal of Physiology and Pharmacology 44: 195-201.
Wright, L.D., Russo, H.F., Skeggs, H.R., Patch, E.A. and Beyer, K.H., 1947, Am. J. Physiol., 149: 130-134.
Yamamoto, H., Aikawa, T., Matsutaka, H., Okuda, T. and Ishikawa, E., 1974, Am. J. Physiol., 226: 1428-1433.
Yoshitomi, K. and Frömter, E., 1985, Pflügers Arch., 405: S121-S126.

AMINO ACID TRANSPORT IN LYMPHOCYTES

George B. Segel

University of Rochester School of Medicine and Dentistry
601 Elmwood Avenue, Box 777, Rochester, New York 14642

INTRODUCTION

Amino acid transport in lymphoid cells involves processes that are conserved in mammalian tissues. The A, L, ASC and y^+ systems for amino acid transport characterized in Ehrlich Ascites cells (Christensen et al, 1973) and Chinese hamster ovary cells (Shotwell et al, 1981) are present in lymphocytes. However, the predominant transport system for any particular amino acid may vary considerably and may be related to the functional requirements of the lymphocyte type. The term "lymphocyte" is generic since the lymphocyte phenotype is heterogenous. T-lymphocytes include T-helper cells, T-suppressor cells and cytotoxic T-cells, and B-lymphocytes mature into immunoglobulin secreting plasma cells of five different isotypes. Differences in amino acid transport have been established between T and B-lymphocyte types and their leukemic counterparts. Less is known of amino acid transport in the T and B-lymphocyte subtypes.

SYSTEM A

The A-system of amino acid transport in lymphoid cells exhibits characteristics similar to A-system transport in other described tissues (Guidotti et al, 1978; Oxender et al, 1985). Amino acids with short, polar or linear side chains such as aminoisobutyric acid (AIB) or methylaminoisobutyric acid (methyl-AIB) are preferred substrates for the A-system. In particular, N-methylated amino acids are transported by the A-system in contrast to the ASC-system (see below). An excess of internal A-system amino acids transinhibits the uptake of methyl-AIB. A-system transport is dependent on a sodium gradient, is sensitive to a decrease in pH and is markedly inhibited below a pH of 6.8. The A-system also is characterized by the process of "adaptation", that is, heightened transport when the cells are exposed to amino acid deprivation (see below).

Amino acid uptake by human lymphocytes has been measured using AIB or methyl-AIB as A-system prototypes for the past 20 years (Mendelsohn et al, 1971; Whitney and Sutherland, 1973). Studies have been conducted in human blood lymphocytes (Segel et al, 1983); rat splenocytes and thymocytes (Wise, 1978); mouse lymphocytic cell lines (Finklestein and Adelberg, 1977); and human lymphoblast cell lines (Bass and Englesberg, 1979). The Michaelis constant (Km) for AIB or methyl-AIB uptake among these cell types is remarkably similar, approximately 500 uM in blood lymphocytes (Segel et al, 1983), a human lymphoblast cell line (Bass and Englesberg, 1979), and a mouse lymphocytic cell line (Finklestein and Adelberg, 1977). The magnitude of the transport rate, however, varies among the cell types. For example, rat thymic as opposed to rat splenic lymphocytes show greater than five times the accumulation rate of methyl-AIB (Wise, 1978).

Mammalian Amino Acid Transport, Edited by M.S. Kilberg and
D. Häussinger, Plenum Press, New York, 1992

Table 1. Kinetics of amino acid transport in human blood lymphocytes[3].

	k_m	v_{max}	K_u
Methyl-AIB (4)	540	28	.004
Proline (3)	1260	75[1]	.031
Alanine (3)	230	395	--[2]
BCH (4)	63	969	.141
Leucine (3)	120	1240	.202

The k_m, v_{max} and K_u were calculated using a computer-assisted multicomponent analysis. The k_m is expressed as µmol/liter, the v_{max} as µmol/L cell water/min, and K_u as min^{-1}. The number of lymphocyte populations is shown in parenthesis.
[1]Indicates sodium-dependent transport only.
[2]Indicates that no nonsaturable transport was measured since the ratio of A_i/A_o was > 1 at the amino acid concentrations that were studied.
[3]Adapted from Segel et al, 1983.

In establishing the pathway of entry of amino acids into lymphocytes we have considered the rate of appearance of an amino acid within the cell as a result of two parallel processes: a carrier mediated process obeying Michaelis Menten kinetics (ν_1), and a nonsaturable process which depends on a membrane constant and the difference in amino acid concentration across the plasma membrane (ν_2) (Segel et al, 1983). The following equation with three unknown constants: v_{max}, k_m, and K_u considers the observed uptake as the sum of the carrier mediated transport and the nonsaturable process.

$$v_T = v_1 + v_2 = \frac{v_{max}}{(k_m/C_{outside}) + 1} + K_u(C_{outside} - C_{inside})$$

The constants are determined from a series of measured velocities as a function of outside and inside concentrations. Curve fitting to the observed velocities can be reduced to a least squares procedure for ν_{max} and K_u by treating $k_m/C_{outside}$ as an independent variable whose possible range is searched to establish the best curve fit (Simon, 1986). For each trial value of $k_m/C_{outside}$, a least squares fit is done to find v_{max} and K_u. The sums of the squared residuals for each trial value are compared to find the value of $k_m/C_{outside}$ which yields the minimum sum of the squared residuals.

The kinetics for transport of various amino acids by human lymphocytes are shown in Table 1. The amino acids other than methyl-AIB are transported only fractionally or not at all (eg. BCH) by system A, and their uptake reflects a composite of several transport pathways (Table 2). Greater than 90% of methyl-AIB traverses the A-system, and >90% of 2-amino-2-carboxybicyclo-heptane (BCH) traverses the L-system. However, alanine and proline use more than one pathway to enter lymphocytes.

Other amino acids such as glycine and glutamic acid are transported at least in part by the A-system (Dantzig et al, 1978). Mouse lymphocyte mutants of the A-system have been produced using a tritium-suicide technique, so that fewer but qualitatively normal A-system transport sites are present (Finklestein and Adelberg, 1977 and Dantzig et al, 1979).

The energy source for A-system transport appears related to the sodium gradient across the plasma membrane as well as to additional less well defined sources such as a redox substrate (Ohsawa et al, 1980 and Heinz et al, 1981). The sodium gradient is maintained by the sodium-potassium transport ATPase to sustain a low (approximately 10 to 15 mM) concentration of intracellular sodium compared to the high extracellular

Table 2. Pathways of natural amino acid-mediated transport in human blood lymphocytes[1].

Amino acid	A-system (%)	ASC-system (%)	L-system (%)
Methyl-AIB	90	--	10
Alanine	10	70	10
Proline	30	30	15
Leucine	7.5	10	70
BCH	--	--	90

[1]Adapted from Segel et al, 1983.

sodium (approximately 140 mM) (Segel et al, 1979). Early studies established that neither glycolysis nor respiration are the direct source of A-system transport energy (Eddy et al, 1967), yet ATP restoration in previously depleted cells is associated with restoration of amino acid uptake, and the involvement of an undefined ATPase remains a possibility (Ohsawa et al, 1980).

The amount of lymphocyte ATP production devoted to maintenance of the Na gradient in an amino acid rich medium is approximately 6 mmol/L cell water/hr when assessed as the ouabain sensitive portion of energy metabolism (Segel et al, 1978). This ouabain sensitive fraction represents about one third of the total ATP production of either resting or activated lymphocytes (Segel et al, 1978) and can be considered closely related to the energy requirement for the Na dependent amino acid transport systems.

Adaptation

The A-system of amino acid transport shows heightened activity when cells are amino acid starved. This phenomenon has been described in chick embryo heart cells (Gazzola et al, 1972) and in other mammalian tissues (Guidotti et al, 1975). A number of studies have shown a similar adaptive process of the A-system in various lymphoid cells. Studies of chronic lymphocytic leukemic B-lymphocytes (CLL) indicate that time dependent increases in both alanine and AIB but not leucine uptake occur when these cells are deprived of environmental amino acids (Frengley et al, 1974). This adaptive increase in A-system transport requires new DNA, RNA and protein synthesis (Frengley et al, 1974, Peck et al, 1976). Additional studies in CLL and acute lymphoblastic leukemia cells (ALL) indicate that the adaptive increase results from a heightened maximal velocity of transport as well as increased affinity for the transported amino acids (decreased k_m) (Frengley et al, 1975a). Further, specific studies eliminate the release of transinhibition as the explanation for the time dependent increase in A-system activity (Frengley et al, 1975a). Interestingly, the adaptation of leukemic lymphocytes (CLL and ALL) is prevented by the presence of cortisol (1 mm), a drug which is a known inhibitor of these lymphoid cells and has no effect on the substantial adaptive increase of myeloid leukemic cells (Frengley et al, 1975b). Studies with rat thymic lymphocytes indicate that adaptation is heightened with increasing pH in the range of 6.5 to 8.0 (Peck et al, 1976). Further, human blood lymphocytes show a clear difference in the mechanism of heightened A-system transport mediated by adaptation when contrasted to the increases mediated by lectin treatment which can be superimposed on the adaptive increase and which affect the ASC and L-systems also (Segel & Lichtman, 1981).

SYSTEM ASC

The ASC System transports nonmethylated neutral amino acids such as alanine, serine and cysteine (Christensen et al, 1967). ASC transport depends on the gradient of sodium and is transstimulated by internal amino acids. The ASC system is insensitive to pH and shows no "adaptation" to amino acid deprivation. There is no amino acid prototype for the ASC system but about 70% of alanine uptake traverses this system in human T-lymphocytes (Segel et al, 1983). The ASC system is particularly prominent in resting porcine lymphocytes where it accounts for the bulk of sodium dependent transport (Borghetti et al, 1979a; Borghetti et al, 1981). The studies of Borghetti and his co-workers in lymphocytes established the sodium dependence, the phenomena of trans-stimulation and a broad specificity of the ASC system for amino acid substrates such as alanine, serine, cysteine and threonine. In resting cells both low (k_m approximately 4 mM) and high (k_m approximately 0.2 mM) affinity alanine transport are observed. Rat splenic lymphocytes also transport amino acids primarily by the ASC system in contrast to rat thymic cells which employ both the A and ASC pathways for sodium dependent transport (Wise, 1978).

Cysteine and Cystine

The uptake of cysteine by the ASC system is related to the maintenance of cellular glutathione in murine lymphocytes (Ishii et al, 1987). Glutathione provides protection from oxidant threats to cellular molecules and membrane lipids (Editorial, Nutritional Reviews, 1989). Cysteine uptake is inhibited in murine lymphocytes more than 80% by a 50 fold excess of glutamine, serine and threonine and is not blocked by methyl-AIB (Ishii et al, 1987). The uptake of cysteine is particularly critical to the maintenance of glutathione since cystine transport is extremely low in both resting and stimulated lymphocytes (Bannai, 1984).

Glutamine

Glutamine, glutamic acid and glutamate are acidic amino acids. The uptake of glutamine involves, in part, the A-system in a mouse lymphocyte line (Dantzig et al, 1978) and the ASC system in rat mesenteric lymphocytes (Ardawi and Newsholme, 1986). Glutamine uptake is sodium dependent and competitively inhibited by serine and by its chloroketone analog, L-2-amino-4-oxo-5 chloropentanoate (Novogrodsky et al, 1979). Methyl-AIB and histidine do not competitively inhibit glutamine uptake in rat mesenteric lymphocytes (Ardawi and Newsholme, 1986) suggesting that systems A and L are not involved. Likewise, glutamine uptake in human blood lymphocytes is mediated by system ASC since neither Methyl-AIB nor BCH inhibit glutamine uptake indicating that there is no measurable transport via the A or L-systems (Koch et al, 1990 and Schroder et al, 1990). The v_{max} for glutamine uptake of 360 μmol/L cell water/min, and the k_m of 142 uM (Schroder et al, 1990) are similar to those of the ASC system in human lymphocytes v_{max} ~400 and k_m ~200 (Segel et al, 1983). Another potential pathway for glutamine uptake has been demonstrated in a mouse lymphocyte line and its transport mutant (Dantzig et al, 1978). In these cells there is sodium independent uptake that is insensitive to inhibition by BCH, leucine and phenylalanine in addition to sodium dependent (A-system) glutamine uptake. The authors suggest that this second system is specific for dicarboxylic acids and may be related to a high affinity transport system described in nerve tissues (Dantzig et al, 1978).

An alternative pathway for glutamate uptake involves the enzyme, gamma glutamyl transpeptidase. This membrane associated enzyme catalyzes glutamate transport when it is incorporated into phospholipid vesicles (Lerner, 1987). The physiologic role for this pathway in lymphocytes is yet unclear, although Novogrodsky et al (1977) showed an association between the level of gamma glutamyl transpeptidase and the initial rate of glutamine uptake in a human lymphoid cell line. Gamma glutamyl transpeptidase activity is markedly decreased in lymphocytes from patients with chronic lymphocytic leukemia, but this decrease does not account for the defect in L-system transport observed in these cells and appears to be a coincidental finding (Woodlock et al, 1990).

SYSTEM L

The L-system transports neutral amino acids with branches or rings in their side chains such as leucine, phenylalanine, and valine. L-system transport is insensitive to the sodium gradient and does not increase its transport rate when the cells are deprived of

Table 3. Sodium dependence of amino acid transport in human blood lymphocytes[1].

	Extracellular sodium concentration		Percentage inhibition
	140 mM	10 mM	
Methyl-AIB	4.7 ± 1	1.2 ± 1	78
Alanine	85.0 ± 14	26.0 ± 6	71
Proline	8.7 ± 1	4.2 ± 1	57
BCH	376.0 ± 51	365.0 ± 48	3
Leucine	612.0 ± 28	539.0 ± 31	12

The data represent v_i, expressed as μmol/L cell water/min. The percent inhibition of uptake by the elimination of the Na gradient is shown. Amino acid uptake was measured at a concentration of 0.1 mM. The data represent the mean ± S.E. of measurements in three lymphocyte populations.
[1]Adapted from Segel et al, 1983.

external L-system amino acids. BCH or 2-amino-2-carboxy-biocyclo (2,2,1) heptane is a synthetic amino acid that is transported primarily by the L-system and has been used to characterize L-system transport (Christensen et al, 1969). L-system transport is prominent in human blood lymphocytes (Table 1) with a maximal velocity of approximately 1,000 μmol/L cell water/min and a k_m of 50 to 100 uM (Segel et al, 1983). Similar kinetic parameters have been found in human tonsillar-B lymphocytes, blood B-lymphocytes and a human B cell line (Segel et al, 1984). A similar k_m was observed for phenylalanine uptake in a human lymphoblast cell line (Bass and Englesberg, 1979), while lower k_m values, 4 to 15 uM, were described in mouse lymphocytic cell lines (Dantzig et al, 1979).

The descriptions of leucine and BCH uptake in various lymphocyte types display the classic characteristics of system L. No evidence for sodium dependence is observed (Table 3), and no time dependent increase in amino acid transport is observed with amino acid deprivation (Segel et al, 1983). Further, transstimulation of BCH uptake by prior BCH loading increases the v_{max} by 1.6 fold (Segel et al, 1988) without a significant change in the k_m. Similar results are seen when leucine rather than BCH is used as the intracellular amino acid. Transinhibition of A-system transport is observed under similar conditions in both human T and B-lymphocytes (Segel et al, 1988).

Leucine influx and efflux have been described in human-mouse myeloma hybrids (Cieplinski et al, 1985). While no difference is observed in leucine uptake between immunoglobulin producing and nonproducing cells, decreased leucine efflux is observed in those hybrids that synthesize immunoglobulin (Cieplinski et al, 1985). Regulatory mutants for L-system transport have been described in chinese hamster ovary cells (Oxender et al, 1985). Mutants with both heightened and diminished L-system transport have been prepared from a temperature resistant CHO-tsH1 cell line.

Chronic Lymphocytic Leukemia (CLL)

B-lymphocytes from patients with chronic lymphocytic leukemia uniformly show diminished L-system transport (Segel and Lichtman, 1982; Segel et al, 1984). A-system transport in CLL B-cells precedes at normal rates, while L-system transport is reduced to approximately 10 to 15% of that observed in normal blood B-lymphocytes as measured by BCH uptake (Segel et al, 1984). When CLL cells are loaded with BCH no increase in ^{14}C BCH uptake (ie. transstimulation) is observed (Segel et al, 1988). Thus, the diminished L-system transport of these cells is not related to a failure of amino acid exchange. Since the maximal velocities of L-system transport in normal B-lymphocytes, tonsillar B-lymphocytes, a B-cell line and T-cell CLL are ten fold that of CLL B-cell preparations, it is likely that the diminished L-system transport in CLL cells is a specific defect associated with this neoplasm (Segel et al, 1988).

Tetradecanoyl phorbol acetate (TPA) treatment of CLL lymphocytes produces a more mature immunoglobulin secreting phenotype, and L-system amino acid uptake is markedly enhanced from eight-fold to fourteen-fold at 16 and 40 hours of exposure to phorbol ester (Woodlock et al, 1988). The maximal velocity of L-system transport in phorbol ester matured CLL cells is similar to that of phorbol ester treated normal B-lymphocytes and parallels the development of plasmacytoid features at 40 hours. The development of heightened L-system transport is dependent on new protein synthesis and likely dependent on new RNA synthesis. The restoration of the selective L-system transport defect in CLL B-lymphocytes coincides with the in vitro maturation of these leukemic cells (Woodlock et al, 1988). Transstimulation likewise is restored after phorbol ester treatment and transmembrane exchange of amino acids is reestablished (Segel et al, 1988). The diminished L-system transport in CLL may reflect the neoplastic origin and/or the immature phenotype of CLL cells. It is improbable that maturational stage alone accounts for the diminished transport. The fundamental importance of amino acid availability for protein synthesis and metabolism of young proliferative, maturing cells makes a profound decrease in amino acid uptake unlikely in normal but immature cells. Unfortunately, sufficient members of normal early B-cells homologous to CLL cells are unavailable to test this hypothesis directly.

Melphalan (4[bis(2 chloroethyl)amino]-L-phenylalanine)

Melphalan is a chemotherapeutic alkylating agent used in the treatment of B-cell neoplasms such as multiple myeloma. Melphalan enters lymphoid cells by two different amino acid carriers in human blood T-lymphocytes (Vistica et al, 1978, Begleiter et al, 1979, Vistica, 1979, Begleiter et al, 1980) and plasmacytoma cells (Goldenberg et al, 1979). These two systems are most similar to the L and ASC-systems described in other cells (Begleiter et al, 1979). The L-system provides the greatest proportion of transport in lymphoblasts (Begleiter et al, 1979) and blood T-lymphocytes (Begleiter et al, 1980). Melphalan uptake also has been studied in cells from CLL patients who are commonly treated with alkylating agents. There are no differences in the kinetic parameters for melphalan uptake in CLL cell populations from untreated, treated and treatment resistant patients (Panasci et al, 1988). The proportion of BCH sensitive melphalan uptake is highest in the cells from treatment resistant CLL patients prompting the authors to suggest that resistance to melphalan is secondary to a mechanism other than a transport defect. When the kinetic parameters for melphalan transport are compared to those for BCH uptake in CLL B-lymphocytes (Segel et al, 1984), they are nearly identical. The v_{max} for the three described CLL cell populations varies from 60 to 120 μmol/L cell water/min and the k_m from 50 to 100 uM (Panasci, 1988). This maximal velocity is less than 10% of that observed for L-system transport in other lymphoid cells (Segel et al, 1984).

SYSTEM y^+

The y^+ system for amino acid transport is independent of the sodium gradient and favors the transport of dibasic cationic amino acids such as ornithine, arginine and lysine, which carry a positively charged group on their side chains (White et al, 1982). The y^+ system has been assessed in rat thymic lymphocytes (Guidotti et al, 1976), human blood T and B-lymphocytes and chronic lymphocytic leukemic B-lymphocytes (Segel et al, 1985). The initial velocity for lysine transport in rat lymphocytes is approximately 60, and in human lymphocytes, 100 μmol/L cell water/min. A kinetic analysis of lysine uptake using a nonlinear least squares computer analysis indicates a maximal velocity for saturable transport of 600 μmol/L cell water/min with a k_m of 340 micromoles per liter and a constant for nonsaturable uptake of .94 min^{-1}. No difference is observed in the kinetic parameters for CLL B-lymphocytes when compared to blood B or T-lymphocytes in contrast to the impaired L-system transport in CLL B-cells (Segel et al, 1985).

Studies of competitive inhibition demonstrate that lysine uptake is markedly inhibited by a 50 fold excess of arginine, while the prototypical substrates for the A and L-systems, methyl-AIB and BCH, show little inhibitory effect (Segel et al, 1985). Fluoronitrophenylazide, which is a photosensitive probe, has been used to study the y^+ system and was originally proposed as a specific probe for this system (Hare, 1976).

Table 4. Amino acid requirements.

Amino Acid	Essential in Vivo	Proliferation in Culture	Not Essential[2] for proliferation	Protein Synthesis
isoleucine	x	x		x
leucine	x	x		x
lysine	x	x		x
methionine	x	x		x
phenylalanine	x	x[1]		x
threonine	x	x		x
tryptophan	x	x		x
valine	x	x		x
tyrosine		x		x
histidine		x		x
glutamine		x		x
cysteine		x		x
arginine		x		x
serine		x[1]		x
alanine		x[1]		
asparagine			x	
aspartic acid			x	
glutamic acid			x	
glycine			x	
hydroxyproline			x	
proline			x	

[1] These amino acids are not essential but required for optimal DNA synthesis.
[2] Lymphocyte proliferation is impaired <20% in the absence of these amino acids.

Further studies of light sensitive fluoronitrophenylazide inhibition of amino acid uptake by human lymphocytes indicate that both Y+ and L-system transport are inhibited when lymphocytes are exposed to the photolysis products of the azide (Segel et al, 1985).

CELL CYCLE AMINO ACID REQUIREMENTS AND TRANSPORT

Although there are eight essential amino acids including isoleucine, leucine, lysine, methionine, phenylalanine, threonine, tryptophan and valine for humans in vivo, cells in culture require fifteen amino acids for optimal survival and proliferation in vitro (Chuang et al, 1990) (Table 4). The additional amino acids include alanine, serine, arginine, cysteine, glutamine, histidine and tyrosine. All except alanine are required for lymphocytes to synthesize proteins (Waithe et al, 1975). The requirements for DNA synthesis were established using experiments with lectin stimulated human lymphocytes and "drop out" medium to define the individual amino acid requirements (Chuang et al, 1990). Twelve of the fifteen amino acids including isoleucine, leucine, lysine, methionine, threonine, tryptophan, valine, tyrosine, histidine, glutamine, cysteine, and arginine are essential for proliferation. Three additional amino acids, serine, phenylalanine and alanine are required for optimal ^{3}H thymidine incorporation into DNA, but substantial DNA synthesis persists in their absence (Chuang et al, 1990) (Table 4). Lymphocyte proliferation is impaired less than 20% in the absence of asparagine, aspartic acid, glutamic acid, glycine, hydroxyproline and proline (Chuang et al, 1990).

As might be predicted from the broad amino acid requirements, heightened amino acid transport occurs when lymphoid cells enter the proliferative cell cycle. Specifically, studies in a lymphoblastoid B-cell line show an increase in L-system (leucine, isoleucine and valine) transport when cells pass from G0 into the proliferative cycle (Glassy and Furlong, 1981). The heightened transport persists through G1 and falls during the latter part of S-phase (Glassy and Furlong, 1981). The fall in S-phase contrasts with the findings in chinese hamster ovary and L-cells in which a gradual rise in amino acid transport occurs during the latter part of S-phase (Sander and Pardee, 1972).

EXTRINSIC REGULATION OF AMINO ACID TRANSPORT

The rate of amino acid uptake is influenced by a variety of external conditions which include cell density, external amino acid concentration and pH. The phenomenon of A-system adaptation in response to amino acid starvation and the regulatory effect of pH are discussed in the System-A section. The binding of polyclonal mitogens such as plant lectins to the cell membrane results in activation of second messenger pathways and increased amino acid transport. Similar effects are seen with specific antigen and sensitized lymphocytes, antibody such as OKT3 and hormones such as insulin or triiodothyronine. Inhibitory effects are seen with glucocorticoids and other toxic agents.

Cell Density

The transport rates of systems A and ASC are accelerated in a Burkitt Lymphoma cell line (SV3T3) at cell densities less than 3×10^5 cells per ml (Piedimonte et al, 1989). This sensitivity to cell density suggests that cell-cell communication either by direct contact or humoral mediators influences the rate of amino acid uptake.

Plant Lectins

Lymphocytes that have been stimulated with phytohemagglutinin (PHA), and concanavalin A (Con A) increase their amino acid transport rates (Mendelsohn et al, 1971; Averdunk, 1972; Whitney and Sutherland, 1973; Segel and Lichtman, 1981). Heightened transport after lectin treatment is observed in pig (Borghetti et al, 1979a; Borghetti et al, 1981), rat (Van Den Berg and Betel, 1971; Van Den Berg and Betel, 1973a; Van Den Berg and Betel, 1973b) and human lymphocytes (Mendelsohn et al, 1971; Whitney and Sutherland, 1973; Greene et al, 1976a; Segel and Lichtman, 1981) and in rabbit thymocyte vesicles (Schmidt-Ullrich et al, 1976). This increased transport reflects an increase in the v_{max} for the A and ASC systems (Segel and Lichtman, 1981; Borghetti et al, 1981) in human and pig lymphocytes respectively. Further, neither the k_m for A, (0.5 mM) (Segel and Lichtman, 1981) nor ASC (0.2 mM) (Segel and Lichtman, 1981; Borghetti et al, 1981) change significantly after PHA treatment. There are, however, species differences. For example, resting pig lymphocytes exhibit little or no A-system transport (Borghetti et al, 1981) when compared to human blood lymphocytes (Segel and Lichtman, 1981) but develop marked A-system enhancement with PHA (Borghetti et al, 1981). L-system transport in human lymphocytes increases little after lectin treatment for four hours (Segel and Lichtman, 1981). However, there are substantial increases in pig lymphocytes, 1.5 fold (Borghetti and Kay, 1979) and human lymphocytes, 2-3 fold (Segel and Lichtman, 1981) if the lymphocytes are exposed to PHA for prolonged periods, for example, 20-24 hours. The effect of lectins on both A (Borghetti et al, 1979b) and L-system transport (Lucivero et al, 1988) parallel the effect on lymphocyte activation as assessed by 3H-uridine incorporation into RNA or 3H-thymidine incorporation into DNA. Wheat germ agglutinin is not mitogenic for human lymphocytes and, in fact, inhibits A-system transport in resting and PHA or Con A treated cells (Greene et al, 1976a).

The heightened amino acid transport observed in lectin treated lymphocytes depends on new protein synthesis and is blocked by the presence of inhibitors such as cycloheximide (Van Den Berg and Betel, 1973a). A secondary effect of the new protein synthesis is an increased stability of A-system transport in lectin treated lymphocytes (Borghetti and Kay, 1979). The half life of A-system decay in the presence of cycloheximide increases from 4 to 8 hours after lectin treatment (Borghetti and Kay, 1979).

The lectin induced increase in A-system transport is independent of transport adaptation to amino acid deprivation and can be superimposed upon it in human lymphocytes (Segel and Lichtman, 1981). Since other amino acid transport pathways, for example ASC and L, also are accelerated by lectins, their enhancement by lectins must occur by a mechanism independent of that responsible for adaptation which is characteristic of system A.

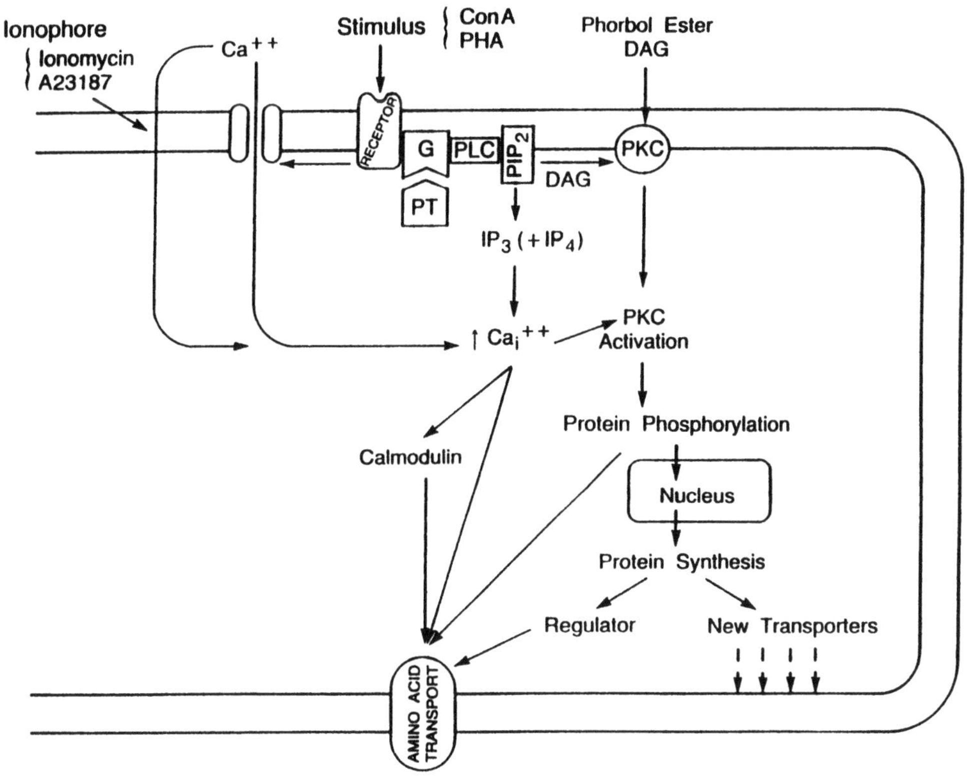

<u>Abbreviations</u>:

G	Guanine nucleotide binding protein
PT	Pertussis Toxin
PLC	Phospholipase C
PIP_2	Phosphatidyl inositol bisphosphate
IP_3	Inositol triphosphate
DAG	Diacylglycerol
PKC	Protein kinase C

Figure 1. Messenger systems in cell activation. Binding of a ligand to surface receptors results in activation of guanine nucleotide binding (G) proteins which are heterotrimers. The subunits G_α and $G_{\beta\gamma}$ then, 1) activate membrane calcium channels, increasing cytoplasmic free calcium and, 2) increase the activity of PLC which produces IP_3 and IP_4, and DAG. IP_3 and IP_4 release calcium from internal stores, and DAG activates PKC which phosphorylates numerous proteins which in turn serve as signals for various effectors. The heightened cytoplasmic free calcium also serves alone and in conjunction with calcium regulator proteins such as calmodulin as a signal for effector processes.

The mechanism whereby PHA or Con A stimulate amino acid transport is unknown. Both agents bind to membrane receptors which may interact with microtubules, microfilaments and the cytoskeleton since colchicine, a microtubule-active agent inhibits AIB uptake by lectin treated human lymphocytes (Resch et al, 1981). In contrast, cytochalasins (A, B, E, and D) heighten further the increases due to PHA and Con A in both DNA synthesis and amino acid uptake (Greene et al, 1976b). This effect of cytochalasins is not due to an alteration in lectin binding to the cells, but may reflect involvement of microfilaments or of calcium uptake which is enhanced by cytochalasins (Greene and Parker, 1975). In this regard, the calcium binding protein, calmodulin, may play a role in Con A stimulated leucine transport in human lymphocytes through calmodulin regulation of calcium dependent potassium channels (Sato et al, 1988). Activation of the potassium channels appears necessary for membrane hyperpolarization and consequent heightened leucine uptake after Con A exposure.

The treatment of lymphocytes with PHA also may result in a shift in the activity of the various transport pathways. For example, methionine transport in human lymphocytes is primarily sodium independent (Scanlon et al, 1983). A four fold to eight fold increase in methionine uptake is observed after the cells are exposed to PHA for 72 hours (Scanlon et al, 1983). However, the increased transport is primarily sodium dependent and appears related to the ASC system.

Activation of Second Messenger Pathways: Protein kinase C, Calcium, Na/H Exchange and Phospholipid Metabolism.

A complex second messenger network is involved in the initiated responses of various cell types including lymphocytes. A summary of this network is shown in Figure 1 to indicate the potential sites of action for plant lectins (see above), phorbol esters such as tetradecanoylphorbol-13-acetate (TPA), diacylglycerol and calcium ionophores such as A23187 or ionomycin. While plant lectins interact with cell surface receptors producing sequential signalling responses leading to new protein synthesis, phorbol esters heighten neutral amino acid transport more directly by binding to protein kinase C (Figure 1). Interaction with protein kinase C results directly in secondary protein phosphorylation and a very prompt increase in amino acid transport. Increased A-system transport is observed in TPA treated bovine, rat and human lymphocytes respectively (Kensler et al, 1979, Novogrodsky et al, 1982, Klip et al, 1986, Woodlock et al, 1988). TPA induced enhancement of L-system transport in human blood B-lymphocytes and rat thymocytes also is documented (Woodlock et al, 1988 and Klip et al, 1986) although longer exposure (~ 8 hours) is required.

When SN-1,2-diacylglycerols (DAG), the natural intracellular ligands for protein kinase C, are used in place of phorbol esters, a time dependent increase in A-system transport is seen beginning as early as one minute after the initial exposure (Woodlock et al, 1989). Such an early response suggests protein alteration such as phosphorylation rather than new protein synthesis in the mechanism of transport control. Supporting this interpretation, neither phorbol ester (Kensler et al, 1979) nor DAG (Woodlock et al, 1989) enhancement of A-system transport is sensitive to inhibition of protein synthesis. However, H-7, an inhibitor of protein kinase C causes substantial reduction of the DAG enhancement of A-system uptake.

Treatment of lymphocytes with calcium ionophores such as A23187 or ionomycin also may initiate lymphocyte mitogenesis (Lichtman et al, 1983). Ionophores facilitate entry of calcium from the extracellular medium to the cytoplasm. The heightened cytoplasmic free calcium (Figure 1) then can interact with effectors directly or through protein kinase C via protein phosphorylation without or with new protein synthesis. Treatment of human T-lymphocytes with ionomycin increases A-system transport after 1 minute and causes a one and one-half to two fold increase in the initial velocity after 30 minutes exposure (Woodlock et al, 1989). This rapid enhancement of A-system transport does not depend on new protein synthesis (Woodlock et al, 1989). In contrast, A23187 blocks AIB transport in rat thymocytes at a concentration of 1 μM to 6 μM (Batzri and Gardner, 1976). This inhibition may reflect a difference in species or result from excess calcium influx as these concentrations are toxic to human T-lymphocytes (Lichtman et al, 1983). If the effect of DAG is compared to that of ionomycin, two relatively separate pathways are suggested for enhancement of A-system amino acid transport in T-lymphocytes (Woodlock et al, 1989). DAG interacts directly with protein kinase C while ionomycin produces an increase in A-system transport of similar magnitude by increasing the cytoplasmic free calcium independent of kinase C. This may reflect a direct interaction of Ca or Ca-calmodulin with the A-system transporter (Figure 1). These pathways likely are complimentary when plant lectins or physiologic stimuli such as antigens incite lymphocyte activation.

Stimulation of Na/H exchange results in cytoplasmic alkalinization which is associated with the action of several mitogenetic growth factors (Grinstein and Rothstein, 1986; Moolenaar, 1986). Na/H exchange is enhanced in lymphocytes stimulated with both mitogens (Hesketh et al, 1985) and phorbol esters (Grinstein et al, 1985). However, the heightened amino acid transport is not sensitive to amiloride inhibition (an inhibitor of Na/H exchange), and increasing the intracellular sodium and cytoplasmic alkalinization do not stimulate amino acid uptake (Klip et al, 1986). Therefore, Na/H exchange does not appear to mediate directly the enhancement of neutral amino acid transport.

Arachidonic acid metabolites such as thromboxanes are suggested as potential mediators of cellular responses (Udey and Parker, 1982). A variety of inhibitors of thromboxane synthesis block A-system transport in both unstimulated and Con-A stimulated human blood lymphocytes (Udey and Parker, 1982). The mechanism whereby thromboxanes influence amino acid transport is unknown, but is not related to direct cytotoxicity. It may involve cyclic AMP or cyclic GMP dependent protein phosphorylation which is dependent on thromboxane A2 (Udey and Parker, 1982).

Other Agents Producing Heightened Transport

Increased transport of the A and L-systems is seen when bovine lymphocytes from animals sensitized to mycobacterium antigens are exposed directly to the antigenic material (Tate, 1982). This antigen specific effect depends upon new protein synthesis and heightens the A and L but not the ASC system. Sodium dependent alanine transport also increases when human lymphocytes are treated with the mitogenic monoclonal antibody, OKT3 (Walls et al, 1984).

Hormones such as insulin and T3 (triiodothyronine) also accelerate neutral amino acid transport. Insulin increases the uptake of AIB by raising the v_{max} and lowering the k_m in rat thymocytes (Goldfine et al, 1972). A-system transport is enhanced by insulin without an effect on L or y^+ transport (Guidotti et al, 1976). The effect of thyroid hormone on AIB and cycloleucine transport in rat thymocytes involves a diminution of amino acid efflux without an effect on amino acid uptake (Goldfine et al, 1974). Hence there is augmented amino acid accumulation in thyroid hormone treated cells.

Other agents such as sodium periodate (Lee et al, 1978) and hypertonic solutions (Klip et al, 1986) also heighten AIB transport into pig splenic lymphocytes and rat thymocytes respectively. The periodate effect involves an increase in the v_{max} with no change in the k_m.

Glucocorticoids and Other Transport Inhibitors

The inhibitory effect of glucocorticoids on lymphoid tissues is well known, and noncompetitive glucocorticoid inhibition of A-system transport has been described in rat thymocytes (Baran et al, 1973) and in human chronic lymphocytic leukemic lymphocytes (Baran et al, 1972). The suppressive effect of cortisol on A-system transport requires new protein synthesis and is not observed in the presence of cycloheximide (Baran et al, 1972). Other agents inhibiting lymphoid A-system amino acid transport include anti-inflammatory drugs (Bayer et al, 1981), superoxide (Kwock et al, 1981), hyperthermia and radiation (Kwock et al, 1978).

ISOLATION OF A TRANSPORTER

Neither the protein nor the gene for the A, ASC or L mammalian amino acid transporters has yet been isolated. Only the gene for y^+ transporter has been identified as a retroviral receptor which mediates transport of dibasic amino acids (Kim et al, 1991; Wang et al, 1991). A variety of innovative approaches have been applied to this problem with tantalizing potential but no other transporter as the prize. The L-system gene has been localized to chromosome #20 using hybrids of temperature sensitive leucyl-tRNA synthetase chinese hamster ovary cell mutants and human white blood cells to restore L-system activity (Oxender et al, 1985). Further, human gene sequences from a cosmid library can correct the L-system transport defect when transformed into a mutant chinese hamster ovary cell line (Oxender et al, 1985).

Studies to isolate an amino acid transport protein have included photolabeling with agents as azidophenylalanine (Noren et al, 1983) and azaserine (Segel et al, 1989). Photolabeling with azidophenylalanine blocks alanine transport, but a specific transport protein candidate could not be identified. Photolabeling of human lymphocytes with azaserine markedly reduces BCH (L-system) transport, and this effect is blocked with excess BCH during photolabeling. Although T-lymphocyte plasma membranes can be photolabeled with radiolabeled azaserine, a specific protein could not be defined (Segel et al, 1989). More sophisticated radio-photolabeling techniques using transport mutants and giant two dimensional gel electrophoresis may yet provide convincing candidates for a transmembrane amino acid carrier.

Even more recent studies use size fractionated mRNA from mammalian liver or kidney injected into xenopus oocytes to conveniently express the mammalian transporter message. A-system transport has been shown in xenopus oocytes with a 2 to 2.5 kilobase fraction of mRNA from rat liver, chinese hamster ovary cells and human placenta (Tarnuzzer et al, 1990). Similarly, L-system and some A-system transport expression are seen in oocytes injected with a 3 to 4 kilobase fraction of mRNA from rat kidney (Tate et al, 1989). A 1.9 to 2.5 kilobase fraction of mRNA from rat liver permitted sodium/alanine transport expression when injected into xenopus oocytes (Palacin et al, 1990). In spite of this advance in localization of the mRNA for the major transport systems, none of the genes have yet been cloned. However, this methodology has been used successfully to clone the sodium/glucose transporter from a 2.3 kilobase fraction of mRNA from rabbit intestine (Hediger et al, 1987), and provides precedent for the success of these studies.

REFERENCES

Ardawi, M.S.M., Newsholme, E.A., 1986, Biochimica. et. Biophysica. Acta., 856:413.
Averdunk, R., 1972, Hoppe-Seyler's Z. Physiol. Chem., 353:79.
Bannai, S., 1984, Biochimica. et. Biophysica. Acta., 779:289.
Baran, D.T., Lichtman, M.A., Peck, W.A., 1972, J. Clin. Invest., 51:2181.
Baran, D.T., Peck, W.A., Frengley, P.A., Lichtman, M.A., 1973, Biochimica. et. Biophysica. Acta., 307:627.
Bass, R., Englesberg, E., 1979, In Vitro, 15:829.
Batzri, S., Gardner, J.D., 1976, J. Biol. Chem., 251:6030.
Bayer, B.M., Almeida, A.P., Beaven, M.A., 1981, J. Pharm. Exper. Thera., 219:752.
Begleiter, A., Lam, H-Y. P., Grover, J., Froese, E., Goldenberg, G.J., 1979, Cancer Res., 39:353.
Begleiter, A., Froese, E.K., Goldenberg, G.J., 1980, Cancer Letters, 10:243.
Borghetti, A.F., Kay, J.E., Wheeler, K.P., 1979a, Biochem. J., 182:27.
Borghetti, A.F., Girit, E.Z., Kay, J.E., 1979b, Biochem. Soc. Trans., 7:1007.
Borghetti, A.F., Kay, J.E., 1979, Biochem. J., 182:237.
Borghetti, A.F., Tramacere, M., Ghiringhelli, P., Severini, A., Kay, J.E., 1981, Biochimica. et. Biophysica. Acta., 646:218.
Christensen, H.N., Liang, M., Archer, E.G., 1967, J. Biol. Chem., 242:5237.
Christensen, H.N., Handlogten, M.E., Lam, I., Tager, H.S., Zand, R., 1969, J. Biol. Chem. 244:1510.
Christensen, H.N., deCepedes, C., Handlogten, M.E., Ronquist, G., 1973, Biochim. Biophys. Acta., 300:487.
Chuang, J.C., Yu, C.L., Wang, S.R., 1990, Clin. Exp. Immunol., 81:173.
Cieplinski, W., Tomicic, T., Schwink, A., Hajjar, J.J., 1985, Cancer Biochem. Biophys., 7:309.
Dantzig, A.H., Finkelstein, M.C., Adelberg, E.A., Slayman, C.W., 1978, J. Biol. Chem., 253:5813.
Dantzig, A.H., Adelberg, E.A., Slayman, C.W., 1979, J. Biol. Chem., 254:8988.
Eddy, A.A., Mulcahy, M.F., Thomson, P.J., 1967, Biochem. J., 103:863.
Editorial, 1989, Nutrition Review, 47:26.
Finkelstein, M.C., Adelberg, E.A., 1977, J. Biol. Chem., 252:7101.
Frengley, P.A., Peck, W.A., Lichtman, M.A., 1974, Exptl. Cell. Res., 88:442.

Frengley, P.A., Peck, W.A., Lichtman, M.A., 1975b, Blood Cells, 1:139.
Frengley, P.A., Lichtman, M.A., Peck, W.A., 1975a, J. Lab. Clin. Med., 86:984.
Gazzola, G.C., Franchi, R., Saibene, V., Ronchi, P., Guidotti, G.G., 1972, Biochim. Biophys. Acta., 226:407.
Glassy, M.C., Furlong, C.E., 1981, J. Cell. Physiol., 107:69.
Goldenberg, G.J., Lam, H-Y.P., Begleiter, A., 1979, J. Biol. Chem., 254:1057.
Goldfine, I.D., Gardner, J.D., Neville, D.M.Jr., 1972, J. Biol. Chem., 247:6919.
Goldfine, I.D., Simons, C.G., Smith, G.J., Ingbar, S.H., 1974, Endo., 96:1030.
Greene, W.C., Parker, C.W., 1975, Biochem. Biophysiol. Res. Commun., 65:456.
Greene, W.C., Parker, C.M., Parker, C.W., 1976a, J. Biol. Chem. 251:4017.
Greene, W.C., Parker, C.M., Parker, C.W., 1976b, Exper. Cell. Res. 103:109.
Grinstein, S., Cohen, S., Goetz, J.D., Rothstein, A., Gelfand, E.W., 1985, Proc. Natl. Acad. Sci. USA, 82:1429.
Grinstein, S., Rothstein, A., 1986, J. Membr. Biol., 90:1.
Guidotti, G.G., Gazzola, G.C., Borghetti, A.F., Franchi-Gazzola, R., 1975, Biochim. Biophys. Acta., 406:264.
Guidotti, G.G., Borghetti, A.F, Gazzola, G.C., Tramacere, M., Dall'Asta, V., 1976, Biochem. J., 160:281.
Guidotti, G.G., Borghetti, A.F., Gazzola, G.C., 1978, Biochimica. et. Biophysica. Acta., 515:329.
Hare, J.D., 1976, Biochimica. et. Biophysica. Acta. 443:485.
Hediger, M.A., Coady, M.J., Ikeda, T.S., Wright, E.M., 1987, Nature, 330:379.
Heinz, A., Jackson, J.W., Richey, B.E., Sachs, G., Schafer, J.A., 1981, J. Membrane Biol. 62:149.
Heskoth, T.R., Moore, J.P., Morris, J.D., Taylor, M.V., Rogers, J., Smith, G.A., Metcalfe, J.C., 1985, Nature, 313:481.
Ishii, T., Sugita, Y., Bannai, S., 1987, J. Cell. Physiol. 133:330.
Kensler, T.W., Wertz, P.W., Mueller, G.C., 1979, Biochimica. et. Biophysica. Acta., 585:43.
Kim, J.W., Closs, E.I., Albritton, M., Cunningham, J.M., 1991, Nature, 352:725.
Klip, A., Mack, E., Cragoe, E.J.Jr., Grinstein, S., 1986, J. Cell. Physiol., 127:244.
Koch, B., Schroder, M.T., Schafer, G., Schauder, P., 1990, J. Cell. Physiol., 143:94.
Kwock, L., Lin, P-S., Hefter, K., Wallach, D.F.H., 1978, Cancer Research, 38:83.
Kwock, L., Lin, P.S., Ciborowski, L., 1981, Cell. Biol. Inter. Reports, 5:83.
Lee, S-H., Nakagawa, H., Kalra, V.K., Brodie, A.F., 1978, Arch. Biochem. Biophy. 187:346.
Lerner, J., 1987, Comp. Biochem. Physiol., 87B:443.
Lichtman, A.H., Segel, G.B., Lichtman, M.A., 1983, Blood 61:413.
Lucivero, G., Surico, G., Mazzini, G., Dell'Osso, A., Bonomo, L., 1988, Mechanisms of Ageing and Development, 43:259.
Mendelsohn, J., Skinner, A., Kornfeld, S., 1971, J. Clin. Invest., 50:818.
Moolinaar, W.H., 1986, Annu. Rev. Physiol., 48:363.
Noren, O., Sjostrom, H., Larsen, P., 1983, Eur. J. Biochem., 134:71.
Novogrodsky, A., Tate, S.S., Meister, A., 1977, Biochem. Biophys. Res. Commun., 78:222.
Novogrodsky, A., Nehring, R.E.Jr., Meister, A., 1979, Proc. Natl. Acad. Sci. USA, 76:4932.
Novogrodsky, A., Ravid, A., Rubin, A.L., Stenzel, K.H., 1982, Proc. Natl. Acad. Sci. USA, 79:1171.
Ohsawa, M., Kilberg, M.S., Kimmel, G., Christensen, H.N., 1980, Biochimica. et. Biophysica. Acta. 599:175.
Oxender, D.L., Collarini, E.J., Shotwell, M.A., Lobaton, C.D., Moreno, A., Campbell, G.S., 1985, Ann. NY Acad. Sci., 456:404.
Palacin, M., Werner, A., Dittmer, J., Murer, H., Biber, J., 1990, Biochem. J., 270:189.
Panasci, L., Henderson, D., Torres-Garcia, S.J., Skalski, V., Caplan, S., Hutchinson, M., 1988, Cancer Res., 48:1972.
Peck, W.A., Rockwell, L.H., Lichtman, M.A., 1976, J. Cell. Physiol., 89:417.
Piedimonte, G., Baginski, I., Silvotti, L, Petronini, P.G., Borghetti, A.F., 1989, Cancer Letters, 47:121.
Resch, K., Wood, T., Northoff, H., Cooper, H.L., 1981, Eur. J. Biochem., 115:659.
Sander, G., Pardee, A.B., 1972, J. Cell. Physiol., 80:267.
Sato, K., Mitsumoto, Y., Mohri, T., 1988, Biochem. Biophys. Res. Commun., 153:570.
Scanlon, K.J., Berkowitz, K., Pallai, M.E., Waxman, S., 1983, Cancer Treatment Reports, 67:631.

Schmidt-Ullrich, R., Mann, N.D., Wallach, D.F.H., 1976, J. Cell. Physiol., 89:775.
Schroder, M.T., Schafer, G., Schauder, P., 1990, J. Cell. Physiol., 145:155.
Segel, G.B., Androphy, E.J., Lichtman, M.A., 1978, J. Cell. Physiol. 97:407.
Segel, G.B., Simon, W., Lichtman, M.A., 1979, J. Clin. Invest., 64:834.
Segel, G.B., Lichtman, M.A., 1981, J. Cell. Physiol., 106:303.
Segel, G.B., Lichtman, M.A., 1982, J. Biol. Chem., 257:9255.
Segel, G.B., Simon, W., Lichtman, M.A., 1983, J. Cell. Physiol., 116:372.
Segel, G.B., Simon, W., Lichtman, M.A., 1984, J. Clin. Invest., 74:17.
Segel, G.B., Tometsko, A.M., Lichtman, M.A., 1985, Arch. Biochem. Biophy. 242:347.
Segel, G.B., Woodlock, T.J., Lichtman, M.A., 1988, J. Cell. Physiol., 134:503.
Segel, G.B., Woodlock, T.J., Murant, F.G., Lichtman, M.A., 1989, J. Biol. Chem., 264:16399.
Shotwell, M.A., Jayme, D.W., Kilberg, M.S., Oxender, D.L., 1981, J. Biol. Chem., 256:5422.
Simon, J., 1979, Biochimica. et. Biophysica. Acta., 585:563.
Simon, W., 1986, Mathematical Techniques for Biology and Medicine, Dover Publications Inc., Mineola, NY, pg. 218.
Tarnuzzer, R.W., Campa, M.J., Qian, N.X., Englesberg, E., Kilberg, M.S., 1990, 265:13914.
Tate, E.H., 1982, J. Cell. Physiol. 113:307.
Tate, S.S., Urade, R., Getchell, T.V., Udenfriend, S., 1989, Arch. Biochem. Biophys., 275:591.
Udey, M.C., Parker, C.W., 1982, Biochem. Pharm. 31:337.
Van Den Berg, K.J., Betel, I., 1971, Exptl. Cell. Res., 66:257.
Van Den Berg, K.J., Betel, I., 1973a, Exptl. Cell. Res., 76:63.
Van Den Berg, K.J., Betel, I., 1973b, FEBS Letters, 29:149.
Vistica, D.T., Toal, J.N., Rabinovitz, M., 1978, Biochem. Pharmacol., 27:2865.
Vistica, D.T., 1979, Biochim. Biophys. Acta., 550:309.
Waithe, W.I., Dauphinais, C., Hathaway, P., Hirschhorn, K., 1975, Cell. Immunol., 17:323.
Walls, E.V., Borghetti, A.F., Benzie, C.R., Kay, J.E., 1984, Cell. Immunol. 89:30.
Wang, H., Kavanaugh, M.P., North, R.A., Kabot, D., 1991, Nature, 352:729.
White, M.F., Gazzola, G.C., Christensen, H.J., 1982, J. Biol. Chem., 257:4443.
Whitney, R.B., Sutherland, R.M., 1973, Biochimica. et. Biophysica. Acta. 298:790.
Wise, W.C., 1978, J. Cell. Physiol., 97:161.
Woodlock, T.J., Segel, G.B., Lichtman, M.A., 1988, J. Clin. Invest., 81:32.
Woodlock, T.J., Segel, G.B., Lichtman, M.A., 1989, J. Cell. Physiol., 141:33.
Woodlock, T.J., Brown, R., Mani, M., Pompeo, L., Hoffman, H., Segel, G.B., Silber, R., 1990, J. Cell. Physiol., 145:217.

TRANSPORT OF CYSTEINYL LEUKOTRIENES

Dietrich Keppler, Michael Müller and Toshihisa Ishikawa

Division of Tumor Biochemistry
Deutsches Krebsforschungszentrum
D–6900 Heidelberg, F.R.G.

INTRODUCTION

The cysteinyl leukotrienes comprise the metabolites formed from the glutathione conjugate leukotriene C_4 (LTC_4) in the mercapturate pathway as well as the ω- and ß-oxidized catabolites derived from the cysteine conjugate LTE_4 and from the N–acetylcysteine conjugate N–acetyl–LTE_4. The cysteinylglycine derivative LTD_4 is the most biologically potent mediator (Lewis and Austen, 1984), whereas the ω– and ß–oxidized catabolites are inactive (Samhoun et al., 1989). The cysteinyl leukotrienes, with the relative potencies LTD_4>LTC_4>LTE_4>N–acetyl–LTE_4, contract at nanomolar concentrations respiratory, vascular, and intestinal smooth muscles, and are involved in inflammatory and anaphylactic reactions (Lewis and Austen, 1984; Samuelsson et al., 1987). The parent molecule LTC_4 is formed in a limited number of cell types including mast cells, basophils, eosinophils, monocytes, and macrophages (Austen and Soberman, 1988). Export of LTC_4 from a leukotriene–synthesizing cell into the extracellular space (Fig. 1) is a prerequisite for interaction with receptors located on the surface of target cells. Moreover, the ectoenzyme–catalyzed removal of the γ–glutamyl moiety from LTC_4 by γ–glutamyltransferase in the extracellular space is a prerequisite for generation of the more potent LTD_4 (Fig. 1).

Transport controls not only the release of LTC_4 after its biosynthesis but also the removal of the cysteinyl leukotrienes from the blood circulation *in vivo*. The liver is the most active organ for uptake, metabolic inactivation, and biliary excretion of cysteinyl leukotrienes (Appelgren and Hammarström, 1982; Keppler et al., 1985; Hagmann et al., 1989; Huber et al., 1989a, 1990; Wettstein et al., 1989). One hour after intravenous injection of labeled LTC_4, LTD_4, or N–acetyl–LTE_4 72-80% of the administered radioactivity were recovered in rat bile (Huber et al., 1987; Keppler et al., 1989). The initial half–lifes of the cysteinyl leukotrienes in the blood circulation range between 30 and 60 seconds. Hepatobiliary elimination predominates over renal excretion in mammalian species, including monkey and man (Huber et al., 1990). A minimum of two transport systems for cysteinyl leukotriene elimination by the liver must be operative: Uptake at the sinusoidal side of the hepatocyte (Ormstad et al., 1982) and excretion across the canalicular membrane of hepatocytes into bile (Huber et al., 1987; Ishikawa et al., 1990). The hereditary defect of cysteinyl leukotriene transport across the canalicular membrane in a mutant rat strain has provided an approach to the molecular characterization of this transport system (Huber et al. 1987). Moreover, this defect indicates that the canalicular secretion may become rate-determining for cysteinyl leukotriene elimination from blood. Additional transport systems for cysteinyl leukotrienes may be involved in renal secretion of LTE_4 and its metabolites (Örning et al., 1985; Huber et al., 1989b; Maltby et al., 1990; Sala et al., 1990) and in the intestinal

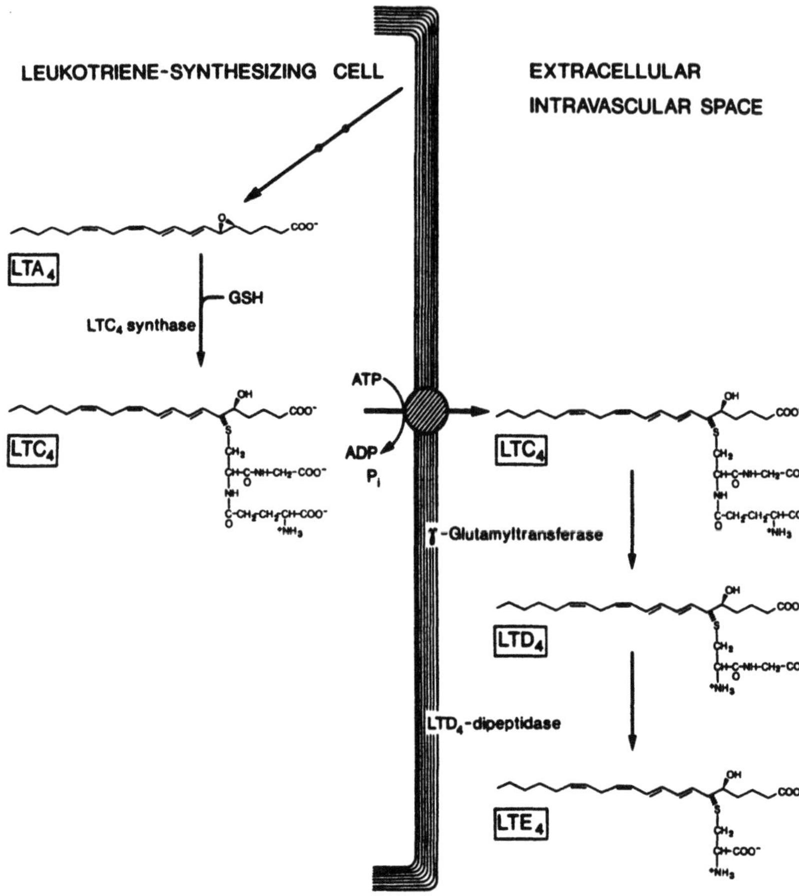

Fig. 1. Biosynthesis, export, and catabolism of leukotriene C_4.

absorption of cysteinyl leukotrienes during enterohepatic circulation (Denzlinger et al., 1986).

TRANSPORT OF LEUKOTRIENE C_4 DURING LEUKOTRIENE RELEASE FROM LEUKOTRIENE–SYNTHESIZING CELLS

The release of LTC_4 has been characterized in cultured human eosinophils incubated with exogenous LTA_4 (Lam et al., 1989). This transport is saturable, temperature–dependent, and progressively inhibited by increasing intracellular concentrations of LTC_5, suggesting a carrier–mediated process. The underlying mechanism for transport of LTC_4, as studied using plasma membrane vesicles prepared from murine mastocytoma cells, has been characterized as a primary active ATP–dependent process with apparent K_m values of 48 μM for ATP and of 110nM for LTC_4 (Schaub et al., 1991). Among the cysteinyl leukotrienes, LTC_4 is the best substrate for the ATP-dependent export carrier. At 10 nM concentration, the relative transport rates for LTC_4, LTD_4, LTE_4, and N–acetyl–LTE_4 are 1.00, 0.31, 0.12, and 0.08, respectively (Schaub et al., 1991). LTC_4 transport is competitively inhibited by the glutathione S-conjugate S–(2,4-dinitrophenyl)glutathione and several other amphiphilic anions including LTD_4 receptor antagonists (see below).

Hepatobiliary Cysteinyl Leukotriene Elimination

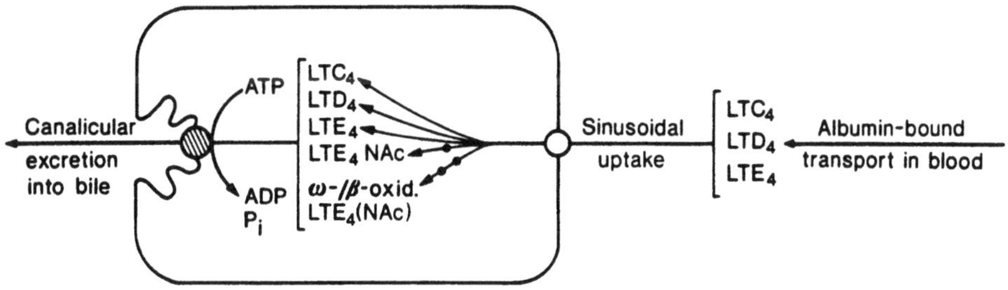

Fig. 2. Transport of cysteinyl leukotrienes through the hepatocyte.

METABOLISM AND TRANSPORT OF CYSTEINYL LEUKOTRIENES IN THE BLOOD CIRCULATION

Direct photoaffinity labeling of human or rat blood serum with [^{3}H]LTE$_4$ and other cysteinyl leukotrienes indicates one predominant protein binding and transporting these leukotrienes in the blood circulation. The molecular weight of 67 kDa and immunoprecipitation have identified the leukotriene–binding protein in blood serum as albumin (Falk et al., 1989). Metabolism within the blood circulation determines whether LTC$_4$, or its metabolite LTD$_4$, or its catabolite LTE$_4$ reach liver and kidney for uptake, intracellular metabolism, and excretion. γ–Glutamyltransferase and LTD$_4$–dipeptidase, predominantly located on cell surfaces within the vascular space, catalyze the conversion of LTC$_4$ to LTD$_4$ and LTE$_4$ within seconds after LTC$_4$ injection into blood (Huber and Keppler, 1988). In the rat, the initial half–life of LTC$_4$ is 12 seconds, and LTE$_4$ is the predominant cysteinyl leukotriene in the blood circulation after 30 seconds. Large differences with respect to degradation of the glutathione moiety of LTC$_4$ in the blood circulation exist between species (Huber and Keppler, 1988). In the guinea pig a very limited conversion of LTD$_4$ to LTE$_4$ occurs and the biologically potent LTD$_4$ is the predominant metabolite both in the blood circulation and in bile (A. Keppler et al., 1987; Guhlmann et al., 1989). The intravascular metabolism of LTC$_4$ thus determines the substrate for the cysteinyl leukotriene uptake system at the liver sinusoidal membrane (Fig. 2) which exhibits the highest affinity and transport rate for LTE$_4$ and the lowest ones for LTC$_4$ transport into the cell (Uehara et al., 1983).

CYSTEINYL LEUKOTRIENE UPTAKE BY HEPATOCYTES

Cysteinyl leukotriene uptake has been demonstrated both by isolated hepatocytes and by the perfused liver (Ormstad et al., 1982; Uehara et al., 1983). Studies using the isolated perfused rat liver demonstrate sodium-dependent and sodium-independent components of the hepatic uptake of cysteinyl leukotrienes (Wettstein et al., 1990). Since these studies were performed in the intact organ, the driving force of the uptake remains to be characterized in freshly isolated hepatocytes and right-side-out membrane vesicles enriched in the basolateral membrane.

Leukotriene-binding polypeptides possibly involved in hepatocellular transport were identified by the method of direct photoaffinity labeling in the deep-frozen state using the [^{3}H]labeled leukotriene itself as the photolabile ligand (Falk et al., 1989; Müller et al., 1991a). Photoaffinity labeling of rat liver membrane subfractions enriched with sinusoidal membranes resulted in the predominant labeling of a polypeptide with an apparent M_r of 48,000. This polypeptide is assumed to be the one also labeled in the particulate fraction by photoaffinity labeling of freshly isolated rat hepatocytes incubated with [^{3}H]LTC$_4$ and [^{3}H]LTE$_4$. Whether this polypeptide is specifically involved in the uptake of cysteinyl leukotrienes or whether it is identical with the polypeptides of similar M_rs, which are labeled in rat hepatocytes by photolabile derivatives of bile salts and other amphiphilic compounds, and are expected to be part of an uptake system of broad specificity (Kurz et al., 1989; Meijer et al., 1990), can only be established after purification of the proteins. Evidence for the function of this leukotriene–binding protein as part of an uptake system for these eicosanoids was obtained by labeling studies with AS-30D ascites hepatoma cells which are deficient in cysteinyl leukotriene uptake (Weckbecker and Keppler, 1986). In these cells, photoaffinity labeling with cysteinyl leukotrienes showed no labeling of 48 kDa polypeptides (Müller et al., 1991a). Corresponding results were obtained when isolated intact human hepatocytes and the human hepatoma cell lines PLC and HepG2 were compared. The 48 kDa polypeptide which was labeled in the plasma membrane of human hepatocytes was not labeled in the dedifferentiated hepatoma cells (Müller et al., 1991a). In addition, no significant labeling of the soluble fraction could be detected which is consistent with the deficient uptake for cysteinyl leukotrienes into hepatoma cells.

In contrast, many labeled polypeptides were found in the soluble fraction of isolated hepatocytes after photoaffinity labeling of intact cells. This indicates that the cysteinyl leukotrienes interact after their uptake into cells with a number of different polypeptides. Among these polypeptides, those with apparent M_rs in the range of 25,000 were identified as subunits of glutathione transferases (Falk et al., 1989). The basic isoenzyme of glutathione transferase (Y_a) might also be strongly associated with membrane fractions from rat liver and exhibits LTC_4-binding properties as detected by photoaffinity labeling (Sun et al., 1987; Ishikawa et al., 1990). Caution is therefore required in the characterization of LTC_4 receptors and carrier proteins in plasma membranes from different tissues which might be contaminated by different intracellular binding proteins.

ATP–DEPENDENT TRANSPORT OF CYSTEINYL LEUKOTRIENES ACROSS THE HEPATOCYTE CANALICULAR MEMBRANE

The mechanism of the hepatobiliary transport has been studied using inside-out vesicles enriched in canalicular and sinusoidal membrane vesicles from rat liver (Ishikawa et al., 1990). Incubation of canalicular membrane vesicles with $[^3H]LTC_4$ in the presence of ATP resulted in an uptake of LTC_4 into vesicles. The initial rate of ATP-stimulated LTC_4 uptake was about 40-fold higher in canalicular than in sinusoidal membrane vesicles, suggesting the existence of an ATP–dependent transport system for LTC_4 in the canalicular membrane. LTD_4, LTE_4, and N–acetyl–LTE_4 were also transported into liver plasma membrane vesicles in an ATP-dependent manner. Mutual competition between the cysteinyl leukotrienes for their ATP-dependent uptake into vesicles indicates that cysteinyl leukotrienes are transported by a common carrier.

A partial hereditary defect in the excretion of several organic anions, such as bilirubin glucuronide and dibromosulfophthalein, from hepatocytes into bile was discovered in a mutant strain of Wistar rats (Jansen et al., 1985, 1987). In these mutant rats excretion of cysteinyl leukotrienes into bile is reduced to less than 2% of normal (Huber et al., 1987). Liver plasma membrane vesicles prepared from these mutant rats lack the ATP-dependent transport of cysteinyl leukotrienes (Ishikawa et al., 1990).

SPECIFICITY OF THE CYSTEINYL LEUKOTRIENE EXPORT CARRIER

Among the cysteinyl leukotrienes LTC_4 is the best substrate for the hepatic canalicular export carrier. Apparent K_m values are 0.25, 1.5, and 5.2 µM for LTC_4, LTD_4, and N–acetyl–LTE_4, respectively, whereas the K_m for LTE_4 is more than 10 µM. LTB_4 is not transported via this transport system (Ishikawa et al., 1990). In addition to N-acetylation, LTE_4 and N–acetyl–LTE_4 undergo ω-oxidation in hepatocytes to give rise to more polar metabolites such as ω-carboxy–N–acetyl–LTE_4 which is also eliminated into bile (Foster et al., 1987; Örning et al., 1988). ω–Carboxy–N–acetyl–LTE_4 is also taken up by inside-out plasma membrane vesicles in an ATP-dependent manner suggesting that ω- and β–oxidized metabolites of N–acetyl–LTE_4 are also transported via the ATP-dependent transport system from hepatocytes into bile (Ishikawa et al., 1990).

Not only cysteinyl leukotrienes but also glutathione S-conjugates (e.g. S–2,4–(dinitrophenyl)–glutathione and glutathione-conjugated bromosulfophthalein) as well as several other organic anions are excreted from hepatocytes into bile (Combes, 1965; Wahllânder and Sies, 1979; Akerboom et al., 1982). Biliary excretion of glutathione S-conjugates has been shown to be dependent on intracellular ATP (Oude Elferink et al., 1990) and canalicular membrane vesicles prepared from rat liver exhibited an ATP-dependent transport of S–(2,4–dinitrophenyl)-glutathione (Kobayashi et al., 1988, 1990; Akerboom et al., 1991). Mutual competition between cysteinyl leukotrienes and S–(2,4–dinitrophenyl)-glutathione for the ATP-dependent transport into canalicular membrane vesicles suggests that both are transported via a common carrier (Ishikawa et al., 1990). Similar ATP–dependent transport systems for glutathione conjugates have been reported in human erythrocytes (Kondo et al., 1980; LaBelle et al., 1986) and rat heart (Ishikawa, 1989a,b). ATP–dependent transport of LTC_4 into plasma membrane vesicles from rat liver and heart was markedly inhibited by glutathione S-conjugates with a long aliphatic carbon chain, e.g. octyl– and decyl–S–glutathione (Ishikawa et al., 1989). The export carrier appears to have high affinities toward the hydrophobic moiety of the glutathione S-conjugates. On the other hand, as represented by the transport of LTD_4, LTE_4, and N–acetyl–LTE_4 via the ATP–dependent carrier, the glutathione moiety is not considered to be a structural determinant but an important property for maximal affinity to the active site of the carrier. It has recently been shown that the canalicular efflux of glutathione S–conjugates can be inhibited by a wide variety of multivalent organic anions, such as 3–sulfotaurolithocholate, 3–sulfoglycolithocholate, lithocholic acid 3–*O*–glucuronide (Oude Elferink et al., 1991), *p*–nitrophenyl glucuronide, testosterone glucuronide (Kobayashi et al., 1991), naphtyl glucuronide and glutathione disulfide (Akerboom et al., 1991). It is, therefore, of interest and importance to define whether the ATP–dependent leukotriene export system consists of a single carrier with a broad substrate specificity or of a cluster of related carrier proteins with overlapping specificities.

INHIBITION OF CYSTEINYL LEUKOTRIENE TRANSPORT ACROSS PLASMA MEMBRANES

Inhibition of LTC_4 uptake into plasma membrane vesicles has been observed in the presence of various glutathione conjugates. In heart sarcolemma and liver plasma membrane vesicles 90% inhibition of LTC_4 transport was induced by S–octylglutathione (10 μM) added at an 800–fold excess over LTC_4 (Ishikawa et al., 1989). S–Hexylglutathione and the glutathione conjugate of 4–hydroxynonenal were also effective inhibitors in this system. More recently, structural analogs of LTD_4 and LTE_4, designed as LTD_4/LTE_4 receptor antagonists and devoid of a glutathione moiety or of a fatty acid side chain, turned out to be potent inhibitors of LTC_4 transport (Schaub et al., 1991) and of the hepatobiliary elimination of cysteinyl leukotrienes *in vivo* (Denzlinger et al., 1991). This may be exemplified by the LTD_4 receptor antagonist MK-571 (previously L–660,711); chemical name, 3–(((3–(2–(7–chloro–2-quinolinyl)ethenyl)-phenyl)((3-dimethylamino-3–oxopropyl)thio)methyl)thio)propanoic acid (Jones et al., 1989). LTC_4 transport, examined in mastocytoma plasma membrane vesicles, was inhibited by 50% in the presence of 1 μM MK–571 added at a 100–fold excess over LTC_4; the corresponding IC_{50} value for S–2,4–(dinitrophenyl)–glutathione was 6.4 μM (Schaub et al., 1991). Interestingly, another amphiphilic anion which is not a LTD_4 receptor antagonist or a structural analog of LTC_4, but an inhibitor of leukotriene biosynthesis, MK–886, chemical name, 3–(1–(4–chlorobenzyl)-3–t–butyl-thio–5–isopropylindol–2–yl)–2,2-dimethylpropanoic acid (Gillard et al., 1989), potently inhibited LTC_4 transport with an IC_{50} value of 1.8 μM (Schaub et al., 1991). Inhibition of carrier–mediated leukotriene export from leukotriene–synthesizing cells provides a novel approach to inhibition of leukotriene synthesis and more potent inhibitors acting at this site may be developed. Moreover, some of these compounds interfere with transport processes involved in the hepatobiliary elimination of cysteinyl leukotrienes as shown by Denzlinger et al. (1991). Studies in the isolated perfused liver have demonstrated inhibition of cysteinyl leukotriene uptake into hepatocytes by the immunosuppressive drug cyclosporin A (Hagmann et al., 1989) and by bromosulfophthalein (Wettstein et al., 1990). Potent inhibition of cysteinyl leukotriene uptake into hepatocytes would be more critical than interference of canalicular excretion into bile since it would extend the half–life of biologically active leukotrienes in the blood circulation. On the other hand,

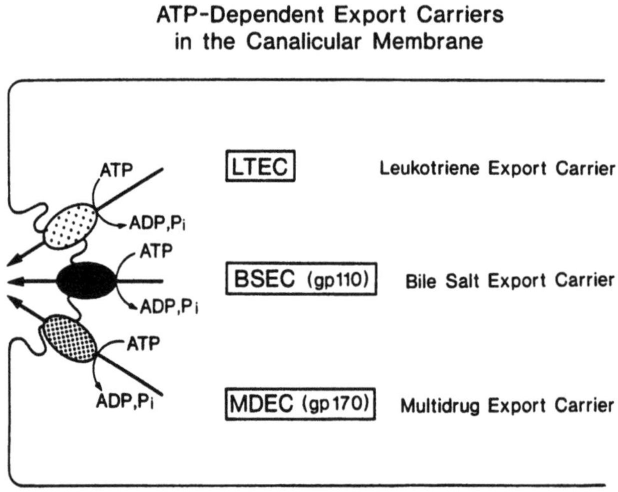

Fig. 3. Primary–active export carriers in the canalicular membrane.

inhibition of canalicular export still allows for inactivation of leukotrienes within the hepatocyte, particularly by ω–oxidation and ß–oxidation from the ω–end.

COMPARISON OF THE ATP-DEPENDENT EXPORT CARRIERS IN THE CANALICULAR MEMBRANE

The driving forces for the transmembrane movement of the molecules transported across the sinusoidal and canalicular membrane seem to be domain-specific. The canalicular domain of the plasma membrane of hepatocytes exhibits highly specialized transport functions that can mediate active canalicular excretion of a variety of structurally unrelated molecules. Recently three ATP-dependent primary-active transport systems which function as vanadate-sensitive export carriers for molecules in the M_r range of300-900 were kinetically characterized using canalicular membrane vesicles from rat liver (Fig. 3): The export system ("multidrug export carrier" or MDEC) for hydrophobic, mostly basic compounds (Kamimoto et al., 1989; Arias, 1990), the leukotriene export carrier (LTEC) for cysteinyl leukotrienes and other amphiphilic anions (Ishikawa et al., 1990; Kitamura et al., 1990; Kobayashi et al., 1990), and the bile salt export carrier (BSEC) which was characterized very recently (Müller et al., 1991b).

Using monoclonal antibodies against the p–glycoprotein (gp170), one of the multidrug resistance gene products overexpressed in several tumor cells (Gottesman and Pastan, 1988; Endicott and Ling, 1989), the presence of a 160 kDa protein species in the canalicular plasma membrane of hepatocytes was shown (Kamimoto et al., 1989; Ishikawa et al., 1990). There is evidence that this protein could have its physiological function as a multidrug export carrier (MDEC) (Kamimoto et al., 1989; Arias, 1990; Fig. 3). On the other hand, the molecular structure of LTEC is not known at present. Whether LTEC is identical at least in part with a 38kDa ATPase which hydrolyzes ATP at a higher rate in the presence of conjugates of bilirubin, bile acids, and glutathione conjugates (Awasthi et al., 1991) remains to be demonstrated, e.g. by kinetic studies using proteoliposomes containing the functionally active, reconstituted purified ATPase.

ATP-dependent taurocholate transport was observed not only in membrane vesicles from normal rats (Müller et al., 1991b; Nishida et al., 1991) but also in mutant rats deficient in the ATP-dependent transport of cysteinyl leukotrienes and related amphiphilic anions. Substrates of the p-glycoprotein (gp170), such as verapamil and doxorubicin, did neither interfere with the ATP-dependent transport of taurocholate nor with the one of cysteinyl leukotrienes, indicating that these transporters are different. Furthermore, it was demonstrated by functional reconstitution of the purified ATP- and bile salt-binding glycoprotein with apparent M_r of 110,000 (gp110) into liposomes that this protein functions as carrier in the ATP-dependent transport of bile salts from the hepatocyte into bile (Müller et al., 1991b). It may be assumed that these different ATP-dependent unidirectional transport systems (Fig. 3) belong to the recently defined superfamily of ATP-binding proteins with homologous ATP-binding cassettes (ABC) in their sequence (Hyde et al., 1990).

REFERENCES

Akerboom, T. P. M., Bilzer, M., and Sies, H., 1982, FEBS Lett., 140:73-76.

Akerboom, T. P. M., Narajanaswami, V., Kunst, M., and Sies, H., 1991, J. Biol. Chem., 266:13147-13152.

Appelgren, L. E., and Hammarström, S., 1982, J. Biol. Chem., 257:531–535.

Arias, I. M., 1990, Hepatology, 12:159-165.

Austen, K. F., and Soberman, R. J., 1988, Ann. N.Y. Acad. Sci., 524:XI–XXV.

Awasthi, Y. C., Singhal, S. S., Gupta, S., Ahmad, H., Zimniak, P., Radominska, A., Lester, R., and Sharma, R., 1991, Biochem. Biophys. Res. Commun., 175:1090-1096.

Combes, B., 1965, J. Clin. Invest., 44:1214-1224.

Denzlinger, C., Guhlmann, A., Scheuber, P. H., Wilker, D., Hammer, D. K., and Keppler, D., 1986, J. Biol. Chem., 261:15601-15606.

Denzlinger, C., Grimberg, M., Kapp, A., Haberl, C., and Wilmanns, W., 1991, Br. J. Pharmacol., 102:865-870.

Endicott, J. A., and Ling, V., 1989, Annu. Rev. Biochem., 58:137-171.

Falk, E., Müller, M., Huber, M., Keppler, D., and Kurz, G., 1989, Eur. J. Biochem., 186:741-747.

Foster, A., Fitzsimmons, B., Rokach, J., and Letts, L. G., 1987, Biochim. Biophys. Acta, 921:486-493.

Gillard, J., Ford-Hutchinson, A. W., Chan, C., Charleson, S., Denis, D., Foster, A., Fortin, R., Leger, S., McFarlane, C. S., Morton, H., Piechuta, H., Reindeau, D., Rouzer, C. A., Rokach, J., Young, R., MacIntyre, D. E., Peterson, L., Bach, T., Eiermann, G., Hopple, S., Humes, J., Hupe, L., Luell, S., Metzger, J., Meurer, R., Miller, D. K., Opas, E., and Pacholok, S., 1989, Can. J. Physiol. Pharmacol., 67:456-464.

Gottesman, M. M., and Pastan, I., 1988, J. Biol. Chem., 263:12163–12166.

Guhlmann, A., Keppler, A., Kästner, S., Krieter, H., Brückner, U. B., Messmer, K., and Keppler, D., 1989, J. Exp. Med., 170:1905-1918.

Hagmann, W., Parthé, S., and Kaiser, I., 1989, Biochem. J., 261:611–616.

Huber, M., Guhlmann, A., Jansen, P. L. M., and Keppler, D., 1987, Hepatology, 7:224-228.

Huber, M., and Keppler, D., 1988, pp. 449-470, in: "Glutathione Conjugation: Mechanisms and Biological Significance", H. Sies and B. Ketterer, eds., Academic Press, Orlando, FL.

Huber, M., Baumert, T., Guhlmann, A., Mayer, D., and Keppler, D., 1989a, pp. 131-140, in: "Hepatic Transport of Organic Substances", E. Petzinger, R. K.-H. Kinne, and H. Sies, eds., Springer-Verlag, Heidelberg.

Huber, M., Kästner, S., Schölmerich, J., Gerok, W., and Keppler, D., 1989b, Eur. J. Clin. Invest., 19:53-60.

Huber, M., Müller, J., Leier, I., Jedlitschky, G., Ball, H. A., Moore, K. P., Taylor, G. W., Williams, R., and Keppler, D., 1990, Eur. J. Biochem., 194:309-315.

Hyde, S. C., Emsley, P., Hartshorn, M. J., Mimmack, M. M., Gileadi, U., Pearce, S. R., Gallagher, M. P., Gill, D. R., Hubbard, R. E., and Higgins, C. F., 1990, Nature, 346:362-365.

Ishikawa, T., 1989a, J. Biol. Chem., 264:17343-17348.

Ishikawa, T., 1989b, FEBS Lett., 246:177-180.

Ishikawa, T., Kobayashi, K., Sogame, Y., and Hayashi, K., 1989, FEBS Lett., 259:95-98.

Ishikawa, T., Müller, M., Klünemann, C., Schaub, T., and Keppler, D., 1990, J. Biol. Chem., 265:19279-19286.

Jansen, P. L. M., Peters, W. H., and Lamers, W. H., 1985, Hepatology, 5:573-579.

Jansen, P. L. M., Groothuis, G. M. M., Peters, W. H., and Meijer, D. K. F., 1987, Hepatology, 7:71-76.

Jones, T. R., Zamboni, R., Belley, M., Champion, E., Charette, L., Ford-Hutchinson, A. W., Frenette, R., Gauthier, J. Y., Leger, S., Masson, P., McFarlane, C. S., Piechuta, H.,

Rokach, J., Williams, H., Young, R. N., Dehaven, R. N., and Pong, S. S., 1989, Can. J. Physiol. Pharmacol., 67:17-28.
Kamimoto, Y., Gatmaitan, Z., Hsu, J., and Arias, I. M., 1989, J. Biol. Chem., 264:11693-11698.
Keppler, A., Örning, L., Bernström, K., and Hammarström, S., 1987, Proc. Natl. Acad. Sci. USA, 84:5903-5907.
Keppler, D., Hagmann. W., Rapp, S., Denzlinger, C., and Koch, H. K., 1985, Hepatology, 5:883-891.
Keppler, D., Huber, M., Baumert, T., and Guhlmann, A., 1989, Advances Enzyme Regul., 28:307-319.
Kitamura, T., Jansen, P., Hardenbrook, C., Kamimoto, Y., Gatmaitan, Z., and Arias, I. M., 1990, Proc. Natl. Acad. Sci. USA, 87:3557-3561.
Kobayashi, K., Sogame, Y., Hayashi, K., Nicotera, P., and Orrenius, S., 1988, FEBS Lett., 240:55-58.
Kobayashi, K., Sogame, Y., Hara, H., and Hayashi, K., 1990, J. Biol. Chem., 265:7737-7741.
Kobayashi, K., Komatsu, S., Nishi, T., Hara, H., and Hayashi, K., 1991, Biochem. Biophys. Res. Commun., 176:622-626.
Kondo, T., Dale, G.L. and Beutler, E., 1980, Proc. Nat. Acad. Sci.., 77:6359.
Kurz, G., Müller, M., Schramm, U., and Gerok, W., 1989, pp. 267-278, in: "Hepatic Transport of Organic Substances", R. Kinne, E. Petzinger, and H. Sies, eds., Springer-Verlag, Berlin.
LaBelle, E. F., Singh, S. V., Srivastava, S. K., and Awasthi, Y. C., 1986, Biochem. J., 238:443-449.
Lam, B. K., Owen, W. F., Austen, K. F., and Soberman, R. J., 1989, J. Biol. Chem., 264:12885-12889.
Lewis, R. A., and Austen, K. F., 1984, J. Clin. Invest., 73:889–897.
Maltby, N. H., Taylor, G. W., Ritter, J. M., Moore, K., Fuller, R. W., and Dollery, C. T., 1990, J. Allergy Clin. Immunol., 85:3-9.
Meijer, D. K. F., Mol, W. E. M., Müller, M., and Kurz, G., 1990, J. Pharmacokin. Biopharm., 18:35-70.
Müller, M., Falk, E., Sandbrink, R., Berger, U., Leier, I., Jedlitschky, G., Huber, M., Kurz, G., and Keppler, D., 1991a, Advances Prostaglandin, Thromboxane, Leukotriene Res., 21:395-398.
Müller, M., Ishikawa, T., Berger, U., Klünemann, C., Lucka, L., Schreyer, A., Kannicht, C., Reutter, W., Kurz, G., and Keppler, D., 1991b, J. Biol. Chem., 266, 18920-18926.
Nishida, T., Gatmaitan, Z., Che, M., and Arias, I. M., 1991, Proc. Natl. Acad. Sci. USA, 88, 6590-6594.
Ormstad, K., Uehara, N., Orrenius, S., Örning, L., and Hammarström, S., 1982, Biochem. Biophys. Res. Commun., 26:1434-1440.
Örning, L., Kaijser, L., and Hammarström, S., 1985, Biochem. Biophys. Res. Commun., 130:214-220.
Örning, L., Keppler, A., Midtvedt, T., and Hammarström, S., 1988, Prostaglandins, 35:493-501.
Oude Elferink, R. P. J., Ottenhoff, R., Liefting, W. G. M., Shoemaker, B., Groen, A. K., and Jansen, P. L. M., 1990, Am. J. Physiol., 258:G699-G706.
Oude Elferink, R. P. J., Ottenhoff, R., Radominska, A., Hofmann, A. F., Kuipers, F., and Jansen, P. L. M., 1991, Biochem. J., 274:281-286.
Sala, A., Voelkel, N., Maclouf, J., and Murphy, R. C., 1990, J. Biol. Chem., 265:21771-21778.
Samhoun, M. N., Conroy, D. M., and Piper, P. J., 1989, Br. J. Pharmacol., 98:1406-1412.
Samuelsson, B., Dahlén, S., Lindgren, J. Å., Rouzer, C. A., and Serhan, C. N., 1987, Science, 237:1171-1176.
Schaub, T., Ishikawa, T., and Keppler, D., 1991, FEBS Lett., 279:83-86.
Sun, F. F., Chau, L.-Y., and Austen, K. F., 1987, Fed. Proc., 46:204-207.
Uehara, N., Ormstad, K., Örning, L., and Hammarström, S., 1983, Biochim. Biophys. Acta, 732:69-74.
Wahlländer, A., and Sies, H., 1979, Eur. J. Biochem., 90:441-446.
Weckbecker, G., and Keppler, D., 1986, Eur. J. Biochem., 154:559–562.
Wettstein, M., Gerok, W., and Häussinger, D., 1989, Eur. J. Biochem., 181:115-124.
Wettstein, M., Gerok, W., and Häussinger, D., 1990, Eur. J. Biochem., 191:251-255.

INTERORGAN GLUTATHIONE TRANSPORT

Theodorus P.M. Akerboom and Helmut Sies

Institut für Physiologische Chemie I
Universität Düsseldorf
Moorenstraße 5
4000 Düsseldorf, Germany

INTRODUCTION

The tripeptide glutathione (GSH, γ-glutamylcysteinylglycine) is the major intracellular low-molecular weight thiol in animals, plants and in most microorganisms. It is involved in a variety of metabolic processes, e.g. the detoxication of xenobiotics, the reduction of hydroperoxides, the synthesis of leukotrienes and prostaglandins, the maintenance of protein and membrane structure, and the regulation of numerous enzyme activities (Meister and Anderson, 1983; Sies and Ketterer, 1988; Dolphin et al., 1989 a,b; Taniguchi et al., 1989; Vina, 1990). These functions are effected by the thiol group participating in redox transitions, thiol exchange reactions, and thioether formation, or as a radical scavenger. The importance of glutathione in the protection of the cell against reactive oxygen species, electrophilic compounds arising from drug metabolism, and free radicals resulting from ionizing radiation is well established (Sies and Ketterer, 1988; Chow, 1988; Sies, 1989).

Most cells possess the capacity to synthesize glutathione from its amino acid precursors, which proceeds in a two-step reaction catalysed by γ-glutamylcysteine synthetase and glutathione synthetase. The intracellular glutathione level is regulated via feedback inhibition of the γ-glutamylcysteine synthetase by glutathione, resulting in intracellular glutathione concentrations usually in the millimolar range. In the cell glutathione is kept in the reduced form by glutathione reductase, so that under normal conditions less than 1% of glutathione is in the disulfide form (GSSG) (Srivastava and Beutler, 1968; Akerboom et al., 1982a). Glutathione may form mixed disulfides with other low molecular weights thiols like CoASH (Crane et al., 1982) or with protein thiol groups, which may be of significance as a means of regulation of a number of enzyme activities (Ziegler, 1985; Brigelius, 1985).

The degradation of glutathione is catalysed by γ-glutamyl transpeptidase and at least two dipeptidases located on the outer surface of the cell membrane. Also glutathione disulfide and glutathione S-conjugates formed within the cell are metabolized via these ecto-enzymes. This implies that the metabolism of the different glutathione species requires transport across the cell membrane. High activities of γ-glutamyl transpeptidase and cysteinylglycine dipeptidase are found in the brush border membranes of the kidney and the intestine, and furthermore in the bile canaliculi of the liver, so that in these organs an active intraorgan turnover of glutathione may occur.

Part of the glutathione exported from cells reaches the blood plasma, mainly in the form of GSH (Bartoli and Sies, 1978; Anderson and Meister, 1980) of which a large

Mammalian Amino Acid Transport, Edited by M.S. Kilberg and
D. Häussinger, Plenum Press, New York, 1992

amount is bound via disulfide linkage to plasma proteins (Lash and Jones, 1985), notably albumin (Inoue et al., 1987). Plasma glutathione may function in the maintenance of the thiol redox state of plasma proteins and in the protection against extracellular oxygen free-radical damage at inflammatory sites. The plasma concentrations of glutathione are in the micromolar range, and differ considerably in the different regions of the blood circulation (Häberle et al., 1979; Anderson et al., 1980). This is consistent with the occurrence of an active interorgan turnover of glutathione, in which the liver is the main organ exporting glutathione, and in which the kidney is effective in extracting glutathione from the circulation. Thus, membrane transport constitutes an essential step in the overall function of glutathione in the organism. In epithelial tissues such as the liver, the cell polarity is an important determinant for the fate of the different forms of glutathione. Glutathione conjugates are preferentially exported from the hepatocyte into the bile canaliculus, whereas hepatic GSH is predominantly released into the sinusoidal space (Sies, 1983). This overview describes current knowledge on transport properties of the different glutathione species in different organs and discusses some physiological implications.

GLUTATHIONE TRANSPORT IN DIFFERENT ORGANS

Liver, sinusoidal

The liver, being equipped with high activities of GSH peroxidases and GSH-S-transferases, is the major organ in detoxification of hydroperoxides and elimination of xenobiotics supplied by the portal vein. Furthermore, the liver plays a central role in the interorgan relationships of glutathione as it exports GSH into plasma at a rate that accounts for nearly all of its hepatic biosynthesis (Bartoli and Sies, 1978; Lauterburg et al., 1984a). The significance of the liver for the supply of systemic glutathione is expressed by the higher plasma levels of glutathione in the hepatic vein (26 μM) as compared to the portal vein (14 μM, Akerboom et al., unpublished) and arterial blood (15 μM) (Anderson et al., 1980). In the perfused rat liver GSH is released into the sinusoidal space at a rate of 14-16 nmol/min x g liver (Akerboom et al., 1982a; Bartoli and Sies, 1978; Ookhtens et al. 1985). Biliary excretion of GSH occurs at a rate of 1-3 nmol/min x g liver (Akerboom et al., 1982a).

The kinetics of sinusoidal efflux of GSH in the perfused rat liver has been studied by manipulating the intracellular GSH level. Efflux is characterized by apparent sigmoidal saturation kinetics with an apparent K_m of 3.2 μmol/g liver and a V_{max} of 20 nmol/min x g liver (Ookhtens et al., 1985). Similar data have been obtained from studies with isolated rat hepatocytes (Aw et al., 1986). Organic anions may competitively inhibit from the inside of the hepatocyte (Ookhtens et al., 1988), and transport is dependent both on the membrane potential and on ATP (Fernandes-Checa et al., 1988), indicating that transport is probably carrier-mediated, normally operating at a near-maximal rate.

In studies with isolated basolateral plasma membrane vesicles, two transport systems have been identified, a low-affinity system with a K_m of 3.9 mM and a V_{max} of 12 nmol/min x mg protein, and a high-affinity system with substantially lower capacity (Inoue et al., 1984b; Aw et al., 1987). Evidence was presented that transport of GSH via the high capacity system is electrogenic and driven by the membrane potential. This would be consistent with the unidirectional properties of GSH transport across the basolateral plasma membrane of the intact liver (Hahn et al., 1978; Sies, 1983). Also hormonal effects on the sinusoidal efflux of GSH from the liver (Sies and Graf, 1985) might in part be explained by changes on the membrane potential (Aw et al., 1987; see below).

Glutathione disulfide and glutathione conjugates formed in the liver are selectively transported into the bile canaliculi (Sies et al., 1978). Only at excessively high rates of formation these glutathione species also appear in the sinusoidal space. Whether this occurs via transport across the basolateral part of the plasma membrane or is due to paracellular backflow is still unclear (Ballatori and Truong, 1989).

In contrast to GSH and GSSG which are not significantly taken up by the intact liver, some glutathione S-conjugates are efficiently extracted from the hepatic sinusoidal space and excreted into bile. The rate of uptake is dependent on the conjugant. Whereas almost no uptake of 2,4-dinitrophenyl-S-glutathione (DNPSG) in the perfused rat liver occurs (Wahlländer and Sies, 1979), the extraction of the sulfobromophthalein glutathione conjugate (BSPSG) amounts to about 35% (Sorrentino et al., 1988). The uptake of BSPSG by isolated hepatocytes is diminished by BSP, suggesting a common uptake system for BSP and its glutathione conjugate (Schwarz et al., 1980). Leukotriene C4 is taken up to about 60% by the liver during a single-pass perfusion (Iwai et al., 1988; see also Keppler et al., this volume).

Liver, canalicular

Determination of the biliary GSH excretion rate is difficult because of rapid autoxidation of GSH in bile (Eberle et al., 1981), conversion by γ-glutamyl transpeptidase located in the biliary tract (Abbott and Meister, 1986; Ballatori et al., 1988) and possible contributions of pancreatic juice (Meister and Anderson, 1983). In the rat, the amount of GSH found in bile is almost doubled after pretreatment of the liver with the γ-glutamyl transpeptidase inhibitor AT 125 (Ballatori et al., 1986). No saturation has been observed for canalicular transport in the intact organ (Kaplowitz et al., 1983). That GSH excretion is carrier-mediated and not representing a simple diffusion process may be inferred from the finding that compounds like sulfobromophthalein (BSP) or phenol-3,6-dibromophthalein (DBSP) are able to inhibit canalicular GSH release (Ballatori and Clarkson, 1985).

Depending on the animal species, glutathione excreted into bile is subject to an extensive metabolism and recycling of its amino acid constituents, for which sodium-coupled transport systems have been identified in purified canalicular plasma membrane vesicles from rat liver (Ballatori et al., 1988). The species differences of interorgan turnover may be related to the hepatic γ-glutamyl transpeptidase activity which is much higher in guinea pigs, rabbits and dogs than in rats. Also other factors like canalicular glutathione excretion rate, bile flow and the size of the biliary space may influence the extent of intrahepatic turnover and explain the absence of detectable glutathione in canine bile.

Biliary excretion of glutathione disulfide is an active process, since the concentration within the hepatocyte is about 20-fold lower than in bile (Akerboom et al., 1982a). In studies with isolated canalicular plasma membrane vesicles from rat liver it has been found recently that the transport of glutathione disulfide (GSSG), and of 2,4-dinitrophenyl-S-glutathione (DNPSG) is substantially stimulated by the addition of ATP (Kobayashi et al., 1990; Akerboom et al., 1991) (Fig. 1). An apparent K_m of 70 μM for DNPSG and a V_{max} of 0.34 nmol/min x mg protein has been estimated for ATP-dependent DNPSG transport (Akerboom et al., 1991) (Fig. 2). It was concluded from experiments with the indicator dye oxonol V, that the stimulation of DNPSG transport by ATP is not due to the generation of a membrane potential. Furthermore, the protonophore FCCP, the ionophore A23187 and inhibitors of cation transport show no influence on the ATP-dependent transport of DNPSG. It is therefore assumed that this transport system is primarily driven by ATP. Stimulation of DNPSG transport by a valinomycin-induced potassium diffusion potential is relatively low in the canalicular plasma membrane vesicles (Inoue et al., 1984a; Kobayashi et al., 1990; Akerboom et al., 1991). Electrogenic transport of DNPSG may play a major role in the basolateral membrane (Kobayashi et al., 1990). Mutual inhibition of GSSG and DNPSG for canalicular transport both in the intact organ (Akerboom et al., 1982b) and in membrane vesicles (Akerboom et al., 1991) indicates that both glutathione species are transported via a common transport system.

Transport of GSH in canalicular plasma membrane vesicles is stimulated by a valinomycin-induced potassium diffusion potential and is probably electrogenic (Inoue et al., 1983). The addition of ATP has no effect (Akerboom et al, 1991). These observations and the lack of inhibition of ATP-dependent DNPSG transport by GSH (Kobayashi et al., 1990; Akerboom et al., 1991) indicate that GSH transport is not related to the active transport of GSSG or glutathione conjugates.

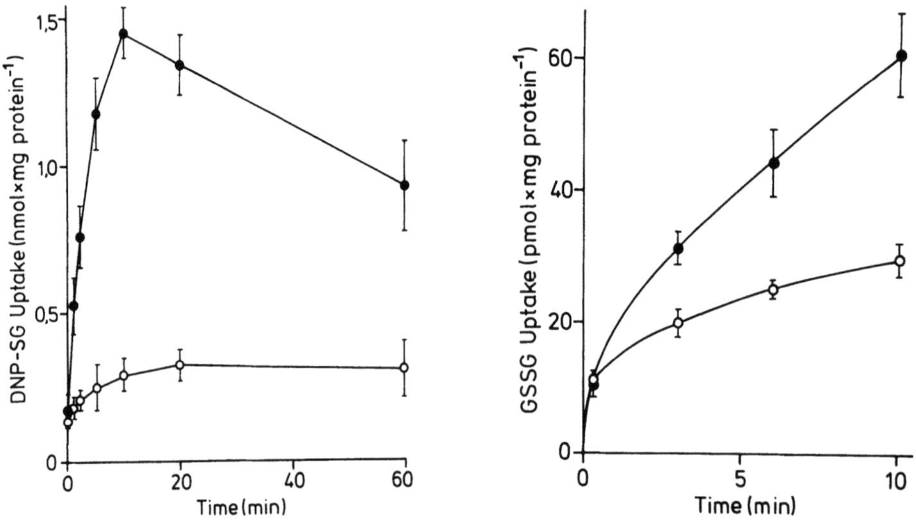

Fig. 1. Uptake of DNPSG (100 μM)(left) and GSSG (5 μM)(right) in rat liver canalicular membrane vesicles in the absence (o) and presence (•) of ATP and an ATP-regenerating system (for details, see Akerboom et al., 1991)

Recently, a 38 kDa protein with DNPSG dependent-ATPase activity has been purified from rat liver membranes which may participate in the canalicular ATP-dependent transport of DNPSG (Awasthi et al., 1991). The protein shows a K_m for DNPSG of 110 μM and is active with a large variety of organic anions including glucuronidated and sulfated lithocholic acid. However, GSSG and taurocholate are not able to stimulate the purified enzyme. Reconstitution studies are required to gain more insight in the transport properties.

Intestine

The intestine is directly exposed to ingested toxic compounds, drugs, food additives and toxins derived from intestinal bacterial flora. Protection by glutathione is essential for intestinal function. In the mouse, glutathione deficiency induced by inhibition of glutathione synthesis leads to severe degeneration of the epithelial cells of jejunum and colon, which can be prevented by oral administration of GSH or GSH monoester (Martensson et al., 1990).

Studies with isolated intestinal epithelial cells showed that externally added GSH is able to protect against intracellular damage by oxidizing agents, and that transport is probably required for its protective effect (Lash et al., 1986). GSH is taken up by these cells in a Na^+-dependent manner.

Transepithelial transport has been investigated using an in situ closed-loop vascular perfusion of rat small intestine (Hagen and Jones, 1987). Addition of radioactively labelled GSH to the luminal space results in the appearance of intact GSH in the perfusate. It was suggested that the transepithelial transport does not involve breakdown and resynthesis of the tripeptides as comparable rates of transport are obtained after pretreatment of the intestine segments with inhibitors of γ-glutamyl transpeptidase and glutathione synthesis. The addition of GSH into the vasculature gives no detectable carrier-mediated transport into the lumen, indicating that the transepithelial transport is undirectional. Regional differences may occur for the rate of uptake along the intestinal tract. The highest uptake of GSH takes place in the upper jejunum, relatively little absorption occurs in the duodenum, and no transport is detectable in the ileum (Hagen et al.,1990).

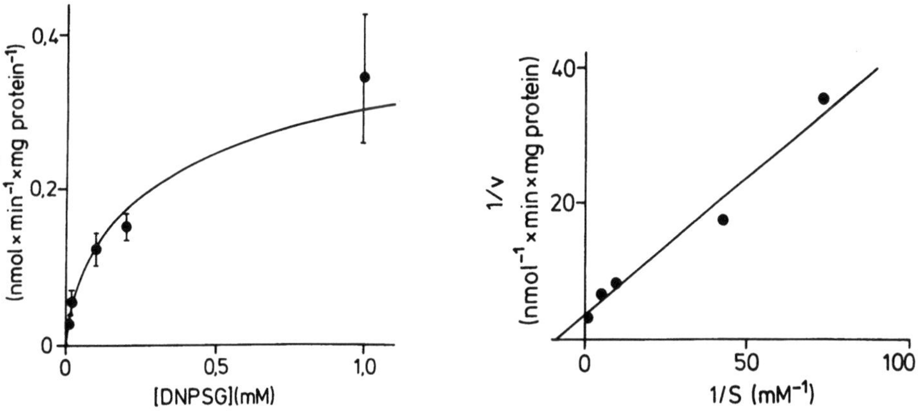

Fig. 2. Effect of substrate concentration on ATP-dependent uptake of DNPSG in rat liver canalicular membrane vesicles (left) and a Lineweaver-Burk plot derived therefrom (right). Conditions as in Fig. 1.

Transepithelial transport of GSH is substantially diminished by replacement of sodium for choline in the lumen (Hagen and Jones, 1987). Also in rabbit brush border membrane vesicles the rate of GSH uptake is higher in the presence of Na^+ (Vincenzini et al., 1988). However, in these studies a similar stimulation of GSH uptake was obtained in the presence of K^+, so that there is no selective requirement for Na^+. Na^+-dependent GSH transport has been reported in brush border membrane vesicles isolated from pig intestine (Linder et al,1986), whereas uptake of GSH in everted sacs of rat small intestine was found to be Na^+-independent (Hunjan and Evered, 1985). Thus, the bioenergetics of luminal GSH transport is not fully elucidated. The interpretation of the results obtained from the studies with the isolated brush border membranes is rendered difficult because of the undefined orientation of the vesicles and possible contributions with basolateral membranes, for which also a Na^+-dependent GSH transport system has been described (Lash et al, 1986).

Luminal GSH transport described in the perfused organ of the rat (Hagen and Jones, 1987) exhibits an apparent K_m of about 1 mM, and is inhibited by ophthalmic acid and γ-glutamylglutamate; the kinetics of the system described in the rabbit brush border membrane vesicles (Vincenzini et al., 1988) is quite different. It shows an apparent K_m of 14 μM, and is not inhibitable by ophthalmic acid, nor by γ-glutamyl-glycine. The latter system is competitively inhibited by glutathione S-conjugates, with apparent K_i values similar to the K_m for GSH; GSSG shows no inhibition. It is proposed that transport of glutathione conjugates and GSH proceed via a common transport system, responsible for the cellular efflux of these compounds (Vincenzini et al., 1991).

No transepithelial transport of GSSG has been detected (Lash et al, 1986), although significant amounts of glutathione disulfide may reach the intestinal tract from food processing or from biliary excretion. Recently it was suggested that a mechanism for reducing GSSG to GSH exists in the lumen of the distal duodenum or proximal jejunum of the rat, the GSH formed being subject to transepithelial transport (Hagen et al, 1991). This would offer an alternative pathway for the utilization of intraluminal GSSG in addition to luminal hydrolysis and intestinal uptake of the amino acid constituents.

Kidney

The kidney is the major organ for glutathione removal from the blood circulation, responsible for at least 50 to 60% of net plasma glutathione turnover (Häberle et al, 1979; Griffith and Meister, 1979). About 80% of both GSH and GSSG is extracted after a single pass through the kidney (Anderson et al.,1980; Häberle et al., 1979). The glomerular filtration capacity amounts to only 20 to 30%, so that peritubular mechanisms must contribute to the plasma disappearance of glutathione.

The kidney possesses an extremely high activity of γ-glutamyl transpeptidase. The enzyme is predominantly located in the brush border membrane of the proximal tubule with the active site oriented towards the tubular lumen (Horiuchi et al., 1978; Silbernagl et al.,1978; Tsao and Curthoys, 1980), but is also present in the basolateral domain of the plasma membrane (Spater et al., 1980). Inhibitors such as acivicin (Griffith and Meister, 1979; Anderson et al., 1980; Rankin and Curthoys, 1982; Inoue et al., 1986) and serine borate (Ormstad et al.,1982) have been used to investigate the role of the enzyme in the renal extraction of glutathione. Pretreatment of the mouse with these inhibitors results in urinary levels of glutathione which increased from 3 μM to 25-35 mM (Meister, 1983). This shows that after glomerular filtration, glutathione is effectively degraded to its constituent amino acids which then are reabsorbed across the brush border membrane.

As the active site of the γ-glutamyl transpeptidase is facing the tubular lumen, its functioning in the hydrolysis of intracellular glutathione as part of an intraorgan turnover requires transport across the brush border membrane (Griffith and Meister, 1979). Indeed, under conditions of inactivated luminal γ-glutamyl transpeptidase and lowered arterial glutathione by intravenous infusion of purified γ-glutamyl transpeptidase, the amount of glutathione excreted exceeds the filtered amount fourty-fold (Scott and Curthoys, 1987). This implies that tubular secretion of glutathione contributes to the luminal catabolism of the tripeptide. The net secretion of glutathione with the final urine in the acivicin-treated animals probably originates from proximal tubules of deep nephrons (Heuner et al., 1989). Transport of glutathione across the brush border membrane as studied with isolated plasma membrane vesicles indicated that GSH transport is electrogenic, exhibits an apparent K_m of 0.21 mM, and is inhibited by S-benzylglutathione (Abbott et al., 1984). Interestingly, these characteristics show similarities to the GSH transport system described for the brush border membrane of the intestine (Vincenzini et al., 1991). Whether these transport systems are identical remains to be established.

Peritubular extraction of glutathione involves degradation via basolateral γ-glutamyl transpeptidase. This is shown by studies with acivicin (Anderson et al., 1980; Abbott et al., 1984; Rankin et al., 1985; Inoue et al., 1986;) and HPLC analysis of venous glutathione metabolites (Abbott et al., 1984; Rankin et al., 1985). However, in the presence of acivicin the renal extraction is still higher than the glomerular filtration capacity, suggesting that transport of intact glutathione across the basolateral membrane contributes to the renal handling of plasma glutathione. Studies using the multiple-dilution indicator method also indicated basolateral transport of intact glutathione (Rankin et al., 1985). Further support for the existence of peritubular GSH transport comes from experiments with renal basolateral plasma membrane vesicles, in which a Na^+-dependent GSH transport system was identified (Lash and Jones, 1984). Transport of glutathione across these membranes is electrogenic and involves the cotransport of at least two Na^+ ions.

Erythrocytes

In erythrocytes, like in other cell types, glutathione is important for protection against oxidative damage. Deficiency of glutathione or defects in the pathway of GSH regeneration as in glucose-6-phosphate dehydrogenase deficiency are known to lead to Heinz-body formation and hemolytic anemia, especially in the presence of oxidative and free radical generating drugs. When human red blood cells are exposed to hydroperoxides in the absence of glucose, they respond by an increased efflux of glutathione disulfide (Srivastava and Beutler, 1969). In the presence of glucose, which regenerates reducing equivalents for the intracellular reduction of GSSG, the rate of transport is extremely low and compares with the turnover rate of glutathione in the erythrocyte; export of GSH from the red blood cells does not occur (Lunn et al., 1979).

Studies with erythrocyte ghosts loaded with GSSG and with inside-out membrane vesicles showed that the transport of GSSG is ATP-dependent and inhibited by fluoride (Prchal et al., 1975; Kondo et al., 1981; Akerboom et al., 1992). Two transport systems for GSSG with different kinetic properties have been identified, one high-affinity system with a K_m of 29 μM and a V_{max} of 3.6 nmol/hr x mg protein, and one low-affinity

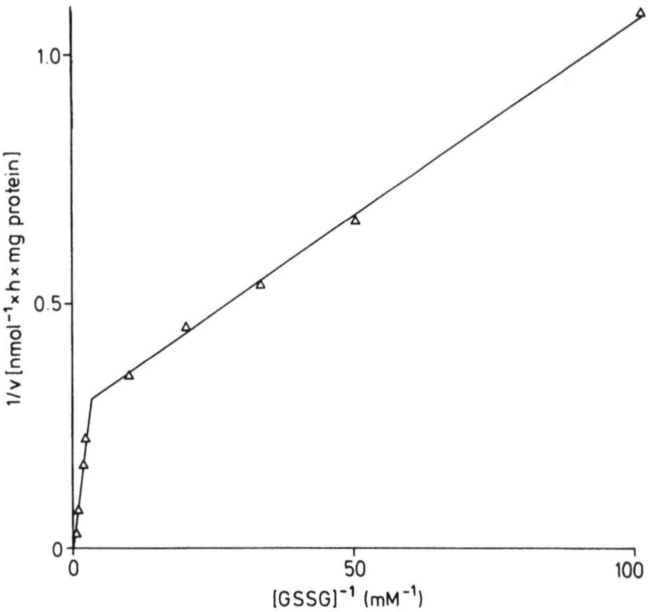

Fig. 3. Lineweaver-Burk plot of GSSG uptake in inside-out vesicles from human erythrocytes (Akerboom et al., in preparation; see also Kondo et al., 1982).

system with a K_m of 5.5 mM and a V_{max} of 3.0 nmol/hr x mg protein (Fig. 3). DNPSG inhibits GSSG transport. The inhibition is competitive with a K_i of 0.55 mM, and it was suggested that DNPSG is transported via the low-affinity GSSG carrier (Kondo et al., 1985).

Treatment of erythrocytes with 1-chloro-2,4-dinitrobenzene gives rise to an efflux of DNPSG formed by intracellular GSH transferases. Analogous to the transport of GSSG it was found that DNPSG transport is sensitive to fluoride and requires ATP (Board, 1981) and, furthermore, that two carrier systems exist (Eckert and Eyer, 1986). The high-affinity transport system exhibits a K_m of 1.4 µM and a V_{max} of 0.9 nmol/min x ml cells, the low affinity system a K_m of 700 µM and a V_{max} of 4.5 nmol/min x ml cells. With inside-out vesicles only one transport system has been described with an apparent K_m ranging from 0.29 mM to 0.94 mM (Kondo et al., 1982; LaBelle et al., 1986a). DNPSG transport is inhibited by thiol reagents and shows similar activity with GTP or ATP. It has been reported that concentrations of GSSG as high as 10 mM do not inhibit DNPSG transport significantly (LaBelle et al., 1986b), suggesting that the transport of glutathione conjugates and GSSG may proceed via separate ways. Recent studies, however, indicate that a common high affinity translocator exists for both glutathione species (Akerboom et al., 1992).

ATPase activities have been described in isolated erythrocyte membranes showing stimulation by GSSG (Kondo et al., 1987, 1989) and DNPSG (LaBelle et al., 1988; Sharma et al., 1990) which are thought to be identical with the transport systems for the respective glutathione species. Elution of solubilized membranes on a S-hexyl-Sepharose 6B affinity column resulted in the resolution of two peaks of GSSG-stimulated ATPase activities. The apparent K_m values observed for the two enzymes (105 µM and 2.0 mM) correspond well with the values observed for GSSG transport. The high-affinity ATPase has been further purified to apparent homogeneity and appeared to consist of two subunits of 85 kDa and 62 kDa. The reconstituted enzyme showed both GSSG-ATPase activity and GSSG transport activity. Interestingly, the enzyme exhibits immunological cross reactivity with the enzyme isolated from human hepatocytes (Kondo et al., 1987).

Also the DNPSG-ATPase activity in erythrocyte membranes showed kinetics similar to the DNPSG transport system (LaBelle et al., 1988). After purification of the enzyme by affinity chromatography a 38 kDa subunit was obtained which showed ATPase activity towards DNPSG but not GSSG (Sharma et al., 1990). Western blot analysis using antibodies against the purified DNPSG-ATPase indicated that the enzyme is also expressed in human liver, lung, placenta and pancreas (Sharma et al., 1990).

PHYSIOLOGICAL IMPLICATIONS

Hormonal control of GSH transport

At least 90% of plasma glutathione is derived from the liver (Bartoli and Sies, 1978; Lauterburg et al., 1984a). Little is known on the factors which regulate the systemic level of glutathione. During fasting there is an increased intraorgan turnover with increased hepatic efflux and plasma levels of GSH (Lauterburg et al., 1984a). Also an initial increase of plasma GSH accompanied by a decrease of the liver GSH content has been observed during physical exercise in the rat (Lew et al., 1984). Hormonal control mechanisms may be involved in the modulation of hepatic glutathione transport. Perivascular nerve stimulation in perfused rat liver leads to increased glutathione release, which is not observed in the presence of α-adrenergic antagonists like phentolamine (Häussinger et al., 1987). Adrenaline, phenylephrine, vasopressin, and angiotensin II stimulate sinusoidal GSH release from the perfused rat liver (Sies and Graf, 1985, 1988), and , to a lesser extent, glucagon and dibutyryl cAMP (Graf and Sies, 1988; Lu et al., 1990).

The hormones produce their stimulatory effect by different mechanisms. For hormones which act through Ca^{2+} signaling, such as phenylephrine and vasopressin, the calcium releasing properties are probably relevant. Their effects can be mimicked by omission of calcium from the perfusate or by the addition of a calcium chelator to the perfusate (Graf and Sies, 1988.) Interestingly, the effects of phenylephrine and vasopressin are not observed in cultured rat hepatocytes (Lu et al., 1990), and may be explained by changes in the hepatic paracellular permeability (Ballatori and Truong, 1990). It has been shown recently that the junctional barrier, which separates the sinusoidal and canalicular spaces, is subject to permeability changes under the influence of the Ca^{2+} mediator hormones (Lowe et al., 1988). The increase in sinusoidal GSH efflux in the perfused rat liver by vasopressin and phenylephrine is completely balanced by decreased biliary excretion, so that these hormones induce a redistribution of exported hepatic GSH in favor of systemic glutathione levels (Ballatori and Truong, 1990).

Dibutyryl cAMP has no effect on the junctional permeability (Lowe et al., 1988). The stimulation of sinusoidal GSH efflux by glucagon and dibutyryl cAMP has been suggested to be due to a stimulation of the Na^+-K^+- ATPase leading to a hyperpolarisation of the hepatocyte plasma membrane (Moule et al., 1987; Lu et al., 1990). Such hyperpolarisation is known to increase GSH release from isolated hepatocytes (Fernandez-Checa et al., 1988). Also the increased GSH release observed after hypotonic exposure of the perfused liver might be explained by hyperpolarization of the hepatocyte plasma membrane, although paracellular backflow of biliary GSH induced by cell swelling could be an alternative explanation (Häussinger et al., 1990).

Interaction of transport with other anions

The specificities of the transport systems for the different glutathione species has been subject of many studies, especially in the liver, and there is increasing evidence that multiple systems with overlapping specificities may exist for the transport of these compounds and a variety of organic anions.

The sinusoidal release of GSH from the perfused rat liver is inhibited by the sulfobromophthalein glutathione conjugate (BSPSG) in a reversible and dose-dependent manner (Ookthens et al., 1985, 1988). The canalicular excretion of GSH from the liver is not affected by BSPSG (Kaplowitz et al., 1983; Lauterburg et al., 1984b; Ballatori and Clarkson 1985). Thus, the sinusoidal transport system for GSH does not seem to be identical to the canalicular system. In the perfused rat liver the organic anions BSP, rose bengal, indocyanine green, and unconjugated bilirubin show potent inhibitory effects on sinusoidal GSH release (Ookthens et al., 1988). The inhibition of bilirubin is overcome by hepatic GSH, presumably by competition for the carrier. In isolated hepatocytes it has been observed that the mono- and diglucuronide conjugates of bilirubin inhibit as well, and that the inhibition occurs from the inside of the cell. The possibility was raised that the efflux of these organic anions is mediated by the GSH carrier, and that under pathological states of hyperbilirubinemia the interorgan turnover of GSH is altered.

Canalicular transport of glutathione conjugates, GSH, and GSSG is defective in mutant Wistar rats with inherited conjugated hyperbilirubinemia (Oude Elferink et al., 1989). These animals lack the capability of biliary secretion of a number of organic anions such as conjugated bilirubin and sulfated and glucuronidated bile acids, but not of taurocholate and cholate (Jansen et al., 1987; Kuipers et al., 1988; Oude Elferink et al., 1989). ATP-dependent transport of BSP and BSPSG is absent in canalicular membrane vesicles isolated from the mutant rat , while ATP-dependent transport of daunomycin, a substrate of the P-glycoprotein multidrug transport system, is normal (Kitamura et al., 1990). The question arises whether a multispecific transport system is operative for a variety of organic anions including the different glutathione species. As discussed above, canalicular efflux of GSH probably occurs via a transport system separate from other glutathione species, whereas GSSG and glutathione conjugates, including LTC4, may share a common transport system (Akerboom et al., 1982b, 1991; Ishikawa et al., 1990). Kinetic studies with canalicular membrane vesicles show that ATP-dependent transport of DNPSG is not only competitively inhibited by other glutathione conjugates, and by GSSG, but also by taurocholate and α-naphthyl-glucuronide (Fig. 4). Moreover, transport of GSSG (Fig. 1) and taurocholate (Fig. 5) (Akerboom et al., 1991) and of p-nitrophenylglucuronide (Kobayashi et al., 1991) is ATP-dependent, and inhibited by DNPSG. This indicates that the canalicular transport of DNPSG occurs via transport system with broad specifity, capable of transporting a variety of amphipatic organic anions. Indeed, in isolated hepatocytes a potent inhibition of DNPSG efflux by a number

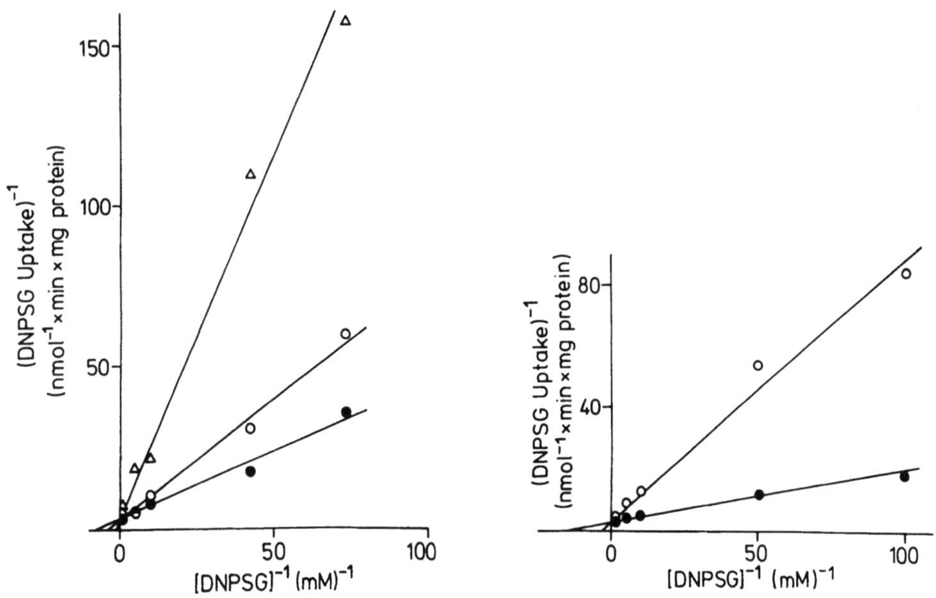

Fig. 4. Competitive inhibition of ATP-dependent uptake of DNPSG (●) by GSSG (△) and taurocholate (○)(left), and by α-naphthyl-glucuronide (○)(right) in rat liver canalicular membrane vesicles (for details, see Akerboom et al., 1991).

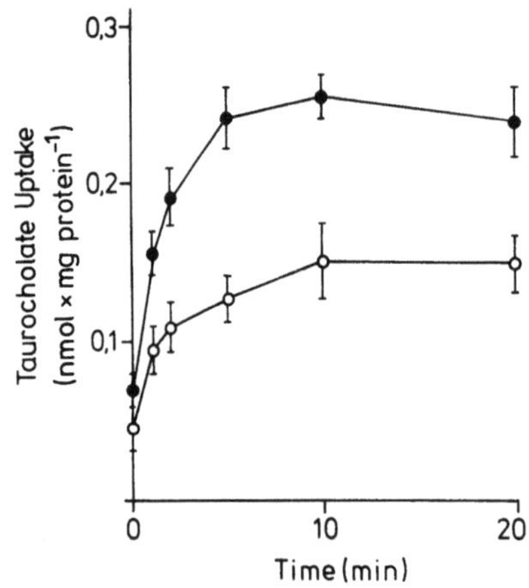

Fig. 5. Uptake of taurocholate (10 μM) in rat liver canalicular membrane vesicles in the absence (○) and presence (●) of ATP and an ATP-regenerating system (for experimental details, see Akerboom et al., 1991).

of conjugates of bilirubin and bile acids was recently observed (Oude Elferink et al., 1990). In this system no inhibition by taurocholate has been detected, possibly related to the high K_i value of this bile acid as observed in canalicular membrane vesicles, and the existence of an other route for taurocholate transport. Thus, although the number of transporters for organic anions and their specificity remain ambiguous, a multispecific organic anion transporter is probably responsible for the transport of GSSG and glutathione conjugates.

ACKNOWLEDGEMENTS

Work in the authors' laboratory was generously supported by the Deutsche Forschungsgemeinschaft, the National Foundation for Cancer Research, the Ernst Jung-Stiftung, and the Fonds der Chemischen Industrie.

REFERENCES

Abbott, W. A., and Meister, A., 1986, Proc. Natl. Acad. Sci. 83:1246
Abbott, W. A., Bridges, R. J., and Meister, A., 1984, J. Biol. Chem. 259:15393
Akerboom, T. P. M., Bartosz, G., and Sies, H., 1992, Biochim. Biophys. Acta 1103:115.
Akerboom, T. P. M., Bilzer, M., and Sies, H., 1982a, J. Biol. Chem. 257:4248
Akerboom, T. P. M., Bilzer, M., and Sies, H., 1982b, FEBS Lett. 140:73
Akerboom, T. P. M., Narayanaswami, V., Kunst, M., and Sies, H., 1991, J. Biol. Chem. 266:13147
Anderson, M. E., and Meister, A., 1980, J. Biol. Chem. 255:9530
Anderson, M. E., Bridges, R. J., and Meister, A., 1980, Biochem. Biophys. Res. Commun. 96:848
Aw, T. Y., Ookthens, M., Ren, C., and Kaplowitz, N., 1986, Am. J. Physiol. 250:G236
Aw, T. Y., Ookthens, M., Kuhlenkamp, J. F., and Kaplowitz, N., 1987, Biochem. Biophys. Res. Commun. 143:377
Awasthi, Y. C., Singhal, S. S., Gupta, S., Ahmad, H., Zimniak, P. Radominska, A., Lester, R., and Sharma, R., 1991, Biochem. Biophys. Res. Commun. 175:1090
Ballatori, N., and Clarkson, T. W., 1985, Am. J. Physiol. 248:G238
Ballatori, N., and Truong, A. T., 1989, J. Pharmacol. Expt. Ther. 251:1069
Ballatori, N., and Truong, A. T., 1990, Mol. Pharmacol. 38:64
Ballatori, N., Jacob, R., and Boyer, J. L., 1986, J. Biol. Chem. 261:7860
Ballatori, N., Jacob, R., Barrett, C., and Boyer, J. L., 1988, Am. J. Physiol. 254:G1
Bartoli, G. M., and Sies, H., 1978, FEBS Lett. 86:89
Board, P., 1981, FEBS Lett. 124:163

Brigelius, R., 1985, in: "Oxidative Stress", H. Sies, ed., pp. 243-272, Academic Press, London.
Chow, C. K., ed., 1988, "Cellular Antioxidant Defense Mechanisms", Vol. II, CRC Press, Boca Raton.
Crane, D., Häussinger, D., and Sies, H., 1982, Eur. J. Biochem. 127:575
Dolphin, D., Poulson, R., and Avramovic, O., eds., 1989a, "Coenzymes and Cofactors", Vol III, Part A, Wiley & Sons, New York.
Dolphin, D., Poulson, R., and Avramovic, O., eds., 1989b, "Coenzymes and Cofactors", Vol III, Part B, Wiley & Sons, New York.
Eberle, D., Clarke, R., and Kaplowitz, N., 1981, J. Biol. Chem. 256:2115
Eckert, K. -G., and Eyer, P., 1986, Biochem. Pharmacol. 35:325
Fernandez-Checa, J. C., Ren, C., Aw, T. Y., Ookthens, M., and Kaplowitz, N., 1988, Am. J. Physiol. 255:G403
Graf, P., and Sies, H., 1988, in: "Calcium-Dependent Processes in the Liver", Heilmann, C., ed., pp. 191-202, MTP Press, Lancaster, England
Griffith, O. W., and Meister, A., 1979, Proc. Natl. Acad. Sci. 76:5606
Häberle, D., Wahllländer, A., and Sies, H., 1979, FEBS Lett. 108:335
Hahn, R., Wendel, A., and Flohé, L., 1978, Biochim. Biophys. Acta 539:324
Hagen, T. M., and Jones, D. P., 1987, Am. J. Physiol. 252:G607
Hagen, T. M., Wierzbicka, G. T., Bowman, B. B., Aw, T. Y., and Jones, D. P., 1990, Am. J. Physiol. 259:G530
Häussinger, D., Stehle, T., Gerok, W., and Sies, H., 1987, Eur. J. Biochem. 163:197
Häussinger, D., Lang, F., Bauers, K., and Gerok, W., 1990, Eur. J. Biochem. 193:891
Heuner, A., Schwegler, J. S., and Silbernagl, S., 1989, Pflügers Arch. 414:551
Horiuchi, S., Inoue, M., and Morino, Y., 1978, Eur. J. Biochem. 87:429
Hunjan, M. K., and Evered, D. F., 1985, Biochim. Biophys. Acta 815:184
Inoue, M., Akerboom, T. P. M., Sies, H., Kinne, R., Tran, T., and Arias, I. M., 1984a, J. Biol. Chem. 259:4998
Inoue, M., Kinne, R., Tran, T., and Arias, I. M., 1983, Eur. J. Biochem. 134:467
Inoue, M., Kinne, R., Tran, T., and Arias, I. M., 1984b, Eur. J. Biochem. 138:491
Inoue, M., Saito, Y., Hirata, E., Morino, Y., and Nagase, S., 1987, J. Prot. Chem. 6:207
Inoue, M., Shinozuka, S., and Morino, Y., 1986, Eur. J. Biochem. 157:605
Ishikawa, T., Müller, M., Klünemann, C., Schaub, T., and Keppler, D., 1990, J. Biol. Chem. 265:19279
Iwai, M., Hagmann, W., Keppler, D., and Jungermann, K., 1988, Biol. Chem. Hoppe-Seyler 369:1131
Jansen, P. L. M., Groothuis, G. M. M., Peters, W. H. M., and Meijer, D. K. F., 1987, Hepatology 7:71
Kaplowitz, N., Eberle, D. E., Petrini, J., Touloukian, J., Corvasce, M. C., and Kuhlenkamp, J., 1983, J. Pharmacol. Expt. Ther. 224:141
Kitamura, T., Jansen, P., Hardenbrook, C., Kamimoto, Y., Gathmaitan, Z., and Arias, I. M., 1990, Proc. Natl. Acad. Sci. 87:3557
Kobayashi, K., Sogame, Y., Hara, H., and Hayashi, K., 1990, J. Biol. Chem. 265:7737
Kobayashi, K., Komatsu, S., Nishi, T., Hara, H., and Hayashi, K., 1991, Biochem. Biophys. Res. Commun. 176:622
Kondo, T., Dale, G. L., and Beutler, E., 1981, Biochim. Biophys. Acta 645:132
Kondo, T., Murao, M., and Taniguchi, N., 1982, Eur. J. Biochem. 125:551
Kondo, T., Kawakami, Y., Taniguchi, N., and Beutler, E., 1987, Proc. Natl. Acad. Sci. 84:7373
Kondo, T., Miyamoto, K., Gasa, S., Taniguchi, N., and Kawakami, Y., 1989, Biochem. Biophys. Res. Commun. 162:1
Kuipers, F., Enserink, M., Havinga, R., van der Steen, A. B. M., Hardonk, M. J., Fevery, J., and Vonk, R. J., 1988, J. Clin. Invest. 81:1593
Kuipers, F., Radominska, A., Zimniak, P., Little, J. M., Havinga, R., Vonk, R. J., and Lester, R., 1989, J. Lip. Res. 30:1835
LaBelle, E. F., Singh, S. V., Srivastava, S. K., and Awasthi, Y. C., 1986a, Biochem. J. 238:443
LaBelle, E. F., Singh, S. V., Srivastava, S. K., and Awasthi, Y. C., 1986b, Biochem. Biophys. Res. Commun. 139:538
LaBelle, E. F., Singh, S. V., Ahmad, H., Wronski, L., Srivastava, S., and Awasthi, Y. C., 1988, FEBS Lett. 228:53

Lash, L. H., and Jones, D. P., 1984, J. Biol. Chem. 259:14508
Lash, L. H., and Jones, D. P., 1985, Arch. Biochem. Biophys. 240:583
Lash, L. H., Hagen, T. M., and Jones, D. P., 1986, Proc. Natl. Acad. Sci. 83:4641
Lauterburg, B. H., Adams, J. D., and Mitchell, J. R., 1984a, Hepatology 4:586
Lauterburg, B. H., Smith, C. V., Hughes, H., and Mitchell, J. R., 1984b, J. Clin. Invest. 73:124
Lew, H., Pyke, S., and Quintanilha, A., 1985, FEBS Lett. 185:262
Linder, M., DeBurlet, G., and Suduka, P., 1984, Biochem. Biophys. Res. Commun. 123:929
Lowe, P. J., Miyai, K., Steinbach, J. H., and Hardison, W. G. M., 1988, Am. J. Physiol. 255:G454
Lu, S. C., Garcia-Ruiz, C., Kuhlenkamp, J., Ookthens, M., Salas-Prato, M., and Kaplowitz, N., 1990, J. Biol. Chem. 265:16088
Lunn, G., Dale, G. L., and Beutler, E., 1979, Blood 54:238
Martensson, J., Jain, A., and Meister, A., 1990, Proc. Natl. Acad. Sci. 87:1715
Meister, A., 1983, in: "Functions of Glutathione: Biochemical, Physiological, Toxicological, and Clinical Aspects", Larsson, A., Orrenius, S., Holmgren, A., and Mannervik, B., eds., pp. 1-22, Raven Press, New York.
Meister, A., and Anderson, M. E., 1983, Ann. Rev. Biochem. 52:711
Moule, S. K., Bradford, M. N., and McGivan, J. D., 1987, Biochem. J. 241:737
Ookthens, M., Hobdy, K., Corvasce, M. C., Aw, T. Y., and Kaplowitz, N., 1985, J. Clin. Invest. 75:258
Ookthens, M., Lyon, I., Fernandez-Checa, J., and Kaplowitz, N., 1988, J. Clin. Invest. 82:608
Ormstad, K., Lastbom, T., and Orrenius, S., 1982, Biochim. Biophys. Acta 700:148
Oude Elferink, R. P. J., Ottenhoff, R., Liefting, W., de Haan, J., and Jansen, P. L. M., 1989, J. Clin. Invest. 84:476
Oude Elferink, R. P. J., Ottenhoff, R., Radominska, A., Hofmann, A. F., Kuipers, F., and Jansen, P. L. M., 1990, Biochem. J. 271:281
Prchal, J., Srivastava, S. K., and Beutler, E., 1975, Blood 46:111
Rankin, B. B., and Curthoys, N. P., 1982, FEBS Lett. 147:193
Rankin, B. B., Wells, W., and Curthoys, N. P., 1985, Am. J. Physiol. 249:F198
Schwarz, L. R., Götz, R., and Klaassen, C. D., 1980, Am. J. Physiol. 239:C118
Scott, R. D., and Curthoys, N. P., 1987, Am. J. Physiol. 252:F877
Sharma, R., Gupta, S., Singh, S. V., Medh, R. D., Ahmad, H., LaBelle, E. f., and Awasthi, Y. C., 1990, Biochem. Biophys. Res. Commun. 171:155
Sies, H., 1983, in: "Glutathione: Storage, Transport and Turnover in Mammals", Sakamoto, Y., Higashi, T., and Tateishi, N., eds., pp. 63-88, Japan Sci. Soc. Press, Tokyo.
Sies, H., 1989, Naturwissenschaften 76:57
Sies, H., and Graf, P., 1985, Biochem. J. 226:545
Sies, H., and Ketterer, B., eds., 1988, "Glutathione Conjugation: Mechanisms and Biological Significance", Academic Press, London.
Sies, H., Wahlländer, A., and Waydhas, Ch., 1978, in: "Functions of Glutathione in Liver and Kidney", Sies, H., and Wendel, A., eds., pp. 120-126, Springer, Berlin
Silbernagl, S., Pfaller, W., Heinle, H., and Wendel, A., 1978, in: "Functions of Glutathione in Liver and Kidney", Sies, H., and Wendel, A., eds., pp. 60-69, Springer, Berlin
Sorrentino, D., Licko, V., and Weisinger, R. A., 1988, Biochem. Pharmacol. 37:3119
Spater, H. W., Poruchynsky, M. S., Quintana, N., Inoue, M., and Novikoff, A. B., 1982, Proc. Natl. Acad. Sci. 79:3547
Srivastava, S. K., and Beutler, E., 1968, Anal. Biochem. 25:70
Srivastava, S. K., and Beutler, E., 1969, J. Biol. Chem. 244:9
Taniguchi, N., Higashi, T., Sakamoto, Y., and Meister, A., eds., 1989, "Glutathione Centennial: Molecular Perspectives and Clinical Implications", Academic Press, San Diego
Tsao, B., and Curthoys, N. P., 1980, J. Biol. Chem. 255:7708
Vincenzini, M. T., Favilli, F., and Iantomasi, T., 1988, Biochim. Biophys. Acta 942:107
Vincenzini, M. T., Favilli, F., Stio, M., and Iantomasi, T., 1991, Biochim. Biophys. Acta 1073:571
Vina, J., ed., 1990, "Glutathione: Metabolism and Physiological Functions", CRC Press, Boca Raton.
Wahlländer, A., and Sies, H., 1979, Eur. J. Biochem. 96:441
Ziegler, D. M., 1985, Ann. Rev. Biochem. 54:305

INTERORGAN AMINO ACID NUTRITION

Halvor N. Christensen

Department of Pediatrics
U.C.S.D. School of Medicine
La Jolla, CA 92037

I should emphasize that by Interorgan Amino Acid Nutrition I mean rather more than the partition of exogenous nutrients among the various tissues, more than a mass-action sharing from the alimentary flow somewhat analogous perhaps to that by a litter of pigs feeding at a common trough. As an example, glutamine, as the reader has already seen, is a vital interorgan nutrient, even when by chance none is being received from an exogenous source. To protect its functional needs, the liver might be said to recycle and partition the glutamine it receives and forms by governing how much it consumes and how much it generates. But these effects are achieved by compartmentalizing these actions, as Häussinger (1987) has shown, by making the first "decision" with regard to NH_4^+ detoxication by ureogenesis in the periportal hepatocytes and the second decision by glutamine synthesis in the perivenous hepatocytes. Although this sequence may appear to be a purely hepatic "decision", note that it allows the liver to allot its glutamine to nourish other organs while protecting its own needs. It seems inevitable that the resulting sequence occasions interacinar flows of glutamine and glutamate, and probably of other amino acids, for which flows must depend upon the presence of and may be regulated by the capacity and the activity of the pertinent amino acid transport systems.

The classical investigations of clinical chemical diagnosis have taught medicine, perhaps all too well, the importance of the concentration of a given circulating metabolite, along with those of the inorganic ions. Only gradually have both the methodology and the perception followed that the flux of a metabolite through the circulation may be more important than its concentration. We still tend to overlook how limited is the information provided us by the concentration alone, because the medical world has in the meantime become so ingenious in deducing a flux abnormality from a concentration abnormality, even in deducing which of several possible fluxes (e.g., liver to kidney) is perturbed. Perhaps to be convinced we need only think of the diagnostic skill that has developed in interpreting a blood sugar determination. One consequence of this history is that we may tend to interpret too narrowly what Walter B. Cannon meant by physiological homeostasis. That term should mean not only the maintenance of, say, the plasma glutamine level between certain limits; it should encompass also the regulation of the glutamine flux, say between liver and kidney. The term, homeorhysis (Christensen, 1981) may help us to remember that sometimes relative constancy of concentration will be sacrificed to maintain constancy of flux, rather than the converse.

A STARTING POINT: AMINO ACID TRANSPORT DURING EMBRYOGENESIS

Some similarities of transport systems of the B family seen in the rodent conceptus to a broad-scope B-type transport system perhaps common to the intestinal and renal

Mammalian Amino Acid Transport, Edited by M.S. Kilberg and
D. Häussinger, Plenum Press, New York, 1992

brush borders raise the interesting possibility of a developmental relation, especially in view of similarities to the blastocyst in the formation of cyst-like cell aggregations during the development of renal epithelia (Wang et al., 1990a,b). The possible ontologic relation has led to the suggestion that the term System B be used for the present generically for Na^+-dependent systems that show a scope much wider than that shown by System A, e.g., to include even the bulky bicyclic amino acids such as BCH and BCO, and possibly also amino acids of net charge. Note that there has been no intent to have the letter B suggest blastocyst, anymore than one would wish it to suggest brush border.

POSSIBLE INTERMEDIATE STATIONS IN A FLOW FROM X TO Y

Transport facilities at the two placental surfaces, maternal and fetal, are not identical. Years ago it was first proposed that the fall of maternal plasma amino acids during human pregnancy arises from the interposition in the organism of yet one more amino acid concentrating organ, the placenta. An early bias tended to regard the placenta as little more than a membrane composed of cells, rather inertly feeding amino acids to the fetus. This view was supported by the early use of difficulty metabolized amino acid analogs, very helpful in describing net effects of transmembrane transport. But here the question arises, to what degree may one or more intermediate metabolic "stations" intervene within what we could instead regard as merely a transporting membrane lying between the main source and the target tissue of an interorgan flow. The term "station" should not incidentally suggest a static compartment. The point seems rather to watch alertly for metabolic diversions changing the composition of the flow in the intermediate compartment. Such metabolic diversions have indeed been encountered in considering transplacental amino acid flows, revealing initial neglect of the role of placental metabolism along with transport of amino acids in the function and growth of that organ. This metabolism has been shown recently by a degree of placental leucine deamination and of glutamate amidation to glutamine in the pregnant sheep (See Marconi et al., 1989, and Loy et al., 1990). Although transport system X^-_{AG} takes up glutamate from both its surfaces (Hoeltzli et al., 1990), it appears to be provided to the fetal lamb via a cycle involving glutamine rather than by net transport of glutamate as such across the placenta. Serine appears to be fed to the placenta mainly from the liver of the fetal lamb, generated there from glycine provided from the ewe by placental transport (Cetin et al., 1991)

Thus, the placenta serves for several of the amino acids not simply and solely for a feeding of them directly to the fetus. In a degree of analogy, on its way from muscle to contribute to hepatic gluconeogenesis, glutamine loses a part of its structure to become alanine in the intestinal wall, thus nourishing that tissue in passing. Evidence for metabolic intermediation has also been obtained during amino acid transport across endothelial cells of the brain microvessels, cells that constitute the blood brain barrier. For example, DeFeudis (1987) has reopened an old question by urging that amino acid metabolizing events intervene within the blood-brain barrier, so that the amino acids presented to the brain are probably not uniformly the same ones departing the general circulation. Endothelial cells produce from arginine the potent vasodialylating agent and proposed neurotransmitter, nitric oxide, the provision of the precursor arginine by membrane transport being subject to inhibition by arginine analogs (Mann et al., 1990). Endothelial cells are also a source of the endothelins, which appear to have one of their actions at a brain receptor, namely by stimulating $^{22}Na^+$ uptake via the Na^+/H^+ antiporter without involving phospholipase C (Vigne et al., 1991), as do other of their actions. A second regulatory barrier is encountered by amino acids on their way to the brain, namely the surface membranes of neurons and glia. The flows, e.g. of glutamate between neurons and astrocytes, play important roles in the control of neurotransmission. For certain parts of the CNS lying outside the protection of the blood-brain barrier (Price et al., 1981), the neuronal surface membrane appears to present the only significant barrier to cytotoxic effects of excess glutamate and its analogs.

AMINO ACID CONTROL OF LYSOSOMAL PROTEOLYSIS

An important aspect of our subject is how large a role is played by membrane transport in determining the balance between the building up of proteins from amino acids in the various cells of the organism, and the breakdown of proteins into amino acids, also the regulatory, nutritional, and pathological influences on that breakdown. Decades ago Whipple and his associates (1948) urged with great influence the view that the ability of tissues to break down serum albumin rapidly was so generally distributed that this newly synthesized protein should serve for a major part of the flow of its contained amino acids from liver to various other tissues. It is now recognized, however, that a disproportionate part of albumin breakdown occurs in the liver where its preceding synthesis took place, namely on its return to the liver, and also in the kidney.

The major although not the sole site of cellular protein breakdown within the organism is the lysosome, particularly in the hepatic parenchyma and in cardiac muscle. From the point of view of this chapter, the most important part of this proteolysis is that of mature, relatively long-lived protein molecules per se. Most proteins with short half-lives are degraded by cytosolic, ATP-dependent pathways. Amino acid release from short-lived sources may also represent the breakdown of temporary components of newly synthesized proteins, for example from signal peptides. The bulk of cellular proteins are relatively long-lived. Just how much of their breakdown occurs by cytosolic and by lysosomal proteolysis varies with the cell type and physiological status. Beyond the macroautophagy (dealing with cellular structural elements) and by microautophagy (dealing with protein molecules per se), a third, much more selective pathway of lysosomal proteolysis is stimulated by the withdrawal of serum growth factors from experimental cell cultures. Cytoplasmic proteins of liver and kidney (and experimentally, of cultured fibroblasts) bearing a specific peptide motif resembling Lys-Phe-Glu-Arg-Gln are preferentially depleted by this pathway on nutritional deficiency, although characteristically early in fasting (Dice, 1990; Wing et al., 1991). Activation of this selective pathway appears not to figure in protein breakdown in muscle during fasting, which appears to occur instead by a non-lysosomal, ATP-dependent pathway. Identification of differences in the routes of proteolysis in liver and muscle, and in their control, will obviously be important in determining the consequences of nutritional or traumatic changes for interorgan amino acid flows.

The best recognized regulation of hepatic microautophagy is exerted by levels of several of the circulating amino acids. However intuitively logical the amino acid economy may seem as a regulator of protein turnover, the physiological advantages of the detailed arrangements are not yet clear. Mortimore et al. (1991), perfused rat liver in situ with blood plasma modified as to amino acid levels. These studies have shown that leucine, glutamine, tyrosine (phenylalanine acts also, but perhaps by its conversion to tyrosine), proline, methionine, histidine, and tryptophan all serve as direct inhibitors of microautophagy, whereas the remaining 12 amino acids as a group as ineffective. The action of the regulatory seven so far seems rather complex. Their actions are seen as mediated by two alternative "mechanisms": one of these, designated L for low is evoked at their low plasma concentrations, with a sharply circumscribed loss of effect at their normal plasma levels. This zonal absence of inhibition is abolished by 0.5 mM alanine, itself noninhibitory, so it is considered to act as a co-regulator at its normal plasma concentration. The inflections describing inhibition and loss of inhibition with rising levels of each of the most effective members of the inhibitory group, namely leucine, glutamine, and tyrosine, all occur respectively at half and 1.25 times their normal plasma levels. The second mode of control is limited to effects at high levels (H) and is less well understood, but is switched on when glucagon is administered along with the regulatory amino acid. It is differentiated from the L mode of regulation by the absence of the inhibition at lowered levels.

Omission from the full mixture of the amino acids in the perfusing plasma of its alanine as a sole co-regulator evokes a proteolytic acceleration. Since this special co-regulatory effect of alanine is retained when its deamination is blocked by aminoxyacetate, alanine appears to be recognized as that amino acid per se. Although these several regulatory effects are attributed to the presence of the evoking amino acid in the perfusing modified plasma, one may well suspect that it occurs at a membrane surface. Tests of possibly interacting amino acid analogs have not yet been made that would justify speculating whether such recognition might occur at the sites serving for their lysosomal uptake.

Recent evidence points to the involvement of cell volume increases in originating anti-proteolytic effects of glutamine or glycine on perfused liver, also producing hormonal effects regulating hepatic proteolysis (Häussinger et al., 1991). These effects, it is suggested, arise from changes in amino acid transport into the cell, although it is not clear yet what intracellular mechanisms may couple cell volume changes to the control of proteolysis. The rates of release of glutamine and alanine from isolated soleus muscle were also modified by changes of cell volume induced by hypo- and hyper-osmotic media, the latter increasing the exodus rates (Parry-Billings et al. 1991). Even though large changes were produced in cell volume, the concentrations of these two amino acids were maintained. When large changes in body hydration occur for two or more tissues involved in an amino acid flux, one wonders what the total metabolic amino acid flow will be.

IS THE INTESTINAL LUMEN INVOLVED ONLY AS A SOURCE BUT NOT AS A TARGET OF AMINO ACID FLOWS?

Perhaps we tend too much to that view. Although it is well accepted now that the enterocytes must in the postabsorptive state receive much of their nutrition from the organism at their basolateral surface via a rather ordinary set of amino acid transport systems, one still notes a tendency to look on the transcellular absorptive function as a purely one-way uptake from the luminal compartment, neglecting the inescapable two-way role of basolateral transporters in that function. To find that simplicity would be unexpected for a complement of transport systems, and its assumption might lead one to a superficial explanation for the obvious differences in the brush-border systems from those of the basolateral plasma membranes.

It was argued years ago that the secretion of protein into the lumen never falls so low as to require an amino acid flux in the outward direction. But what evidence do we have for that neat balance, amino acid by amino acid, and hence for an unusual unidirectional rigidity of transport? The reversibility of transmucosal transport was shown nearly three decades ago with the metabolism-resistant transport models 2-aminoisobutyric acid (AIB) and 2-aminocyclopentanecarboxylic acid (cycloleucine) (Figure 1). Virtually the same distribution ratio (about 12:1 for AIB, and 90:1 for cycloleucine) was reached across the mucosa of isolated rat ileal loops, whether the principal source of the labeled test amino acid was the blood stream or the lumen, always as expected yielding a gradient characteristic of that amino acid in favor of the blood plasma, regardless of its source. One could still argue that an amino acid that might ordinarily arise from proteolysis in the gut would never encounter the mass-law call for its net release into the gut; but the potentiality is there, leaving no necessity to invoke a special irreversibility and hence a unique one-way character for this intercompartmental flow.

We recognize the converse situation: at the highest digestive activity, the amino acid (and peptide) gradients may face strongly in the other direction, and difficultly saturated modes of transmembrane migration may come into dominant play. The situation calls strongly for the development of what are so far lacking, methods that reliably distinguish simple diffusion from transport by unusually high-Km mediations. The different membrane permeabilities for molecules of similar size and charge suggest that diffusion may be invoked all too freely. Likewise, only with improved methods can we expect to confirm that simple diffusion of amino acids is permitted more characteristically at the basolateral surface than across the brush border. Wilbrandt and Rosenberg pointed out in their classical review in 1961 the described difficulty in reliably differentiating a component of simple diffusion from mediated fluxes across a biomembrane.

The career-long studies of Matthews, as well as studies by others describe the transmural absorption of di- and tri-peptides, and demonstrate situations where flows of amino acids in small peptide form are nutritionally advantageous (see Matthews' 1991 monograph). Many but not all of these peptide molecules are split to amino acids in the enterocyte. Let us open the question here, what if any target tissues might depend on flows of amino acids arising from the gut in the form of more or less random small

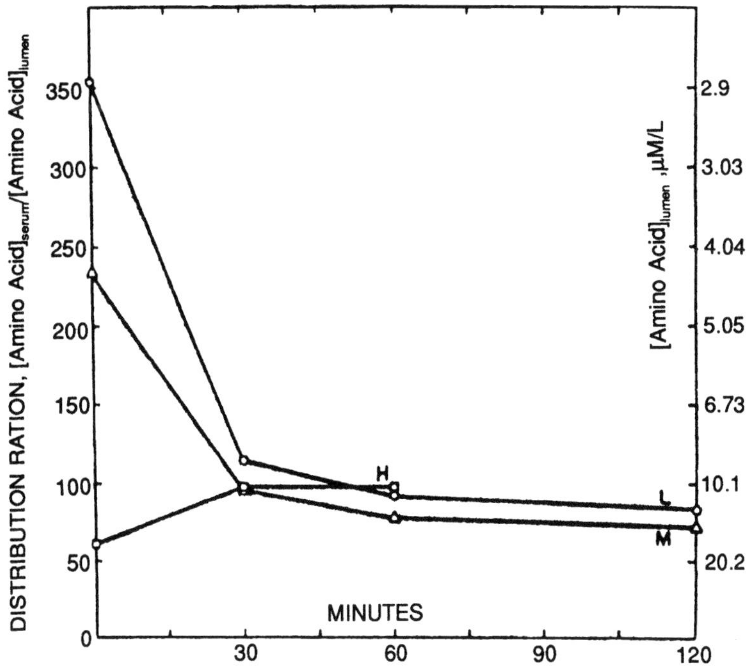

Figure 1. Approach to similar steady-state distributions for 2-aminocyclopentanecarboxylic acid between serum and the contents of ileal loops, from either above or below. Scale at right shows concentration of the amino acid within the ileum, the serum concentration being 1010 µM. The letters L, M, and H indicate relative position of the ileal loops, whether lowest, second, or highest of three in the same animal. This behavior has been confirmed repeatedly also with AIB. (Reproduced with permission from Christensen et al., 1963)

peptides surviving enterocyte peptidase action? Do we face only salvage processes for recycling the contained amino acids as is proposed to be the case for circulating glutathione? Does a basis arise for reassessing any general advantage for administering products of the incomplete hydrolysis of proteins for parenteral nutrition?

LIVER, MUSCLE, KIDNEY, INTESTINE, A CENTRAL SET OF FLOWS?

The amino acid transport processes in the liver and in skeletal muscle have been shown to determine the pertinent metabolic relations within each of these organs. We come now to consider metabolic relations among a group of dominant organs including these two, and on their close relation to the metabolism of kidney and intestine, without neglecting the parallelism of interorgan relations to the brain, the embryo, and other body compartments.

The liver participates in such flows also as a recipient of amino acids. It removes the largest part of the amino acids brought to it by the portal blood, with the exception of the branched-chain amino acids, the latter to be removed mainly by muscle. The liver is both a recipient and a donor of glutamine. We are familiar with the extensive interconversions among the amino acids catalyzed in the liver, and by the net degradation of amino acids produced in the organ, often via glucongeogensis. Hence the quantities reaching the liver are much larger than those departing it. Glutamine is a most important amino acid in its being the main vehicle for the transfer of nitrogen between organs. It is the main substrate for ureagenesis and is even more important in that its fate determines the balance between the nitrogen that goes to urea and that having other fates. Krebs pointed out that most amino acids have multiple functions, but that glutamine appears to be the most versatile of them all. That versatility depends on the abilities of its interorgan fluxes to respond to mass-action and to regulatory influences.

One of the most striking cases of the compartmentation of the whole body economy of an amino acid appears to be presented by arginine. This amino acid is available, from renal argininosuccinase action on argininosuccinate, to the extrahepatic compartment; but because of the inappreciable y^+ transport into the hepatocyte of the rat (White, Gazzola, and Christensen, 1982) the liver appears not to share in the arginine economy of the rest of the body. Instead it depends largely on its own arginine synthesis from ornithine. Because hepatic synthesis cannot accordingly be relied upon to meet the whole-body needs; hence the organism can suffer from arginine deficiency despite the intensity of its concurrent hepatic synthesis. The high y^+ activity of the hepatoma cell line HTC (White and Christensen, 1982), indicates that transformation can cause the appearance or intensification of hepatocyte y^+ activity.

Muscle is similarly important in supplying and modulating the flow of fuel and nitrogen between the periphery and the viscera, both diurnally and during starvation, injury, and disease. This function includes the important supplying of gluconeogenic substrates, and the mounting of the immune and the wound-healing responses. Muscle serves as the principal organ determining the balance between whole body amino acid catabolism and anabolism. Glutamine and alanine together provide 60 to 80% of the amino acids released from muscle, glutamine accounting for over 25%. The formation of glutamine, necessarily from other free amino acids, serves as a vehicle for removal of ammonium ion excesses. The nitrogen of ^{15}N-leucine provided about 9% of the muscle alpha amino nitrogen released as glutamine (Darmaun and Dechelotte, 1991) in experiments in which it provided 13% of that released as alanine. The latter shows itself as an apparently conditionally indispensable amino acid when the branched-chain amino acids therapeutically are largely omitted from the diet in maple-syrup urine disease (Nyhan et al., 1991).

The discovery by Kilberg et al. (1980) of the hepatic transport System N as a major and more specific and distinctly regulatable glutamine transport system than Systems ASC, A, and L, followed by the finding of a similar transport system called N^m in skeletal muscle, has set up an anxiety to discover what differences in these two systems may determine the very different glutamine fluxes in which these two organs participate. Tests that might have revealed differences in substrate specificity between rat liver sinusoidal and sarcolemmal membrane vesicles (Low et al., 1991) unfortunately did not isolate the Systems N and N^m, respectively, from other components of glutamine transport. Indeed no analog relatively specific to just one or two of these systems has yet been found. Hundal et al. (1987, 1991) have, however, made a more promising beginning toward the desired discrimination by finding that System N^m lacks the high pH specificity of System N. This means that these two transporter molecules must differ as to whether or not they bear dissociable groups on amino acid residues occupying critical positions, to account for this difference in pH sensitivity, which is likely in turn to help us understand the necessary functional differences. Furthermore the greater insulin sensitivity of System N than that observed for N^m by Hundal et al. (1991) is significant. Other differences in regulatory effects are likely to be found. The identification of the transport protein for System N using two monoclonal antibodies (Tammarapoo et al., 1991) should accelerate the discrimination of the structures of transporters of the N type. Gulve et al. (1991) have reported adaptive regulation for System A of rat skeletal muscle.

Jepson et al. (1988) have presented charts to show that declines in the free glutamine of muscle produced by endotoxin treatment of rats or by their fasting or protein deprivation are closely correlated with declines in muscle protein synthesis. Warner et al. (1990) sought evidence that the decreased muscle amino acid uptake in experimental endotoxemia in the rat is probably unrelated to the associated hypotension in experiments in which endotoxemic plasma was shown to decrease AIB uptake by isolated muscle by 22%. Although the result with the widely overworked transport model AIB is interesting, one wishes a metabolism-resistant glutamine analog had by now been identified for that test.

GENERAL CONSIDERATION OF CENTRAL FLUXES INVOLVING GLUTAMINE AND BRANCHED-CHAIN AMINO ACIDS

The administration of transport-competitive analogs of the branched-chain amino acids continue to be considered for producing therapeutic changes in their fluxes, for example to the brain in hepatic encephalopathy (Fischer, 1990; May et al., 1991).

A series of diagrams by Souba (1991) appears useful to show how such fluxes of glutamine may be changed in pathophysiologic states. The beakers at the left of these diagrams, here appearing as Figs. 2 to 4, are intended to indicate changes in the portal blood glutamine level. Fig. 2 shows how in metabolic acidosis the stimulated flow of glutamine from the liver (arrow at the lower right), in fact to the kidney, lowers the glutamine available for intestinal nutrition and at the same time for portal uptake by the liver. Regulation of the glutamine flux to the kidney has been discussed in an earlier chapter. As emphasized by Häussinger, hepatic ammonia disposition is thus diverted to the high-affinity synthetase pathway of the perivenous hepatocytes, allowing the ultimate renal disposition of the excess hydrogen ion in the form of the H^+-carrying ammonium ion in place of urea, a form of nitrogen not carrying hydrogen ion.

Fig. 3 proposes that sepsis with endotoxemia shifts glutamine utilization from the gut, normally the major glutamine consumer, to the liver, the organ most flexible in its role in glutamine fluxes. In the case of Fig. 3, the impairment is attributed to the gut, although the consequent hepatic glutamine uptake serves to support hepatic and glutathione biosynthesis, as well as of course ureagenesis and gluconeogenesis. The sharing also by immune cells of the glutamine released by muscle is emphasized in a review by Newsholme (1989).

Fig. 4 represents the influence of advanced malignant disease in rats bearing tumors that are fast growing and avid glutamine consumers. As the tumor with time

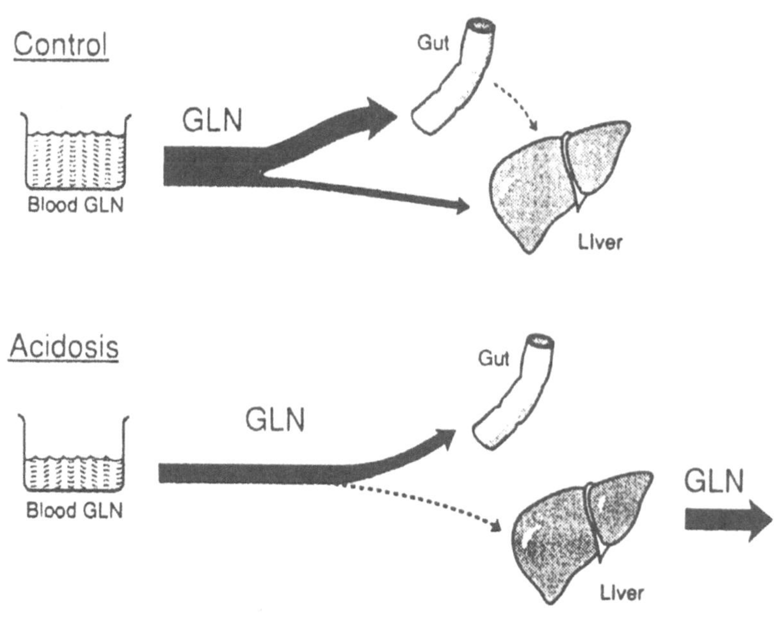

Figure 2. Changes in splanchnic glutamine flux during acidosis. A high portal ammonia load coupled with a relatively low portal glutamine level (such as seen during metabolic acidosis) decreases urea synthesis (and therefore hydrogen ion production) but increases glutamine synthesis and release, which can help support renal ammoniagenesis. Reproduced by permission from Souba, 1991.

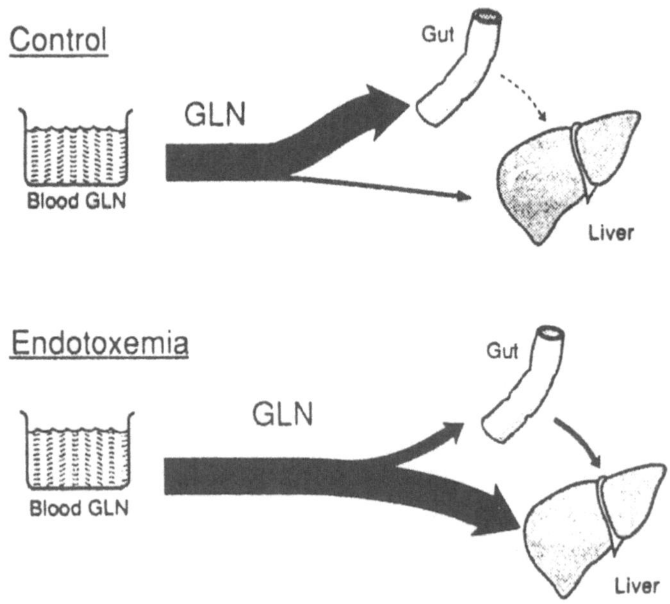

Figure 3. Glutamine utilization by the gut and liver during endotoxemia. Normally the gut is the major glutamine consumer, exceeding hepatic glutamine uptake by threefold. During endotoxemia, intestinal glutamine consumption falls and hepatic glutamine utilization increases tenfold, such that the liver becomes the major organ of glutamine uptake. The huge increase in hepatic glutamine uptake is due to an increase in the rates of hepatic glutamine delivery and intracellular metabolism rather than to a change in hepatocyte plasma membrane activity. The accelerated hepatic glutamine uptake appears to support nucleotide biosynthesis and glutathione biosynthesis as well as ureagenesis and gluconeogenesis. Reproduced with permission from Souba, 1991.

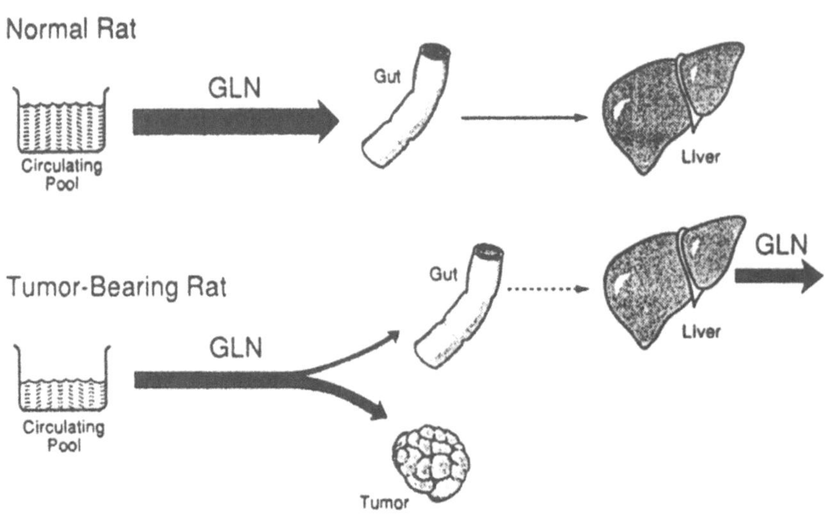

Figure 4. Splanchnic glutamine metabolism in advanced malignant disease. In tumor-bearing rats, intestinal glutamine uptake falls and the liver reverses to an organ of net glutamine release. With time the tumor becomes the major organ of glutamine uptake. Reproduced with permission from Souba, 1991.

becomes the body's main glutamine consumer, the liver reverses to an organ of net glutamine release, although as indicated for a different reason than in endotoxemia. Unpublished results of the same laboratory showed an associated interesting 30% increase in hepatic System L activity, which was considered, despite the complexity of that situation, to assist as a net the transport of glutamine from hepatocytes. Some non-transport adjustments were also proposed in explanation of some of the changes of glutamine flux in Figures 2 to 4.

On the basis of his review of glutamine nutrition of the splanchnic bed, Souba summarized advantages of providing supplemental amounts of this amino acid, both for enteral and parenteral feeding. The immune function of the gut can be sustained by oral dietary supplements of glutamine. For parenteral supplementation, certain glutamine dipeptides may be advantageous because of their stability relative to this particular free amino acid (Adibi, 1989).

CONCLUDING COMMENT

Manifestly, to study the biological regulation of interorgan amino acid flows, once we know what metabolic steps must occur where and in a coordinated way, we need to learn what transport system or systems can mediate the transmembrane flows of a given amino acid. Beyond that, for a biologically important amino acid and for its associated analogs, over biologically pertinent concentration ranges, we need to know which transport systems actually do participate to a significant extent, also whether in the inward or outward direction as a net. We encounter in studies of important flows such amino acids as glutamine which might perhaps migrate simultaneously by several systems. Hence special care will be necessary before we can settle the main issues that determine to what degree modulation of membrane transport processes, on the one hand, and enzymatic reactions on the other hand, control the important interorgan flows. This chapter has provided the privilege of trying to examine critically the incompleteness of the present approach to these targets.

REFERENCES

Adibi, S.A., 1989, Metabolism 38:89.
Cetin, I., Fennessey, P.V., Quick, A.N. Jr., Marconi, A.M., Meschia, G., Battaglia, F.C., and Sparks, J.W., 1991, Am. J. Physiol. 260:E371.
Christensen, H.N., 1985, Perspect. Biol. Med. 30:186.
Christensen, H.N., 1990, Physiol. Rev. 70:43.
Christensen, H.N., Feldman, B.H., and Hastings, A.B. 1963, Am. J. Physiol. 205:255.
Curran, R.D., Ferrari, F.K., Kispert, P.H., Stadler, J., Stuehr, D.J., Simmonds, R.I., and Billier, T.R., 1991, FASEB J. 5:2095.
Darmaun, D.D., and Dechelotte, P., 1991, Am. J. Physiol. 260:E326.
Dice, J.F., 1990 Trends Biochem. Sci. 15:305.
DeFeudis, F.V., in: "Amino acids in Health and Disease, New Perspectives", S. Kaufman, ed., Liss, New York, 1987, 135.
Fischer, J.E., 1990, J. Parenter. Enteral Nutr. 14:2499.
Gulve, E.A., Cartee, G.D., Youn, J.H., and Hollozy, 1991, Am. J. Physiol. 260:C82.
Häussinger, D. 1983 Eur. J. Biochem. 133:269.
Häussinger, D., Hallbrucker, C., vom Dahl, S., Decker, S., Schweizer, U., Lang, F., and Gerok, K. 1991, FEBS Lett. 283:470.
Hoeltzli, S.D., Kelley, L.K., Moe, A.J. and Smith, C.H. 1990, Am. J. Physiol. C47.
Hundal, H.S., Rennie, M.J., and Watt, P.W., 1987, J. Physiol. 393:283.
Hundal, H.S., Babij, P., Taylor, P.M., Watt, P.W., and Rennie, M.J., 1991, Biochim. Biophys. Acta. 1092:376.
Kilberg, M.S., Handlogten, M.E., and Christensen, H.N., 1980, J. Biol. Chem. 255:4011.
Low, S.Y., Taylor, P.M., Ahmed, A., Pogson, C.I., and Rennie, M.J., 1991, Biochem. J. (London), 278:185.
Loy, G.L., Quick, A.N. Jr., Meschia, G., Battaglia, F.C., and Fennessey, P.V., 1990, Am. J. Physiol. 259:E492.

Madden, S.C., Woods, R.R., Schull, F.W., Remington, J.H., and Whipple, G., 1945, J. Exp. Med. 81:439.
Mann, G.E., Sheriff, C.J., and Pearson, J.D. 1989, in: "Nitric oxide from L-arginine, a bioregulatory system", S. Moncada and E.A. Higgs, eds., Excerpta Med., Amsterdam 1990, pp.331.
Marconi, A.M., Battaglia, F.C., Meschia, G., and Sparks, J.W. 1989 Am. J. Physiol. 257:E909.
Matthews, D.M., 1991, Protein Absorption. Wiley-Liss, N.Y., pp.414.
May, R.C., Piepenbrock, N., Kelley, R.A., and Mitch, W.E., 1991, J. Nutr. 121:293.
Mortimore, G.E., Hkurana, K.K., and Miotto, G., 1991, J. Biol. Chem. 266, 1021.
Newsholme, E.A., Newsholme, P., Curi, R., Crabtree, B., and Ardawi, S.M., 1989, Perspect. Clin. Nutr., J.M. Kinney and J.R. Borum, eds., pp.71-78.
Nyhan, W.L., Rice-Asaro, M., and Acosta, P. 1991, in: "Treatment of Genetic Diseases", R.J. Desnick, ed., Chap. 3, Churchill Livingstone, 1991.
Parry-Billings, M., Bevan, S.J., Opara, E., and Newsholme, E.A. 1991, Biochem. J. 276:559.
Price, M.T., Olney, J.W., Lowry, O.H., and Buchsbaum, S. 1981, J. Neurochem. 36:1774.
Souba, W.W., 1991, Annu. Rev. Nutr. 11:285.
Tammarapoo, B.K., Handlogten, M.E., Laine, R.O., Dugan, J., and Kilberg, M.S., 1991, J. Biol. Chem. in press.
Vigne, P., Ledoux, A., and Frelin, C., 1991 J. Biol. Chem. 266:3925.
Wing, S.S., Chiang, H.L., Goldberg, A.L., and Dice, J.F., 1991, Biochem. J. (London) 275:165.
Wang, A.Z., Ojakian, G.K., and Nelson, W.J., 1990a, J. Cell Sci. 95:137.
Wang, A.Z., Ojakian, G.K., and Nelson, W.J., 1990b, J. Cell Sci. 95:153.
Warner, B.W., Hasselgren, P.O., James, J.H., Hummel, R.P., III, Rigel, D.F., and Fischer, J.E., 1990, Ann. Surg. 211:323.
Wilbrandt, W., and Rosenberg, T., 1961, Pharmacol. Rev. 13:109.

INDEX

MIX
Papier aus verantwortungsvollen Quellen
Paper from responsible sources
FSC® C105338

If you have any concerns about our products,
you can contact us on
ProductSafety@springernature.com

In case Publisher is established outside the EU,
the EU authorized representative is:
Springer Nature Customer Service Center GmbH
Europaplatz 3, 69115 Heidelberg, Germany

Printed by Libri Plureos GmbH
in Hamburg, Germany